Probability and Mathematical Statistics (Continued)

LEHMANN • Testing Statistical Hypotheses, *Second Edition*
LEHMANN • Theory of Point Estimation
MATTHES, KERSTAN, and MECKE • Infinitely Divisible Point Processes
MUIRHEAD • Aspects of Multivariate Statistical Theory
OLIVER and SMITH • Influence Diagrams, Belief Nets and Decision Analysis
PRESS • Bayesian Statistics: Principles, Models, and Applications
PURI and SEN • Nonparametric Methods in General Linear Models
PURI and SEN • Nonparametric Methods in Multivariate Analysis
PURI, VILAPLANA, and WERTZ • New Perspectives in Theoretical and Applied Statistics
RANDLES and WOLFE • Introduction to the Theory of Nonparametric Statistics
RAO • Asymptotic Theory of Statistical Inference
RAO • Linear Statistical Inference and Its Applications, *Second Edition*
ROBERTSON, WRIGHT and DYKSTRA • Order Restricted Statistical Inference
ROGERS and WILLIAMS • Diffusions, Markov Processes, and Martingales, Volume II: Îto Calculus
ROHATGI • Statistical Inference
ROSS • Stochastic Processes
RUBINSTEIN • Simulation and The Monte Carlo Method
RUZSA and SZEKELY • Algebraic Probability Theory
SCHEFFE • The Analysis of Variance
SEBER • Linear Regression Analysis
SEBER • Multivariate Observations
SEBER and WILD • Nonlinear Regression
SEN • Sequential Nonparametrics: Invariance Principles and Statistical Inference
SERFLING • Approximation Theorems of Mathematical Statistics
SHORACK and WELLNER • Empirical Processes with Applications to Statistics
STOYANOV • Counterexamples in Probability
STYAN • The Collected Papers of T. W. Anderson 1943—1985, Volumes 1 and 2
WHITTAKER • Graphical Models in Applied Multivariate Statistics
YANG • The Construction Theory of Denumerable Markov Processes

Applied Probability and Statistics

ABRAHAM and LEDOLTER • Statistical Methods for Forecasting
AGRESTI • Analysis of Ordinal Categorical Data
AGRESTI • Categorical Data Analysis
AICKIN • Linear Statistical Analysis of Discrete Data
ANDERSON and LOYNES • The Teaching of Practical Statistics
ANDERSON, AUQUIER, HAUCK, OAKES, VANDAELE, and WEISBERG • Statistical Methods for Comparative Studies
ARTHANARI and DODGE • Mathematical Programming in Statistics
ASMUSSEN • Applied Probability and Queues
*BAILEY - The Elements of Stochastic Processes with Applications to the Natural Sciences
BARNETT • Interpreting Multivariate Data
BARNETT and LEWIS • Outliers in Statistical Data, *Second Edition*
BARTHOLOMEW • Stochastic Models for Social Processes, *Third Edition*
BARTHOLOMEW and FORBES • Statistical Techniques for Manpower Planning
BATES and WATTS • Nonlinear Regression Analysis and Its Applications
BECK and ARNOLD • Parameter Estimation in Engineering and Science

*Now available in a lower priced paperback edition in the Wiley Classics Library

Applied Probability and Statistics (Continued)

BELSLEY • Conditioning Diagnostics: Collinearity and Weak Data in Regression

BELSLEY, KUH, and WELSCH • Regression Diagnostics: Identifying Influential Data and Sources of Collinearity

BHAT • Elements of Applied Stochastic Processes, *Second Edition*

BHATTACHARYA and WAYMIRE • Stochastic Processes with Applications

BLOOMFIELD • Fourier Analysis of Time Series: An Introduction

BOLLEN • Structural Equations with Latent Variables

BOX • R. A. Fisher, The Life of a Scientist

BOX and DRAPER • Empirical Model-Building and Response Surfaces

BOX and DRAPER • Evolutionary Operation: A Statistical Method for Process Improvement

BOX, HUNTER, and HUNTER • Statistics for Experimenters: An Introduction to Design, Data Analysis, and Model Building

BROWN and HOLLANDER • Statistics: A Biomedical Introduction

BUCKLEW • Large Deviation Techniques in Decision, Simulation, and Estimation

BUNKE and BUNKE • Nonlinear Regression, Functional Relations and Robust Methods: Statistical Methods of Model Building

BUNKE and BUNKE • Statistical Inference in Linear Models, Volume I

CHAMBERS • Computational Methods for Data Analysis

CHATTERJEE and HADI • Sensitivity Analysis in Linear Regression

CHATTERJEE and PRICE • Regression Analysis by Example

CHOW • Econometric Analysis by Control Methods

CLARKE and DISNEY • Probability and Random Processes: A First Course with Applications, *Second Edition*

COCHRAN • Sampling Techniques, *Third Edition*

COCHRAN and COX • Experimental Designs, *Second Edition*

CONOVER • Practical Nonparametric Statistics, *Second Edition*

CONOVER and IMAN • Introduction to Modern Business Statistics

CORNELL • Experiments with Mixtures: Designs, Models and The Analysis of Mixture Data, Second Edition

COVER and THOMAS · Elements of Information Theory

COX • A Handbook of Introductory Statistical Methods

COX • Planning of Experiments

DANIEL • Applications of Statistics to Industrial Experimentation

DANIEL • Biostatistics: A Foundation for Analysis in the Health Sciences, *Fourth Edition*

DANIEL and WOOD • Fitting Equations to Data: Computer Analysis of Multifactor Data, *Second Edition*

DAVID • Order Statistics, *Second Edition*

DAVISON • Multidimensional Scaling

DEGROOT, FIENBERG and KADANE • Statistics and the Law

DEMING • Sample Design in Business Research

DILLON and GOLDSTEIN • Multivariate Analysis: Methods and Applications

DODGE • Analysis of Experiments with Missing Data

DODGE and ROMIG • Sampling Inspection Tables, *Second Edition*

DOWDY and WEARDEN • Statistics for Research

DRAPER and SMITH • Applied Regression Analysis, *Second Edition*

DUNN • Basic Statistics: A Primer for the Biomedical Sciences, *Second Edition*

DUNN and CLARK • Applied Statistics: Analysis of Variance and Regression, *Second Edition*

ELANDT-JOHNSON and JOHNSON • Survival Models and Data Analysis

FLEISS • The Design and Analysis of Clinical Experiments

FLEISS • Statistical Methods for Rates and Proportions, *Second Edition*

FLURY • Common Principal Components and Related Multivariate Models

FRANKEN, KÖNIG, ARNDT, and SCHMIDT • Queues and Point Processes

(***cardplate continues in back of book***)

The Construction Theory of Denumerable Markov Processes

The Construction Theory of Denumerable Markov Processes

XIANG-QUN YANG
Xiangtan University, Hunan,
People's Republic of China

With a Foreword by
D. G. KENDALL

HUNAN SCIENCE AND TECHNOLOGY PUBLISHING HOUSE
Changsha

JOHN WILEY & SONS
Chichester · New York · Brisbane · Toronto · Singapore

First published as Keliemaerkefuguochenggouzaolun
by the Hunan Science and Technology Publishing House, Changsha.

Other Wiley Editorial Offices

John Wiley & Sons, Inc., 605 Third Avenue,
New York, NY 10158–0012, USA

Jacaranda Wiley Ltd, G.P.O. Box 859, Brisbane,
Queensland 4001, Australia

John Wiley & Sons (Canada) Ltd, 22 Worcester Road,
Rexdale, Ontario M9W 1L1, Canada

John Wiley & Sons (SEA) Pte Ltd, 37 Jalan Pemimpin 05–04,
Block B, Union Industrial Building, Singapore 2057

Library of Congress Cataloging-in-Publication Data:

Yang, Hsiang-ch' ün.
[K' o lieh Ma-erh-k' o-fu kuo ch' eng kou tsao lun. English]
The construction theory of denumerable Markov processes / Xiang-qun Yang ; with a foreword by D. G. Kendall.
p. cm.
Translation of: K' o lieh Ma-erh-k' o-fu kuo ch' eng kou tsao lun.
Includes bibliograhical references.
ISBN 0 471 92490 3
1. Markov processes. I. Title.
QA274.7.Y3613 1990
519.2′ 33—dc20 89–22671

British Library Cataloguing in Publication Data:

Yang, Xiang-qun
The construction theory of denumerable Markov processes.
1. Markov processes
I. Title
519.2′33

ISBN 0 471 92490 3

Typeset by Thomson Press (India) Ltd, New Delhi
Printed and bound in Great Britain by
Biddles Ltd, Guildford and King's Lynn

Contents

PART IV PATH STRUCTURE OF DENUMERABLE MARKOV PROCESSES

Foreword to the English Edition

In this book my friend Xiang-qun Yang presents the theory of temporally-homogeneous countable-state Markov chains developing in continuous time. Systems of this kind with a finite set of states have been studied for many years. Indeed, the second volume of Fréchet's book on probability, published in 1938, gives a detailed treatment. The interest and difficulty both increase dramatically when the set of states becomes countably infinite. The first detailed theoretical treatments of this 'infinite' case were given in two papers published in 1951 and written by Kolmogorov and Lévy respectively, and these have subsequently generated an enormous literature. The book by Kai-lai Chung (1967) at once became the standard reference for the subject, but now it by no means covers all that has been done—even by Chung himself.

In recent years mathematicians in the People's Republic of China have taken up the subject with enthusiasm, and there are two especially strong centres of Markovian studies in the Province of Hunan in which Professor Yang lives—Xiangtan University (of which he is the President), and the Chinese Institute of Railways in Changsha (of which Professor Zhen-ting Hou is the Vice-President).

David Williams and I, accompanied by Sheila Williams, spent two weeks at Xiangtan University in 1983. A few lines from my diary attempt to sketch the exotic environment.

'Xiangtan University is set quite by itself on a plateau of red clay standing out in a vast landscape of rice fields. These are arranged in tiers so that the water flowing out of one field forms a source for the next, and if one follows the flow backwards then the system forks as the valleys fork in an complicated network—almost a branching process.... This was well shown when we visited and climbed (in a typhoon) the holy mountain Nan Yue. There the "branching process" ascends the mountain in a succession of ever-decreasing rice-fields (ultimately just of handkerchief size) until the forest takes over as a sort of Martin boundary.'

Our task was to give two weeks of advanced lectures to a group of some forty Chinese probabilists drawn from widely different parts of the People's Republic. I lectured on archaeological seriation, and on the statistical theory of shape, both novel topics for that audience. David Williams was not so fortunate. He set out armed with his recently published and now famous book *Diffusions, Markov Processes, and Martingales* as a text, but was astonished to see the members of the audience pour into the lecture hall, each clutching a well-thumbed xerox copy of it. 'What we want to know', they said, 'is *what happens next*'. So they received an impromptu and highly original set of lectures from him.

That incident impressed in our minds how seriously mathematical probability is taken in China—dramatic advances in Markov chain theory are exactly what one would expect. And dramatic advances there have been.

The present book first sets out the basic framework of the theory of such Markov chains, and then in the latter chapters develops more recent advanced work by Professor Yang himself and by some other Chinese writers. The author made his own translation into English, and in the earlier chapters this has been carefully checked by Professor G. E. H. Reuter whose own work forms a basis for much of the subsequent developments. But the latter chapters of the book are essentially as the author himself wrote them.

The reader should be alerted to a special feature of Chinese mathematical writing in the English language. This arises from the fact that in different traditions there are different conventions about what must be mentioned, and what need not be mentioned. Anyone who has tried translating Chinese mathematics into English will have noticed this. Sometimes, for example, algebraic manipulations will be completely omitted when we would expect them to be included. Conversely, Chinese readers must find our own mathematical writing to be extraordinarily prolix, and packed with references to apparently irrelevant matters, such as cricket. Readers from both language traditions have to learn to expect and to be patient with this. It is tempting to enlarge on this theme, but I will stop here, and wish the reader an interesting journey.

DAVID KENDALL

Foreword to the Original Edition

Markov processes occupy an extremely important position in random processes. The denumerable Markov processes form a very active and theoretically fairly complete branch of knowledge of Markov processes, which is universally applicable in many fields of science and technology. Quite a few famous probability scholars, such as A. N. Kolmogorov, J. L. Doob, W. Feller, K. L. Chung, etc., have done a lot of work in this area of knowledge for quite some time and have made important contributions. For more than two decades, Chinese workers engaged in probability theory have conducted extensive and profound studies on this subject. So far there have appeared three monographs published respectively under the following titles: *Birth–Death Processes and Markov Chains, Homogeneous and Denumerable Markov Processes* and *Reversible Markov Processes.*

In 1958 Zi-kun Wang, Professor of Nankai University (now President of Beijing Normal University), began to publish his research results in this field. Then he, together with his students and colleagues, further probed and developed his studies on the subject. Because of their long, unremitting and painstaking efforts, they have achieved excellent results on several principal problems closely related to this subject. This book is a treatise by Professor Xiang-qun Yang. It gives a comprehensive summary of his research results on the construction theory of denumerable Markov processes, which he has studied for the past 20 years. Some of the achievements contained in this book are made known for the first time.

The construction problem is a central one in denumerable Markov processes. Xiang-qun Yang's researches are mainly concentrated on the construction of birth–death processes in construction theory, and of the Q processes with finite boundaries, on the probability method in the construction theory, and on the relationship between the probability method and the analytical method. For these aspects he has obtained very good results, which come up to the most advanced level in the world.

Quite a number of people in the world have studied birth–death processes. S. Karlin gave the integral representation of the minimal solution. W. Feller constructed all birth–death processes by the analytical method, which satisfy simultaneously the systems of both backward and forward equations. Zi-kun Wang succeeded in constructing all honest birth-death processes by means of the probability method. Xiang-qun Yang has constructed all the birth–death processes by using these two methods and, moreover, has meticulously investigated the properties of birth–death processes. Therefore, the construction problem of birth–death processes has been completely solved.

The general construction problem of denumerable Markov processes is a very difficult one. In 1957 Feller constructed all Q processes satisfying the system of forward equations under the conditions that Q is conservative, finite exit and finite entrance. Latter, Xiang-qun Yang constructed all Q processes under the same conditions. D. Williams and Kai-Lai Chung derived all Q processes respectively in the case that Q is conservative and finite exit. Xiang-qun Yang has recently made further advances on this problem. He has derived all Q processes under the more extensive conditions that the non-conservative state and exit boundary of Q are both finite.

The two methods in the construction theory, i.e. the probability method and the analytical method, each have their own merits and demerits. By applying either of the two methods we should achieve certain results. But the results generated by these two methods are quite different in form. Xiang-qun Yang has found the relatonship between the two methods with regard to birth–death processes and combined these two kinds of results. Hence, he succeeds not only in giving a clear, definitive probability meaning to the results of analysis but also in expressing the probability results in equally concise forms. The work in this aspect pioneered by Xiang-qun Yang is to be further studied because it is a branch of knowledge with a broad, bright future for development.

This book is not a simple collection of the author's research achievements, but a treatise in which his research results are elaborately put into good order and carefully compiled. I believe that the reader can read this book through without much difficulty and reach the forefront of research in this field if he or she has finished reading the first three chapters of the treatise entitled *Birth–Death Processes and Markov Chains* by Zi-kun Wang. Consequently, the publication of this work will certainly give an impetus to the development of probability theory in our country.

At a time when this work is about to be published, I have expressed my views briefly on it. On account of my limited knowledge, my personal remarks cannot be without something inappropriate. Colleagues are sincerely requested to be kind enough to make comments or criticisms.

ZHEN-TING HOU
Changsha Railway Institute
3 March, 1980

Preface to the English Edition

I feel greatly honoured and very happy that my treatise *The Construction Theory of Denumerable Markov Processes* is to be available outside China in the form of this English edition. Readers abroad will be able to read the book and my research results on denumerable Markov processes will be introduced to the UK and the world.

Comprehension of this book presupposes one being familiar with the basic knowledge of random processes and the fundamental theory of Markov chains, which can be found in the book *Markov Chains with Stationary Transition Probabilities* by K. L. Chung and in the treatise *The Birth–Death Processes and Markov Chains* by Zi-kun Wang and Xiang-qun Yang. These two books were published by Springer-Verlag. For the readers' convenience, to this English edition I have added a chapter entitled 'Theoretical background', most of which is devoted to an introduction to the basic concepts and fundamental conclusions of random processes and, especially, to these of Markov chains, which are necessary for reading this book.

I sincerely thank Professor D. G. Kendall, who recommended my book to John Wiley & Sons Ltd. I am most grateful to John Wiley & Sons Ltd and to the Hunan Science and Technology Publishing House for having decided to publish the English edition of this book; likewise I would like to extend my heartfelt thanks to the Editor, Mrs Charlotte Farmer, and Associate Editor, Mr Hai-qing Hu, for their kind cooperation and great efforts. In particular, I feel very much indebted to Professor David Kendall, who has been kind enough to write a foreword to the English edition of this book. The first draft of the English version of this book was mainly done by Dr Shou-jun Luo and Mr Ying-qiu Li, and Madam Dong-ya Zou and Mr Wei-guo Tan translated the forward and introduction and some difficult paragraphs. Finally, the entire English manuscript was examined and approved by myself. Mrs Jun-fang Peng typed out the approved manuscript. Therefore, the English version of this book was born of joint efforts. To all those mentioned above, I would like to give my hearty thanks.

XIANG-QUN YANG
Xiangtan University
February 1989

Preface to the Second Edition

The second edition differs from the first one in that some major revisions and additions have been made in the following aspects: first, a concise proof of the uniqueness criterion is given; secondly, the theory of the Martin boundary is elaborated and verified, and therefore considerably enriched; thirdly, those sections about the construction of the bifinite (finite non-conservative and finite exit) Q processes are rewritten; fourthly, the approximating Markov chains and approximating minimal processes are supplemented, and DV-type and (DV)*-type extension processes and generalized DV-type and (DV)*-type extension processes are also supplemented. In addition, a large number of minor changes have been made.

The book has been enthusiastically received and heartily supported by a good many colleagues, who have expressed their interest in it since its first edition came out in 1981. Thus, the author has been greatly encouraged. What is particularly exciting is that quite a number of colleagues have suggested valuable improvements. To all these colleagues I extend my hearty thanks. My heartfelt thanks also go to the Hunan Science and Technology Publishing House for offering me the opportunity of having the second edition of this book published.

XIANG-QUN YANG
Xiangtan University
5 April 1984

Introduction

This book is a summary of the research results of the author on the construction theory of denumerable Markov processes. With an introduction to the analytical basis of the construction theory in the first chapter, special topics are presented progressively.

Construction theory is a central subject in the theory of Markov processes. It is aimed at constructing Markov processes on the basis of some known conditions. In other words, we describe the Markov processes one by one. Thus we can study the properties of each Markov process according to its general and specific characters. For example, by basing our approach on the construction theory, we may select with relative ease those processes possessing inversibility from the different types of constructed Markov processes just as does the author of the treatise entitled *Inversible Markov Processes.*

Up to now there have been two methods for dealing with the construction theory. One is the analytical method, and the other is the probability method (both of which have their merits and demerits). Certain results have been achieved by means of one or other of these two methods. We can find a brief account of this aspect in this book.

The present work has six parts, basically dealing with three major aspects. In Parts I–III, the analytical method and results of the construction theory are discussed. In Parts IV and are expounded the probability method and results of the construction theory, and the relationship between these two methods. In Part VI a discussion is given on some properties of Markov processes related to the construction theory. Considerable importance is attached to the birth–death processes, not only because they have their own important theoretical significance and application value but also because they are often the source that gives the ideas and methods for solving general problems.

The analytical basis of the construction theory is discussed in Part I. First, a study is made of the analytical propertis of the Q process as a transition probability, of the construction and properties of the minimal solution, and of the general form of the Q process; then Q processes in simple cases are directly constructed; finally, the uniqueness problem for the Q processes is discussed.

Part II is devoted to an exposition of the construction theory of birth–death processes. We have succeeded in constructing all bilateral birth–death processes and all unilateral birth–death processes, the results being complete and enlightening. A good grasp of the construction theory of birth–death processes is extremely conductive to a thorough understanding of the general construction of the Q processes.

Part III is concentrated on a study of the Martin boundary of Q processes and of its application in the construction theory. First of all, we apply the Martin boundary theory of discrete parameter Markov chains to the Q processes, conducting an extensive and profound discussion. Next we introduce the Martin exit boundary of the minimal Q process and further describe the general form of Q processes in the light of the exit boundary. Finally, we construct all the finite non-conservative and finite exit Q processes. That is to say, all the Q processes in the 'double finite' case are constructed.

In Part IV emphasis is laid upon analysing the path structures of probabilistic Q processes. First the concepts of W transformation and strong limit are introduced. General Q processes can be changed into various processes with simpler path structures by W transformation. Thus it is convenient to study the paths of processes from various aspects. The strong limit theorems demonstrate that the paths of relatively complicated Q processes can be approached by those of Q processes with simpler structures. Secondly, the concept of leaping intervals is introduced to study the exit and entrance of Q processes. A discussion is given on how leaping intervals and leaping points are related to the system of Kolmogorov equations, and the entrance decomposition theorem is derived.

Finally, the extension of processes is studied and the sample paths of processes with simpler structures are directly constructed. We mainly consider the D-type extension. In order that the Q matrix of the extension process remains unchanged, we also take into account the D^*-type extension.

Part V exclusively treats construction of the probability method for birth–death processes. As these processes are particular, the results obtained are relatively profound. Each birth–death process is not only the strong limit of a sequence of Doob processes but also corresponds to a characteristic sequence. We have established the relations between the birth–death processes constructed by the analytical method and those constructed by the probability method, so that the processes constructed by the analytical method have clear probability structure and the processes constructed by the probability method have succinct analytical expression. In this way the virtues of each method can be brought out. With simultaneous use of the two methods, the results are more remarkable.

In Part VI some properties of Markov processes, mainly recurrence and ergodic properties, are investigated. These properties depend closely on the construction theory.

I extend my hearty thanks to my teacher, Professor Zi-kun Wang, for his patient guidance, without which it would have been impossible to attain these research results. I have also greatly benefited by numerous discussions with

Professor Zhen-ting Hou. Moreover, he has read through the manuscript of this book very carefully and suggested many valuable improvements. Among those who had discussions with me and rendered me much support and encouragement are Associate Professor Qing-feng Guo and colleagues Rong Wu, Wen-chuan Mo and Mu-fa Chen. To all these I express my sincere thanks.

CHAPTER 1

Theoretical Background

1.1 INTRODUCTION

A number of basic concepts and major conclusions of random processes and Markov chains are given briefly in this chapter, but the proofs of conclusions are omitted. These concepts and conclusions will be used often in this book.[1]

1.2 RANDOM PROCESSES

1.2.1 Definitions

One or finitely many random trials are usually considered in elementary probability theory. These random trials can be described by one or finitely many random variables. In the law of large numbers and the central limit theorems, a sequence of random variables is involved, but it is assumed that the variables are independent. However, in practice, we have to study the development and process of a random phenomenon, and the events considered must be concerned with infinitely many (not necessarily denumerable) random variables. Consequently, we need to study all the random variables involved in the development and process of the random phenomenon. Thus we can depict the whole statistical law and the development of a random phenomenon. Hence we usually call a family of random variables a random process.

Example 1

Let $T = \{1, 2, \ldots, N\}$ or $T = \{0, 1, 2, \ldots\}$ and $X(i)$ be a random variable for each $i \in T$. Then $X = \{X(i), i \in T\}$ is just a random process. But at this time we call it a random vector or a random sequence.

Example 2

In order to investigate how much service is performed at a telephone service counter, we calculate it by starting from a certain time $t = 0$ and by letting $X(t)$

[1] *A note on notation.* 'Equation (1)' and 'Theorem 1', 'equation (2.1)' and 'Theorem 2.1', 'equation (3.2.1)' and 'Theorem 3.2.1', etc., refer respectively to equation (1) and Theorem 1 in the same section, in section 2 of the same chapter, and in section 2 of Chapter 3.

denote the number of telephone calls received at the service counter up to time t. Thus $X = \{X(t), t \in [0, \infty)\}$ is a random process.

Example 3

When considering the storing capacity of a warehouse, we denote by $X(t)$ the amount or quantity of some material stored in the warehouse at time t. Then $\{X(t), t \in [0, \infty)\}$ is a random process.

Example 4

When considering the problem of how far and wide an infectious disease has been spread, we use $X(t)$ to denote the number of people suffering from the disease in some region at time t. Then $\{X(t), t \in [0, \infty)\}$ is a random process.

Definition 1. Let a set T of parameters, a measurable space $(E, \mathscr{B})$ and a probability space $(\Omega, \mathscr{F}, P)$ be given. For each $t \in T$, there exists a measurable mapping from $(\Omega, \mathscr{F})$ into $(E, \mathscr{B})$ (the image of ω for the mapping $X(t)$ is denoted by $X(t, \omega)$). Then $\{X(t), t \in T\}$ is called a random process defined on the probability space $(\Omega, \mathscr{F}, P)$ and taking values in $(E, \mathscr{B})$.

An element ω of Ω is called a sample point while an element of $\mathscr{F}$ is called an event. The E or $(E, \mathscr{B})$ is called the state space; an element of E is called a state.

The parameter t is usually understood to stand for time; of course, the parameter t can also be understood to refer to something else, for instance the points in a plane or in a space. The state space $(E, \mathscr{B})$ is in general an abstract space. The one often used is $(E_0, \mathscr{B}_0)$ or $(R_1, \mathscr{B}_1)$, where E_0 is a set of all non-negative integers, $\mathscr{B}_0$ is the Borel field composed of all subsets of E_0, $R_1 = (-\infty, +\infty)$ and $\mathscr{B}_1$ is the Borel field composed of all Borel sets in R_1.

According to the definition of a random process, $X(t, \omega)$ $(t \in T,\ \omega \in \Omega)$ is a bivariate function taking values in E. When t is fixed, $X(t, \cdot)$ is a random variable; when ω is fixed, $X(\cdot, \omega)$ is a function defined on T, and is called the sample function or the path corresponding to the sample point ω.

From now on, if we make no special statement, $(\Omega, \mathscr{F}, P)$ always refers to a complete probability space, i.e. a probability space with the property that any subset of a null-probability event is also an event. T always refers to $\{0, 1, 2, \ldots\}$ or $[0, \infty)$; $(E, \mathscr{B})$ always refers to $(E_0, \mathscr{B}_0)$ or $(R_1, \mathscr{B}_1)$.

1.2.2 Family of finite-dimensional distribution functions and existence theorem

Just as the probabilistic feature of a random variable is represented by the distributions of the random variable in elementary probability theory, so we use the family of finite-dimensional distribution functions in the case of random processes.

Definition 2. Let $X = \{X(t), t \in T\}$ be a random process. For any positive integer n and $t_i \in T$, $1 \leqslant i \leqslant n$, denote the n-dimensional joint distribution function of $X(t_1), X(t_2), \ldots, X(t_n)$ by $F_{t_1,t_2,\ldots,t_n}$, i.e.

$$F_{t_1,t_2,\ldots,t_n}(\lambda_1, \lambda_2, \ldots, \lambda_n) = P\{X(t_i) \leqslant \lambda_i, 1 \leqslant i \leqslant n\} \qquad \lambda_i \in R_1, \quad 1 \leqslant i \leqslant n \quad (1)$$

The family of distribution functions

$$F = \{F_{t_1,t_2,\ldots,t_n}: t_i \in T, 1 \leqslant i \leqslant n, n = 1, 2, \ldots\} \qquad (2)$$

is called a family of finite-dimensional distribution functions of the random process X.

Obviously, the family F of finite-dimensional distribution functions of a random process satisfies the following consistency conditions (a) and (b):

(a) Assume that $(\alpha_1, \alpha_2, \ldots, \alpha_n)$ is any permutation of $(1, 2, \ldots, n)$. Then

$$F_{t_1,t_2,\ldots,t_n}(\lambda_1, \ldots, \lambda_n) = F_{t_{\alpha_1},t_{\alpha_2},\ldots,t_{\alpha_n}}(\lambda_{\alpha_1}, \lambda_{\alpha_2}, \ldots, \lambda_{\alpha_n})$$

(b) Assume $m < n$. Then

$$F_{t_1,\ldots,t_m}(\lambda_1, \ldots, \lambda_m) = F_{t_1,\ldots,t_m,t_{m+1},\ldots,t_n}(\lambda_1, \ldots, \lambda_m, \infty, \ldots, \infty)$$

Theorem 1. Given a set T of parameters and a family (2) of finite-dimensional distribution functions satisfying the consistency conditions (a) and (b), then there exists a probability space $(\Omega, \mathscr{F}, P)$ on which is defined a random process $X = \{X(t), t \in T\}$ whose family of finite-dimensional distribution functions coincides with the given F.

1.2.3 Separability

In researching into random processes, non-denumerable union ($\cup$) and intersection ($\cap$) of events are often involved. For instance, sometimes we have to consider the set A of those sample points for which the corresponding sample functions of the random process $X = \{X(t), t \in T\}$ are bounded by C in the interval $[0, 1]$, i.e.

$$A = \{\omega: X(\cdot, \omega) \text{ is bounded by } C \text{ in } [0, 1]\} = \bigcap_{0 \leqslant t \leqslant 1} \{\omega: |X(t, \omega)| \leqslant C\} \qquad (3)$$

Although each $\{\omega: |X(t, \omega)| \leqslant C\} \in \mathscr{F}$, the set A, as a non-denumerable intersection of sets of $\mathscr{F}$, does not necessarily belong to $\mathscr{F}$. Consequently we have to introduce the concept of separability.

In the discussion of separability, we allow the random process to take the value ∞, but for each $t \in T$,

$$P\{X(t) = \infty\} = 0 \tag{4}$$

Definition 3. A random process $X = \{X(t), t \in T\}$ is called separable if there exist a denumerable dense subset R of T and a null-probability event N such that, for every $\omega \notin N$, the sample function $X(\cdot, \omega)$ possesses the following property: for any $t \in T$ there exists a sequence of points $r_n \in R$ such that

$$\lim_{n \to \infty} r_n = t \qquad \lim_{n \to \infty} X(r_n, \omega) = X(t, \omega) \tag{5}$$

The set R is said to be a separability set and the set N to be an exceptional set (relative to R).

If the process is separable, then for the set A in (3), the following holds:

$$A_R \supset A \supset A_R \cap N^c$$

where $N^c = \Omega - N$ and

$$\begin{aligned} A_R &= \{\omega: X(\cdot, \omega) \text{ is bounded by } C \text{ in } [0, 1] \cap R\} \\ &= \bigcap_{\substack{0 \leqslant r \leqslant 1 \\ r \in R}} \{\omega: |X(r, \omega)| \leqslant C\} \in \mathscr{F} \end{aligned}$$

Since $(\Omega, \mathscr{F}, P)$ is complete, $A \in \mathscr{F}$.

Definition 4. Let $X = \{X(t), t \in T\}$ be a random process. If there exist a denumerable dense subset R of T and a null-probability ω set N, such that for any closed set Λ in $[-\infty, +\infty]$ and any open interval I, the following holds:

$$\{\omega: X(r, \omega) \in \Lambda, r \in IR\} - \{\omega: X(t, \omega) \in \Lambda, t \in IT\} \subset N \tag{6}$$

then the process X is called separable.

Theorem 2. The two definitions in Definitions 3 and 4 are equivalent.

In order to emphasize the separability set, we say that the separable process X is separable relative to R. If X is separable relative to any denumerable dense set R of T, then X is called well separable.

Definition 5. Two random processes $X = \{X(t), t \in T\}$ and $Y = \{Y(t), t \in T\}$ defined on the same probability space $(\Omega, \mathscr{F}, P)$ are said to be stochastically equivalent if, for any $t \in T$, it is true that

$$P\{X(t) = Y(t)\} = 1$$

In that case, X is called a modification or a version of Y.

Theorem 3. Let $X=\{X(t), t\in T\}$ be a real-valued random process. Then there always exists a separable process $Y=\{Y(t), t\in T\}$ stochastically equivalent to X.

Y is called a separable version of X. Notice that Y take the value ∞ but $P\{Y(t)=\infty\}=0$, $t\in T$.

Definition 6. A random process $X=\{X(t), t\in T\}$ is called stochastically continuous if, for any $t_0\in T$, we have

$$P\lim_{t\to t_0} X(t)=X(t_0) \tag{7}$$

Here the limit is the limit in probability.

Theorem 4. Assume that a separable process X is stochastically continuous; then X is well separable.

1.2.4 Measurability

In order to study the measurability of a sample function $X(\cdot,\omega)$, we need to define measurable random processes.

Let $T=[0,\infty)$, $\mathscr{B}_T$ be the Borel field composed of all Borel sets in T, L be the Lebesgue measure on $\mathscr{B}_T$, $\mu=L\times P$ be the product measure defined on $\mathscr{B}_T\times\mathscr{F}$ and $\overline{\mathscr{B}_T\times\mathscr{F}}$ be the completion of $\mathscr{B}\times\mathscr{F}$ relative to μ.

Definition 7. A random process $X=\{X(t), t\in T\}$ is called Borel-measurable if, for any $c\in R_1$, we have

$$\{(t,\omega): X(t,\omega)\leqslant c\}\in\mathscr{B}_T\times\mathscr{F} \tag{8}$$

X is called measurable if, for any $c\in R_1$, we have

$$\{(t,\omega): X(t,\omega)\leqslant c\}\in\overline{\mathscr{B}_T\times\mathscr{F}}. \tag{9}$$

Theorem 5. A right-continuous process (i.e. a process whose sample functions are all right-continuous) is Borel-measurable. A process almost all of whose sample functions are right-continuous is measurable.

Theorem 6. A stochastically continuous process must have a measurable version.

1.2.5 Sample functions: continuity and step functions

Theorem 7. Let $X=\{X(t),\ t\in[a,b]\}$ be a separable random process. Assume that there exist three real numbers $\alpha>0$, $\varepsilon>0$ and $c\geqslant 0$ such that for arbitrary

$t \in [a, b]$, $t + \Delta \in [a, b]$, we have

$$E\{|X(t+\Delta) - X(t)|^{\alpha}\} \leqslant c|\Delta|^{1+\varepsilon} \tag{10}$$

Then almost all sample functions of $\{X(t), t \in [a, b]\}$ are continuous.

Definition 8. Let $\{X(t), t \in [a, b]\}$ be a real-valued function, and suppose that there is a finite subdivision $a = t_0 < t_1 < \cdots < t_n = b$ such that

(i) $X(t) = c_i$ for $t_i < t < t_{i+1}$, $i = 0, 1, \ldots, n-1$, so that the limits $X(t_i - 0) = c_{i-1}$ $(0 < i \leqslant n)$ and $X(t_i + 0) = c_i$ $(0 \leqslant i < n)$ exist.

(ii) Either $X(t_i) = X(t_i - 0)$ $(0 < i \leqslant n)$, in which case $X(\cdot)$ is left-continuous on $(a, b]$, or $X(t_i) = X(t_i + 0)$ $(0 \leqslant i < n)$, in which case $X(\cdot)$ is right-continuous on $[a, b)$.

Then $\{X(t),\ t \in [a, b]\}$ is said to be a step function defined on $[a, b]$; if $X(t_i - 0) \neq X(t_i + 0)$, we call t_i a jumping point.

A function $\{X(t), t \in [0, \infty)\}$ is called a step function if $\{X(t), t \in [0, b_n]\}$ is step function for a sequence $b_n \uparrow \infty$.

Theorem 8. Let $\{X(t), t \in [a, b]\}$ be a separable random process. If there exists a constant $c \geqslant 0$ such that, for arbitrary $t \in [a, b]$, $t + \Delta \in [a, b]$, the following holds:

$$P\{X_{t+\Delta} \neq X_t\} \leqslant c|\Delta| \tag{11}$$

then almost all sample functions of $\{X(t), t \in [a, b]\}$ are step functions.

1.2.6 Classification

According to whether the set T of parameters and the state space E are denumerable or not, we can simply classify random processes into four classes:

(i) T and E are denumerable;

(ii) T is denumerable whereas E is not;

(iii) T and E are not denumerable at all;

(iv) T is not denumerable, but E is.

A random process with a denumerable set of parameters is called a random sequence or a time series. A random process with a denumerable state space is called a denumerable process.

The classifications above are only in form. They are not concerned with the inherent probability relation of a random process. If we classify random processes according to the inherent probability relation of a process, then a lot of important classes of processes can be obtained. For example, consider the following.

Definition 9. Let $X(t), t\in T$, be independent random variables. Then $X = \{X(t), t\in T\}$ is said to be an independent process.

Definition 10. Suppose that, for arbitrary $t_i \in T$, $t + t_i \in T$, $1 \leqslant i \leqslant n$,

$$X(t_1), X(t_2), \ldots, X(t_n)$$

and

$$X(t_1 + t), X(t_2 + t), \ldots, X(t_n + t)$$

have the same joint distribution function. Then the random process $X = \{X(t), t\in T\}$ is called stationary.

An extremely important class of random processes is that of Markov processes, which will be introduced below.

1.3 MARKOV PROCESSES

1.3.1 Definitions

The intrinsic probability structure of an independent random process is simple. Let $T = [0, \infty)$; the parameter is understood as time, and $X(t)$ is understood as the position of a particle making stochastic movement at time t. We consider the time s as the 'present'; then the time interval $[0, s)$ is the 'past', and the time interval (s, ∞) is the 'future'. Suppose that we want to predict the 'future' of the random process if we have information about the 'past' and the 'present', or about the 'present', or know nothing about the 'past' and 'present'. For an independent process, the results predictable are the same in the three cases above. But for a Markov process, the result predictable are the same in the first and second cases above; roughly speaking, under the condition that the 'present' is known, the 'future' and the 'past' are conditionally independent. This property is said to be the Markov property.

Examples in reality possessing the Markov property are numerous. For instance, if we leave out secondary factors, population growth is characterized by the Markov property. The number of a population in future is relative only to the present population base and not relative to the number of people in the past. The process in example 2 in section 1.2 is also characterized by the Markov property.

Definition 1. A random process $X = \{X(t), t\in T\}$ is called a Markov process if, for arbitrary $0 \leqslant t_1 < t_2 < \cdots < t_n$ and $\Gamma \in \mathscr{B}$, we have

$$P\{X(t_n)\in\Gamma \mid X(t_1), \ldots, X(t_{n-1})\} = P\{X(t_n)\in\Gamma \mid X(t_{n-1})\} \qquad \text{a.e.} \qquad (1)$$

The property (1) is called the Markov property. The Markov property has many equivalent forms. For instance:

(a) For arbitrary $0 \leqslant s \leqslant t$, $\Gamma \in \mathscr{B}$,

$$P\{X(t) \in \Gamma \mid X(u), 0 \leqslant u \leqslant s\} = P\{(t) \in \Gamma \mid X(s)\} \qquad \text{a.e.} \tag{2}$$

(b) For arbitrary $0 \leqslant s \leqslant t$, and N^s-measurable random variable $\eta, E|\eta| < \infty$,

$$E\{\eta \mid X(u), 0 \leqslant u \leqslant s\} = E\{\eta \mid X(s)\} \qquad \text{a.e.} \tag{3}$$

where $N^s = \sigma\{X(u), u \geqslant s\}$ is the Borel field generated by $\{X(u), u \geqslant s\}$.

1.3.2 Transition functions and homogeneity

Definition 5. A Markov process $\{X(t), t \in T\}$ is called homogeneous if there exists a regular transition probability function (simply, transition function) if the following four conditions are satisfied:

(i) for fixed s, x, t, $P(s, x; t, \cdot)$ is a probability measure on $\mathscr{B}$;

(ii) for fixed $s, t, \Gamma, P(s, \cdot; t, \Gamma)$ is a $\mathscr{B}$-measurable function;

(iii) for any $s \in T$

$$P(s, x; s, \Gamma) = I_\Gamma(x) \tag{4}$$

where I_Γ is the indicator of Γ;

(iv) the Chapman–Kolmogorov equation holds: for arbitrary $0 \leqslant s \leqslant t \leqslant u$, there is

$$P(s, x; u, \Gamma) = \int_E P(s, x; t, \mathrm{d}y) P(t, y; u, \Gamma) \tag{5}$$

The condition (iii) is called the regularity condition.

Definition 3. A transition function $P(s, x; t, \Gamma)$ is said to be homogeneous if for fixed x and Γ the values of the function only depend on $t - s$. That is, there exists a three-variate function $P(t, x, \Gamma)$ such that

$$P(s, x; t, \Gamma) = P(t - s, x, \Gamma) \tag{6}$$

In that case (i)–(iv) in Definition 2 become:

(a) for fixed t and $x, P(t, x, \cdot)$ is a probability measure on $\mathscr{B}$;

(b) for fixed t and $\Gamma, P(t, \cdot, \Gamma)$ is a $\mathscr{B}$-measurable function;

(c) $P(0, x, \Gamma) = I_\Gamma(x)$;

(d) for arbitrary $s, t \geqslant 0$, we have

$$P(s + t, x, \Gamma) = \int_E P(s, x, \mathrm{d}y) P(t, y, \Gamma) \tag{7}$$

The three-variate function $P(t, x, \Gamma)$ satisfying (a)–(d) above is said to be a homogeneous transition functions.

Example 1

(Wiener transition function).

$$P(t,x,\Gamma)=\begin{cases}\dfrac{1}{\sigma(2\pi t)^{1/2}}\displaystyle\int_\Gamma \exp\left(-\frac{(y-x)^2}{2\sigma^2 t}\right)\mathrm{d}y & t>0\\ I_\Gamma(x) & t=0\end{cases}$$

Here $\sigma>0$ is a constant.

Example 2

(Poisson transition function). Let $E=\{0,1,2,\ldots\}$. Put

$$p_{ij}(t)=\begin{cases}\dfrac{(\lambda t)^{j-i}}{(j-i)!}\mathrm{e}^{-\lambda t} & \text{if } t>0,\ j\geqslant i\\ 0 & \text{if } t>0,\ j<i\\ \delta_{ij} & \text{if } t=0\end{cases}$$

Here $\lambda>0$ is a constant. And then set

$$P(t,i,\Gamma)=\sum_{j\in\Gamma} p_{ij}(t) \qquad t\geqslant 0$$

Definition 4. A Markov process is said to have a transition function if there exists a transition function $P(s,x;t,\Gamma)$ such that

$$P\{X(t)\in\Gamma\,|\,X(s)\}=P(s,X(s);t,\Gamma) \qquad \text{a.e.} \tag{8}$$

Definition 5. A Markov process $\{X(t),t\in T\}$ is called homogeneous if there exists a homogeneous transition function $P(t,x,\Gamma)$ such that

$$P\{X(s+t)\in\Gamma\,|\,X(s)\}=P\{t,X(s),\Gamma\} \qquad \text{a.e.} \tag{9}$$

1.3.3 Family of finite-dimensional distributions and existence theorem

Theorem 1. Let $\{X(t),t\in T\}$ be a random process. In order that $\{X(t),t\in T\}$ is a homogeneous Markov process, it is necessary and sufficient that there exists a homogeneous transition function $P(t,x,\Gamma)$ such that for any $0\leqslant t_1<t_2<\cdots<t_n$ and $\Gamma_i\in\mathscr{B}$, $1\leqslant i\leqslant n$, the finite-dimensional distributions are given by

$$P(X(t_i)\in\Gamma,1\leqslant i\leqslant n)$$

$$=\int_E \mu(\mathrm{d}y_0)\int_{\Gamma_1}P(t_1,y_0,\mathrm{d}y_1)\int_{\Gamma_2}P(t_2-t_1,y_1,\mathrm{d}y_1)\cdots\int_{\Gamma_n}P(t_n-t_{n-1},y_{n-1},\mathrm{d}y_n) \tag{10}$$

where $\mu(\Gamma)=P(X(0)\in\Gamma)$ is the initial distribution.

Theorem 2 (Existence theorem). Let a homogeneous transition function $P(t, x, \Gamma)$, $t \geqslant 0$, $x \in R_1$, $\Gamma \in \mathscr{B}_1$, and a probability measure $\mu(\Gamma), \Gamma \in \mathscr{B}$, be given. Then there exists a probability space $(\Omega, \mathscr{F}, P)$ on which is defined a homogeneous Markov process $\{X(t), t \geqslant 0\}$ with the given transition function $P(t, x, \Gamma)$ and the given initial distribution $\mu(\Gamma)$.

1.3.4 Semigroup property and Feller transition functions

Let B be the collection of all bounded $\mathscr{B}$-measurable functions defined on E. For $f \in B$ we define a norm $\| f \| = \sup_{x \in E} | f(x) |$. Then B is a Banach space. For $f \in B$ we define a function by

$$T_t f(x) = \int_E f(y) P(t, x, \mathrm{d}y) \tag{11}$$

Obviously, $T_t f \in B$, and moreover, from (a) and (b) for a homogeneous transition function (Definition 3) we have

$$\| T_t \| \leqslant 1 \qquad T_{s+t} = T_s T_t \tag{12}$$

That is, $\{T_t, t \geqslant 0\}$ is a family of contraction operators, forming a semigroup. Whence we can determine its strong (or weak) infinitesimal operator A (or $\tilde{A}$) and strong (or weak) resolvent operators R_λ (or $\tilde{R}_\lambda$), $\lambda > 0$. Therefore we may research into the tansition function of Markov processes by means of the theory of semigroups in functional analysis.

Assume that the state space E is endowed with a topology and C denotes the collection of all bounded continuous functions defined on E. If $T_t f \in C$ for any $f \in C$, the transition function is called the Feller transition function.

1.3.5 Strong Markov property

Definition 6. Let $X = \{X(t), t \in T\}$ be a random process. We say that X is progressively measurable if, for any $t \in T$ and $\Gamma \in \mathscr{B}$, it is valid that

$$\{(s, \omega): 0 \leqslant s \leqslant t, X(s, \omega) \in \Gamma\} \in \mathscr{B}_{[0,t]} \times N_t \tag{13}$$

where $\mathscr{B}_{[0,t]}$ is the Borel field composed of all Borel sets in $[0, t] \cap T$, and N_t is the Borel field generated by $\{X(s), 0 \leqslant s \leqslant t, s \in T\}$, which is called the pre-t field.

Theorem 3. A random process whose sample functions all are right- (or left-) continuous is progressively measurable.

Definition 7. Let τ be a non-negative (including $+\infty$) $\mathscr{F}$-measurable function defined on Ω. If for any $t \in T$ the following holds:

$$(\tau \leqslant t) \in N_t \tag{14}$$

τ is called a stopping time of the process $\{X(t), t\in T\}$. For a stopping time τ, the Borel field

$$N_\tau = \{A: A\in\mathscr{F}, \text{ and moreover, } A\cap(\tau\leqslant t)\in N_t \text{ for each } t\geqslant 0\} \tag{15}$$

is called the pre-τ field.

Definition 8. Let $X = \{X(t), t\in T\}$ be a homogeneous Markov process. Assume that X is progressively measurable, and moreover, for any stopping time τ, it is valid that

$$P\{X(\tau+t)\in\Gamma \mid N_\tau\} = P(t, X(\tau), \Gamma) \qquad \text{a.e. } \Omega_\tau \tag{16}$$

where a.e. Ω_τ indicates that it is valid for almost all $\omega\in\Omega_\tau = (\tau<\infty)$. X is called a strong Markov process and (16) is said to be the strong Markov property.

Theorem 4. A homogeneous Markov process $X = \{X(n), \; n = 0, 1, 2, \ldots\}$ possesses the strong Markov property.

Theorem 5. A right-continuous Feller process is a strong Markov process.

1.4 MARKOV CHAINS

1.4.1 Definitions and transition matrices

A homogeneous Markov chain is a special homogeneous Markov process. When $T = \{0, 1, 2, \ldots\}$ and E is denumerable, we might as well set $E = \{0, 1, 2, \ldots\}$, a homogeneous Markov process is said to be a homogeneous Markov chain, which is called a Markov chain for short. Of course, we can define a Markov chain directly.

Definition 1. $\{X(n), n = 0, 1, 2, \ldots\}$ is called a Markov chain if, for arbitrary $m\geqslant 0$, $n\geqslant 0$, $i_0, i_1, \ldots, i_{m-1}, i, j\in E$, the following equality holds:

$$\begin{aligned} P\{X(m+n) = j \mid X(0) = i_0, X(1) = i_1, \ldots, X(m-1) = i_{m-1}, X(m) = i\} \\ = P\{X(m+n) = j \mid X(m) = i\} \end{aligned} \tag{1}$$

and moreover, the right side is independent of m, provided the conditional probability is defined.

We may as well assume that for each $i\in E$, there exists a parameter $m\in T$ such that $P\{X(m) = i\} > 0$. Otherwise, i is a non-essential state and i may be combed out of E. We can consider the state space $E - \{i\}$ instead of E.

Denote the right in (1) by $p_{ij}^{(n)}$, which is called the n-step transition probability from i to j. The matrix $P^{(n)} = (p_{ij}^{(n)}, \; i, j\in E)$ is called the n-step transition matrix. When $n = 1$, we write $P = P^{(1)}$, $p_{ij} = p_{ij}^{(1)}$. The matrix P is called the transition matrix of the chain. Owing to the Chapman–Kolmogorov equations, it follows that

$$P^{(n)} = P^n \tag{2}$$

Consequently, the transition law of the Markov chain is determined by its one-step transition matrix. Therefore, sometimes we call P the Markov chain.

For a Markov chain $\{X(n),\ n=0,1,2,\ldots\}$, the distribution $\gamma=(\gamma_i)$, where $\gamma_i = P(X(0)=i)$, is called the initial distribution of the chain. Moreover, the family of distributions

$$P(X(0)=i_0, X(1)=i_1,\ldots, X(n)=i_n) = \gamma_{i_0} P_{i_0 i_1} \cdots P_{i_{n-1} i_n}, \qquad i_0, i_1, \ldots, i_n \in E$$

and the family of finite-dimensional distributions

$$P(X(n_1)=i_1, X(n_2)=i_2,\ldots, X(n_k)=i_k), \qquad n_1 < n_2 < \cdots < n_k,$$

are determined by the initial distribution and the transition matrix P. Conversely, given a distribution $\gamma=(\gamma_i)$ and a nonnegative matrix $P=(P_{ij})$ with $\sum_j P_{ij}=1$ for every i, by the existence theorem, there exists a probability space $(\Omega, \mathscr{F}, P)$, on which is defined a Markov chain $\{X(n),\ n \geqslant 0\}$. This chain has the initial distribution $\gamma=(\gamma_i)$ and transition matrix $P=(P_{ij})$. The probability measure P depends on the initial distribution $\gamma=(\gamma_i)$, and we also denote it by P_γ instead of the probability measure P. The measure P_γ is said to be the measure generated by the initial distribution $\gamma=(\gamma_i)$ and the matrix $P=(P_{ij})$. When $\gamma_i=1,\ \gamma_j=0\ (j\neq i)$, we denote it by $p_\gamma = p_i$. For general γ we have $p_\gamma = \sum_i \gamma_i p_i$.

Example 1

(Random Walk). Assume that $Z(0)=0$ is a constant, $Z(i)\ (i=1,2,\ldots)$ are independent and of identical distribution, and, moreover,

$$P\{Z(i)=1\}=p \qquad P\{Z(i)=-1\}=q \qquad P\{Z(i)=0\}=r$$

put

$$p+q+r=1 \qquad 0<p, \qquad 0<q, \qquad 0\leqslant r<1.$$

$$X(n)=Z(0)+Z(1)+\cdots+Z(n)$$

Then $\{X(n),\ n=0,1,2,\ldots\}$ is a Markov chain, its state space is $\{\cdots, -2, -1, 0, 1, 2, \ldots\}$, and its transition probabilities are

$$P_{i,i+1}=p \qquad P_{i,i-1}=q \qquad P_{ii}=r \tag{3}$$

$\{X(n), n=0,1,2,\ldots\}$ is called a simple random walk, and a symmetric random walk when $r=0$ and $p=q=\frac{1}{2}$.

The example above can be explained intuitively as follows. A particle moves stochastically in a number axis. It starts from the origin. Then it moves once per unit time. It shifts a unit distance either towards the left with probability

q or towards the right with probability p; it may stand still with probability r. $X(n)$ denotes the position of the particle after it has moved n times.

Example 2

(Random walk with barriers). Let the state space be $\{0, 1, 2, \ldots, b\}$. Where the particle is at i $(0 < i < b)$, it makes the simple random walk. Let the particle be at 0. If the particle goes to 1 after unit time with probability 1, i.e. $p_{01} = 1$, the barrier 0 is called a reflecting barrier; if the particle stays at 0 after unit time with probability 1, then 0 is called an absorbing barrier. If the particle stays at 0 with probability r_0 $(0 < r_0 < 1)$ and goes to 1 with probability $p_0 = 1 - r_0$, the barrier 0 is called an elastic barrier. If we place variously different barriers on 0 and b, we will obtain variously different random walks. For instance, the matrix

$$P = \begin{pmatrix} 1 & 0 & 0 & \cdots & 0 & 0 & 0 \\ q & r & p & \cdots & 0 & 0 & 0 \\ 0 & q & r & \cdots & 0 & 0 & 0 \\ \vdots & \vdots & \vdots & & \vdots & \vdots & \vdots \\ 0 & 0 & 0 & \cdots & q & r & p \\ 0 & 0 & 0 & \cdots & 0 & 1 & 0 \end{pmatrix}$$

represents a random walk with an absorbing barrier 0 and a reflecting barrier b.

Example 3

(Biological chains). Let the state space be $\{0, 1, 2, \ldots\}$. If the particle is at i, when the particle walks stochastically to the left or to the right, or stands motionless its probabilities are related to the position i. For example, the probabilities are $q_i, p_i, r_i, q_i + p_i + r_i = 1$, $q_0 = 0$, respectively, that is,

$$P = \begin{pmatrix} r_0 & 1 - r_0 & 0 & 0 & \cdots \\ q_1 & 1 - (p_1 + q_1) & p_1 & 0 & \cdots \\ 0 & q_2 & 1 - (p_2 + q_2) & p_2 & \cdots \\ 0 & 0 & q_3 & 1 - (p_3 + q_3) & \cdots \\ \vdots & \vdots & \vdots & \vdots & \end{pmatrix}$$

A chain with the transition matrix P above is called a birth–death chain. The intuitive meaning goes as follows:

Assume that there is a biological group, the number of whose individuals is changeable. Suppose that at time n the number of individuals of the group is denoted by i. Then at time $n + 1$, the biological group increases to $i + 1$ individuals with probability p_i, reduces to $i - 1$ individuals with probability q_i and keeps i individuals with probability r_i.

1.4.2 Classification of states

Let $P=(p_{ij}, i,j\in E)$ be the transition matrix of a Markov chain $\{X(n),\ n=0, 1,2,\ldots\}$.

Defintion 2. Assume that the state i is fixed and the set $\{n{:}n\geqslant 1,\ p_{ii}^{(n)}>0\}$ is non-empty. Let $d=d(i)$ be the greatest common divisor of the set. The number d is called the period of the state i. The state i is called periodic when $d>1$ and is non-periodic when $d=1$.

Let $\tau=\tau(j)$ denote the first time of reaching the state j:

$$\tau=\begin{cases}\min\{n\geqslant 1{:}X(n)=j\} & \text{if the set is non-empty}\\ +\infty & \text{otherwise}\end{cases}$$

Then

$$f_{ij}^{(n)}=P\{\tau=n\,|\,X(0)=i\}\qquad n\geqslant 1$$

is the probability that the Markov chain will first visit the state j at time n, given that it starts from state i. Furthermore,

$$f_{ij}^{*}=\sum_{n=1}^{\infty} f_{ij}^{(n)}$$

is the probability that the Markov chain will first visit the state j after finitely many steps, given that it starts from the state i, and

$$\mu_{ij}=\sum_{n=1}^{\infty} n f_{ij}^{(n)}$$

is the mean number of steps that the Markov chain visits j first, given that it starts from state i. Write $f_i^{*}=f_{ii}^{*}$, $\mu_i=\mu_{ii}$.

Definition 3. (a) The state i is called recurrent if $f_i^{*}=1$ and transient if $f_i^{*}<1$. (b) Let the state i be recurrent. The state i is called positive-recurrent if $\mu_i<\infty$ and null-recurrent if $\mu_i=\infty$. (c) The state i is said to be ergodic if it is non-periodic and positive-recurrent.

Theorem 1. For arbitrary $i,j\in E$ and $n\geqslant 1$, it is true that

$$p_{ij}^{(n)}=\sum_{k=1}^{n} f_{ij}^{(k)}p_{jj}^{(n-k)}=\sum_{k=0}^{n-1} p_{ij}^{(k)}f_{jj}^{(n-k)}$$

where $p_{ij}^{(0)}=\delta_{ij}, f_{jj}^{(0)}\doteq 1$.

Set

$$G_{ij}=\sum_{n=0}^{\infty} p_{ij}^{(n)}$$

Theorem 2. It is true that

$$G_{ii} = 1/(1 - f_i^*)$$

$$G_{ij} = f_{ij}^* G_{jj} \qquad (i \neq j)$$

Theorem 3. For the state i to be recurrent it is necessary and sufficient that $G_{ii} = \infty$.

Theorem 4. Let the state i be recurrent.

(a) i is null-recurrent if and only if $\lim_{n \to \infty} p_{ii}^{(n)} = 0$.

(b) i is ergodic if and only if $\lim_{n \to \infty} p_{ii}^{(n)} = 1/\mu_i > 0$.

1.4.3 Decomposition of the state space

Definition 4. We say that the state i leads to the state j if there exists an integer $n \geqslant 1$ such that $p_{ij}^{(n)} > 0$. In this case we write $i \to j$. Also we write $i \leftrightarrow j$ to mean that i and j communicate, i.e. $i \to j$ and $j \to i$.

Definition 5. A subset C of E is called closed if for any $i \in C$ and $k \bar{\in} C$ it is true that $p_{ik} = 0$. If two arbitrary states in a closed set C communicate, the set C is called an irreducible closed set.

Theorem 5 (Decomposition theorem). The state space has a unique decomposition

$$E = E_0 \cup \bigcup_{a \in \mathscr{A}} E_a$$

where $\mathscr{A}$ is empty, or the finite set $\{1, 2, \ldots, b\}$, or the denumerable set $\{1, 2, 3, \ldots\}$; E_0 is the set lf all transient states, which may be empty; for each $a \in \mathscr{A}$, E_a is a irreducible and recurrent class; the states in E_a are all null, or all positive, or all ergodic; finally, all E_a, $a \in \{0\} \cup \mathscr{A}$, are mutually disjoint.

1.5 DENUMERABLE MARKOV PROCESSES

1.5.1 Definitions

A denumerable Markov process is a Markov process with a continuous time and a discrete state space. We usually take $T = [0, \infty)$, $E = \{0, 1, 2, \ldots\}$.

Definition 1. Let the state space E of the random process $X = \{X(t),\ t \geqslant 0\}$ be denumerable. X is said to be a Markov process if for arbitrary $n \geqslant 1$,

$0 \leqslant t_1 < \cdots < t_n < s < t$, $i_1, \ldots, i_n$, $i, j \in E$, we have

$$P\{X(s+t)=j \mid X(t_1)=i_1, \ldots, X(t_n)=i_n, X(s)=i\} = P\{X(s+t)=j \mid X(s)=i\} \quad (1)$$

provided that

$$P\{X(t_1)=i_1, \ldots, X(t_n)=i_n, X(s)=i\} > 0$$

If the right side of (1) depends on only $t-s$ and does not depend on s, the Markov process X is called homogeneous, or a denumerable Markov process.

Denoting the right side of (1) by $p_{ij}(s,t)$, then the matrices $P(s,t) = \{p_{ij}(s,t), i,j \in E\}$, $0 \leqslant s \leqslant t$, are called the transition matrices of X. Write $E_+(s) = \{k: P\{X(s)=k\} > 0\}$, $E_0(s) = E - E_+(s)$. If $E_+(s) = E$ for each $s \geqslant 0$, then the transition matrices $\{P(s,t),\ 0 \leqslant s \leqslant t\}$ of X exist, and satisfy the following conditions:

(a) $p_{ij}(s,t) \geqslant 0 \qquad \sum_j p_{ij}(s,t) = 1$

(b) $p_{ij}(s,t) = \sum_k p_{ik}(s,u) p_{kj}(u,t) \qquad 0 \leqslant s \leqslant u \leqslant t$

(c) $p_{ij}(s,s) = \delta_{ij}$

But if $E_+(s) \neq E$ for some s, $p_{ij}(s,t)$ is defined for $i \in E_+(s)$ and undefined for $i \in E_0(s)$. Does X possess the transition matrices satisfying the above condition (a), (b), (c) in the general case? The books that I have read affirm the truth of the conclusion but the proof is not given. My postgraduate student Yu-Quan Xie gives a proof of this below.

Theorem 1. Each Markov process X possesses the transition matrices $P(s,t)$ satisfying the above conditions (a), (b), (c), such that when $i \in E_+(s)$, the right side of (1) is $p_{ij}(s,t)$.

Proof. For $i \in E_+(s)$, we define $p_{ij}(s,t)$ by the right side of (1).

Let $i \in E_0(s)$. Take arbitrary non-negative numbers $u_{ij}(s)\,(j \in E)$ such that $u_{ij}(s) = 0$ for $j \in E_0(s)$ and $\sum_{j \in E_+(s)} u_{ij}(s) = 1$. Set

$$p_{ij}(s,t) = \sum_{k \in E_+(s)} u_{ik}(s) p_{kj}(s,t) \qquad 0 \leqslant s \leqslant t \cdot \quad (2)$$

By a direct verification, it easily follows that $P(s,t) = \{p_{ij}(s,t),\ i,j \in E\}$, $0 \leqslant s \leqslant t$, are just what we want. QED

In the following X is always assumed to be a homogeneous denumerable Markov process. Then $p_{ij}(s,t) = p_{ij}(t-s)$ is independent of s, and moreover, (a), (b), (c) become

(α) $p_{ij}(t) \geqslant 0 \qquad \sum_j p_{ij}(t) = 1$

(β) $p_{ij}(s+t) = \sum_k p_{ik}(s)p_{kj}(t)$

(γ) $p_{ij}(0) = \delta_{ij}$

Furthermore, we assume the condition of standardness:

(δ) $\lim_{t \downarrow 0} p_{ij}(t) = p_{ij}(0)$

1.5.2 Properties of sample functions

A family of matrices $P(t) = \{p_{ij}(t),\ i,j \in E\}$, $t \geqslant 0$, satisfying the above conditions (α)–(δ) are called standard transition matrices.

Theorem 2. For standard transition matrices there exist right derivatives at $t = 0$,

$$q_{ij} = p'_{ij}(0) = \lim_{t \downarrow 0} \frac{p_{ij}(t) - p_{ij}(0)}{t} \tag{3}$$

and moreover,

$$0 \leqslant q_{ij} < \infty \qquad (i \neq j) \qquad 0 \leqslant -q_{ii} \leqslant \infty$$

$$\sum_{j \neq i} q_{ij} \leqslant -q_{ii} \tag{4}$$

From (3) it follows that

$$p_{ij}(t) = q_{ij}t + o(t) \qquad i \neq j \tag{5}$$

if $q_i < \infty$ then

$$p_{ii}(t) = 1 - q_i t + o(t) \tag{6}$$

The matrix $Q = (q_{ij}, i,j \in E)$ is called the Q matrix of the process X or of the transition matrix. It can be seen from (3) that Q is determined by the values of $P(t)$ in a small interval $[0. \varepsilon)$. In practice Q is easier to calculate than $P(t)$. Therefore how to calculate from Q constitutes the central problem discussed in this book, that is, the construction problem.

Definition 2. Write $q_i = -q_{ii}$. The state i is said to be stable if $q_i < \infty$, to be absorbing if $q_i = 0$, and to be instaneous if $q_i = +\infty$.

Notice that $X(t, \omega)$ may be $+\infty$ if we consider a separable process $X = \{X(t), t \geqslant 0\}$, but $P\{X(t) = \infty\} = 0$ for fixed $t \geqslant 0$.

Theorem 3. Let X be separable and measurable, and i be an instantaneous state.

Then for almost all $\omega \in \Omega$, the set

$$S_i(\omega) = \{t : t \geqslant 0, X(t, \omega) = i\}$$

is nowhere dense in $[0, \infty)$.

Theorem 4. Let X be separable and measurable, and $t \geqslant 0$ be fixed. Then for allmost all $\omega \in \Omega$, the following conclusions hold:

(i) $\lim_{s \to t} X(s, \omega) = i$ if $X(t, \omega) = i$ and i is stable.

(ii) $X(s, \omega)$ has exactly two limiting values i and ∞ as $s \downarrow t$ (or $s \uparrow t$) if $X(t, \omega) = i$ and i is instantaneous.

(iii) $X(t, \omega) \neq \infty$.

Theorem 5. Assume that X is separable and measurable. Then for almost all $\omega \in \Omega$, the sample functions $X(\cdot, \omega)$ have the following property.

Property (A): for any generic $t \geqslant 0$, as $s \downarrow t$ (or $s \uparrow t$) $X(\cdot, \omega)$ has at most one finite limiting value. There are three posibilities:

(a) $X(s, \omega) \to i$, where i is stable;

(b) $X(s, \omega)$ has exactly two limiting values i and ∞, where i is instantaneous;

(c) $X(s, \omega) \to \infty$.

Furthermore

(α) if $X(t, \omega) = i$ where i is stable, then (a) is true (with the same i) as $s \to t$ from at least one side;

(β) if $X(t, \omega) = i$ where i is instantaneous, then (b) is true (with the same i) as $s \to t$ from at least one side.

1.5.3 Canonical processes

Definition 3. A process $X = \{X(t),\ t \geqslant 0\}$ is called right lower semicontinuous if, for all $\omega \in \Omega$, we have

$$\varliminf_{s \downarrow t} X(s, \omega) = X(t, \omega) \qquad \text{all } t \geqslant 0 \tag{4}$$

A separable and measurable process that is right lower semicontinuous is said to be a canonical process.

Theorem 6. Given a denumerable Markov process X, there always exists a canonical process stochastically equivalent to X.

Theorem 7. A canonical process possesses the strong Markov property. That is, for any stopping time τ and $\Lambda \in N_\tau$, arbitrary $0 \leqslant t_0 < t_1 < \cdots < t_N$, $j_0, j_1, \ldots, j_N \in E$, the following equality holds:

$$P\{\Lambda, X(\tau + t_v) = j_v, 0 \leqslant v \leqslant N\} = P\{\Lambda, X(\tau + t_0) = j_0\} \prod_{v=0}^{N-1} p_{i_v, i_{v+1}}(t_{v+1} - t_v) \quad (7)$$

Moreover,

$$P\{X(\tau + t) = \infty \mid \Delta\} = 0 \qquad t > 0 \quad (8)$$

where $\Delta = \{\tau < \infty\}$ and $P\{X(\tau) = \infty \mid \Delta\}$ may be positive.

1.5.4 Special examples

(i) Poisson process. Let $\lambda > 0$ be a constant. The process $X = \{X(t), t \geqslant 0\}$ with the Q matrix

$$Q = \begin{pmatrix} -\lambda & \lambda & 0 & 0 & \cdots \\ 0 & -\lambda & \lambda & 0 & \cdots \\ 0 & 0 & -\lambda & \lambda & \cdots \\ \vdots & \vdots & \vdots & \vdots & \end{pmatrix}$$

is called a Poisson process.

The Poisson process is a special birth process. Assume that there is a biological group, whose individuals are deathless and may give birth to new individuals. Let $X(t)$ denote the content of the biological group (i.e. the number of all individuals) at time t. Assume that Δt is very small, and the probability that the content increases by one in the interval $[t, t + \Delta t]$ is $\lambda \Delta t + o(\Delta t)$, the probability that the content increases by two and over in the interval $[t, t + \Delta t]$ is $o(\Delta t)$, and the probability that the content is invariable in $[t, t + \Delta t]$ is $1 - \lambda \Delta t + o(t)$. Such a process $\{X(t), t \geqslant 0\}$ is just a Poisson process.

The transition probabilities of the Poisson process must be

$$p_{ij}(t) = \begin{cases} 0 & \text{if } j < i \\ e^{-\lambda t} \dfrac{(\lambda t)^{j-i}}{(j-i)!} & \text{if } j \geqslant i \end{cases} \quad (9)$$

(ii) Birth processes. A denumerable Markov process $X = \{X(t), t \geqslant 0\}$ with the Q matrix

$$Q = \begin{pmatrix} -\lambda_0 & \lambda_0 & 0 & 0 & \cdots \\ 0 & -\lambda_1 & \lambda_1 & 0 & \cdots \\ 0 & 0 & -\lambda_2 & \lambda_2 & \cdots \\ \vdots & \vdots & \vdots & \vdots & \end{pmatrix} \quad (10)$$

is called a birth process. Here $\lambda_i > 0\,(i = 0, 1, 2, \ldots)$ also. Let $X(t)$ be the content of the biological group at time t, and the individuals be deathless and give birth to new individuals. Suppose $X(t)=i$. The probability that $X(t+\Delta t)=i+1$ is $\lambda_i\Delta t+o(\Delta t)$, the probability that $X(t+\Delta t)>i+1$ is $o(\Delta t)$, and the probability that $X(t+\Delta t)=i$ is $1-\lambda_i\Delta t+o(\Delta t)$. Such a denumerable Markov process $\{X(t), t\geqslant 0\}$ is a birth process. The transition matrix $P(t)$ with its Q matrix (10) may not be unique.

(iii) Birth–death processes. A denumerable Markov process $X=\{X(t), t\geqslant 0\}$ with Q matrix

$$Q=\begin{pmatrix} -(a_0+b_0) & b_0 & 0 & 0 & \cdots \\ a_1 & -(a_1+b_1) & b_1 & 0 & \cdots \\ 0 & a_2 & -(a_2+b_2) & b_2 & \cdots \\ \vdots & \vdots & \vdots & \vdots & \end{pmatrix}$$

is called a birth–death process. Here $a_0\geqslant 0$, $b_0>0$, $a_i>0$, $b_i>0\,(i=1,2,\ldots)$.

Let $X(t)$ be the content of the biological group whose individuals may give birth to new individuals and may die. Assume $X(t)=i$. For $i>0$ the probability that $X(t+\Delta t)=i-1$ is $a_i\Delta t+o(\Delta t)$. For $i\geqslant 0$ the probability that $X(t)=i+1$ is $b_i\Delta t+o(\Delta t)$, and the probability that $X(t)=i$ is $1-(a_i+b_i)\Delta t+o(\Delta t)$. The Markov process $X=\{X(t),\ t\geqslant 0\}$ is a birth–death process.

PART I GENERAL CONSTRUCTION THEORY

CHAPTER 2

Introduction to Construction Theory

2.1 INTRODUCTION

In this chapter the fundamental results of the construction theory are introduced. They are summarized as follows: the analytical properties of processes, such as continuity, existence of Q matrices, differentiability; the conditions under which the Kolmogorov equations are satisfied; the construction and properties of the minimal solution and the general form of the Q processes. The greater part of this chapter is derived from Reuter (1957). Sections 2.4 and 2.5 and Theorem 6.2 are obtained from Chung (1967) and the conclusions in section 2.8 from Reuter (1959, 1962) and Feller (1957a). Sections 2.10–2.12 are derived from Xiang-qun Yang (1981a).

2.2 NOTATION AND DEFINITIONS

Let E be a denumerable set of indices. We call it the state space. We denote by m the Banach space that is composed of the bounded column vectors (or bounded functions) on E. Let the norm of $f \in m$ be defined by $\| f \| = \sup_{i \in E} |f_i|$. We denote by l the Banach space that is composed of summable row vectors on E and let the norm of $g \in l$ be defined by $\| g \| = \sum_{j \in E} |g_i|$. If $f \in m, g \in l$, their inner product will be defined by

$$[g, f] = \sum_{i \in E} g_i f_i \tag{1}$$

Matrix notation will be used and the limit of matrices will be defined element by element. From the analytical point of view, a Markov process (abbreviated to process) is a family of real matrices $P(t) = \{p_{ij}(t)\}$ $(i, j \in E, t \geqslant 0)$ which satisfy the following conditions:

(A) $P(t) \geqslant 0 \qquad P(t)\mathbf{1} \leqslant \mathbf{1}$

(B) $P(t + s) = P(t)P(s)$

(C) $\lim_{t \downarrow 0} P(t) = P(0) = \boldsymbol{I}$

where 0 represents the zero matrix and sometimes the zero column (or zero row) vector; **1** represents the unit column vector whose components are 1. ***I*** represents the unit matrix.

In terms of the elements of the matrix $P(t)$, conditions (A), (B) and (C) become: for arbitrary $i, j \in E$, $s, t \geqslant 0$, we have

(A) $p_{ij}(t) \geqslant 0 \qquad \sum_j p_{ij}(t) \leqslant 1$

(B) $p_{ij}(t+s) = \sum_k p_{ik}(t) p_{kj}(s)$

(C) $\lim_{t \downarrow 0} p_{ij}(t) = p_{ij}(0) = \delta_{ij} = \begin{cases} 1 & i = j \\ 0 & i \neq j \end{cases}$

Here the sums are taken over E. (B) is usually called the Chapman–Kolmogorov equation.

We call

$$d_i(t) = 1 - \sum_j p_{ij}(t) \tag{2}$$

the stopping function of the process $P(t)$.

Lemma 1. $d_i(t)$ is a non-decreasing function of t. If $d_i(t) = 0$ holds for some $t > 0$ and all $i \in E$, then $d_i(t) = 0$ holds for all $t \geqslant 0$ and $i \in E$.

Proof. For $s, t > 0$, by (B), (C),

$$\begin{aligned} d_i(s+t) &= 1 - \sum_j p_{ij}(s+t) \\ &= 1 - \sum_j \sum_k p_{ik}(s) p_{kj}(t) \\ &= 1 - \sum_k p_{ik}(s) \sum_j p_{kj}(t) \geqslant 1 - \sum_k p_{ik}(s) \\ &= d_i(s) \end{aligned}$$

From this we see that $d_i(t)$ is non-decreasing. From the above expression we find out that if $\sum_j p_{kj}(t) = 1$ for some $t > 0$ and all k, then $d_i(s+t) = 0$ $(s > 0)$. Since $d_i(t) = 0$, an easy induction gives $d_i(nt) = 0$ for all position integers n. For any $u \geqslant 0$, choose n so that $u \leqslant nt$, then $d_i(u) \leqslant d_i(nt) = 0$ because $d_i(\cdot)$ is non-decreasing, and so $d_i(u) = 0$. QED

When $d_i(t) = 0$ $(i \in E)$ for some (hence all) $t > 0$, namely,

(D) $P(t)\mathbf{1} = \mathbf{1} \qquad t > 0$

or equivalently

(D) $\sum_j p_{ij}(t) = 1 \qquad i \in E, t > 0$

then the process $P(t)$ is said to be an honest process; otherwise it is called a stopping process.

For a stopping process $P(t)$ we can get an honest process $\tilde{P}(t)$ by enlarging the state space as follows: take an arbitrary index $\Delta \notin E$, and write $\tilde{E} = E \cup \{\Delta\}$; set

$$\tilde{p}_{ij}(t) = p_{ij}(t) \qquad \tilde{p}_{i\Delta}(t) = d_i(t) \qquad i, j \in E$$
$$\tilde{p}_{\Delta j}(t) = \begin{cases} 1 & \text{if } j = \Delta \\ 0 & \text{if } j \in E \end{cases} \tag{3}$$

We then have the following lemma.

Lemma 2. Let $P(t) = \{p_{ij}(t)\}$ $(i, j \in E, t \geqslant 0)$ be a stopping process. Then $\tilde{P}(t) = \{\tilde{p}_{ij}(t)\}$ $(i, j \in \tilde{E}, t \geqslant 0)$ will be an honest process.

The class of all processes $P(t)$ is denoted by $\mathscr{P}$.

It will be proved latter (Theorems 5.1 and 5.2) that for each $P(t) \in \mathscr{P}$, its right derivative $P'(0)$ at $t = 0$ exists, that is, the limits

$$q_{ij} = p'_{ij}(0) = \lim_{t \downarrow 0} \frac{p_{ij}(t) - p_{ij}(0)}{t} \tag{4}$$

exist, and

$$0 \leqslant q_{ij} < \infty \qquad (i \neq j)$$
$$\sum_{j \neq i} q_{ij} \leqslant -q_{ii} \equiv q_i \leqslant \infty \tag{5}$$

We call the matrix $Q = (q_{ij})$ $(i, j \in E)$ the Q matrix of the process $P(t)$.

When $q_i < \infty$, we say that the state i is stable. From now on we only study the process whose states are all stable, that is, the process has a finite Q matrix. We denote by $\mathscr{P}_s$ the class of such processes, that is, their Q matrices satisfy the following conditions:

$$0 \leqslant q_{ij} < \infty \qquad (i \neq j)$$
$$q_i \equiv -q_{ii} < \infty \qquad d_i \equiv q_i - \sum_{j \neq i} q_{ij} \geqslant 0 \tag{6}$$

In order to emphasize the relation (4) between $P(t) \in \mathscr{P}_s$ and the finite matrix Q, that is,

$$P'(0) = Q \tag{7}$$

we say that $P(t)$ is a Q process. It should be emphatically pointed out that in this book all the states of the Q process $P(t)$ are supposed to be stable. The class of all Q processes with the same Q matrix will be denoted by $\mathscr{P}_s(Q)$.

2.3 THE PROBLEM OF CONSTRUCTION

The problem of construction is a converse one. Except for Williams (1976), so far consideration of the problem of construction has been limited almost entirely to Q matrices satisfying equation (2.6).

Let Q be a fixed matrix satisfying (2.6). We call $d=(d_i, i\in E)$ a non-conservative column vector of Q. If $d_i=0$, we call i a conservative state. We call

$$H=\{i|d_i>0\} \tag{1}$$

the set of non-conservative states. If H is empty, we call Q itself conservative.

The formulation of the problem of construction is as follows. Suppose a fixed Q matrix satisfying (2.6) is given. Problem 1: is there a process $P(t)$ satisfying (2.7)? In other words, does there exist such a Q process? Problem 2: if there exists such a Q process, is it unique? Problem 3: if such a Q process is not unique, how can we construct all Q processes?

The problem of construction was first proposed by Kolmogorov (1931), and he was the first to write down the system of backward differential equations

$$\text{(KB)} \quad p'_{ij}(t)=\sum_k q_{ik}p_{kj}(t) \qquad i,j\in E, t\geqslant 0$$

and the system of forward differential equations

$$\text{(KF)} \quad p'_{ij}(t)=\sum_k p_{ik}(t)q_{kj} \qquad i,j\in E, t\geqslant 0$$

Feller (1940) proved that the Q process always exists. Moreover, he constructed the minimal Q process. So problem 1 is settled.

Doob (1945) proved that for a conservative Q either there is only one Q process, i.e. the minimal Q process, or there exist infinitely many Q processes. For a conservative Q Reuter (1957) found out the necessary and sufficient condition under which the Q process is unique. Therefore for a conservative Q matrix, problem 2 was solved. For a general Q, Zhen-ting Hou (1974) obtained the uniqueness criterion for Q processes, and thus problem 2 is solved completely.

As far as problem 3 is concerned, a complete solution seems still far away. There are two approaches to this problem at present. One is the analytical approach, as used in Reuter (1957, 1959, 1962, 1976), Feller (1940, 1945, 1956, 1957a, b, 1958, 1971), Williams (1964, 1966, 1976), Zhen-Zu Sun (1962), Xiang-qun Yang (1964b, 1965a, 1966a) and Di-he Hu (1965, 1966, 1983). They mainly make use of analytical tools and methods to find solutions satisfying Kolmogorov's backward or forward equations, which are Q processes, or to get the infinitesimal operators of the contraction semigroups derived from the Q processes, or to find the resolvent operators of the Q processes. The other approach is the probability method, i.e. the limit transition method. This method was given by Zi-kun Wang (1958), resulting in a very successful solution to the construction problem for birth–death processes (Zi-kun Wang, 1962; Zi-kun Wang and Xiang-qun Yang 1978, 1979). The basic idea of this method is to approach the sample paths of a Q process by those of Doob processes which have simpler structures. That is, the sample paths of a Q process are the strong limits of the sample paths of a sequence of Doob processes. This method was latter improved by Zhen-ting Hou (1975) and Xiang-qun Yang (1978, 1979, 1980a, b, c).

The solution to problem 3 by means of the analytical approach goes as follows. Suppose that Q is conservative. Reuter (1959) and Zhen-zu Sun (1962) constructed all Q processes in the case that Q is single exit. Feller (1957a), in the case of finite exit and finite entrance, derived all Q processes that satisfy simultaneously Kolmogorov's backward and forward equations. Xiang-qun Yang (1966a), with the same hypotheses as Feller's, obtained all Q processes. Williams (1964, 1966) and Chung (1963, 1966), with the finite exit hypothesis, found all Q processes.

Since each Q process must satisfy Kolmogorov's backward equations when Q is conservative but need not satisfy the backward equations when Q is non-conservative, it follows that few books or articles are concerned with the problem of the construction of a general Q.

As for the birth–death matrix Q, which is conservative except perhaps at the state 0, Feller (1957b, 1971) got all birth–death processes that satisfy both Kolmogorov's backward equations and forward equations. Xiang-qun Yang (1965a) has constructed all birth–death processes, that is, the birth–death processes that satisfy one of the two systems of equations, or niether. In Chapter 4, for general Q in the case when Q is single exit, we have constructed all the Q processes that satisfy the system of backward equations, and in the case when Q is single entrance we have constructed all the Q processes that satisfy the forward equations. Especially, we have derived all Q processes in the case of null exit when there exists only one non-conservative state. In Chapter 8, we find all the Q processes when Q is finite exit and finite non-conservative.

It can be seen in the bibliography of this work that much work on the subject of the construction theory has been carried out by Chinese writers.

2.4 CONTINUITY

Theorem 1. If $P(t)\in\mathscr{P}$, then for arbitrary $i, j\in E$ and $h>0$,

$$|p_{ij}(t\pm h)-p_{ij}(t)|\leqslant 1-p_{ii}(h) \tag{1}$$

$$|d_i(t\pm h)-d_i(t)|\leqslant 1-p_{ii}(h) \tag{2}$$

and, furthermore, the stopping function $d_i(t)$ and $p_{ij}(t)$ are uniformly continuous in $[0,\infty)$.

Proof. It suffices to prove the theorem in the case of the plus sign. By condition

(B) in section 2.2[1],

$$p_{ij}(t+h)-p_{ij}(t)=\sum_k p_{ik}(h)p_{kj}(t)-p_{ij}(t)$$
$$=[p_{ii}(h)-1]p_{ij}(t)+\sum_{k\neq i} p_{ik}(h)p_{kj}(t) \tag{3}$$

The first term is non-positive and $\geqslant -(1-p_{ii}(h))$; the second term is non-negative, and $\leqslant \sum_{k\neq i}p_{ik}(h)\leqslant 1-p_{ii}(h)$; hence (1) follows. Extend $P(t)$ to $\tilde{P}(t)$ according to (2.3) and apply (1) to $\tilde{p}_{i\Delta}(t)$, and we obtain (2). The proof is terminated. QED

Theorem 2. The series $\sum_j p_{ij}(t)$ is uniformly convergent in any finite interval $[0,b]$.

Proof. By Theorem 1, both $p_{ij}(t)$ and $1-d_i(t)$ are continuous functions; hence it suffices to quote the Dini theorem (Titchmarsh, 1939). QED

Theorem 3. Let $h>0$. The sum

$$\sum_j |p_{ij}(t+h)-p_{ij}(t)| \tag{4}$$

is non-increasing when t increases and is uniformly convergent to zero for $t\geqslant\delta>0$ as $h\to 0$. In particular, for every $\delta>0$, $p_{ij}(t)$ is uniformly continuous in $[\delta,\infty]$.

Proof. Let $0\leqslant s<t$. By conditions (A) and (B) is section 2.2[1],

$$\sum_j |p_{ij}(t+h)-p_{ij}(t)|=\sum_j\left|\sum_k [p_{ik}(s+h)-p_{ik}(s)]p_{kj}(t-s)\right|$$
$$\leqslant\sum_k |p_{ik}(s+h)-p_{ik}(s)|\sum_j p_{kj}(t-s)$$
$$\leqslant\sum_k |p_{ik}(s+h)-p_{ik}(s)| \tag{5}$$

Thus, we have proved the first conclusion. Then on account of Theorem 1, $p_{ij}(t)$ is continuous; hence the above expression can be integrated for $s\in[0,\delta]$. So if $t\geqslant\delta>0$,

$$\sum_j |p_{ij}(t+h)-p_{ij}(t)|\leqslant\sum_k (1/\delta)\int_0^\delta |p_{ik}(s+h)-p_{ik}(s)|\,\mathrm{d}s$$

But if $0\leqslant h\leqslant\delta$, the second series is dominated by the series

$$\sum_k (2/\delta)\int_0^{2\delta} p_{ik}(s)\,\mathrm{d}s$$

[1] Shortened in the following to (2.A) and (2.B) etc.

and, therefore, uniformly continuous in $h \in [0, \delta]$. But according to a theorem in Titchmarsh (1939), for every k

$$\lim_{h \to 0} \int_0^\delta |p_{ik}(s+h) - p_{ik}(s)| \, ds = 0 \tag{6}$$

It follows that uniformly for $t \geqslant \delta$,

$$\overline{\lim_{h \to 0}} \sum_j |p_{ij}(t+h) - p_{ij}(t)| \leqslant \sum_k (1/\delta) \lim_{h \to 0} \int_0^\delta |p_{ik}(s+h) - p_{ik}(s)| \, ds$$
$$= \sum_k (1/\delta) 0 = 0$$

The proof is complete. QED

Theorem 4. We have $p_{ii}(t) > 0$ in $(0, \infty)$; for $i \neq j$, we have $p_{ij}(t) = 0$ or $p_{ij}(t) > 0$ in $(0, \infty)$.

Proof. (i) By (2.B), for arbitrary $t > 0$, we have

$$p_{ii}(t) \geqslant [p_{ii}(t/n)]^n \qquad \text{for all } n \tag{7}$$

From (2.C) it follows that $p_{ii}(t) > 0$.

(ii) Suppose $i \neq j$. By (2.B)

$$p_{ij}(t+s) \geqslant p_{ii}(s) p_{ij}(t) \tag{8}$$

Hence if we have $p_{ij}(t_1) > 0$ for some $t_1 > 0$, then $p_{ij}(t) > 0$ for all $t \geqslant t_1$.

(iii) Suppose that $t_0 > 0$ exists so that

$$p_{il}(t) = 0 \qquad (0 < t \leqslant t_0)$$
$$p_{il}(2t_0) = c > 0 \tag{9}$$

To simplify things, we may suppose that E is the set of all non-negative integers. By Theorem 2 there exists a positive integer N so that

$$\sum_{j > N} p_{ij}(t) < c/4 \qquad 0 < t \leqslant 2t_0 \tag{10}$$

Set $s = t_0/(2N)$, and define

$$A_m = \{k \mid p_{ik}(ms) > 0\} \qquad m \geqslant 1 \tag{11}$$

By (ii), $A_m \subset A_{m+1}$. Write $B_1 = A_1$, $B_m = A_m - A_{m-1}$ $(m \geqslant 2)$. If $k \notin A_n$, then

$$0 = p_{ik}(ms) = \sum_j p_{ij}[(m-1)s] p_{jk}(s)$$
$$= \sum_{j \in A_{m-1}} p_{ij}[(m-1)s] p_{jk}(s) \tag{12}$$

Therefore,

$$p_{jk}(s) = 0 \qquad j \in A_{m-1}, k \notin A_m \tag{13}$$

If for some m, $1 < m \leqslant 2N$, we have $A_m = A_{m-1}$, then by the above expression we obtain

$$p_{ik}[(m+1)s] = \sum_{j \in A_m} p_{ij}(ms) p_{jk}(s) = 0 \qquad k \notin A_m$$

Hence $A_{m+1} = A_m$. Repeating this proof we get $A_n = A_m$ $(n > m)$ and in particular we have $A_{2N} = A_{4N}$. But by (9), $l \notin A_{2N}$ and $l \in A_{4N}$. This contradiction shows that all B_m $(1 \leqslant m \leqslant 2N)$ are non-empty and mutually disjoint.

Let $1 \leqslant m \leqslant 2N$, then $A_m \subset A_{2N}$. If $k \notin A_m$, then by (13), for every $n \geqslant 1$, we have

$$p_{ik}[(n+1)s] = \left(\sum_{j \notin A_m} + \sum_{j \in B_m} + \sum_{j \in A_{m-1}} \right) p_{ij}(ns) p_{jk}(s)$$

By (13) the third sum is zero, so

$$\sum_{j \notin A_m} p_{ik}[(n+1)s] \leqslant \sum_{j \notin A_m} p_{ij}(ns) + \sum_{j \in B_m} p_{ij}(ns)$$

Summing n from 1 to $4N - 1$, we have

$$\sum_{k \notin A_m} p_{ik}(4Ns) \leqslant \sum_{n=1}^{4N} \sum_{j \in B_m} p_{ij}(ns)$$

Because $l \notin A_{2N}$ the left-hand side at least equals $p_{il}(4NS) = c$, and therefore

$$c \leqslant \sum_{n=1}^{4N} \sum_{j \in B_m} p_{ij}(ns) \qquad 1 \leqslant m \leqslant 2N \tag{14}$$

Since $B_1, B_2, \ldots, B_{2N}$ are non-empty and mutually disjoint, there exists at least N of B_N $(1 \leqslant n \leqslant 2N)$, whose union is denoted by B, which is disjoint from the set $(1, 2, \ldots, N)$. Thus

$$Nc \leqslant \sum_{n=1}^{4N} \sum_{j \in B} p_{ij}(ns) \tag{15}$$

On the other hand, we have by (10)

$$\sum_{j \in B} p_{ij}(ns) \leqslant \sum_{j > N} p_{ij}(ns) < c/4$$

Consequently the right-hand side of (15) is strictly less than $4N(c/4) = Nc$. This contradiction implies that (9) cannot hold. The proof is concluded. QED

Corollary

For any fixed i, $d_i(t)$ is identically zero in $(0, \infty)$ or never zero.

Proof. If suffices to combine Lemma 2.2 with Theorem 4.4

2.5 EXISTENCE OF Q MATRICES

Theorem 1. Let $P(t)\in\mathscr{P}$, then for each $i\in E$

$$-p'_{ii}(0)=\lim_{t\downarrow 0}\frac{1-p_{ii}(t)}{t} \tag{1}$$

exists, but may be infinite.

Proof. Put

$$\Phi(t)=-\ln p_{ii}(t) \tag{2}$$

Then $\Phi(t)$ is non-negative and finite by Theorem 5.4. By (2.A) and (2.B) we have

$$p_{ii}(s+t)\geqslant p_{ii}(s)p_{ii}(t) \tag{3}$$

From the foregoing expression it follows that

$$\Phi(s+t)\leqslant\Phi(s)+\Phi(t) \tag{4}$$

Denote

$$q_i\equiv\sup_{t>0}\,[\Phi(t)/t] \tag{5}$$

If $q_i<\infty$, then for arbitrary $\varepsilon>0$, there exists $t_0>0$ so that $\Phi(t_0)/t_0>q_i-\varepsilon$. But for every $t>0$, we can write $t_0=nt+\delta$ $(0\leqslant\delta<t)$, so

$$q_i-\varepsilon\leqslant\frac{\Phi(t_0)}{t_0}\leqslant\frac{n\Phi(t)+\Phi(\delta)}{t_0}=\frac{nt}{t_0}\frac{\Phi(t)}{t}+\frac{\Phi(\delta)}{t_0}$$

$nt/t_0\to 1$ when $t\to 0$. By (2.C) we have $\Phi(\delta)\to 0$. It follows that

$$q_i-\varepsilon\leqslant\varliminf_{t\downarrow 0}\frac{\Phi(t)}{t}\leqslant\varlimsup_{t\downarrow 0}\frac{\Phi(t)}{t}\leqslant q_i$$

Since ε is arbitrary, $\lim_{t\downarrow 0}\Phi(t)/t=q_i$. If $q_i=\infty$, upon replacing $q_i-\varepsilon$ by the arbitrarily large positive number M, we still get $\lim_{t\downarrow 0}\Phi(t)/t=\infty$. However,

$$\lim_{t\downarrow 0}\frac{\Phi(t)}{t}=\lim_{t\downarrow 0}\frac{-\ln\{1-[1-p_{ii}(t)]\}}{t}$$

$$=\lim_{t\downarrow 0}\frac{1-p_{ii}(t)}{t}$$

Thus, the theorem is proved. QED

Theorem 2. For arbitrary $i\neq j$,

$$p'_{ij}(0)=\lim_{t\downarrow 0}\frac{p_{ij}(t)}{t} \tag{6}$$

exists and is non-negative but finite.

Proof. Let $i \neq j$ and $h > 0$. Define ${}_jp_{ii}^0(h) = 1$

$$ {}_jp_{ii}^n(h) = \sum p_{ik_1}(h)p_{k_1k_2}(h)\cdots p_{k_{n-1}i}(h) \tag{7} $$

$$ f_{ij}^n(h) = \sum p_{ik_1}(h)p_{k_1k_2}(h)\cdots p_{k_{n-1}j}(h) \tag{8} $$

where the sum is taken over those $k_1 \neq j, k_2 \neq j, \ldots, k_{n-1} \neq j$. Although we define ${}_jp_{ii}^n(h)$ and $f_{ij}^n(h)$ by analytical expressions, in actual fact they are respectively the probability of going from i to i in n steps without visiting j and the probability of the first visit to j from i in n steps for a Markov chain whose one-step transition probability matrix is $\{p_{ij}(h)\}$.

From (2.A) and (2.B), we can deduce that

$$ p_{ij}(nh) \geqslant \sum_{m=0}^{n-1} {}_jp_{ii}^m(h)p_{ij}(h)p_{jj}[(n-m-1)h] \tag{9} $$

$$ {}_jp_{ii}(mh) = {}_jp_{ii}^m(h) + \sum_{a=1}^{m-1} f_{ij}^a(h)p_{ji}[(m-a)h] \tag{10} $$

$$ p_{ij}(mh) = \sum_{a=1}^{m} f_{ij}^a(h)p_{jj}[(m-a)h] \tag{11} $$

From the probability point of view, the above relations are even clearer. By (2.C), for $\varepsilon < \frac{1}{2}$, there exists $t_0 > 0$ so that

$$ \max_{0 \leqslant t \leqslant t_0} p_{ij}(t) < \varepsilon \qquad \max_{0 \leqslant t \leqslant t_0} p_{ji}(t) < \varepsilon $$

$$ \min_{0 \leqslant t \leqslant t_0} p_{ii}(t) > 1 - \varepsilon \qquad \min_{0 \leqslant t \leqslant t_0} p_{jj}(t) > 1 - \varepsilon \tag{12} $$

From this and (11), we obtain

$$ \sum_{a=1}^{m} f_{ij}^a(h)(1-\varepsilon) < \varepsilon $$

thus

$$ \sum_{a=1}^{m} f_{ij}^a(h) \leqslant 1 \tag{13} $$

Hence, by (10), we get

$$ {}_jp_{ii}^m(h) \geqslant p_{ii}(mh) - \max_{1 \leqslant a \leqslant m} p_{ji}[(m-a)h] $$

From this, if $nh < t_0$, then ${}_jp_{ii}^m(h) > 1 - 2\varepsilon$. So by (9), we find

$$ p_{ij}(nh) > (1-2\varepsilon)\sum_{m=0}^{n-1} p_{ij}(h)(1-\varepsilon) \geqslant (1-3\varepsilon)np_{ij}(h) $$

$$ \frac{p_{ij}(nh)}{nh} > (1-3\varepsilon)\frac{p_{ij}(h)}{h} \qquad \text{if } nh < t_0 \tag{14} $$

Set

$$q_{ij} = \lim_{t \downarrow 0} \frac{p_{ij}(t)}{t}$$

Then we have $q_{ij} < \infty$ by (14), and moreover, there exists $t_1, 0 < t_1 < t_0/2$, so that

$$p_{ij}(t_1)/t_1 < q_{ij} + \varepsilon$$

The left-hand side is continuous in t_1; hence, there exists $h_0 > 0$ so that

$$p_{ij}(t)/t < q_{ij} + 2\varepsilon \qquad |t - t_1| < h_0 \tag{15}$$

For arbitrary $h \in (0, \min(h_0, t_1))$, there exists n so that $t_1 \leqslant nh < t_1 + h < t_0$. By (14) and (15), we get

$$(1 - 3\varepsilon)\frac{p_{ij}(h)}{h} < \frac{p_{ij}(nh)}{nh} < q_{ij} + 2\varepsilon$$

Since ε is arbitrary, we obtain $\lim_{t \downarrow 0} p_{ij}(t)/t \leqslant q_{ij}$. QED

Corollary

For all $i \in E$, the right derivative

$$D_i \equiv d_i'(0) = \lim_{t \downarrow 0} \frac{d_i(t)}{t} \geqslant 0 \tag{16}$$

exists and is finite.

Proof. By Lemma 2.2, to get (16) it suffices to apply Theorem 2 to $\tilde{p}_{i\Delta}(t) = d_i(t)$. QED

Theorem 3. For arbitrary $i \in E$, we have

$$\sum_{j \neq i} p_{ij}'(0) + d_i'(0) \leqslant -p_{ii}'(0) \tag{17}$$

Proof. Because

$$\sum_{j \neq i} \frac{p_{ij}(t)}{t} + \frac{d_i(t)}{t} = \frac{1 - p_{ii}(t)}{t}$$

(17) follows by using the Fatou lemma. QED

2.6 DIFFERENTIABILITY

Let $P(t) \in \mathscr{P}$. $Q = (q_{ij}) = \{p_{ij}'(0)\}$ is its Q matrix, and $q_i = -q_{ii}$, $D_i = d_i'(0)$.

Theorem 1. If $q_i < \infty$[1] for fixed i, then $p_{ij}(t)$ (for all $j \in E$) and $d_i(t)$ have finite and continuous derivatives in $[0, \infty)$. Moreover

$$|p_{ij}(t+h) - p_{ij}(t)| \leqslant q_i h \qquad t \geqslant 0, h \geqslant 0 \tag{1}$$

$$|d_i(t+h) - d_i(t)| \leqslant q_i h \qquad t \geqslant 0, h \geqslant 0 \tag{2}$$

$$\sum_j |p'_{ij}(t)| + d'_i(t) \leqslant 2q_i \qquad t > 0 \tag{3}$$

$$\sum_j p'_{ij}(t) + d'_i(t) = 0 \qquad t > 0 \tag{4}$$

$$p'_{ij}(t_1 + t_2) = \sum_k p'_{ik}(t_1) p_{kj}(t_2) \qquad t_1 > 0, t_2 \geqslant 0 \tag{5}$$

Proof. (i) By (4.7) and the definition of q_i, we have

$$p_{ii}(t) \geqslant e^{-q_i t} \geqslant 1 - q_i t \tag{6}$$

From this, (4.1) and (4.2), (1) and (2) follow.

Inequalities (1) and (2) show that $p_{ij}(t)$ and $d_i(t)$ satisfy Lipschitz's condition; hence, they are absolutely continuous. It follows that $p'_{ij}(t)$ and $d'_i(t)$ exist for almost all $t \geqslant 0$. Furthermore,

$$p_{ij}(t) = \delta_{ij} + \int_0^t p'_{ij}(u)\,du \tag{7}$$

$$d_i(t) = \int_0^t d'_i(u)\,du \tag{8}$$

(ii) Set

$$\Delta_{ij}(t, t+s) = \frac{p_{ij}(t+s) - p_{ij}(t)}{s} \qquad t \geqslant 0, s > 0 \tag{9}$$

$$\Delta_i(t, t+s) = \frac{d_i(t+s) - d_i(t)}{s} \qquad t \geqslant 0, s > 0 \tag{10}$$

Since $\sum_j p_{ij}(t) + d_i(t) = 1$ and $d_i(t)$ are non-decreasing, we have

$$\sum_j \Delta_{ij}(t, t+s) + \Delta_i(t, t+s) = 0 \tag{11}$$

$$\Delta_i(t, t+s) \geqslant 0 \tag{12}$$

[1] When $q_i = \infty$, it follows also that $p_{ij}(t)$ and $d_i(t)$ have finite and continuous derivatives in $(0, \infty)$, and (5) holds for $t_1 > 0$, $t_2 \geqslant 0$. See Zi-kun Wang (1980, §2.2, Theorem 2) or Chung (1967a, II.12, Theorem 8).

By (4.8) and (6), it follows that

$$\Delta_{ij}(t,t+s) \geqslant -\frac{1-p_{ii}(s)}{s}p_{ij}(t) \geqslant -q_i p_{ij}(t) \tag{13}$$

Hence for arbitrary set $A \subset E$, we have

$$\sum_{j\in A}\Delta_{ij}(t,t+s) \geqslant -q_i \sum_{j\in A} p_{ij}(t) \geqslant -q_i \tag{14}$$

Starting from this and noticing (12), by (11), we get

$$\sum_{j\in A}\Delta_{ij}(t,t+s) = -\sum_{j\in E-A}\Delta_{ij}(t,t+s) - \Delta_i(t,t+s) \leqslant q_i \tag{15}$$

So for arbitrary $A \subset E$,

$$\left|\sum_{j\in A}\Delta_{ij}(t,t+s)\right| \leqslant q_i \tag{16}$$

Moreover

$$0 \leqslant \Delta_i(t,t+s) = -\sum_j \Delta_{ij}(t,t+s) \leqslant q_i \tag{17}$$

Take $A = \{j|\Delta_{ij} \geqslant 0\}$ in (14) and (15). We obtain

$$\sum_j |\Delta_{ij}| = \sum_{j\in A}\Delta_{ij} + \sum_{j\in E-A}(-\Delta_{ij}) \leqslant 2q_i \tag{18}$$

From this it follows that

$$\sum_j |p'_{ij}(t)| \leqslant 2q_i \tag{19}$$

so long as the derivatives involved exist. In particular the foregoing equation holds for almost all $t \geqslant 0$.

Sum (7) over j and add it to (8); it follows by (19) that the order of summation and integration can be interchanged, therefore

$$\begin{aligned}\sum_j p_{ij}(t) + d_i(t) &= 1 + \sum_j \int_0^t p'_{ij}(u)\,\mathrm{d}u + \int_0^t d'_i(u)\,\mathrm{d}u \\ &= 1 + \int_0^t \left(\sum_j p'_{ij}(u) + d'_i(u)\right)\mathrm{d}u\end{aligned}$$

and it follows that

$$\int_0^t \left(\sum_j p'_{ij}(u) + d'_i(u)\right)\mathrm{d}u = 0 \qquad t \geqslant 0$$

Hence for almost all $t \geqslant 0$, we have

$$\sum_j p'_{ij}(t) + d'_i(t) = 0 \tag{20}$$

(iii) Suppose that $p'_{ij}(t)$ and $d'_i(t)$ exist for some t and (20) holds. We prove

$$\Sigma(s) \equiv \sum_j |\Delta_{ij}(t, t \pm s) - p'_{ij}(t)| \to 0 \qquad s \downarrow 0 \tag{21}$$

First we prove the case for the plus sign. Given arbitrary $\varepsilon > 0$, by (19) we can choose a finite set $A \subset E$ such that

$$q_i \sum_{j \notin A} p_{ij}(t) + \sum_{j \notin A} |p'_{ij}(t)| < \varepsilon \tag{22}$$

Hence,

$$\Sigma(s) \leqslant \sum_{j \in A} |\Delta_{ij}(t, t+s) - p'_{ij}(t)| + \sum_{j \notin A} |\Delta_{ij}(t, t+s)| + \varepsilon$$

Use $\sum'$ to denote the summation of index j for $\Delta_{ij} < 0$. Then by (14) and (22), we have

$$\begin{aligned} \sum_{j \notin A} |\Delta_{ij}| &= \sum_{j \notin A} \Delta_{ij} - 2 \sum_{j \notin A}{}' \Delta_{ij} \\ &\leqslant \sum_{j \notin A} \Delta_{ij} + 2q_i \sum_{j \notin A} p_{ij}(t) \\ &\leqslant \sum_{j \notin A} \Delta_{ij} + 2\varepsilon \end{aligned} \tag{23}$$

Hence by (11) we have

$$\Sigma(s) \leqslant \sum_{j \notin A} |\Delta_{ij}(t, t+s) - p'_{ij}(t)| - \sum_{j \in A} \Delta_{ij} - \Delta_i + 3\varepsilon$$

By (20) and (22), we obtain

$$\begin{aligned} \overline{\lim_{s \downarrow 0}}\, \Sigma(s) &\leqslant 0 - \sum_{j \in A} p'_{ij}(t) - d'_i(t) + 3\varepsilon \\ &= \sum_{j \notin A} p'_{ij}(t) + 3\varepsilon < 4\varepsilon \end{aligned}$$

Thus we have proved (21) where the plus sign appears.

In the case of the minus sign the above proof is still valid after a few revisions. Inequality (22) will be replaced by

$$q_i \sum_{j \notin A} p_{ij}(t-s) + \sum_{j \notin A} |p'_{ij}(t)| < \varepsilon \qquad (s \leqslant t) \tag{24}$$

The above expression holds by Theorem 4.2. Inequality (23) still holds after replacing $p_{ij}(t)$ by $p_{ij}(t-s)$.

(iv) Suppose that $p'_{ij}(t)$ and $d'_i(t)$ exist for some $t > 0$ and (20) holds. Then for arbitrary $u > t$, by (21)

$$\begin{aligned} \sum_j \left| \Delta_{ij}(u, u+s) - \sum_k p'_{ik}(t) p_{kj}(u-t) \right| &= \sum_j \left| \sum_k [\Delta_{ik}(t, t+s) - p'_{ik}(t)] p_{kj}(u-t) \right| \\ &\leqslant \sum_k |\Delta_{ik}(t, t+s) - p'_{ik}(t)| \to 0 \qquad s \downarrow 0 \end{aligned} \tag{25}$$

From this, it follows that

$$\left|\Delta_i(u, u+s) - \sum_j \left(-\sum_k p'_{ik}(t) p_{kj}(u-t) \right)\right| \leqslant \sum_k |\Delta_{ik}(t, t+s) - p'_{ik}(t)| \to 0 \qquad s \downarrow 0$$

Hence the right derivatives of $p_{ij}(u)$ and $d_i(u)$ exist. It is analogous to prove that the left derivatives also exist. Hence $p'_{ij}(u)$ and $d'_i(u)$ exist in (t, ∞). Moreover

$$p'_{ij}(u) = \sum_k p'_{ik}(t) p_{kj}(u-t) \tag{26}$$

$$d'_i(u) = -\sum_k p'_{ik}(u) \tag{27}$$

From (19) and the two above equations it follows that $p'_{ij}(u)$ and $d'_i(u)$ are continuous functions for $u \in (t, \infty)$. Since by (i) it follows that the above t can be arbitrarily small, hence $p'_{ij}(t)$ and $d'_i(t)$ exist and are continuous functions in $(0, \infty)$; moreover, (4) and (5) hold. Because (19) holds for all $t > 0$, according to Lemma 2.2 and by applying conclusion (19) to $\tilde{P}(t)$, we obtain (3).

(v) We are going to prove

$$\lim_{t \downarrow 0} p'_{ij}(t) = p'_{ij}(0) \qquad \lim_{t \downarrow 0} d'_i(t) = d'_i(0) \tag{28}$$

We only need to prove the first statement, because according to Lemma 2.2 we can get the second statement by applying the first statement to $\tilde{P}(t)$.

By (6)

$$p_{ij}(t+h) - p_{ij}(t) \geqslant [p_{ii}(h) - 1] p_{ij}(t) \geqslant -q_i h p_{ij}(t) \tag{29}$$

Therefore

$$R_{ij}(t) \equiv p'_{ij}(t) + q_i p_{ij}(t) \geqslant 0 \tag{30}$$

Moreover, by (5) and (2.B), we have

$$R_{ij}(t+s) = \sum_k R_{ik}(t) p_{kj}(s) \qquad s > 0, t > 0 \tag{31}$$

Hence

$$R_{ij}(t+s) \geqslant R_{ij}(t) p_{jj}(s)$$

Since $p'_{ij}(t)$ is continuous in $(0, \infty)$, we obtain

$$R_{ij}(s) \geqslant \overline{\lim_{t \downarrow 0}} R_{ij}(t) p_{jj}(s) \qquad \underline{\lim_{t \downarrow 0}} R_{ij}(s) \geqslant \overline{\lim_{t \downarrow 0}} R_{ij}(t)$$

Hence $\lim_{t \downarrow 0} R_{ij}(t)$ exists, i.e. $\lim_{t \downarrow 0} p'_{ij}(t)$ exists. And by the mean value theorem of differential calculus, we have

$$\lim_{t \downarrow 0} p'_{ij}(t) = p'_{ij}(0)$$

and the proof is complete. QED

Corollary

If $q_i < \infty$, then for arbitrary $\delta > 0$

$$\lim_{s \to 0} \sum_j \left| \frac{p_{ij}(u+s) - p_{ij}(u)}{s} - p'_{ij}(u) \right| = 0 \tag{32}$$

holds uniformly for $u \geqslant \delta$.

Proof. The conclusion follows from (25) and (5). QED

Theorem 2. Suppose $q_j < \infty$ for fixed j, then for all $i \in E$ there exist finite and continuous derivatives of $p_{ij}(t)$ in $[0, \infty)$. Moreover

$$p'_{ij}(s+t) = \sum_k p_{ik}(t) p'_{kj}(s) \qquad t \geqslant 0, s > 0 \tag{33}$$

Proof. By (4.8) and (6)

$$p_{ij}(t+h) - p_{ij}(t) \geqslant p_{ij}(t)[p_{jj}(h) - 1] \geqslant -p_{ij}(t) q_j h$$

Thus

$$D[p_{ij}(t) e^{q_j t}] = [D p_{ij}(t) + p_{ij}(t) q_j] e^{q_j t} \geqslant 0$$

where D represents the right lower derivative. Therefore $p_{ij}(t) e^{q_j t}$ is non-decreasing when t increases and $D p_{ij}(t)$ are almost everywhere finite.

Set

$$v_{ij}(t) = D p_{ij}(t) + p_{ij}(t) q_j \geqslant 0 \tag{34}$$

Rewriting (2.B) we have

$$p_{ij}(s+t) e^{q_j (t+s)} = e^{q_j t} \sum_k p_{ik}(t) p_{kj}(s) e^{q_j s}$$

By differentiating the above equation for s and using Fubini's theorem on differentiation[1] it follows that for every $t \geqslant 0$ and almost all s we have

$$v_{ij}(s+t) = \sum_k p_{ik}(t) v_{kj}(s) \tag{35}$$

If we use Fatou's lemma, then for all $s \geqslant 0$ and $t \geqslant 0$ we have

$$v_{ij}(s+t) \geqslant \sum_k p_{ik}(t) v_{kj}(s) \tag{36}$$

In particular, for almost all t_0 and all $s \leqslant t_0$,

$$\infty > v_{ij}(t_0) \geqslant p_{ii}(t_0 - s) v_{ij}(s)$$

holds. So $v_{ij}(s)$ is bounded in any finite interval. By Fubini's theorem it follows

[1] See Saks (1937; p. 117)

that (35) holds for $s \notin Z$ and $t \notin Z_s$, where the measures of Z and Z_s are zero. For some $s_0 \notin Z$, suppose that

$$v_{ij}(t+s_0) > \sum_k p_{ik}(t) v_{ki}(s_0) \tag{37}$$

holds for some t, then for $t' > t$ we have

$$\begin{aligned} v_{ij}(t'+s_0) &\geqslant \sum_l p_{il}(t'-t) v_{lj}(t+s_0) \\ &> \sum_l p_{il}(t'-t) \sum_k p_{lk}(t) v_{kj}(s_0) \\ &= \sum_k p_{ik}(t') v_{kj}(s_0) \end{aligned}$$

The above inequality cannot hold, because (35) holds for almost all t when $s_0 \notin Z$. Thus Z_s is empty when $s \notin Z$. Also let $s > 0$ be arbitrary, $0 < s' < s$ and $s' \notin Z$. Then

$$\begin{aligned} v_{ij}(t+s) &= v_{ij}(t+s-s'+s') \\ &= \sum_k p_{ik}(t+s-s') v_{kj}(s') \\ &= \sum_k \sum_l p_{il}(t) p_{lk}(s-s') v_{kj}(s') = \sum_l p_{il}(t) v_{lj}(s) \end{aligned}$$

and it follows that Z is also empty. Hence (35) holds for all $t \geqslant 0$ and $s > 0$.

Let $t \geqslant 0$ and moreover let $t_n \downarrow 0$, $t'_n \downarrow 0$ so that $v_{ij}(t+t_n) \to a_{ij}$, $v_{ij}(t+t'_n) \to a'_{ij}$. We might as well suppose $t'_n < t_n$. By (35),

$$v_{ij}(t+t_n) \geqslant p_{ii}(t_n - t'_n) v_{ij}(t+t'_n)$$

Hence $a_{ij} \geqslant a'_{ij}$. By symmetry it follows that $a_{ij} = a'_{ij}$. That is, $v_{ij}(t+0)$ $(t \geqslant 0)$ exists. Similarly, we can prove that $v_{ij}(t-0)$ $(t > 0)$ exists. Moreover, from $v_{ij}(t+t_n) \geqslant p_{ii}(t_n) v_{ij}(t)$ it follows that $v_{ij}(t+0) \geqslant v_{ij}(t)$ $(t \geqslant 0)$. From $v_{ij}(t) \geqslant p_{ii}(t_n) v_{ij}(t-t_n)$ it follows that $v_{ij}(t) \geqslant v_{ij}(t-0)$ $(t>0)$. Therefore, $v_{ij}(t+0) \geqslant v_{ij}(t) \geqslant v_{ij}(t-0)$. For $t > 0$ we can take $0 < s < t$, since

$$v_{ij}(t-t_n) = \sum_k p_{ik}(t-s-t_n) v_{kj}(s)$$

$$v_{ij}(t-0) \geqslant \sum_k p_{ik}(t-s) v_{kj}(s) = v_{ij}(t)$$

Thus $v_{ij}(t)$ is continuous in $(0, \infty)$. Hence $p_{ij}(t)$ has the continuous Dini's derivative $Dp_{ij}(t)$ in $(0, \infty)$, and in fact there exists the continuous derivative $p'_{ij}(t)$ in $(0, \infty)$ (Saks 1937, p. 204). As just stated, $v_{ij}(0+)$ exists, namely, $p'_{ij}(0+) = \lim_{t \downarrow 0} p'_{ij}(t)$ exists. By the mean value theorem we get $p'_{ij}(0+) = p'_{ij}(0)$. So $p'_{ij}(t)$ is continuous in $[0, \infty)$, and (35), which holds for all $t \geqslant 0$ and $s > 0$, becomes (33). The proof is complete. QED

2.7 THE KOLMOGOROV EQUATIONS

Let $P(t)\in\mathscr{P}$. Notice that (6.5) and (6.33) do not necessarily hold for $t_1=0$ and $s=0$ respectively. That is, when $q_i<\infty$,

$$(\mathrm{KB}_i)\quad p'_{ij}(t)=\sum_k q_{ik}p_{kj}(t) \qquad t\geqslant 0, j\in E$$

does not necessarily hold, and when $q_j<\infty$,

$$(\mathrm{KF}_j)\quad p'_{ij}(t)=\sum_k p_{ik}(t)q_{kj} \qquad t\geqslant 0, i\in E$$

does not necessarily hold. But from

$$\Delta_{ij}(t,t+s)=\sum_k \Delta_{ik}(0,s)p_{kj}(t)=\sum_k p_{ik}(t)\Delta_{kj}(0,s)$$

and Fatou's lemma it follows that when $q_i<\infty$

$$p'_{ij}(t)\geqslant\sum_k q_{ik}p_{kj}(t) \qquad t\geqslant 0, j\in E \tag{1}$$

holds and when $q_j<\infty$

$$p'_{ij}(t)\geqslant\sum_k p_{ik}(t)q_{kj} \qquad t\geqslant 0, i\in E \tag{2}$$

holds. We call (1) and (2) backward inequalities and forward inequalities respectively. We call the system of equations (KB_i) $(i\in E)$, i.e. (3.KB), the system of backward equations; and (KF_j) $(j\in E)$, i.e. (3.KF), is said to be the system of forward equations.

Applying (1) to $\tilde{P}(t)$ of Lemma 2.2, we obtain

$$d'_i(t)\geqslant\sum_k q_{ik}d_k(t)+D_i \tag{3}$$

where $D_i=d'_i(0)\geqslant 0$.

If $P(t)\in\mathscr{P}$, and moreover (KB_i) or (KF_j) holds, then q_i or q_j is finite and also $p'_{ij}(0)=q_{ij}$ ($j\in E$ or $i\in E$). In particular, if $P(t)$ satisfies the system of backward equations or forward equations, then $P(t)$ is a Q process, i.e. $P(t)\in\mathscr{P}_s(Q)$. In fact, by (KB_i),

$$p'_{ii}(t)=q_{ii}p_{ii}(t)+\sum_{k\neq i}q_{ik}p_{ki}(t)$$

Hence q_{ii} is finite. Similarly, by (KF_j) it follows that q_{jj} is finite. Taking $t=0$ in (KB_i) or (KF_j), we get $p'_{ij}(0)=q_{ij}$.

Naturally we may put forward the question: What are the conditions under which the Q process $P(t)\in\mathscr{P}_s(Q)$ satisfies the backward or forward equations? We shall discuss this problem now.

Lemma 1. Let $P(t)\in\mathscr{P}$. If (KB_i) holds for almost all t, then (KB_i) holds for all $t\geqslant 0$.

Proof. By hypothesis it follows that q_{ii} is finite. Moreover

$$p_{ij}(t)=\delta_{ij}+\int_0^t\left(\sum_k q_{ik}p_{kj}(u)\right)du \qquad t\geqslant 0 \tag{4}$$

The integrand expression is continuous by (2.6) and (KB_i) holds for all $t\geqslant 0$ by differentiating with respect to t. QED

Theorem 2. Assume that $P(t)\in\mathscr{P}$. For fixed i, the necessary and sufficient condition under which (KB_i) holds is that q_i is finite and moreover

$$\sum_j q_{ij}+D_i=0 \tag{5}$$

where $D_i=d_i'(0)$.

Proof. Suppose (KB_i) holds, then q_i is finite and

$$\sum_j p_{ij}'(t)=\sum_k q_{ik}\sum_j p_{kj}(t)$$

By (6.4) it follows that the above equation is equivalent to

$$-d_i'(t)=\sum_k q_{ik}[1-d_k(t)]$$

By Theorem 6.1 it follows that, setting $t\rightarrow 0$, we obtain

$$-D_i=\sum_k q_{ik}$$

That is (5).

Conversely if q_i is finite and (5) holds, then by (1), (3) and (6.4) we find

$$\begin{aligned}0&=\sum_j p_{ij}'(t)+d_i'(t)\\&\geqslant\sum_k q_{ik}[1-d_k(t)]+\sum_k q_{ik}d_k(t)+D_i\\&=\sum_k q_{ik}+D_i=0\end{aligned}$$

Hence equality must hold in (1), i.e. (KB_i) holds. QED

When i is a conservative state, that is, when

$$q_i=\sum_{j\neq i}q_{ij}<\infty \tag{6}$$

by Theorem 5.3 it necessarily follows that $D_i=0$. Therefore (5) holds. Hence we have as a corollary of Theorem 2, the following.

Theorem 3. If (6) holds, then (KB_i) holds. In particular, when Q is conservative, each Q process satisfies the system of backward equations.

Consequently when Q is conservative, the problem of construction is simpler, and the problem of deriving all the Q processes becomes the problem of deriving all the Q processes that satisfy the system of backward equations.

Theorem 4. Let $P(t)\in\mathscr{P}_s(Q)$. A necessary and sufficient condition under which the system of backward equations holds is that, for arbitrary $t\geqslant 0$,

$$\lim_{h\to 0}\sum_j\left|\frac{p_{ij}(t+h)-q_{ij}(t)}{h}-\sum_k q_{ik}p_{kj}(t)\right|=0 \tag{7}$$

And a necessary and sufficient condition under which the system of forward equations holds is that, for arbitrary $t\geqslant 0$,

$$\lim_{h\to 0}\sum_j\left|\frac{p_{ij}(t+h)-p_{ij}(t)}{h}-\sum_k p_{ik}(t)q_{kj}\right|=0 \tag{8}$$

Proof. The necessity is derived from the corollary of Theorem 6.1. The sufficiency is obvious.

Theorem 5. Let $P(t)\in\mathscr{P}$. If the system of forward equations (3.KF) holds for almost all $t\geqslant 0$, then it must hold for all $t\geqslant 0$.

Proof. Suppose Z is a null set and (3.KF) holds if $t\notin Z$. From this it follows that

$$p_{ij}(t)=\delta_{ij}+\int_0^t\left(\sum_k p_{ik}(u)q_{kj}\right)\mathrm{d}u$$

$$\geqslant\delta_{ij}+q_{ij}\int_0^t p_{ii}(u)\,\mathrm{d}u+(1-\delta_{ij})q_{jj}\int_0^t p_{ij}(u)\,\mathrm{d}u$$

$$\frac{p_{ij}(t)-\delta_{ij}}{t}\geqslant q_{ij}\frac{1}{t}\int_0^t p_{ii}(u)\,\mathrm{d}u+(1-\delta_{ij})q_{jj}\frac{1}{t}\int_0^t p_{ij}(u)\,\mathrm{d}u$$

By (2.C) and Theorem 5.1 and 5.2, we get

$$\bar{q}_{ij}\equiv p'_{ij}(0)\geqslant q_{ij}+(1-\delta_{ij})q_{jj}\delta_{ij}=q_{ij} \tag{9}$$

Therefore $\bar{q}_{ii}\geqslant q_{ii}>-\infty$, and hence all $\bar{q}_{ij}$ are finite; furthermore, by forward inequalities (2) we have

$$p'_{ij}(t)\geqslant\sum_k p_{ik}(t)\bar{q}_{kj}\qquad t\geqslant 0$$

Combining (3.KF) of $t\notin Z$, we obtain

$$\sum_k p_{ik}(t)(\bar{q}_{kj}-q_{kj})\leqslant 0\qquad t\notin Z,\ i,j\in E$$

Since every term is non-negative, in particular, we have $p_{ii}(t)(\bar{q}_{ij}-q_{ij})=0\ (t\notin Z)$. It follows that $\bar{q}_{ij}=q_{ij}$, that is, $p'_{ij}(0)=q_{ij}$. Hence (3.KF) holds for $t=0$.

Now let $u>0$ and moreover $u\in Z$. We can find $t>0$ so that $t\notin Z$, $0<u-t\notin Z$. So by Theorem 6.2 we have

$$\begin{aligned} p'_{ij}(u) &= \sum_l p_{il}(t)p'_{lj}(u-t) \\ &= \sum_l p_{il}(t)\sum_k p_{lk}(u-t)q_{kj} \\ &= \sum_k\left(\sum_l p_{il}(t)p_{lk}(u-t)\right)q_{kj} \\ &= \sum_k p_{ik}(u)q_{kj} \end{aligned}$$

QED

Theorem 6. Let $P(t)\in\mathscr{P}$. If the system of forward equations holds, then $\sum_k p_{ik}(t)q_{kj}$ is uniformly convergent in any finite interval $[0, T]$.

Proof. By Theorems 4.1 and 6.2 it follows that both $p'_{ij}(t)$ and $p_{ik}(t)q_{kj}$ are continuous; then using Dini's theorem we can prove this conclusion. QED

It is to be pointed out that if we only construct the Q processes that satisfy the system of backward equations, we can change the non-conservative case into the conservative case.

Theorem 7. Assume that $P(t)\in\mathscr{P}_s(Q)$ and $P(t)$ satisfies the system of backward equations, and $\tilde{P}(t)=\{\tilde{p}_{ij}(t)\}$ $(i,j\in\tilde{E})$ is determined by (2.3). Then the matrix $\tilde{Q}=\tilde{P}'(0)$ is conservative and $\tilde{P}(t)$ satisfies the system of backward equations.

Proof. Obviously

$$\tilde{Q}=\begin{pmatrix} Q & D \\ 0 & 0 \end{pmatrix}$$

where $D=\{D_i\}_{i\in E}$ is a column vector. By Theorem 2 it follows that $\tilde{Q}$ is conservative. By (3.KB) we can see that

$$\tilde{p}'_{ij}(t)=\sum_{k\in\tilde{E}}\tilde{q}_{ik}\tilde{p}_{kj}(t) \qquad i,j\in\tilde{E} \tag{10}$$

is valid for $i,j\in E$. For $i=\Delta$ and $j\in\tilde{E}$, the above equation is obviously valid. For $i\in E$, $j=\Delta$, (10) becomes

$$d'_i(t)=\sum_k q_{ik}d_k(t)+D_i \tag{11}$$

When $t=0$, the above equation obviously holds. And when $t>0$, it can be derived from (3.KB) and (6.4), and the proof is complete. QED

2.8 RESOLVENT OPERATORS

Let $P(t)\in\mathscr{P}$. Consider the Laplace transform of $P(t)$, $\psi(\lambda)=\{\psi_{ij}(\lambda)\}$ $(i,j\in E, \lambda>0)$:

$$\psi_{ij}(\lambda)=\int_0^\infty e^{-\lambda t}p_{ij}(t)\,dt \qquad \lambda>0 \tag{1}$$

$\psi(\lambda)(\lambda>0)$ are called resolvent operators of the process $P(t)$.

By (2.A)–(2.C) it follows that for arbitrary $i,j\in E, \lambda,\mu>0$,

$$\psi_{ij}(\lambda)\geqslant 0 \qquad \lambda\sum_j\psi_{ij}(\lambda)\leqslant 1 \tag{2}$$

$$\psi_{ij}(\lambda)-\psi_{ij}(\mu)+(\lambda-\mu)\sum_k\psi_{ik}(\lambda)\psi_{kj}(\mu)=0 \tag{3}$$

$$\lim_{\lambda\to\infty}\lambda\psi_{ij}(\lambda)=\delta_{ij} \tag{4}$$

Here the deduction of (3) from (2.B) requires the calculation of the integral $\int_0^\infty\int_0^\infty p_{ij}(t+s)e^{-\lambda t-\mu s}dt\,ds$, for which see Feller (1971, XIII.8, Ex, (a), p. 452–3). We call (2) the norm condition, (3) the resolvent equation and (4) the continuity condition. The matrix forms of (2), (3) and (4) are as follows:

$$\psi(\lambda)\geqslant 0 \qquad \lambda\psi(\lambda)\mathbf{1}\leqslant\mathbf{1} \tag{5}$$

$$\psi(\lambda)-\psi(\mu)+(\lambda-\mu)\psi(\lambda)\psi(\mu)=0 \tag{6}$$

$$\lim_{\lambda\to\infty}\lambda\psi(\lambda)=\boldsymbol{I} \tag{7}$$

From (2.D) we obtain

$$\lambda\sum_j\psi_{ij}(\lambda)=1 \tag{8}$$

or

$$\lambda\psi(\lambda)\mathbf{1}=\mathbf{1} \tag{9}$$

when $P(t)$ is an honest process.

The systems of backward and forward inequalities (7.1) and (7.2) imply that

$$\lambda\psi_{ij}(\lambda)-\delta_{ij}\geqslant\sum_k q_{ik}\psi_{kj}(\lambda) \tag{10}$$

$$\lambda\psi_{ij}(\lambda)-\delta_{ij}\geqslant\sum_k\psi_{ik}(\lambda)q_{kj} \tag{11}$$

From Theorem 7.3 we have

$$\lambda\psi_{ij}(\lambda) - \sum_k q_{ik}\psi_{kj}(\lambda) = \delta_{ij} \qquad i \in E - H \tag{12}$$

where H is the set of non-conservative states.

Theorem 1. In order that $\psi(\lambda)(\lambda > 0)$ are the resolvent operators of a process $P(t) \in \mathscr{P}$, the necessary and sufficient conditions are that the norm condition, resolvent equation and continuity condition hold. The process $P(t)$ is honest if and only if (9) holds.

Proof. The necessity has been proved already. We are now going to prove the sufficiency. Take $\psi(\lambda)$ to be a linear operator that operates on Banach space l described at the beginning of section 2.2: for $g \in l, g\psi(\lambda) \in l$,

$$[g\psi(\lambda)]_j = \sum_i g_i \psi_{ij}(\lambda) \tag{13}$$

By the norm condition it follows that $\psi(\lambda)$ is a non-negative linear operator from l to l, and its norm is bounded by λ^{-1}. By the resolvent equation it follows that in the uniform topology of operators,

$$\left(-\frac{d}{d\lambda}\right)^n \psi(\lambda) = n!\,[\psi(\lambda)]^{(n+1)} \tag{14}$$

Hence

$$0 \leqslant \left(-\frac{d}{d\lambda}\right)^n \psi_{ij}(\lambda) \leqslant \frac{n!}{\lambda^{n+1}} \tag{15}$$

$$0 \leqslant \left(-\frac{d}{d\lambda}\right)^n \sum_j \psi_{ij}(\lambda) \leqslant \frac{n!}{\lambda^{n+1}} \tag{16}$$

Using the theory of complete monotonic functions (as can be seen in Feller (1971, pp. 415–18)), by (15) it follows that $\psi_{ij}(\lambda)$ is the Laplace transform of some measurable function $f_{ij}(t)$. From (15) and (16) we obtain the following two inequalities:

$$0 \leqslant f_{ij}(t) \leqslant 1 \tag{17}$$

$$0 \leqslant \sum_j f_{ij}(t) \leqslant 1 \tag{18}$$

for almost all $t \geqslant 0$. Modify the values of $f_{ij}(t)$ on a set t whose Lebesgue measure is zero, so that (17) and (18) hold for all $t > 0$. We assume that the values have already been modified.

Then we shall prove that, according to Lebesgue measure in the plane,

$$f_{ij}(s+t) = \sum_k f_{ik}(s) f_{kj}(t) \tag{19}$$

holds for almost all non-negative s and t. By the uniqueness theorem of the double Laplace transform it suffices to prove that the double Laplace transform

$$\int_0^\infty \int_0^\infty \cdots e^{-\lambda s - \nu t} ds\, dt$$

for both sides of (12) are equal. When $\lambda \neq \nu$, the double Laplace transform for both sides of (19) are

$$(\nu - \lambda)^{-1}[\psi_{ij}(\lambda) - \psi_{ij}(\nu)] \qquad \text{and} \qquad \sum_k \psi_{ik}(\lambda)\psi_{kj}(\nu)$$

According to the resolvent equation, they are equal. Letting $\lambda \to \nu$, from (14) it follows that the double Laplace transforms for both sides of (19) also are equal in the case $\lambda = \nu$.

Now by modifying values of $f_{ij}(t)$ in a null measure set we shall obtain $p_{ij}(t)$ such that $p_{ij}(t)$ will become a Markov process.

Define $p_{ij}(0) = \delta_{ij}$ for $t = 0$ and define

$$\begin{aligned} p_{ij}(t) &= t^{-1} \sum_k \int_0^t f_{ik}(u) f_{kj}(t-u) du \\ &= t^{-1} \int_0^t \left(\sum_k f_{ik}(u) f_{kj}(t-u) \right) du \end{aligned} \tag{20}$$

for $t > 0$.

Since (19) holds for almost all (s, t), for almost all $t > 0$ the integrated expression in (14) is equal to $f_{ij}(t)$ for almost all $u \in (0, t)$. Hence

$$p_{ij}(t) = f_{ij}(t) \qquad \text{for almost all } t > 0$$

and the function

$$g_k(t) = \int_0^t f_{ik}(u) f_{kj}(t-u) du \tag{21}$$

as a convolution of two bounded measurable functions f_{ik} and f_{kj}, is continuous. However, the series $\sum_k g_k(t)$ is dominated by the series

$$\sum_k \int_0^t f_{ik}(u) du \tag{22}$$

term by term. In addition, each term of series (22) is continuous. Therefore by Dini's theorem it follows that the series (22) is uniformly convergent in any finite interval $[0, T]$ so that series $\sum_k g_k(t)$ is uniformly convergent in $[0, T]$, and hence $p_{ij}(t) = t^{-1} \sum_k g_k(t)$ is continuous in $(0, T)$. It follows that

$$p_{ij}(t) \text{ is continuous in } (0, \infty) \tag{23}$$

By (21) and (23), from (17) we have

$$0 \leqslant p_{ij}(t) \leqslant 1 \qquad \text{for all } t \geqslant 0 \tag{24}$$

By (18) and (21) we obtain that

$$\sum_j p_{ij}(t) \leqslant 1 \tag{25}$$

is valid for almost all $t > 0$. By (23), using Fatou's lemma it follows that the above expression holds for all $t > 0$.

We have already proved that $P(t) = \{p_{ij}(t)\}$ satisfies (2.A) and (2.C). We proceed to prove that $P(t)$ satisfies (2.B). Considering that (21) and (19) hold for almost all (s, t), we have

$$p_{ij}(s+t) = \sum_k p_{ik}(s) p_{kj}(t) \qquad \text{for almost all } (s, t) \tag{26}$$

By (23), the left-hand side of the above expression is continuous for $s > 0, t > 0$. Hence in order to prove that (26) holds for all $s > 0, t > 0$, we only need to prove that for fixed $s > 0$, the right-hand side of (26) is continuous for $t > 0$; and for fixed $t > 0$, the right-hand side of (26) is continuous for $s > 0$. By (23) and (25) it follows that the first conclusion is right. In order to prove the second conclusion, we only need to prove that for fixed $t > 0$, the series

$$\sum_k p_{ik}(s) p_{kj}(t)$$

is uniformly convergent in any interval $[a, b]$ $(0 < a < b)$. By (24) it suffices to prove that the series $\sum_k p_{ik}(s)$ is uniformly convergent in $[a, b]$. In fact, by (20)

$$\begin{aligned} \sum_j p_{ij}(t) &= t^{-1} \int_0^t \left(\sum_k \sum_j f_{ik}(u) f_{kj}(t-u) \right) \mathrm{d}u \\ &= t^{-1} \sum_k \int_0^t f_{ik}(u) \left(\sum_j f_{kj}(t-u) \right) \mathrm{d}u \end{aligned}$$

By (18) it follows that the above series $\sum_k$ is dominated by series (22), and it has been pointed out that series (22) is uniformly convergent in any finite interval $[0, T]$. Hence the series $\sum_j p_{ij}(t)$ converges uniformly in $[a, b]$.

We have already proved that the elements of $P(t) = \{p_{ij}(t)\}$ are measurable functions, and that (2.A) and (2.B) hold, namely, $P(t) = \{p_{ij}(t)\}$ is a measurable generalized transition matrix. Therefore the limit $\lim_{t \to 0+} p_{ij}(t) = u_{ij}$ exists (see Zi-kun Wang 1980, Theorem 1 in section 2.1) or Chung (1967, II.1, Theorem 3). According to Abel's property of Laplace transforms, with which we are familiar, we have

$$\lim_{\lambda \to \infty} \lambda \psi_{ij}(\lambda) = \lim_{t \to 0+} p_{ij}(t)$$

By noticing also the continuity condition (7) of $\psi(\lambda)$ it follows that (2.C) holds. Thus $P(t)=\{p_{ij}(t)\}$ is a Markov process. Its resolvent matrix is the given $\psi(\lambda)$. The sufficiency is proved.

Let $P(t)$ be honest; obviously (9) holds. Conversely, assume that (9) holds. If $\sum_j p_{ij}(t)<1$ for some $t>0$, then by the corollary of Theorem 1.4.4 $\sum_j p_{ij}(t)<1$ for all $t>0$, so that (9) cannot possibly hold. Thus (9) implies that $P(t)$ is honest. The proof is complete. QED

Theorem 2. Let the derivative matrix of $P(t)\in\mathscr{P}$ be $Q=(q_{ij})$, and let its resolvent operators be $\psi(\lambda)$, then

$$q_{ij}=\lim_{\lambda\to\infty}\lambda[\lambda\psi_{ij}(\lambda)-\delta_{ij}] \tag{27}$$

Proof. When q_{ij} is finite, for any arbitrarily given $\varepsilon>0$, there exists $\delta>0$ so that, if $t>\delta$,

$$\left|\frac{p_{ij}(t)-\delta_{ij}}{t}-q_{ij}\right|<\varepsilon$$

Therefore

$$|\lambda[\lambda\psi_{ij}(\lambda)-\delta_{ij}]-q_{ij}|=\left|\lambda^2\int_0^\infty e^{-\lambda t}[p_{ij}(t)-\delta_{ij}-q_{ij}t]\,dt\right|$$

$$\leqslant\lambda^2\int_0^\delta e^{-\lambda t}\varepsilon t\,dt+\lambda^2\int_\delta^\infty e^{-\lambda t}(2+|q_{ij}|t)\,dt$$

$$=\varepsilon[-e^{-\lambda\delta}(\lambda\delta+1)+1]+2\lambda e^{-\lambda\delta}+|q_{ij}|e^{-\lambda\delta}(\lambda\delta+1)$$

When $\lambda\to\infty$ we have

$$\overline{\lim_{\lambda\to\infty}}\,|\lambda[\lambda\psi_{ij}(\lambda)-\delta_{ij}]-q_{ij}|\leqslant\varepsilon$$

Because ε is arbitrary, (27) follows.

When q_{ii} is infinite, for arbitrary $N>0$, there exists $\delta>0$, if $t<\delta$, such that

$$[p_{ii}(t)-1]/t<-N$$

Consequently

$$\lambda[\lambda\psi_{ii}(\lambda)-1]=\lambda^2\int_0^\infty e^{-\lambda t}[p_{ii}(t)-1]\,dt\leqslant\lambda^2\int_0^\delta e^{-\lambda t}(-Nt)\,dt$$

$$=-N[-e^{-\lambda\delta}(\lambda\delta+1)+1]\to-N\qquad(\lambda\to\infty)$$

Because N is arbitrary, $\lim_{\lambda\to\infty}\lambda[\lambda\psi_{ij}(\lambda)-1]=-\infty$. Thus the theorem is proved. QED

When Q is finite, we call condition (27) the Q condition.

Theorem 3. In order that $\psi(\lambda)\,(\lambda > 0)$ are the resolvent operators of a Q process $P(t)\in\mathscr{P}_s(Q)$, the necessary and sufficient conditions are that the norm condition, resolvent equation and Q condition are valid.

Proof. This is quite obvious because the Q condition implies the continuity condition. QED

Hereafter, the resolvent operators $\psi(\lambda)\,(\lambda > 0)$ of a process or a Q process will be directly called a process or a Q process, and denoted by $\psi(\lambda)\in\mathscr{P}$ or $\psi(\lambda)\in\mathscr{P}_s(Q)$.

Theorem 4. Assume that Q is finite. In order that $\psi(\lambda)\in\mathscr{P}_s(Q)$ and $\psi(\lambda)$ satisfies the Kolmogorov backward equations (3.KB), the necessary and sufficient conditions are that the norm condition, the resolvent equation and the backward (B) condition

$$(\lambda I - Q)\psi(\lambda) = I \qquad \lambda > 0 \tag{28}$$

are valid. The expression of the B condition with elements is

$$\lambda\psi_{ij}(\lambda) - \sum_k q_{ik}\psi_{kj}(\lambda) = \delta_{ij} \qquad \lambda > 0, i, j\in E \tag{29}$$

Proof. Taking the Laplace transform of (2.KB) and noticing the initial condition (2.4) we obtain the B condition (29). The necessity is proved.

We now consider sufficiency. By the resolvent equations we get $\psi_{ij}(\lambda)\downarrow(\lambda\uparrow)$. By (27) we have

$$\psi_{ij}(\lambda)\downarrow 0 \qquad \lambda\uparrow\infty \tag{30}$$

Also by (29) we obtain

$$(\lambda + q_i)\psi_{ij}(\lambda)\downarrow\delta_{ij} \qquad \lambda\uparrow\infty \tag{31}$$

Hence

$$\lambda\psi_{ij}(\lambda)\to\delta_{ij} \qquad \lambda\to\infty \tag{32}$$

i.e. the continuity condition holds. Again by (29) we have

$$\lambda[\lambda\psi_{ij}(\lambda) - \delta_{ij}] = \sum_k q_{ik}\lambda\psi_{kj}(\lambda) \tag{33}$$

By (32) and the dominated convergence theorem it follows that the Q condition holds. As a result $\psi(\lambda)\in\mathscr{P}_s(Q)$.

According to Theorems 4.1 and 6.1, both $p'_{ij}(t)$ and $\sum_k q_{ik}p_{kj}(t)$ are continuous functions of t, whereas (29) shows that they have the same Laplace transform. Hence they are equal, i.e. the backward equations (3.KB) hold. QED

Theorem 5. Let Q be finite. In order that $\psi(\lambda)\in\mathscr{P}_s(Q)$ and $\psi(\lambda)$ satisfies the system of Kolmogorov forward equations (3.KF) the necessary and sufficient conditions are that the norm condition, the resolvent equation and the forward (F) condition

$$\psi(\lambda)(\lambda I - Q) = I \qquad \lambda > 0 \tag{34}$$

hold. The expression of the F condition with elements is

$$\lambda\psi_{ij}(\lambda) - \sum_k \psi_{ik}(\lambda)q_{kj} = \delta_{ij} \qquad \lambda > 0, i, j \in E \tag{35}$$

Proof. Taking Laplace transforms of both sides of (3.KF) and noticing the initial condition (2.C) the F condition (35) follows. The necessity is proved.

We prove next the sufficiency. By (35) it follows that (30) still holds. Again by (35) we have

$$\psi_{ij}(\lambda)(\lambda + q_j) \downarrow \delta_{ij} \qquad \lambda \uparrow \infty \tag{36}$$

So we find that (32) still holds, too. By Theorem 1, $\psi(\lambda)(\lambda > 0)$ are the resolvent operators of a process $P(t) \in \mathscr{P}$.

$$\lambda[\lambda\psi_{ij}(\lambda) - \delta_{ij}] = \sum_k \lambda\psi_{ik}(\lambda)q_{kj} \tag{37}$$

Likewise the Q condition holds, i.e. $\psi(\lambda) \in \mathscr{P}_s(Q)$.

By (35) it follows that the functions $p'_{ij}(t)$ and $\sum_k p_{ik}(t)q_{kj}$ have the same Laplace transform. Therefore they are equal for almost all t, i.e. (3.KF) holds for almost all $t \geqslant 0$. According to Theorem 7.5, the system of forward equations (3.KF) holds. The proof is complete. In the proof of Theorem 7.5, it is proved directly that $p'_{ij} = q_{ij}$, i.e. $p(t) \in \mathscr{P}_s(Q)$, $\psi(\lambda) \in \mathscr{P}_s(Q)$. QED

Definition 1. $\psi(\lambda)$, which satisfies (28) or (35), is said to be B-type or F-type respectively.

2.9 FELLER'S EXISTENCE THEOREM

Does a Q process exist for a fixed matrix Q that satisfies (2.6)? In other words, is the $\mathscr{P}_s(Q)$ empty? Feller (1940) solved this problem: he constructed the minimal solution.

Define $f^n_{ij}(t)\,(t \geqslant 0)$ as follows:

$$\begin{aligned} &f^0_{ij}(t) \equiv 0 \\ &f^{n+1}_{ij}(t) = \delta_{ij}\mathrm{e}^{-q_i t} + \mathrm{e}^{-q_i t}\int_0^t \left(\sum_{k \neq i} q_{ik} f^n_{kj}(u)\right)\mathrm{e}^{q_i u}\,\mathrm{d}u \end{aligned} \tag{1}$$

or equivalently

$$\begin{aligned} &f^0_{ij}(t) \equiv 0 \\ &f^{n+1}_{ij}(t) = \delta_{ij}\mathrm{e}^{-q_i t} + \mathrm{e}^{-q_j t}\int_0^t \left(\sum_{k \neq j} f^n_{ik}(u)q_{kj}\right)\mathrm{e}^{q_j u}\,\mathrm{d}u \end{aligned} \tag{2}$$

Set

$$f_{ij}^n(t)\uparrow f_{ij}(t) \qquad \text{as } n\uparrow\infty \tag{3}$$

To show that (1) is equivalent to (2), consider the Laplace transforms $\phi_{ij}^n(\lambda)$ $(\lambda>0)$ and $\phi_{ij}(\lambda)$ of $f_{ij}^n(t)$ and $f_{ij}(t)$

$$\begin{aligned}\phi_{ij}^0(\lambda)&=0\\ \phi_{ij}^{n+1}(\lambda)&=\frac{1}{\lambda+q_i}\delta_{ij}+\sum_{k\neq i}\frac{q_{ik}}{\lambda+q_i}\phi_{kj}^n(\lambda)\end{aligned} \tag{4}$$

or

$$\begin{aligned}\phi_{ij}^0(\lambda)&=0\\ \phi_{ij}^{n+1}(\lambda)&=\frac{1}{\lambda+q_j}\delta_{ij}+\sum_{k\neq j}\phi_{ik}^n(\lambda)\frac{q_{kj}}{\lambda+q_j}\end{aligned} \tag{5}$$

Condition (3) becomes

$$\phi_{ij}^n(\lambda)\uparrow\phi_{ij}(\lambda) \qquad n\uparrow\infty \tag{6}$$

It remains only to prove that (4) is equivalent to (5). Using matrix notation is simpler. Denote

$$\begin{aligned}\Pi&=(\Pi_{ij})\\ \Pi_{ij}&=\begin{cases}(1-\delta_{ij}q_{ij})/q_i & \text{as } q_i>0\\ \delta_{ij} & \text{as } q_i=0\end{cases}\end{aligned} \tag{7}$$

Set

$$\begin{aligned}\Pi(\lambda)&=\{\Pi_{ij}(\lambda)\}\\ \Pi_{ij}(\lambda)&=\frac{q_i}{\lambda+q_i}\Pi_{ij}\end{aligned} \tag{8}$$

Define diagonal matrix

$$\lambda I+q=\operatorname{diag}(\lambda+q_i) \tag{9}$$

Then (4) becomes

$$\begin{aligned}\phi^0(\lambda)&=0\\ \phi^{n+1}(\lambda)&=\Pi^0(\lambda)(\lambda I+q)^{-1}+\Pi(\lambda)\phi^n(\lambda)\end{aligned}$$

By induction we have

$$\phi^{n+1}(\lambda)=\sum_{a=0}^{n}\Pi^a(\lambda)(\lambda I+q)^{-1} \qquad n\geqslant 0 \tag{10}$$

From (5) it follows that

$$\phi^0(\lambda)=0$$

$$\phi^{n+1}(\lambda)=\Pi^0(\lambda)(\lambda I+q)^{-1}+\phi^n(\lambda)(\lambda I+q)\Pi(\lambda)(\lambda I+q)^{-1}$$

By induction we still have (10). Hence (6) becomes

$$\phi^{n+1}(\lambda)=\sum_{a=0}^{n}\Pi^a(\lambda)(\lambda I+q)^{-1}\uparrow\phi(\lambda)\qquad \lambda\to\infty \tag{11}$$

Theorem 1. $f(t)\in\mathscr{P}_s(Q)$ and moreover $f(t)$ satisfies both Kolmogorov's backward and forward equations. Moreover, $f(t)$ is the minimal Q process, that is, for arbitrary $P(t)\in\mathscr{P}_s(Q)$, we have

$$p_{ij}(t)\geqslant f_{ij}(t)\qquad t\geqslant 0, i,j\in E \tag{12}$$

Theorem 1 can be expressed by the resolvent operators as follows:

Theorem 2. $\phi(\lambda)\in\mathscr{P}_s(Q)$ and moreover $\phi(\lambda)$ satisfies both the backward and forward equations. Furthermore, $\phi(\lambda)$ is the minimal Q process; that is, for any $\psi(\lambda)\in\mathscr{P}_s(Q)$, we have

$$\psi_{ij}(\lambda)\geqslant\phi_{ij}(\lambda)\qquad \lambda>0, i,j\in E \tag{13}$$

Proof. It suffices to prove Theorem 2. While clearly (12) implies (13), the converse implication is not obvious. But suppose that (13) holds, i.e. that $\psi(\lambda)\geqslant\phi(\lambda)$ for $\lambda>0$, and look back to the proof of Theorem 8.1. There (8.14) tells us that

$$(-\mathrm{d}/\mathrm{d}\lambda)^n(\psi(\lambda)-\phi(\lambda))=n![\psi(\lambda)^{n+1}-\phi(\lambda)^{n+1}]$$

whence easily

$$0\leqslant(-\mathrm{d}/\mathrm{d}\lambda)^n(\psi_{ij}(\lambda)-\phi_{ij}(\lambda))\leqslant n!/\lambda^{n+1}$$

so that $\psi_{ij}(\lambda)-\phi_{ij}(\lambda)$ is the Laplace transform of a function $g(\cdot)$ satisfying $0\leqslant g(t)\leqslant 1$ for almost all $t>0$. Hence $p_{ij}(t)\geqslant f_{ij}(t)$ for almost all $t>0$, so for all $t\geqslant 0$ by continuity.

For a more elegant argument, see Feller (1971), Chapter XIII.10, Theorem 3, p. 426; this leads to the result that $[(n/t)\psi(n/t)]^n\to p(t)$ and likewise $[(n/t)\phi(n/t)]^n\to f(t)$, so that obviously $\psi(\lambda)\geqslant\varphi(\lambda)$ implies $p(t)\geqslant f(t)$, we need to prove that $\phi(\lambda)$ satisfies the norm condition, the resolvent equation, the B condition and the F condition, and that (13) holds.

The non-negativity of $\phi(\lambda)$ is clear. By induction we have $\lambda\sum_j\phi^n_{ij}(\lambda)\leqslant 1$ for all n; thus the norm condition is satisfied.

To prove that the resolvent equation of $\phi(\lambda)$ holds, we only need to prove that for all n

$$\Pi^n(\lambda)(\lambda I+q)^{-1}-\Pi^n(\mu)(\mu I+q)^{-1}$$
$$=(\mu-\lambda)\sum_{a=0}^{n}\Pi^a(\lambda)(\lambda I+q)^{-1}\pi^{n-a}(\mu)(\mu I+q)^{-1} \tag{14}$$

holds. By summing over n in the above expression, the resolvent equation follows.

To prove (14), denoting the right-hand side of (14) as A_n we have

$$\Pi(\lambda)A_n=A_{n+1}-(\mu-\lambda)(\lambda I+q)^{-1}\Pi^{n+1}(\mu)(\mu I+q)^{-1} \tag{15}$$

Thus (14) holds for $n=0$. Suppose (14) holds for some n. Operating $\Pi(\lambda)$ on the left-hand side of (14) we have

$$\begin{aligned}\Pi(\lambda)A_n&=\Pi^{n+1}(\lambda)(\lambda I+q)^{-1}-\Pi(\lambda)\Pi^n(\mu)(\mu I+q)^{-1}\\&=\Pi^{n+1}(\lambda)(\lambda I+q)^{-1}-\Pi^{n+1}(\mu)(\mu I+q)^{-1}\\&\quad-(\mu-\lambda)(\lambda I+q)^{-1}\Pi^{n+1}(\mu)(\mu I+q)^{-1}\end{aligned}$$

By substituting the above expression into (15), it follows that (14) holds where n is replaced by $n+1$. Hence (14) holds for all n.

By (4) and (5) we obtain

$$(\lambda+q_i)\phi_{ij}^{n+1}(\lambda)=\delta_{ij}+\sum_{k\neq i}q_{ik}\phi_{kj}^n(\lambda) \tag{16}$$

$$\phi_{ij}^{n+1}(\lambda)(\lambda+q_j)=\delta_{ij}+\sum_{k\neq j}\phi_{ik}^n(\lambda)q_{kj} \tag{17}$$

Letting $n\to\infty$, the B condition and the F condition of $\phi(\lambda)$ are satisfied.

Finally, let $\psi(\lambda)\in\mathscr{P}_s(Q)$. It is clear that $\psi_{ij}(\lambda)\geqslant\phi_{ij}^0(\lambda)$. From (8.10) and (4), by induction we can easily see $\psi_{ij}(\lambda)\geqslant\phi_{ij}^n(\lambda)$ for all n. Hence we obtain (13), and the proof is complete. QED

2.10 PROPERTIES OF THE MINIMAL SOLUTION

Lemma 1. Let a column vector $f\geqslant 0$ and a row vector $g\geqslant 0$. Then when $n\uparrow\infty$,

$$\xi^n\uparrow\phi(\lambda)f \tag{1}$$

$$\eta^n\uparrow g\phi(\lambda) \tag{2}$$

where ξ^n is determined by

$$\xi_i^0=0$$

$$\xi_i^{n+1}=\frac{f_i}{\lambda+q_i}+\sum_{k\neq i}\frac{q_{ik}}{\lambda+q_i}\xi_k^n \tag{3}$$

and η^n is defined as

$$\eta_j^0 = 0$$

$$\eta_j^{n+1} = \frac{g_j}{\lambda + q_j} + \sum_{k \neq j} \eta_k^n \frac{q_{kj}}{\lambda + q_j} \tag{4}$$

Proof. It suffices to set $\xi^n = \phi^n(\lambda) f$ and $\eta^n = g\phi^n(\lambda)$. Then (1) and (2) hold, and by (9.16) and (9.17) it follows that (3) and (4) hold. QED

Lemma 2. Assume that a column vector $f \geqslant 0$, while u satisfies

$$(\lambda + q_i)u_i = \sum_{i \neq j} q_{ij} u_j + f_i \qquad i \in E$$

$$0 \leqslant u_i \leqslant \infty \tag{5}$$

Then $u \geqslant \phi(\lambda) f$.

Suppose that a row vector $g \geqslant 0$, whereas v satisfies

$$v_j(\lambda + q_j) = \sum_{i \neq j} v_i q_{ij} + g_j \qquad j \in E$$

$$0 \leqslant v_i \leqslant \infty \tag{6}$$

Then $v \geqslant g\phi(\lambda)$.

Proof. It suffices to prove the first case. Because $u \geqslant \xi_i^0 = 0$, owing to (5) and (3) and by induction, we obtain $u \geqslant \xi^n$; hence $u \geqslant \phi(\lambda) f$. The proof is complete. QED

Set

$$A(\mu, \lambda) = I + (\mu - \lambda)\phi(\lambda) \qquad \lambda, \mu > 0 \tag{7}$$

Lemma 3. $A(\mu, \lambda)$ is bounded[1], and for arbitrary $\lambda, \mu, v > 0$,

$$A(\mu, \lambda)A(\lambda, v) = A(\mu, v) \tag{8}$$

$$\phi(\mu)A(\mu, \lambda) = A(\mu, \lambda)\phi(\mu) = \phi(\lambda) \tag{9}$$

In particular,

$$A(\mu, \lambda)A(\lambda, \mu) = I \tag{10}$$

That is, for $A(\mu, \lambda)$, there exists a bounded left inverse matrix $A(\lambda, \mu)$ and a bounded right inverse matrix $A(\lambda, \mu)$.

[1] A matrix $A = (a_{ij})$ is said to be bounded if $\sup_i \sum_j |a_{ij}| < \infty$.

Proof. By using the norm condition and the resolvent equation of $\phi(\lambda)$ it can be easily obtained that $A(\mu, \lambda)$ is bounded and (8) and (9) hold. QED

Denote the equations

$$(U_\lambda) \qquad \lambda u_i - \sum_j q_{ij} u_j = 0 \qquad i \in E \tag{11}$$

The class composed of all solutions $u \in m$ to equations (U_λ) will be denoted by $\mathscr{M}_\lambda$. The class composed of all non-negative solutions $u \in m$ is denoted by $\mathscr{M}_\lambda^+$. $\mathscr{M}_\lambda^+(K)$ denotes a subset of $\mathscr{M}_\lambda^+$ whose elements are bounded by K. Denote the equations

$$(V_\lambda) \qquad \lambda v_j - \sum_i v_i q_{ij} = 0 \qquad j \in E \tag{12}$$

The collection of all solutions $v \in l$ to equations (V_λ) will be denoted by $\mathscr{L}_\lambda$. The collection of all non-negative solutions $v \in \mathscr{L}_\lambda$ is denoted by $\mathscr{L}_\lambda^+$.

Lemma 4. If $f \in \mathscr{M}_\mu$ or $\mathscr{M}_\mu^+$, then $A(\mu, \lambda) f \in \mathscr{M}_\lambda$ or $\mathscr{M}_\lambda^+$. If $g \in \mathscr{L}_\lambda$ or $\mathscr{L}_\lambda^+$, then $g A(\mu, \lambda) \in \mathscr{L}_\lambda$ or $\mathscr{L}_\lambda^+$.

Proof. If $f \in \mathscr{M}_\mu$, then by the B condition of $\phi(\lambda)$,

$$\begin{aligned} QA(\mu, \lambda) f &= Qf + (\mu - \lambda) Q \phi(\lambda) f \\ &= \mu f + (\mu - \lambda)[\lambda \phi(\lambda) f - f] \\ &= \lambda A(\mu, \lambda) f \in m \end{aligned}$$

That is $A(\mu, \lambda) f \in \mathscr{M}_\lambda$.[1]

Let $f \in \mathscr{M}_\mu^+$. When $\lambda \leqslant \mu$, obviously

$$A(\mu, \lambda) f = f + (\mu - \lambda) \phi(\lambda) f \geqslant 0 \tag{13}$$

When $\lambda > \mu$, we have

$$(\lambda I - Q) f = (\lambda - \mu) f \geqslant 0$$

By Lemma 2, $f \geqslant \phi(\lambda)(\lambda - \mu) f$, that is, $A(\mu, \lambda) f \geqslant 0$.

Similarly, we can prove that $g A(\mu, \lambda) \in \mathscr{L}_\lambda$ or $\mathscr{L}_\lambda^+$. QED

Theorem 5. $\lambda \phi(\lambda) \mathbf{1} = \mathbf{1} - \phi(\lambda) \boldsymbol{d} - \bar{X}(\lambda)$, that is,

$$\lambda \sum_j \phi_{ij}(\lambda) = 1 - \sum_{a \in H} \phi_{ia}(\lambda) d_a - \bar{X}_i(\lambda) \tag{14}$$

[1] The above calculation needs careful justification, because $Q = (q_{ij})$ may fail to be bounded so that we cannot appeal to the associative law of multiplication to obtain

$$(\lambda \mathbf{I} - Q)[\phi(\lambda) f] = f \tag{X}$$

from $(\lambda \mathbf{I} - Q) \phi(\lambda) = I$ (the B condition for $\phi(\lambda)$). A strict proof of (X) will probably involve interchanging the order of repeated summations: see the proof of Lemma 11.6 below.

The reader will notice that this point occurs repeatedly later on.

where H is the set of non-conservative states, $\boldsymbol{d}=(d_i)$ is the non-conservative column vector of Q and $\bar{X}(\lambda)$ is the maximal solution of

$$(\lambda I-Q)u=0 \qquad 0\leqslant u\leqslant 1 \tag{15}$$

Moreover, $u^n=\Pi^n(\lambda)\mathbf{1}\downarrow\bar{X}(\lambda)$. We define u^n as follows:

$$u_i^0\equiv 0 \qquad u_i^{n+1}=\sum_{k\neq i}\frac{q_{ik}}{\lambda+q_i}u_k^n \tag{16}$$

Proof. By the B condition of $\phi(\lambda)$

$$(\lambda I-Q)\phi(\lambda)\boldsymbol{d}=\boldsymbol{d} \tag{17}$$

$$(\lambda I-Q)[\mathbf{1}-\lambda\phi(\lambda)\mathbf{1}]=(\lambda I-Q)\mathbf{1}-\lambda I=-Q\mathbf{1}=\boldsymbol{d} \tag{18}$$

By Lemma 2

$$\mathbf{1}-\lambda\phi(\lambda)\mathbf{1}\geqslant\phi(\lambda)\boldsymbol{d}$$

Thus by (17) and (18) we find that

$$\bar{X}(\lambda)=\mathbf{1}-\lambda\phi(\lambda)\mathbf{1}-\phi(\lambda)\boldsymbol{d}$$

is the solution of equation (15), and also we obtain (14).

By Lemma 1 it follows that $1-\bar{X}(\lambda)=\phi(\lambda)(\lambda\mathbf{1}+\boldsymbol{d})$ is the limit of the increasing sequence ξ^n, where

$$\xi_i^0=0 \qquad \xi_i^{n+1}=\frac{\lambda+d_i}{\lambda+q_i}+\sum_{k\neq i}\frac{q_{ik}}{\lambda+q_i}\xi_k^n \tag{19}$$

Thus $\bar{X}(\lambda)$ is the limit of the decreasing sequence $u^n=1-\xi^n$. By the above expression (16) follows.

Finally, suppose that u is the solution of equation (15). So by $u\leqslant\mathbf{1}$ we have $u=\Pi(\lambda)\ u\leqslant\Pi(\lambda)\mathbf{1}$. Therefore $u\leqslant\Pi^n(\lambda)\mathbf{1}$, and hence $u\leqslant\lim_{h\to\infty}\Pi^n(\lambda)\mathbf{1}=\bar{X}(\lambda)$, that is, $\bar{X}(\lambda)$ is the maximal solution of equation (15).

Lemma 6. If $u\in\mathscr{M}_\lambda$ and $|u|\leqslant 1$, then

$$-\bar{X}(\lambda)\leqslant u\leqslant\bar{X}(\lambda) \tag{20}$$

Proof. The right inequality has been proved in the proof of Theorem 5. The left inequality can be proved similarly. QED

Lemma 7. If $\bar{X}(\lambda)\neq 0$, then

$$\sup_i\bar{X}_i(\lambda)=1 \qquad \inf_i\sum_{a\in H}\phi_{ia}(\lambda)d_a=0 \tag{21}$$

$$\inf_i[\phi(\lambda)u]_i=0 \qquad u\in m \tag{22}$$

Proof. Let $\sup_i \bar{X}_i(\lambda) = a$, then $0 < a \leqslant 1$. Since $a^{-1}\bar{X}(\lambda) \in \mathcal{M}_\lambda^+(1)$, by the maximum of $\bar{X}(\lambda)$ we have $a^{-1}\bar{X}(\lambda) \leqslant \bar{X}(\lambda)$, so $1 \leqslant a$; hence $a = 1$, namely $\sup_i \bar{X}_i(\lambda) = 1$. From this and (14) we obtain $\inf_i \sum_{a\in H} \phi_{ia}(\lambda) d_a = 0$ and $\inf_i [\lambda\phi(\lambda)\mathbf{1}]_i = 0$.Thus we get $\inf_i [\phi(\lambda)u]_i = 0$ for $u \in m$. The proof is complete.

QED

2.11 EXIT FAMILY AND ENTRANCE FAMILY

Definition 1. $(\xi(\lambda), \lambda > 0)$ is called an exit family if $0 \leqslant \xi(\lambda) \in m$ and

$$\xi(\mu) = A(\lambda, \mu)\xi(\lambda) \qquad \lambda, \mu > 0 \tag{1}$$

An exit family $(\xi(\lambda), \lambda > 0)$ is said to be harmonic if $\xi(\lambda) \in \mathcal{M}_\lambda^+$.

$(\eta(\lambda), \lambda > 0)$ is called an entrance family if $0 \leqslant \eta(\lambda) \in l$ and

$$\eta(\mu) = \eta(\lambda)A(\lambda, \mu) \qquad \lambda, \mu > 0 \tag{2}$$

An entrance family $(\eta(\lambda), \lambda > 0)$ is said to be harmonic if $\eta(\lambda) \in \mathcal{L}_\lambda^+$.

Obviously, for an exit family $\xi(\lambda)$ or an entrance family $\eta(\lambda)$ we have

$$\xi(\lambda)\downarrow \qquad \eta(\lambda)\downarrow \qquad (\lambda \uparrow \infty) \tag{3}$$

Moreover if $\xi(\lambda) = 0$ or $\eta(\lambda) = 0$ for some λ, then they hold for all $\lambda > 0$.

Definition 2. We call

$$\xi = \lim_{\lambda \downarrow 0} \xi(\lambda) \tag{4}$$

the standard image of the exit family $\xi(\lambda)$ $(\lambda > 0)$. We call

$$\eta = \lim_{\lambda \downarrow 0} \eta(\lambda) \tag{5}$$

the standard image of the entrance family $\eta(\lambda)$ $(\lambda > 0)$.

Lemma 1. Let ξ be the standard image of exit family $\xi(\lambda)$, then[1]

$$\xi = \xi(\mu) + \mu\Gamma\xi(\mu) \qquad \mu > 0 \tag{6}$$

$$\xi = \xi(\lambda) + \lambda\phi(\lambda)\xi \qquad \lambda > 0 \tag{7}$$

where

$$\Gamma = \lim_{\lambda \downarrow 0} \phi(\lambda) = \sum_{n=0}^{\infty} \Pi^n q^{-1} \tag{8}$$

Let η be the standard image of entrance family $\eta(\lambda)$, then

$$\eta = \eta(\mu) + \mu\eta(\mu)\Gamma \qquad \mu > 0 \tag{9}$$

$$\eta = \eta(\lambda) + \lambda\eta\phi(\lambda) \qquad \lambda > 0 \tag{10}$$

[1] Agree on $0.\infty = \infty.0 = 0$, $1/0 = \infty$.

Proof. Notice that

$$\xi(\lambda) = \xi(\mu) + (\mu - \lambda)\phi(\lambda)\xi(\mu) \tag{11}$$

$$\xi(\mu) = \xi(\lambda) + (\lambda - \mu)\phi(\lambda)\xi(\mu) \tag{12}$$

If $\sum_j \Gamma_{ij}\xi_j(u) < \infty$, (6) follows by letting $\lambda \downarrow 0$ in (11). If $\sum_j \Gamma_{ij}\xi_j(u) = \infty$, let $\lambda \downarrow 0$ in (11); by Fatou's lemma we have

$$\xi_i \geqslant \xi_i(\mu) + \mu \sum_j \Gamma_{ij}\xi_j(\mu) = \infty$$

Therefore (6) still holds. In the same way, let $\mu \downarrow 0$ in (12), and we obtain (7). Similarly, we can prove (9) and (10), and the proof is complete. QED

Corollary

Given a fixed $i, \xi_i < \infty$ if and only if for some (hence for all) $\lambda > 0$ we have

$$[\lambda\phi(\lambda)\xi]_i = [\lambda\Gamma\xi(\lambda)]_i < \infty$$

In that case, we have

$$[\lambda\phi(\lambda)\xi]_i = \xi_i - \xi_i(\lambda) \qquad \lambda > 0$$

Similarly, given a fixed $j, \eta_j < \infty$ if and only if for some (hence for all) $\lambda > 0$ we have

$$[\lambda\eta\phi(\lambda)]_j = [\lambda\eta(\lambda)\Gamma]_j < \infty$$

In that case, we have

$$[\lambda\eta\phi(\lambda)]_j = \eta_j - \eta_j(\lambda) \qquad \lambda > 0$$

Lemma 2. Denote $X^0 = \lim_{\lambda\downarrow 0} \lambda\phi(\lambda)\mathbf{1}$. Then

$$\sum_{a\in H} X^a + \bar{X} + X^0 = 1 \tag{13}$$

where

$$X_i^a = \Gamma_{ia} d_a \qquad a \in H \tag{14}$$

is the standard image of the exit family

$$X_i^a(\lambda) = \phi_{ia}(\lambda) d_a \qquad a \in H, \quad \lambda > 0 \tag{15}$$

Thus

$$\lambda\phi(\lambda)X^a = X^a - X^a(\lambda) \qquad a \in H \tag{16}$$

$\bar{X}$ is the standard image of the exit family $\bar{X}(\lambda)$. Hence

$$\lambda\phi(\lambda)\bar{X} = \bar{X} - \bar{X}(\lambda) \qquad \lambda > 0 \tag{17}$$

$\bar{X}$ is a solution of the equation

$$\Pi u = u \qquad 0 \leqslant u \leqslant 1 \tag{18}$$

while

$$\lambda\phi(\lambda)X^0 = X^0 \tag{19}$$

and X^0 is the maximal one of the solutions that satisfy the condition $\lambda\phi(\lambda)u = u$ and equation (18).

Proof. From the resolvent equation of $\phi(\lambda)$ it follows that $X^a(\lambda)$ $(a\in H)$ is an exit family whose standard image is

$$X_i^a = \lim_{\lambda\downarrow 0} X_i^a(\lambda) = \lim_{\lambda\downarrow 0} \phi_{ia}(\lambda)d_a = \Gamma_{ia}d_a \tag{20}$$

On account of (10.14) $\sum_{a\in H} X^a(\lambda) \leqslant 1$ and therefore $\sum_{a\in H} X^a \leqslant 1$. Hence (16) follows from (7).

By (10.14) and (16) we obtain

$$\lambda\phi(\lambda)\left(\mathbf{1} - \sum_{a\in H} X^a\right) = \mathbf{1} - \sum_{a\in H} X^a - \bar{X}(\lambda) \tag{21}$$

From this and the resolvent equation of $\phi(\lambda)$ it follows that $\bar{X}(\lambda)$ $(\lambda > 0)$ is an exit family. And since $\bar{X}(\lambda) \leqslant 1$, we know its standard image $\leqslant 1$. Hence by (7) we obtain (17). Also because

$$(\lambda I - Q)\bar{X}(\lambda) = 0 \qquad 0 \leqslant \bar{X}(\lambda) \leqslant \mathbf{1}$$

and consequently, letting $\lambda\downarrow 0$, we find that $\bar{X}$ is a solution of equation (18).

By (10.14), $\sum_{a\in H} X^a(\lambda) + \bar{X}(\lambda) \leqslant 1$. Letting $\lambda\downarrow 0$, we obtain $\lambda\phi(\lambda)\mathbf{1}\downarrow X^0$ and

$$X^0 = \mathbf{1} - \sum_{a\in H} X^a - \bar{X} \geqslant 0$$

From this, (13) follows. By (17) and (21) we obtain (19). By the B condition of $\phi(\lambda)$ we have

$$(\lambda I - Q)\lambda\phi(\lambda)\mathbf{1} = \lambda \tag{22}$$

and $\lambda\phi(\lambda)\mathbf{1}\downarrow X^0$ $(\lambda\downarrow 0)$. In the foregoing expression, let $\lambda\downarrow 0$, and we obtain $QX^0 = 0$, that is, X^0 satisfies (18). Suppose that u satisfies (18) and $\lambda\phi(\lambda)u = u$. Then $u = \lambda\phi(\lambda)\, u \leqslant \lambda\phi(\lambda)\mathbf{1}$, hence $u \leqslant X^0$.

Definition 3. $\bar{X}$ is called the maximal exit solution of matrix Q, and X^0 is called the maximal passive solution of matrix Q.

Lemma 3. (i) $\eta(\lambda)$ $(\lambda > 0)$ is an entrance family if and only if there exists a Riesz decomposition as follows:

$$\eta(\lambda) = \alpha\phi(\lambda) + \bar{\eta}(\lambda) \tag{23}$$

where the row vector $\alpha \geqslant 0$, and there exists some (hence all) $\lambda > 0$ so that

$\alpha\phi(\lambda)\in l$, that is

$$\left[\alpha, 1-\sum_{a\in H} X^{a}(\lambda)-\bar{X}(\lambda)\right]<\infty \tag{24}$$

$\bar{\eta}(\lambda)\in\mathscr{L}_{\lambda}^{+}$ is a harmonic entrance family. The vector α and hence $\bar{\eta}(\lambda)$ are uniquely determined by $\eta(\lambda)$:

$$\eta(\lambda)(\lambda I-Q)=\alpha \tag{25}$$

$$\eta(\lambda)\downarrow 0 \qquad \lambda\eta(\lambda)\to\alpha \qquad (\lambda\uparrow\infty) \tag{26}$$

(ii) $\xi(\lambda)$ $(\lambda>0)$ is an exit family if and only if there exists a Riesz decomposition as follows:

$$\xi(\lambda)=\phi(\lambda)\beta+\bar{\xi}(\lambda) \tag{27}$$

where the column vector $\beta\geqslant 0$, and there exists some (hence all) $\lambda>0$ so that $\phi(\lambda)\beta\in m$. $\bar{\xi}(\lambda)\in\mathscr{M}_{\lambda}^{+}$ is a harmonic exit family. Vector β and hence $\bar{\xi}(\lambda)$ are uniquely determined by $\xi(\lambda)$:

$$(\lambda I-Q)\xi(\lambda)=\beta \tag{28}$$

$$\xi(\lambda)\downarrow \mathbf{0} \qquad \lambda\xi(\lambda)\to\beta \qquad (\lambda\uparrow\infty) \tag{29}$$

Proof. We first prove (i) and proceed to prove the necessity. Since $(\eta(\lambda),\lambda>0)$ is an entrance family it follows that for arbitrary $v>0$, $\lambda>0$, we have

$$\begin{aligned}
\eta(v)+(v-\lambda)\eta(v)\phi(\lambda)&=\eta(\lambda)\geqslant\mathbf{0}\\
\eta(v)&\geqslant(\lambda-v)\eta(v)\phi(\lambda)\\
\eta_j(v)&\geqslant(\lambda-v)\eta_j(v)\lambda^{-1}+(\lambda-v)\sum_i\eta_i(v)[\phi_{ij}(\lambda)-\lambda^{-1}\delta_{ij}]\\
v\eta_j(v)&\geqslant(1-v\lambda^{-1})\sum_i\eta_i(v)\lambda[\lambda\phi_{ij}(\lambda)-\delta_{ij}]
\end{aligned}$$

In the summation on the right-hand side of the above expression, the terms for $i\neq j$ are non-negative. Hence according to the Q condition of $\phi(\lambda)$ and Fatou's lemma, let $\lambda\to\infty$, and we obtain

$$v\eta_j(v)\geqslant\sum_i\eta_i(v)q_{ij}$$

Hence there exists a finite non-negative row vector $\alpha(v)$ so that

$$\eta(v)(vI-Q)=\alpha(v) \tag{30}$$

By Lemma 10.2, $\eta(v)\geqslant\alpha(v)\phi(v)$, therefore $\alpha(v)\phi(v)\in l$ is valid for all $v>0$. From the F condition of $\phi(\lambda)$ it follows that for any non-negative row vector α, provided $\alpha\phi(v)\in l$, there exists

$$\alpha\phi(v)(vI-Q)=\alpha \tag{31}$$

Hence

$$\eta(v) = \alpha(v)\phi(v) + \bar{\eta}(v) \tag{32}$$

where $\bar{\eta}(v) \in \mathscr{L}_\lambda^+$. Multiplying the above expression from the right by $A(v, \lambda)$ and noticing (10.9), and (2) as well, we obtain

$$\eta(\lambda) = \alpha(v)\phi(\lambda) + \bar{\eta}(v)A(v, \lambda) \tag{33}$$

According to Lemma 10.4, $\bar{\eta}(v)A(v, \lambda) \in \mathscr{L}_\lambda^+$. Substituting (33) into (30) and noticing (31) we obtain $\alpha(v) = \alpha(\lambda)$, that is, $\alpha(\lambda) = \alpha$ is independent of λ. Thus $\bar{\eta}(v)A(v, \lambda) = \bar{\eta}(\lambda)$. In this way (32) becomes (23), where there exists non-negative row vector α so that $\alpha\phi(\lambda) \in l$ for all $\lambda > 0$, and $(\bar{\eta}(\lambda), \lambda > 0)$ is a harmonic entrance family. From (3) and (25) we derive the first expression of (26). From this and the dominated convergence theorem, letting $\lambda \to \infty$ in (25), we obtain the second expression of (26). The necessity is proved. The sufficiency is obvious.

Now we are going to prove (ii). Suppose that $(\xi(\lambda), \lambda > 0)$ is an exit family. Noticing (2.6), for arbitrary $v > 0$, the column vector

$$\beta(v) \equiv (vI - Q)\xi(v) \tag{34}$$

is finite. On account of (2.6) the B condition of $\phi(\lambda)$ is

$$\begin{aligned}(vI - Q)\xi(v) &= (vI - Q)\{[I + (\lambda - v)\phi(v)]\xi(\lambda)\} \\ &= \{(vI - Q)[I + (\lambda - v)\phi(v)]\}\xi(\lambda) \\ &= [vI - Q + (\lambda - v)I]\xi(\lambda) \\ &= (\lambda I - Q)\xi(\lambda)\end{aligned}$$

Therefore, $\beta(\lambda) = \beta$ is independent of $\lambda > 0$, and hence (28) holds. From (3) and (28) we derive the first expression of (29). Because of this and the dominated convergence theorem, let $\lambda \to \infty$ in (28) and we obtain the second expression of (29). The limit $\beta \geqslant 0$ in the second expression of (29) because $\xi(\lambda) \geqslant 0$. According to Lemma 10.2, from (28) we get $\xi(v) \geqslant \phi(v)\beta$, hence $\phi(v)\beta \in m$ is valid for all $v > 0$. By the B condition of $\phi(\lambda)$ it follows that for any non-negative column vector β, provided $\phi(v)\beta \in m$, we have

$$(vI - Q)\phi(v)\beta = \beta \tag{35}$$

Hence from the above expression, (28) and lemma 10.2, we have

$$\xi(v) = \phi(v)\beta + \bar{\xi}(v) \tag{36}$$

where $\xi(v) \in \mathscr{M}_v^+$. Multiplying the above expression from the left by $A(v, \lambda)$ and noticing (10.9) and (1) we obtain

$$\xi(\lambda) = \phi(\lambda)\beta + A(v, \lambda)\bar{\xi}(v) \tag{37}$$

Therefore $\bar{\xi}(\lambda) = A(v, \lambda)\bar{\xi}(v)$. Hence $(\bar{\xi}(\lambda), \lambda > 0)$ is a harmonic exit family. Thus the necessity is proved. The sufficiency is obvious. QED

Corollary

When $\lambda \uparrow \infty$,

$$X^a(\lambda) \downarrow 0 \qquad \lambda X_i^a(\lambda) \to \delta_{ia} d_a \qquad a \in H \tag{38}$$

$$\bar{X}(\lambda) \downarrow 0 \qquad \lambda \bar{X}(\lambda) \to 0 \tag{39}$$

Lemma 4. If $(\eta(\lambda), \lambda > 0)$ is an entrance family, then $\sigma^0 = \lambda[\eta(\lambda), X^0] < \infty$ is independent of $\lambda > 0$. If $(\xi(\lambda), \lambda > 0)$ is an exit family and its standard image $\xi \in m$, then

$$(\lambda - \mu)[\eta(\lambda), \xi(\mu)] = \lambda[\eta(\lambda), \xi] - \mu[\eta(\mu), \xi] \tag{40}$$

$$\lambda[\eta(\lambda), \xi] \uparrow V \leqslant \infty \tag{41}$$

In particular, when $a \in H$

$$V_\lambda^a = \lambda[\eta(\lambda), X^a] \uparrow V^a < \infty \tag{42}$$

where

$$V^a = V_\mu^a + \eta_a(\mu) d_a \qquad \text{is independent of } \mu \tag{43}$$

Suppose that η denotes the standard image of the entrance family $(\eta(\lambda), \lambda > 0)$. Then

$$[\eta(\lambda), \xi] = [\eta, \xi(\lambda)] \tag{44}$$

If $[\eta, X^0] < \infty$, then

$$[\eta, X^0] = 0 \tag{45}$$

Proof. By (2), (7) and $\xi \in m$,

$$\begin{aligned}
\lambda[\eta(\lambda), \xi] &= \lambda[\eta(\mu) A(\mu, \lambda), \xi] \\
&= \lambda[\eta(\mu), A(\mu, \lambda)\xi] \\
&= \lambda[\eta(\mu), \xi] + (\mu - \lambda)[\eta(\mu), \lambda\phi(\lambda)\xi] \\
&= \lambda[\eta(\mu), \xi] + (\mu - \lambda)[\eta(\mu), \xi - \xi(\lambda)] \\
&= \mu[\eta(\mu), \xi] + (\lambda - \mu)[\eta(\mu), \xi(\lambda)]
\end{aligned} \tag{46}$$

From this, (40) follows, and therefore (41) is obtained. Similarly, by using (19), it can be btained that σ^0 is independent of λ.

When $a \in H$, $\xi(\lambda) = X^a(\lambda)$, (40) becomes

$$V_\lambda^a = V_\mu^a + (\lambda - \mu)[\eta(\mu), X^a(\lambda)] \tag{47}$$

Since $\lambda X^a(\lambda) \leqslant d_a$, by (37) and the dominated convergence theorem,

$$[\eta(\mu), \lambda X^a(\lambda)] \to \eta_a(\mu) d_a \qquad \lambda \uparrow \infty, \quad a \in H \tag{48a}$$

Hence let $\lambda \to \infty$ in (47), and (42) and (43) follow.

When $\mu \to 0$,

$$(\lambda - \mu)[\eta(\lambda), \xi(\mu)] = \lambda[\eta(\lambda), \xi(\mu)] - \mu[\eta(\lambda), \xi(\mu)] \to \lambda[\eta(\lambda), \xi]$$

From this, let $\mu \to 0$ in (40) and we have

$$\mu[\eta(\mu), \xi] \to 0 \qquad \text{if} \quad \mu \to 0 \tag{48b}$$

Thus let $\lambda \to 0$ in (40), and we obtain (44).

By (39), if $[\eta, X^0] < \infty$, then $\sigma^0 \leqslant \lambda[\eta, X^0] \to 0$, as $\lambda \to 0$. Hence $\sigma^0 = 0$, $[\eta(\lambda), X^0] = 0$. Letting $\lambda \to 0$, we obtain (45). QED

Lemma 5. If Q is non-conservative, then

$$\lambda \sum_{a \in H} X^a(\lambda) \to \boldsymbol{d} \qquad \lambda \to \infty \tag{49}$$

Proof. We have

$$Z(\lambda) = \mathbf{1} - \lambda\phi(\lambda)\mathbf{1} = \sum_{a \in H} X^a(\lambda) + \bar{X}(\lambda)$$

is an exit family, and moreover $(\lambda I - Q)Z(\lambda) = d$. Hence by Lemma 3, $\lambda Z(\lambda) \to d$. Noticing (38), (49) follows. QED

Lemma 6. If, for the row vector α, $\alpha\phi(\lambda) = 0$ (that is, $\sum_i \alpha_i \phi_{ij}(\lambda) = 0$ for every j, and moreover the series is absolutely convergent), then $\alpha = 0$.

If, for the column vector β, $\phi(\lambda)\beta = 0$ (that is, $\sum_j \phi_{ij}\beta_j = 0$ for every i, and furthermore the series is absolutely convergent), then $\beta = 0$.

Proof. Since $\phi(\lambda)$ satisfies the system of backward equations, it follows that

$$\lambda\phi_{ij}(\lambda) = \delta_{ij} + \sum_k q_{ik}\phi_{kj}(\lambda) \tag{50}$$

Fix i. Multiplying both sides by β_j and summing the resulting expression for j, we obtain

$$0 = \beta_i + \sum_j \sum_k q_{ik}\phi_{kj}(\lambda)\beta_j = \beta_i + \sum_k q_{ik}\left(\sum_j \phi_{kj}(\lambda)\beta_j\right) = \beta_i$$

The order of summation can be interchanged because

$$\begin{aligned}\sum_j\left(\sum_k |q_{ik}|\phi_{kj}(\lambda)|\beta_j|\right) &= \sum_j |\beta_j|\left(\sum_k |q_{ik}|\phi_{kj}(\lambda)\right) = \sum_j |\beta_j|\left(\sum_k q_{ik}\phi_{kj}(\lambda) + 2q_i\phi_{ij}(\lambda)\right)\\ &= \sum_j |\beta_j|[\lambda\phi_{ij}(\lambda) - \delta_{ij} + 2q_i\phi_{ij}(\lambda)] \leqslant (\lambda + 2q_i)\sum_i \phi_{ij}(\lambda)|\beta_j| < \infty\end{aligned}$$

Applying the F condition of $\phi(\lambda)$,

$$\lambda\phi_{ij}(\lambda)=\delta_{ij}+\sum_k \phi_{ik}(\lambda)q_{kj} \tag{51}$$

we can prove similarly that $\alpha=0$. QED

2.12 GENERAL FORM OF Q PROCESSES

If $\psi(\lambda)\in\mathscr{P}_s(Q)$ then $\psi(\lambda)-\phi(\lambda)\geqslant 0$. As $\psi(\lambda)$ satisfies the system of backward inequalities (8.10) and (8.12), whereas $\phi(\lambda)$ satisfies the B conditon, it follows that $(\lambda I-Q)[\psi(\lambda)-\phi(\lambda)]\geqslant 0$. Moreover if $i\in E-H$, then equality holds. More precisely, for fixed $j, u_i=v_{ij}(\lambda)-\phi_{ij}(\lambda)$ satisfies

$$\lambda u_i-\sum_k q_{ik}u_k=d_iF_j^i(\lambda) \tag{1}$$

where $F_j^i(\lambda)\geqslant 0$ and d is the non-conservative quantity. By Lemma 10.2

$$B_{ij}(\lambda)=\psi_{ij}(\lambda)-\phi_{ij}(\lambda)-\sum_{a\in H}\phi_{ia}(\lambda)d_aF_j^a(\lambda)\geqslant 0$$

If j is fixed, $B_{\cdot j}(\lambda)\in\mathscr{M}_\lambda^+(1/\lambda)$. Thus the first part of the following theorem is derived.

Theorem 1. Any Q process $\psi(\lambda)\in\mathscr{P}_s(Q)$ must take the following form:

$$\psi_{ij}(\lambda)=\phi_{ij}(\lambda)+\sum_{a\in H}X_i^a(\lambda)F_j^a(\lambda)+B_{ij}(\lambda) \tag{2}$$

where $X_i^a(\lambda)=\phi_{ia}(\lambda)d_a$ $(a\in H)$. If i is fixed, then $0\leqslant B_{i\cdot}(\lambda)\in l$; if j is fixed, $B_{\cdot j}(\lambda)\in\mu_\lambda^+(1/\lambda)$. Moreover,

$$F^a(\lambda)\geqslant 0 \qquad \lambda[F^a(\lambda),1]\leqslant 1 \qquad a\in H \tag{3}$$

If $\psi(\lambda)$ satisfies the system of backward equations, and Q is non-conservative, then $F^a(\lambda)=0$ $(a\in H)$.

If $\psi(\lambda)$ satisfies the system of forward equations, then when i is fixed, $B_{i\cdot}(\lambda)\in\mathscr{L}_\lambda^+$. And if Q is non-conservative, then $F^a(\lambda)\in\mathscr{L}_\lambda^+$ $(a\in H)$.

Proof. Since $\psi_{i\cdot}(\lambda)\in l$ when i is fixed, so $F^a(\lambda)\in l$ $(a\in H)$, $B_{i\cdot}(\lambda)\in l$.

Suppose $\psi(\lambda)$ satisfies the system of backward equations, and Q is non-conservative. By Theorem 8.4,

$$\sum_{a\in H}X_i^a(\lambda)F_j^a(\lambda)=\sum_{a\in H}\phi_{ia}(\lambda)d_aF_j^a(\lambda)=0$$

By Lemma 11.6, this is equivalent to $F^a(\lambda)=0$ $(a\in H)$.

If $\psi(\lambda)$ satisfies the system of forward equations, since both $\psi(\lambda)$ and $\phi(\lambda)$

satisfy the F condition, it follows that if we set $G(\lambda)=\{d_iF^i_j(\lambda)\}$ $(i, j\in E)$, then

$$[\phi(\lambda)G(\lambda)+B(\lambda)](\lambda I-Q)=0$$

Multiplying from the left by $(\lambda I-Q)$ we obtain

$$G(\lambda)(\lambda I-Q)=0$$

Hence $B(\lambda)(\lambda I-Q)=0$. That is, for fixed $i, B_{i\cdot}(\lambda)\in\mathscr{L}^+_\lambda$, $F^a(\lambda)\in\mathscr{L}^+_\lambda$ $(a\in H)$.

Now let us prove (3). In section 7.18 we shall also prove (3). Here we shall give another proof.

Operating $\lambda I-Q$ on both sides of (2) and summing (2) over j, we obtain

$$(\lambda I-Q)(\lambda\psi(\lambda)\mathbf{1})_i=\lambda+\sum_{a\in H}\delta_{ia}d_a\lambda[F^a(\lambda),\mathbf{1}] \tag{4}$$

If $i=a$,

$$(\lambda I-Q)(\lambda\psi(\lambda)\mathbf{1})_a=\lambda+d_a\lambda[F^a(\lambda),\mathbf{1}] \tag{5}$$

Taking Laplace transforms on both sides of (7.3) we have

$$(\lambda I-Q)\lambda\psi(\lambda)\mathbf{1}\leqslant\lambda\mathbf{1}+\boldsymbol{d} \tag{6}$$

By (5) and (6), (3) follows. QED

Theorem 2. Each Q process $\psi(\lambda)\in\mathscr{P}_s(Q)$ must have the following form:

$$\psi_{ij}(\lambda)=\phi_{ij}(\lambda)+\sum_k H_{ik}(\lambda)\phi_{kj}(\lambda)+C_{ij}(\lambda) \tag{7}$$

where $H(\lambda)=\{H_{ik}(\lambda)\}\geqslant 0$ and $C(\lambda)=\{C_{ij}(\lambda)\}\geqslant 0$. When i is fixed, $C_{i\cdot}(\lambda)\in\mathscr{L}^+_\lambda$; when j is fixed, $C_{\cdot j}(\lambda)\in m$, $H_{\cdot j}(\lambda)\in m$.

$\psi(\lambda)$ satisfies the system of backward equations if and only if $H_{\cdot j}(\lambda)\in\mathscr{M}^+_\lambda$, $C_{\cdot j}(\lambda)\in\mathscr{M}^+_\lambda(1/\lambda)$.

$\psi(\lambda)$ satisfies the system of forward equations if and only if $H(\lambda)=0$.

Proof. By the system of forward inequalities (8.11) and the F condition of $\phi(\lambda)$, we have

$$[\psi(\lambda)-\phi(\lambda)](\lambda I-Q)=H(\lambda)\geqslant 0$$

Therefore, by Lemma 10.2

$$C(\lambda)\equiv\psi(\lambda)-\phi(\lambda)-H(\lambda)\phi(\lambda)\geqslant 0$$

and $C_{i\cdot}(\lambda)\in\mathscr{L}^+_\lambda$; hence we have (7). By (7) we obtain $C_{\cdot j}(\lambda)\in m$. Since

$$H_{\cdot j}(\lambda)\lambda\phi_{jj}(\lambda)\leqslant\lambda\psi_{\cdot j}(\lambda)\leqslant 1$$

it follows that $H_{\cdot j}(\lambda)\in m$.

The B condition of $\psi(\lambda)$ is equivalent to

$$(\lambda I-Q)[H(\lambda)\phi(\lambda)+C(\lambda)]=0$$

Multiplying the above expression form the right by $(\lambda I - Q)$, we have $(\lambda I - Q)H(\lambda) = 0$. Thus by the above expression we have $(\lambda I - Q)C(\lambda) = 0$. Hence the B conditionis equivalent to $C_{\cdot j}(\lambda) \in \mathscr{M}_\lambda^+(1/\lambda)$, $H_{\cdot j}(\lambda) \in \mathscr{M}_\lambda^+$.

The F condition of $\psi(\lambda)$ is equivalent to

$$[H(\lambda)\phi(\lambda) + C(\lambda)](\lambda I - Q) = 0$$

That is, $H(\lambda) = 0$. The proof is complete. QED

CHAPTER 3

Construction of Q Processes in Simple Cases

3.1 INTRODUCTION

Q is assumed to be conservative in many works on construction of Q processes such as Reuter (1959), Feller (1940, 1945, 1957), Doob (1945), Chung (1963, 1966), Williams (1964, 1966) and Xiang-qun Yang (1966a) because in that case any Q process satisfies the system of backward equations. Only a small number of articles like Reuter (1962), Feller (1957b, 1971) and Xiang-qun Yang (1965a) are concerned with non-conservative Q. If we just construct the Q processes satisfying the system of backward equations, we may do so in principle by using Theorem 2.7.7 for $P(t)$. But this construction is not an immediate one. Moreover this way of construction does not suit the construction of Q processes satisfying the system of forward equations.

In this chapter we consider construction of Q processes in simple cases, and it is not necessary to assume Q to be conservative. If Q is conservative and in single exit, all Q processes have been constructed in Reuter (1959). As for non-conservative Q, a class of honest Q processes that do not satisfy the system of backward equations are constructed in Reuter (1962). In section 3.2 all Q processes satisfying the system of backward equations are directly constructed in the case of single exit. A class of processes containing the results in Reuter (1962) are constructed in section 3.3 for non-conservative Q. In particular, when Q is single non-conservative and in null exit, we construct all Q processes. And in section 3.4 all Q processes satisfying the system of forward equations are constructed when the minimal solution is a stopping process and furthermore in single exit. The results of this chapter are seen in Xiang-qun Yang (1981a).

3.2 SINGLE EXIT CASE: CONSTRUCTION OF Q PROCESSES SATISFYING THE BACKWARD EQUATIONS

By Lemma 2.10.4, the dimension m^+ of the solution space $\mathscr{M}_\lambda^+$ of the system of equations

$$(\lambda I - Q)u = 0 \qquad 0 \leqslant u \in m \tag{1}$$

is independent of λ. When $m^+ = 0$, that is, if $\mathscr{M}_\lambda^+$ constains only the zero element, Q is called null exit. When $m^+ = 1$, Q is called single exit. And if m^+ is finite, Q is called finite exit.

Suppose that Q is single exit. By Lemma 2.10.6, $\bar{X}(\lambda) \neq 0$; on account of Theorem 2.10.5, the minimal solution $\phi(\lambda)$ is a stopping process. According to Theorem 2.12.1, every Q process $\psi(\lambda)$ satisfying the system of backward equations possesses the following form:

$$\psi_{ij}(\lambda) = \phi_{ij}(\lambda) + \bar{X}_i(\lambda)F_j(\lambda) \tag{2}$$

Under the conditions of $m^+ > 0$ we shall determine the conditions under which $F(\lambda)$ makes $\psi(\lambda) \in \mathscr{P}_s(Q)$, that is, $\psi(\lambda)$ satisfies the norm condition and resolvent equation because the B condition is always satisfied for $\psi(\lambda)$ defined by (2).

Theorem 1. Let $m^+ > 0$. For $\psi(\lambda)$ defined by (2) to be a Q process it is necessary and sufficient that either $\psi(\lambda) = \phi(\lambda)$ or $\psi(\lambda)$ can be obtained as follows:

Take a row vector $\alpha \geqslant 0$ such that $\alpha\phi(\lambda) \in l$, and take a harmonic entrance family $\bar{\eta}(\lambda) \in \mathscr{L}_\lambda^+$. Moreover

$$\eta(\lambda) = \alpha\phi(\lambda) + \bar{\eta}(\lambda) \neq 0 \tag{3}$$

Take a constant c such that

$$[\alpha, X^0] + \bar{\sigma}^0 + \sum_{a \in H} ([\alpha, X^a] + \bar{V}^a) \leqslant c \tag{4}$$

is satisfied, where $X^0, \bar{X}, X^a (a \in H)$ are defined by Lemma 2.11.2, while

$$\bar{\sigma}^0 = \lambda[\bar{\eta}(\lambda), X^0] < \infty \qquad \text{is independent of } \lambda \tag{5}$$

$$\bar{V}_\lambda^a = \lambda[\bar{\eta}(\lambda), X^a] \uparrow \bar{V}^a < \infty \qquad \lambda \uparrow \infty, a \in H \tag{6}$$

$$\bar{V}^a = \bar{V}_\lambda^a + \bar{\eta}_a(\lambda) d_a \qquad \text{is independent of } \lambda \tag{7}$$

Finally set

$$\psi_{ij}(\lambda) = \phi_{ij}(\lambda) + \bar{X}_i(\lambda) \frac{\sum_k \alpha_k \phi_{kj}(\lambda) + \bar{\eta}_j(\lambda)}{c + [\alpha, \bar{X} - \bar{X}(\lambda)] + \lambda[\bar{\eta}(\lambda), \bar{X}]} \tag{8}$$

The process $\psi(\lambda)$ is honest if and only if

$$Q \text{ is conservative and, moreover, } [a, X^0] + \sigma^0 = c \tag{9}$$

The process $\psi(\lambda)$ satisfies the system of backward equations. The necessary and sufficient condition under which this process satisfies the system of forward equations is that $\alpha = 0$.

When $m^+ = 1$ the above-stated processes include all the Q processes that satisfy the system of backward equations. When Q is conservative and, furthermore, $m^+ = 0$, the processes above include all Q processes.

Proof. By minimality of $\phi(\lambda), \psi(\lambda) \geqslant 0$ is equivalent to $F(\lambda) \geqslant 0$. Upon noting (2.10.14), the norm condition is equivalent to

$$F(\lambda) \geqslant 0 \qquad \bar{X}(\lambda)\lambda[F(\lambda), \mathbf{1}] \leqslant \bar{X}(\lambda) + \sum_{a \in H} X^a(\lambda) \tag{10}$$

Because of (2.10.14),

$$\sum_{a \in H} X^a(\lambda) = \phi(\lambda)\mathbf{d} = \sum_{n=0}^{\infty} \Pi^n(\lambda)(\lambda + q)^{-1}\mathbf{d} \leqslant 1 \tag{11}$$

Therefore,

$$\Pi^n(\lambda)\left(\sum_{a \in H} X^a(\lambda)\right) = \sum_{a=n}^{\infty} \Pi^a(\lambda)(\lambda I + q)^{-1}\mathbf{d} \to 0 \qquad n \to \infty \tag{12}$$

Multiplying (10) from the left by $\Pi^n(\lambda)$ and taking its limit, we obtain

$$\bar{X}(\lambda)\lambda[F(\lambda), \mathbf{1}] \leqslant \bar{X}(\lambda)$$

Hence $\lambda[F(\lambda), 1] \leqslant 1$. Consequently, the norm condition is equivalent to

$$F(\lambda) \geqslant 0 \qquad \lambda[F(\lambda), 1] \leqslant 1 \tag{13}$$

$\psi(\lambda)$ is honest if and only if

$$Q \text{ is conservative and, moreover, } \lambda[F(\lambda), \mathbf{1}] = 1 \tag{14}$$

Since $\phi(\lambda)$ satisfies the resolvent equation, substituting $\psi(\lambda)$ in the resolvent equation, and observing that $\bar{X}(\lambda)$ is an exit family, we obtain that the resolvent equation of $\psi(\lambda)$ is equivalent to

$$F(\lambda)A(\lambda, \mu) = \{1 + (\mu - \lambda)[F(\lambda), \bar{X}(\mu)]\}F(\mu) \tag{15}$$

Or by (2.10.10), upon multiplying the above formula by $A(\mu, \lambda)$ from the right, we find that (15) is equivalent to

$$F(\lambda) = \{1 + (\mu - \lambda)[F(\lambda), \bar{X}(\mu)]\}F(\mu)A(\mu, \lambda) \tag{16}$$

If, for some $\mu > 0, F(\mu) = 0$, then from the formula above we know that for all $\lambda > 0, F(\lambda) = 0$. Hence $\psi(\lambda) = \phi(\lambda)$.

Otherwise, for all $\mu > 0, F(\mu) \neq 0$. Since $\lambda[F(\lambda), \bar{X}(\mu)] \leqslant \lambda[F(\lambda), \mathbf{1}] \leqslant 1$, it follows that $1 + (\mu - \lambda)[F(\lambda), \bar{X}(\mu)] > 0$. From (16) it can be seen that $F(\mu)A(\mu, \lambda) \geqslant 0$. Therefore, if one $\mu > 0$ is fixed, then $\eta(\lambda) = F(\mu)A(\mu, \lambda)$ is an entrance family. Thus, (16) is equivalent to

$$F(\lambda) = m_\lambda \eta(\lambda) \qquad m_\lambda > 0, \eta(\lambda) \neq 0 \tag{17}$$

where the quantity m_λ satisfies

$$m_\lambda = m_\mu + (\mu - \lambda)m_\lambda[\eta(\lambda), \bar{X}(\mu)]m_\mu \tag{18}$$

where $\eta(\lambda)$ is a non-zero entrance family. According to Lemma 2.11.3, $\eta(\lambda)$ has Riesz representation (3).

Dividing (18) on both sides by $m_\lambda m_\mu$, we obtain

$$m_\mu^{-1} = m_\lambda^{-1} + (\mu - \lambda)[\eta(\lambda), \bar{X}(\mu)] \tag{19}$$

But by (2.11.40),

$$(\mu - \lambda)[\eta(\lambda), \bar{X}(\mu)] = \mu[\eta(\mu), \bar{X}] - \lambda[\eta(\lambda), \bar{X}] \tag{20}$$

Consequently (19) becomes

$$m_\lambda^{-1} - \lambda[\eta(\lambda), \bar{X}] = c \qquad \text{(constant)} \tag{21}$$

Hence by (2.11.17),

$$m_\lambda = \frac{1}{c + \lambda[\eta(\lambda), \bar{X}]} = \frac{1}{c + \lambda[\alpha\phi(\lambda), \bar{X}] + \lambda[\bar{\eta}(\lambda), \bar{X}]}$$

$$= \frac{1}{c + [\alpha, \bar{X} - \bar{X}(\lambda)] + \lambda[\bar{\eta}(\lambda), \bar{X}]} \tag{22}$$

Furthermore, every deduction from (18) to (22) can be inverted.

Substituting (17) (22) in (13), we have

$$\lambda[\eta(\lambda), \mathbf{1} - \bar{X}] \leqslant c \tag{23}$$

But by (2.11.13), (2.11.16)–(2.11.19) and Lemma 2.11.4,

$$\lambda[\eta(\lambda), 1 - \bar{X}] = \lambda[\alpha\phi(\lambda), \mathbf{1} - \bar{X}] + \lambda[\bar{\eta}(\lambda), \mathbf{1} - \bar{X}]$$

$$= \lambda\left[\alpha\phi(\lambda), X^0 + \sum_{a \in H} X^a\right] + \lambda\left[\bar{\eta}(\lambda), X^0 + \sum_{a \in H} X^a\right]$$

$$= \left[\alpha, \lambda\phi(\lambda)\left(X^0 + \sum_{a \in H} X^a\right)\right] + \bar{\sigma}^0 + \sum_{a \in H} \bar{V}_\lambda^a$$

$$= [\alpha, X^0] + \bar{\sigma}^0 + \sum_{a \in H} ([\alpha, X^a - X^a(\lambda)] + \bar{V}_\lambda^a)$$

$$\uparrow [\alpha, X^0] + \bar{\sigma}^0 + \sum_{a \in H} ([\alpha, X^a] + \bar{V}^a) \qquad \lambda \uparrow \infty \tag{24}$$

The last step is arrived at on account of (2.11.37). Therefore, the norm condition (23) becomes (4) whereas (14) becomes (9). Operating on (8) from the right by $(\lambda I - Q)$, we find that the F condition holds if and only if $\alpha = 0$. The other conclusions of the theorem are quite clear, and the proof is complete. QED

In the proof of Theorem 1, if, in (2), we replace $\bar{x}(\lambda)$ by a solution $\bar{\xi}(\lambda)$ to the equation (2.10.15), which is a non-null harmonic exit family, and slightly modify it, the theorem remains valid. For this we have the following theorem.

Theorem 2. Let $m^+ > 0$. Let $\bar{\xi}(\lambda) \in \mathcal{M}^+_\lambda(1)$ be a non-null harmonic exit family and let $\bar{\xi}$ be the standard image of $\bar{\xi}(\lambda)$, $\text{Sup}_i\, \bar{\xi}_i(\lambda) = 1$. For $\psi(\lambda) \in \mathcal{P}_s(Q)$ defined by

$$\psi_{ij}(\lambda) = \phi_{ij}(\lambda) + \bar{\xi}_i(\lambda) F_j(\lambda) \tag{25}$$

it is necessary and sufficient that either $\psi(\lambda) = \phi(\lambda)$ or $\psi(\lambda)$ can be obtained as follows: Take a row vector $\alpha \geqslant 0$ so that $\alpha\phi(\lambda) \in l$; take a harmonic entrance family $\bar{\eta}(\lambda) \in \mathcal{L}^+_\lambda$ so that (3) holds and satisfies

$$[\alpha, \bar{X} - \bar{\xi}] < \infty$$

$$W_\lambda \equiv \lambda[\bar{\eta}(\lambda), \bar{X} - \bar{\xi}] \uparrow W < \infty \qquad \lambda \uparrow \infty \tag{26}$$

Take a constant c such that

$$[\alpha, X^0) + \bar{\sigma}^0 + [\alpha, \bar{X} - \bar{\xi}] + W + \sum_{a \in H} ([\alpha, X^a] + \bar{V}^a) \leqslant c \tag{27}$$

is satisfied, where the notation is the same as in Theorem 1. Finally, set

$$\psi_{ij}(\lambda) = \phi_{ij}(\lambda) + \bar{\xi}_i(\lambda) \frac{\sum_k \alpha_k \phi_{kj}(\lambda) + \bar{\eta}_j(\lambda)}{c + [\alpha, \bar{\xi} - \bar{\xi}(\lambda)] + \lambda[\bar{\eta}(\lambda), \bar{\xi}]} \tag{28}$$

The process $\psi(\lambda)$ satisfies the system of backward equations. For the process to be honest, it is necessary and sufficient that Q is conservative, and

$$\bar{\xi}(\lambda) = \bar{X}(\lambda), [\alpha, X^0] + \sigma^0 = c \tag{29}$$

3.3 NULL EXIT CASE: CONSTRUCTION OF Q PROCESSES WITH ONE NON-CONSERVATIVE STATE

Suppose that Q is non-conservative, and moreover,

$$Z(\lambda) = 1 - \lambda\phi(\lambda)\mathbf{1} = \sum_{a \in H} X^a(\lambda) + \bar{X}(\lambda) \tag{1}$$

Obviously $Z(\lambda)$ is a non-zero exit family, whose standard image is

$$Z = \sum_{a \in H} X^a + \bar{X} = 1 - X^0 \tag{2}$$

Now let us find the necessary and sufficient conditions on $F(\lambda)$ under which $\psi(\lambda) \in \mathcal{P}_s(Q)$ can be defined by

$$\psi_{ij}(\lambda) = \phi_{ij}(\lambda) + Z_i(\lambda) F_j(\lambda) \tag{3}$$

Since $Z(\lambda) \notin \mathcal{M}^+_\lambda$ it follows that the B condition does not hold. Thus we need to consider the norm condition, resolvent equation and Q condition.

Theorem 1. For $\psi(\lambda)$ defined by (3) to be a non-minimal Q process it is necessary and sufficient that $\psi(\lambda)$ can be obtained as follows:

Take a row vector $\alpha \geqslant 0$ such that $\alpha\phi(\lambda) \in l$, and take a harmonic entrance family $\bar{\eta}(\lambda) \in \mathscr{L}_\lambda^+$; furthermore, (2.3) holds, and also it is necessary to satisfy

$$[\alpha, Z] + U = \infty \qquad \text{if } \alpha \neq 0 \tag{4}$$

or, equivalently,

$$[\alpha, \mathbf{1}] + Y = \infty \qquad \text{if } \alpha \neq 0 \tag{5}$$

where

$$U_\lambda = \lambda[\bar{\eta}(\lambda), Z] \uparrow U \qquad \lambda \uparrow \infty \tag{6}$$

$$Y_\lambda = \lambda[\bar{\eta}(\lambda), \mathbf{1}] \uparrow Y \qquad \lambda \uparrow \infty \tag{7}$$

And take a constant c so that

$$[\alpha, X^0] + \bar{\sigma}^0 \leqslant c \tag{8}$$

is satisfied, where σ^0 is the same as in (2.5). Finally, set

$$\psi_{ij}(\lambda) = \phi_{ij}(\lambda) + Z_i(\lambda)\frac{\sum_k \alpha_k \phi_{kj}(\lambda) + \bar{\eta}_i(\lambda)}{c + [\alpha, Z - Z(\lambda)] + \lambda[\bar{\eta}(\lambda), Z]}$$

The process $\psi(\lambda)$ does not satisfy the system of backward equations. It satisfies the system of forward equations if, and only if, $\alpha = 0$. For the process $\psi(\lambda)$ to be honest, it is necessary and sufficient that

$$[\alpha, X^0] + \bar{\sigma}^0 = c \tag{10}$$

Proof. Following Theorem 2.1, in order that the norm condition and resolvent equation may hold, it is necessary and sufficient that (2.17) holds, where $\eta(\lambda)$ processes representation (2.3), and m_λ is determined by (2.22) (by replacing $\bar{X}, \bar{X}(\lambda)$ with $Z, Z(\lambda)$). Substituting (2.17) and replacing (2.22) in the norm condition (2.13), we have

$$\lambda[\eta(\lambda), \mathbf{1} - Z] \leqslant c \tag{11}$$

But

$$\lambda[\eta(\lambda), \mathbf{1} - Z] = \lambda[\eta(\lambda), X^0] = [\alpha, X^0] + \bar{\sigma}^0 \tag{12}$$

Consequently (11) becomes (8).

For $\psi(\lambda)$ in (9) to be a Q process, the Q condition is yet to be checked, that is,

$$\lim_{\lambda \to \infty} \lambda Z_i(\lambda)\frac{\lambda\eta_j(\lambda)}{c + \lambda[\eta(\lambda), Z]} = 0 \tag{13}$$

Noting (2.11.38), from (2) and Lemma 2.11.5 we have

$$\lambda Z_i(\lambda) \to d_i \qquad (\lambda \to \infty) \tag{14}$$

By Lemma 2.11.3, $\lambda\eta(\lambda)\to\alpha$. Again by $Z_i(\lambda)\downarrow 0$ it follows that

$$\lambda[\eta(\lambda), Z] = [\alpha, Z - Z(\lambda)] + \lambda[\bar{\eta}(\lambda), Z]$$
$$\uparrow [\alpha, Z] + U \qquad \lambda\uparrow\infty \tag{15}$$

Hence (13) becomes

$$d_i \frac{\alpha_j}{c + [\alpha, Z] + U} = 0 \tag{16}$$

and this is equivalent to (4). Since $Z = 1 - X^0$, by (2.11.39),

$$\sigma^0 = \lambda[\eta(\lambda), X^0] = [\alpha, X^0] + \bar{\sigma}^0 < \infty \tag{17}$$

It follows that (4) is equivalent to (5), and the proof is concluded. QED

According to Lemma 2.10.4, the dimension n^+ of the solution space $\mathscr{L}_\lambda^+$ of the equation

$$v(\lambda I - Q) = 0 \qquad 0 \leqslant v \in l \tag{18}$$

is independent of λ. When $n^+ = 0$, the matrix Q is called null entrance. If $n^+ = 1$, the matrix Q is called single entrance. If n^+ is finite, the matrix Q is called finite entrance.

Suppose Q is non-conservative and non-null entrance, then we may take $\alpha = 0$ and the non-zero harmonic entrance family $\bar{\eta}(\lambda) \in \mathscr{L}_\lambda^+$ in Theorem 1. Hence the non-minimal Q process $\psi(\lambda)$ in Theorem 1 exists.

If Q is non-conservative and null entrance, then for the non-minimal Q process $\psi(\lambda)$ in Theorem 1 to exist, we must have $\alpha \geqslant 0$ so that $\alpha\phi(\lambda) \in l$ and, moreover, $[\alpha, 1] = \infty$. This condition can be given by the following lemma (which is also found in the work by Zhen-ting Hou (1974, Lemma 12.2.4)).

Lemma 2. There exists $\alpha \geqslant 0$ so that $\alpha\phi(\lambda) \in l$ while $[\alpha, 1] = \infty$ if, and only if, for some (hence all) $\lambda > 0$,

$$\inf_i \lambda \sum_j \phi_{ij}(\lambda) = 0 \tag{19}$$

Proof. Suppose that there exists $\alpha \geqslant 0$ such that for some $\lambda > 0$, we have $\alpha\phi(\lambda) \in l$ and, moreover, $[\alpha, 1] = \infty$. From the resolvent equation and the norm condition of $\phi(\lambda)$ we know that $\alpha\phi(\lambda) \in l$ holds for all $\lambda > 0$. Since

$$\lambda[\alpha\phi(\lambda), \mathbf{1}] = [\alpha, \lambda\phi(\lambda)\mathbf{1}]$$
$$\geqslant [\alpha, \mathbf{1}] \inf_i \lambda \sum_i \phi_{ij}(\lambda)$$

it follows that (19) is valid for all $\lambda > 0$.

Assume that (19) holds. Since $\lambda\phi_{ii}(\lambda) > 0$, it follows that $\lambda\sum_j \phi_{ij}(\lambda) > 0$. Hence it is deduced from (19) that there exist mutually different $i_k \in E, k = 1, 2, 3, \ldots,$ such that

$$\lambda \sum_j \phi_{i_k j}(\lambda) < 1/2^k$$

Take

$$\alpha_j = \begin{cases} 1 & \text{if } j \in \{i_1, i_2, \ldots\} \\ 0 & \text{if } j \notin \{i_1, i_2, \ldots\} \end{cases}$$

and we have

$$\lambda[\alpha\phi(\lambda), \mathbf{1}] = [\alpha, \lambda\phi(\lambda)\mathbf{1}] < \sum_{k=1}^{\infty} 1/2^k = 1$$

$$[\alpha, \mathbf{1}] = \sum_{k=1}^{\infty} \alpha_{i_k} = \sum_{k=1}^{\infty} 1 = \infty$$

The lemma is proved. QED

The condition (19) is equivalent to

$$\sup_i Z_i(\lambda) = 1 \tag{20}$$

If $\bar{X}(\lambda) \neq 0$, then by Lemma 2.10.7,

$$\sup_i \bar{X}_i(\lambda) = 1 \tag{21}$$

Therefore (20) is satisfied.

Suppose $\bar{X}(\lambda) = 0$; then (20) becomes

$$\sup_i \sum_{a \in H} X_i^a(\lambda) = 1 \tag{22}$$

If Q is single non-conservative, that is, H contains only one state a, then (22) becomes

$$\sup_i X_i^a(\lambda) = 1 \tag{23}$$

or, equivalently

$$\sup_i \phi_{ia}(\lambda) = 1/d_a \tag{24}$$

Theorem 3. Assume Q to be null exit and, furthermore, non-conservative only in one state a. Then

$$Z_i(\lambda) = \phi_{ia}(\lambda) d_a \qquad Z_i = \Gamma_{ia} d_a = 1 - X_i^0 \tag{25}$$

where Γ is defined by (2.10.8). If Q is null entrance, and also

$$\sup_i \phi_{ia}(\lambda) < 1/d_a$$

then the Q process is unique.

If Q is non-zero entrance, or if Q is null entrance and, moreover, (24) holds, then the Q process is not unique. In this case every non-minimal Q process can be obtained by means of Theorem 1.

3.4 SINGLE ENTRANCE CASE: CONSTRUCTION OF Q PROCESSES SATISFYING THE FORWARD EQUATIONS

In this section Q is not required to be conservative, but we assume that the minimal solution is a stopping process, and that $n^+ > 0$. Then we can select a non-zero harmonic entrance family $\bar{\eta}(\lambda) \in \mathscr{L}_\lambda^+$. If the Q process $\psi(\lambda)$ has the form

$$\psi_{ij}(\lambda) = \phi_{ij}(\lambda) + F_i(\lambda)\bar{\eta}_j(\lambda) \tag{1}$$

then $\psi(\lambda)$ satisfies the system of forward equations. Conversely, if $n^+ = 1$, then every Q process $\psi(\lambda)$ satisfying the system of forward equations must take the form (1). Under the condition $n^+ > 0$ we shall determine $F(\lambda)$ so that $\psi(\lambda)$ in (1) is a Q process. As the F condition is always satisfied, we only need to investigate the norm condition and the resolvent equation.

Theorem 1. Suppose that the minimal solution $\phi(\lambda)$ is a stopping process and $n^+ > 0$. For $\psi(\lambda)$ defined by (1) to be a Q process it is necessary and sufficient that either $\psi(\lambda) = \phi(\lambda)$ or $\psi(\lambda)$ can be derived as follows: Take a constant $\delta \geqslant 0$ and a harmonic exit family $\bar{\xi}(\lambda) \in \mathscr{N}^t_\lambda(1)$, whose standard image is $\bar{\xi}$. If $\delta > 0$, it is also demanded that $\sup_i \bar{\xi}_i(\lambda) = 1$ is satisfied. If Q is non-conservative, then for each $a \in H$, take a quantity $\beta^a \geqslant 0$ such that $\sum_{a\in H} \beta^a X^a(\lambda) \in m$, and, moreover,

$$\xi(\lambda) = \sum_{a\in H} \beta^a X^a(\lambda) + \delta\xi(\lambda) \neq 0 \tag{2}$$

and

$$\begin{aligned} &\lambda[\bar{\eta}(\lambda), \xi] < \infty \\ &\bar{W}_\lambda = \lambda[\bar{\eta}(\lambda), \bar{X} - k\delta\bar{\xi}] \uparrow \bar{W} < \infty \qquad \lambda \uparrow \infty \end{aligned} \tag{3}$$

where[1]

$$\xi = \sum_{a\in H} \beta^a X^a + \delta\bar{\xi} \neq 0$$

$$k = \inf\{1/\delta, 1/\beta^a, (a \in H)\} \tag{4}$$

[1] Agree on $1/0 = \infty$.

Take a constant c such that

$$\bar{\sigma}^0 + \bar{W} + \sum_{a\in H} (1 - k\beta^a)\bar{V}^a \leqslant kc \tag{5}$$

is satisfied, where $\bar{\sigma}^0$, $\bar{V}^a$ are defined by (2.5) and (2.6). Finally, let us set

$$\psi_{ij}(\lambda) = \phi_{ij}(\lambda) + \frac{(\sum_{a\in H}\beta^a X_i^a(\lambda) + \delta\bar{\xi}_i(\lambda))\bar{\eta}_j(\lambda)}{c + \sum_{a\in H}\beta^a\lambda[\bar{\eta}(\lambda), X^a] + \delta\lambda[\bar{\eta}(\lambda), \bar{\xi}]} \tag{6}$$

The process $\psi(\lambda)$ is honest if, and only if,

$$\bar{\xi} = \bar{X} \qquad \beta^a = \delta \qquad (a\in H) \qquad \delta^{-1}c = \bar{\sigma}^0 \tag{7}$$

The process $\psi(\lambda)$ satisfies the system of forward equations. For the process $\psi(\lambda)$ to satisfy the system of backward equations, it is necessary and sufficient that $\beta^a = 0$ $(a\in H)$. When $n^+ = 1$, the processes obtained above include all the Q processes satisfying the system of forward equations.

Proof. On account of (2.10.14), the norm condition is equivalent to

$$0 \leqslant F(\lambda) \qquad F(\lambda)\lambda[\bar{\eta}(\lambda), \mathbf{1}] \leqslant \sum_{a\in H} X^a(\lambda) + \bar{X}(\lambda) \tag{8}$$

In analogy with (2.15), the resolvent equation is equivalent to

$$A(\mu, \lambda)F(\mu) = F(\lambda) + (\lambda - \mu)F(\lambda)[\bar{\eta}(\lambda), F(\mu)] \tag{9}$$

or, equivalently,

$$F(\mu) = \{1 + (\lambda - \mu)[\bar{\eta}(\lambda), F(\mu)]\}A(\lambda, \mu)F(\lambda) \tag{10}$$

so that if $F(\lambda) = 0$ for some $\lambda > 0$; then $F(\lambda) = 0$ for all $\lambda > 0$, therefore $\psi(\lambda) = \phi(\lambda)$. Otherwise (10) is equivalent to

$$F(\lambda) = m_\lambda \xi(\lambda), \qquad m_\lambda > 0, \xi(\lambda) \neq 0 \tag{11}$$

where the quantity $m_\lambda > 0$ satisfies

$$m_\mu = m_\lambda\{1 + (\lambda - \mu)[\bar{\eta}(\lambda), \xi(\mu)]m_\mu\} \tag{12}$$

whereas $\xi(\lambda)$ is a non-zero exit family. According to Lemma 2.11.3, $\xi(\lambda)$ has Riesz representation

$$\xi(\lambda) = \phi(\lambda)\beta + \tilde{\xi}(\lambda) \neq 0 \tag{13}$$

where the column vector $\beta \geqslant 0$ is such that $\phi(\lambda)\beta \in m$, and the exit family $\tilde{\xi}(\lambda) \in \mathcal{M}_\lambda^+$. Using Lemmas 2.10.6 and 2.10.7, it is easy to prove that $\delta = \sup_i \tilde{\xi}_i(\lambda)$ is independent of $\lambda > 0$. If $\delta = 0$, set $\bar{\xi}(\lambda) = 0$; if $\delta > 0$, set $\bar{\xi}(\lambda) = \delta^{-1}\tilde{\xi}(\lambda)$. Then $(\bar{\xi}(\lambda), \lambda > 0)$ is a harmonic exit family. Let its standard image be $\bar{\xi}$; hence

$$\sup_i \bar{\xi}(\lambda) = 1 \qquad \text{if } \delta > 0 \tag{14}$$

comparing this with (2.12.2), we obtain $\beta_j = 0\ (j \in E - H)$. Upon setting $\beta^a = \beta_a/d_a$ $(a \in H)$, (13) becomes

$$\xi(\lambda) = \sum_{a \in H} \beta^a X^a(\lambda) + \delta\bar{\xi}(\lambda) \neq 0 \tag{15}$$

whose standard image is

$$\xi = \sum_{a \in H} \beta^a X^a + \delta\bar{\xi} \neq 0 \tag{16}$$

We are going to prove $\lambda[\bar{\eta}(\lambda), \xi] < \infty$. If not, upon dividing (12) on both sides by $m_\lambda m_\mu$, we have

$$m_\lambda^{-1} = m_\mu^{-1} + (\lambda - \mu)[\bar{\eta}(\lambda), \xi(\mu)]$$

Setting $\mu \downarrow 0$, we obtain

$$m_\lambda^{-1} \geqslant \lim_{\mu \downarrow 0} (\lambda - \mu)[\bar{\eta}(\lambda), \xi(\mu)] = \lambda[\bar{\eta}(\lambda), \xi] = \infty$$

Thus $m_\lambda = 0$, which contradicts (11).

By using (2.11.40),

$$(\mu - \lambda)[\bar{\eta}(\lambda), \xi(\mu)] = \mu[\bar{\eta}(\mu), \xi] - \lambda[\bar{\eta}(\lambda), \xi] \tag{17}$$

Therefore, upon dividing (12) by $m_\lambda m_\mu$, we find that there exists a constant c such that

$$m_\lambda = \{c + \lambda[\bar{\eta}(\lambda), \xi]\}^{-1} \tag{18}$$

Upon substituting (11) and (15) in (1) and comparing the result with (2.12.2), we obtain $F^a_{(\lambda)} = \beta^a m_\lambda \bar{\eta}(\lambda)$.

Observing (2.12.3), we have

$$\beta^a m_\lambda \lambda[\bar{\eta}(\lambda), \mathbf{1}] \leqslant 1 \qquad a \in H. \tag{19}$$

By substituting (15) and (11) in the norm condition (8), we find

$$\left(\sum_{a \in H} \beta^a X^a(\lambda) + \delta\bar{\xi}(\lambda) \right) m_\lambda \lambda[\bar{\eta}(\lambda), \mathbf{1}] \leqslant \sum_{a \in H} X^a(\lambda) + \bar{X}(\lambda) \tag{20}$$

Operating on the formula above by $\Pi^n(\lambda)$, setting $n \to \infty$, and considering (2.12), we obtain

$$\delta\bar{\xi}(\lambda) m_\lambda \lambda[\bar{\eta}(\lambda), \mathbf{1}] \leqslant \bar{X}(\lambda). \tag{21}$$

Hence it can be seen that the norm condition (20) is equivalent to (19) and (21).

By (14) and

$$\bar{\xi}(\lambda) \leqslant \bar{X}(\lambda) \tag{22}$$

we obtain that (21) is equivalent to

$$\delta m_\lambda \lambda[\bar{\eta}(\lambda), \mathbf{1}] \leqslant 1 \tag{23}$$

Thus the norm conditions (21) and (19) become

$$m_\lambda \lambda[\bar{\eta}(\lambda), \mathbf{1}] \leqslant k \tag{24}$$

By (16) and (4), $0 < k < \infty$. Upon substituting (15) and (18) in (24) we find

$$\lambda[\bar{\eta}(\lambda), X^0] + \sum_{a\in H} \lambda[\bar{\eta}(\lambda), X^a] + \lambda[\bar{\eta}(\lambda), \bar{X}]$$

$$\leqslant kc + k \sum_{a\in H} \beta^a \lambda[\bar{\eta}(\lambda), X^a] + k\delta\lambda[\bar{\eta}(\lambda), \bar{\xi}]$$

that is,

$$\bar{\sigma}^0 + \bar{W}_\lambda + \sum_{a\in H} (1 - k\beta^a)\bar{V}^a_\lambda \leqslant kc \tag{25}$$

Setting $\lambda \uparrow \infty$, we find that the norm condition is equivalent to (5).

If $\psi(\lambda)$ is honest, we must establish an equality in the second formula in (8), that is to say, the equalities in (21) and (19) hold. From this it follows that $\beta^a = \delta$ $(a\in H)$, $\bar{\xi}(\lambda) = \bar{X}(\lambda)$, i.e. $\bar{\xi} = \bar{X}$. Hence $k = \delta^{-1}$, and the equality in (5) becomes $\bar{\sigma}^0 = \delta^{-1}c$. Therefore, for $\psi(\lambda)$ to be honest, the necessary and sufficient condition is (7). The other conclusions to the theorem are easily seen, and the proof is complete. QED

CHAPTER 4

Uniqueness

4.1 INTRODUCTION

Given that Q satisfies (2.2.6), there always exist Q processes. This chapter discusses the problem of uniqueness. Section 4.2 provides the necessary and sufficient conditions for uniqueness of Q processes satisfying the system of backward equations. Section 4.3 gives those for uniqueness of Q processes satisfying the system of forward equations. The content of these two sections is derived from Reuter (1957). The proof of the Hou–Reuter theorem in section 4.4, that is, the criterion for uniqueness of Q processes, is obtained from Reuter (1976), but now it has been simplified and improved. Thus, the problem of uniqueness in construction theory has been completely solved.

4.2 UNIQUENESS THEOREM: THE SYSTEM OF BACKWARD EQUATIONS

When we consider the problem of uniqueness of the Q processes satisfying the system of backward equations, the equation

$$(U_\lambda) \qquad \lambda u - Qu = 0 \qquad (1)$$

will play a prominent role. Recall that the set of solutions $u \in m$ of the equation (U_λ) is denoted by $\mathcal{M}_\lambda$, the set of non-negative solutions $u \in m$ is written as $\mathcal{M}_\lambda^+$, and the set of solutions in $\mathcal{M}_\lambda^+$ which are bounded by K is denoted by $\mathcal{M}_\lambda^+(K)$. The dimension of $\mathcal{M}_\lambda^+$ is denoted by m^+.

Theorem 1. The following conditions are equivalent:

(i) The Q process satisfying the system of backward equations is unique.

(ii) For some $\eta > 0$ (hence all $\lambda > 0$), $\mathcal{M}_\lambda$ consists of the zero solution only.

(iii) For some $\lambda > 0$ (hence all $\lambda > 0$), $\mathcal{M}_\lambda^+$ consists of the zero solution only.

If one of the above-stated conditions does not hold, then there are infinitely many Q processes satisfying the system of backward equations. If Q is conservative, there exist infinitely many honest Q processes satisfying the system of backward equations; if Q is non-conservative, then all Q processes satisfying the system of backward solutions are stopping.

Proof. (i)$\Rightarrow$(ii). Suppose (ii) does not hold. By Lemma 2.10.6, $\bar{X}(\lambda)\neq 0$. The process $\psi(\lambda)$ obtained by taking $\bar{\eta}(\lambda)=0, \alpha\geqslant 0, [\alpha, 1]=1$, and $c=[\alpha, X^0]$ in (3.2.8) satisfies the system of backward equations. If Q is conservative, such $\psi(\lambda)$ is still honest. But there are infinitely many choices of α, which therefore is in contradiction with (i). Hence (ii) is valid.

(ii)$\Rightarrow$(iii) need not be proved. We are going to prove (iii)$\Rightarrow$(i). Assume that (iii) holds. Since $\bar{X}(\lambda)\in\mathscr{M}_\lambda^+(1)$, it follows that $\bar{X}(\lambda)=0$. And since $\bar{X}(\lambda)$ is an exit family, it follows that, for all $\lambda>0, \bar{X}(\lambda)=0$.

If $\psi(\lambda)$ satisfies the system of backward equations according to Theorem 2.12.1, it follows that $F^a(\lambda)=0\,(a\in H)$ in (2.12.2). Nevertheless when j is fixed, $u_i=\lambda B_{ij}(\lambda)\in\mathscr{M}_\lambda^+(1)$. By maximality of $\bar{X}(\lambda), \lambda B_{ij}(\lambda)\leqslant\bar{X}_i(\lambda)$, hence $B_{ij}(\lambda)=0$. Thus, (2.12.2) becomes $\psi(\lambda)=\phi(\lambda)$, that is, (i) is obtained.

If there exists an honest Q process $\psi(\lambda)$ satisfying the system of backward equations, then by summing for j in the B condition of $\psi(\lambda)$, we find that Q is conservative, and the proof is complete. QED

4.3 UNIQUENESS THEOREM: THE SYSTEM OF FORWARD EQUATIONS

When the problem of uniqueness of Q processes satisfying the system of forward equations is taken into consideration, the system of equations

$$(V_\lambda) \qquad \lambda v - vQ = 0 \qquad (1)$$

will play an important role. Let us recall that the set of solutions $v\in l$ of the equation (V_λ) is written as $\mathscr{L}_\lambda$, all non-negative solutions $v\in l$ are denoted by $\mathscr{L}_\lambda^+$, and the dimension of $\mathscr{L}_\lambda^+$ is denoted by n^+.

Theorem 1. (i) If the minimal solution is honest, or if it is a stopping process but $n^+=0$, then the Q process satisfying the system of forward equations is unique.

(ii) Suppose that the minimal solution is a stopping process and $n^+=1$. Then we have infinitely many Q processes that satisfy the system of forward equations, and only one of them is honest.

(iii) If the minimal solution is a stopping process and $n^+>1$, there exist infinitely many Q processes satisfying the system of forward equations, of which infinitely many are honest.

Proof. (i) When the minimal solution is honest, uniqueness is obvious. Let the minimal solution be stopping and moreover let $n^+=0$. If $\psi(\lambda)$ is a Q process satisfying the system of forward equations, by Theorem 2.12.1, it follows that $F^a(\lambda)\in\mathscr{L}_\lambda^+$ $(a\in H)$ in (2.12.2), $u_j = B_{ij}(\lambda)\in\mathscr{L}_\lambda^+$, hence $F^a(\lambda)=0\,(a\in H)$, $B_{ij}(\lambda)=0$. Equation (2.12.2) turns into $\psi(\lambda)=\phi(\lambda)$, that is, the Q process satisfying the system of forward equations is unique.

(ii) Theorem 3.4.1 has described the construction.

(iii) The preceding part has already been answered by Theorem 3.4.1. Since $n^+ > 1$, it follows that we may select the exit family $\bar{\eta}a(\lambda)\in\mathscr{L}_\lambda^+$ $(a = 1, 2)$, so that $\bar{\eta}^1(\lambda), \bar{\eta}^2(\lambda)$ become linearly independent. Arbitrarily take two constants $p^a \geqslant 0$ $(a = 1, 2)$ such that

$$\bar{\eta}(\lambda) = p^1\bar{\eta}^1(\lambda) + p^2\bar{\eta}^2(\lambda) \neq 0$$

For $\bar{\eta}(\lambda)$, there exists an honest Q process satisfying the system of forward equations according to Theorem 3.4.1. But there are infinitely many ways of selecting $p^a\,(a = 1, 2)$ so that $\bar{\eta}(\lambda)$ is made different (constant factors not being considered). Therefore there exist infinitely many honest Q processes that satisfy the system of forward equations, and the proof is terminated. QED

4.4 CRITERION FOR UNIQUENESS: THE HOU–REUTER THEOREM

The criterion for uniqueness of Q processes is given by Zhen-ting Hou (1974) and summarized in his book (Zhen-ting Hou, 1982). In addition these discuss the existence of combinations of various cases and the problem of uniqueness, namely the so-called qualitative theory. Reuter (1976) has simplified the proof in Zhen-ting Hou (1974). Here the simplified proof of Reuter is adopted and further improved.

Theorem 1. Let a given matrix Q satisfy (2.2.6). Then for the Q process to be unique it is necessary and sufficient that the minimal Q process $\phi(\lambda)$ is honest, or that it is stopping and moreover satisfies the two conditions below:

(i)
$$\inf_i \lambda \sum_j \phi_{ij}(\lambda)_\lambda > 0 \qquad \lambda > 0 \tag{1}$$

(ii) $n^+ = 0$, that is, the equation (V_λ) has no non-zero and non-negative solution $v\in l$.

We point out that condition (i) implies $m^+ = 0$. In fact, by (2.10.14), (1) becomes

$$\sup_i \left(\sum_{a\in H} X_i^a(\lambda) + \bar{X}_i(\lambda) \right) < 1 \qquad \lambda > 0 \tag{2}$$

since, if $\bar{X}(\lambda) \neq 0$, (2.10.21) holds. From (2) it necessarily follows that $\bar{X}(\lambda) = 0$, and hence $m^+ = 0$.

Thus the condition (i) is equivalent to the following two conditions:

(i_1) $$m^+ = 0$$

(ii_2) $$\sup_i \sum_{a\in H} \phi_{ia}(\lambda)d_a < 1 \qquad \lambda > 0$$

Proof of the theorem. Necessity: Assume that the minimal Q process is stopping and moreover the Q process is unique; hence a Q process of F type is unique. By Theorem 3.1, $n^+ = 0$. Now let us assume that

$$\inf_i \lambda \sum_j \phi_{ij}(\lambda) = 0 \tag{3}$$

According to Lemma 3.2 there exists a row vector $\alpha \geqslant 0$ such that $[\alpha, 1] = \infty$ and furthermore $\alpha\phi(\lambda)\in l$, and for this α, by Theorem 3.3.1,

$$\psi_{ij}(\lambda) = \phi_{ij}(\lambda) + Z_i(\lambda)\frac{\sum_k \alpha_k \phi_{kj}(\lambda)}{c + [\alpha, Z - Z(\lambda)]} \tag{4}$$

is a Q process where $Z(\lambda) = 1 - \lambda\phi(\lambda)1 \neq 0$, $c \geqslant [\alpha, X^0]$. Since the selection of constant c may not be unique, it follows that the Q process is not unique. This is in contradiction with the hypothesis of the necessary conditions. Hence (i) is valid.

Sufficiency: Let (i) and (ii) hold and let $\psi(\lambda)$ be a Q process. From (i_1) it follows that, in Theorem 2.12.1, $B(\lambda) = 0$; hence (2.12.2) becomes

$$\psi_{ij}(\lambda) = \phi_{ij}(\lambda) + \sum_{a\in H} X_i^a(\lambda)F_j^a(\lambda) \tag{5}$$

If H is empty, from the formula above we know $\psi(\lambda) = \phi(\lambda)$, consequently the Q process is unique. In what follows we suppose $H \neq \varnothing$. Substituting the formula (5) in the resolvent equation of $\psi(\lambda)$, noting that $\phi(\lambda)$ satisfies the resolvent equation, and that because of Lemma 2.11.6, $X^a(\lambda)$, $a\in H$ are linearly independent, we obtain

$$F^a(\lambda)A(\lambda, \mu) = F^a(\mu) + (\mu - \lambda)\sum_{b\in H} [F^a(\lambda), X^b(\mu)]F^b(\mu) \qquad a\in H \tag{6}$$

Since $F^a(\lambda) \geqslant 0, \lambda[F^a(\lambda), \mathbf{1}] \leqslant 1$, it can be seen from the above formula that for arbitrary $\lambda, \mu > 0$, $F^a(\lambda)A(\lambda, \mu)\in l$. Hence fix a and $\lambda > 0$ as well, and by (2.10.8), $\eta(\mu) = F^a(\lambda)A(\lambda, \mu)\,(\mu > 0)$ is an entrance family. According to Lemma 2.11.3,

$$\eta(\mu) = \alpha\phi(\mu) + \bar{\eta}(\mu)$$

where $\alpha \geqslant 0$ is independent of μ so that $\alpha\phi(\mu)\in l$ and $(\bar{\eta}(\mu), \mu > 0)$ is a harmonic entrance family. By (ii) $n^+ = 0$, therefore $\bar{\eta}(\mu) = 0$. Again since α depends on a and λ, it follows that $\alpha = \alpha^a(\lambda)$, i.e.

$$F^a(\lambda)A(\lambda, \mu) = \alpha^a(\lambda)\phi(\mu) \tag{7}$$

In particular when $\mu = \lambda$,

$$F^a(\lambda) = \alpha^a(\lambda)\phi(\lambda) \tag{8}$$

On account of condition (i)

$$\begin{aligned} 1 \geqslant \lambda[F^a(\lambda), \mathbf{1}] &= \lambda[\alpha^a(\lambda)\phi(\lambda), \mathbf{1}] \\ &= [\alpha^a(\lambda), \lambda\phi(\lambda)\mathbf{1}] \geqslant \eta_\lambda[\alpha^a(\lambda), \mathbf{1}] \end{aligned}$$

so that

$$[\alpha^a(\lambda), \mathbf{1}] \leqslant 1/\eta_\lambda. \tag{9}$$

Upon substituting (8) in (6) and observing (2.10.8) and Lemma 2.11.6, we have

$$\alpha^a(\lambda) = \alpha^a(\mu) + (\mu - \lambda) \sum_{b \in H} [\alpha^a(\lambda), \phi(\lambda)X^b(\mu)]\alpha^b(\mu) \tag{10}$$

or owing to the fact that $(X^a(\lambda), \lambda > 0)$ is an exit family, we have

$$\alpha^a(\lambda) = \alpha^a(\mu) + \sum_{b \in H} [\alpha^a(\lambda), X^b(\lambda) - X^b(\mu)]\alpha^b(\mu) \tag{11}$$

Because of (10), $\alpha^a(\lambda)$ decreases as λ increases. Now let us proceed to prove that

$$\alpha^a(\lambda) \downarrow 0 \qquad \lambda \uparrow \infty \tag{12}$$

Actually since both $\psi(\lambda)$ and $\phi(\lambda)$ satisfy Q conditions it follows by (5) and (8) that

$$\lim_{\lambda \to \infty} \sum_{a \in H} \lambda X_i^a(\lambda)[\lambda\alpha^a(\lambda)\phi(\lambda)]_j = 0$$

Hence

$$\lim_{\lambda \to \infty} \lambda X_i^a(\lambda)\alpha_j^a(\lambda)\lambda\phi_{jj}(\lambda) = 0 \qquad a \in H$$

By the continuity condition of $\phi(\lambda)$, that is,

$$\delta_{ia} d_a \lim_{\lambda \to \infty} \alpha_j^a(\lambda)\delta_{jj} = 0$$

taking $i = a$ we obtain (12).

Because, if $\lambda > \mu$, by (9),

$$\begin{aligned} \sum_{b \in H} [\alpha^a(\lambda), X^b(\lambda)]\alpha^b(\mu) &\leqslant \sum_{b \in H} [\alpha^a(\lambda), X^b(\mu)]\alpha^b(\mu) \\ &\leqslant \sum_{b \in H} [\alpha^a(\mu), X^b(\mu)]\alpha^b(\mu) \leqslant \sum_{b \in H} [\alpha^a(\mu), X^b(\mu)]1/\eta_\mu \\ &\leqslant \left[\alpha^a(\mu), \sum_{b \in H} X^b(\mu)\right]1/\eta_\mu \leqslant [\alpha^a(\mu), \mathbf{1}]1/\eta_\mu \leqslant 1/\eta_\mu^2 < \infty. \end{aligned}$$

Therefore (11) may be written as

$$\alpha^a(\lambda) + \sum_{b\in H} [\alpha^a(\lambda), X^b(\mu)]\alpha^b(\mu) = \alpha^a(\mu) + \sum_{b\in H} [\alpha^a(\lambda), X^b(\lambda)]\alpha^b(\mu)$$

and, moreover, when $\lambda \to \infty$, by using the theorem of dominated convergence we have

$$0 + \sum_{b\in H} [0, X^b(\mu)]\alpha^b(\mu) = \alpha^a(\mu) + \sum_{b\in H} [0, 0]\alpha^b(\mu)$$

Therefore $\alpha^a(\mu) = 0$ $(a\in H, \mu > 0)$. Thus $F^a(\lambda) = 0$ $(a\in H, \lambda > 0)$. Hence $\psi(\lambda) = \phi(\lambda)$, and so the Q process is unique. The theorem is proved. QED

PART II CONSTRUCTION THEORY OF BIRTH–DEATH PROCESSES

In order to study further the construction theory of processes, it is very useful to study the construction of two kinds of special processes, i.e. bilateral birth–death processes and unilateral birth–death processes. The bilateral birth–death process that we consider is conservative; for the unilateral birth–death process there possibly exists a non-conservative state. The dimension of the solution space $\mathscr{M}_\lambda^+$ for the bilateral birth–death process is $m^+ \leqslant 2$; for the unilateral birth–death process, it is $m^+ \leqslant 1$. The birth–death processes are important because they have very important theoretical significance and application value. By studying them, many varied and deep results can be derived. More importantly, the study of birth–death processes is often the source that gives the ideas and methods for solving problems of general processes.

CHAPTER 5

Bilateral Birth–Death Processes

5.1 INTRODUCTION

If the state space E is taken to be the set of all integers, and $Q=(q_{ij})$ has the following form:

$$\begin{aligned} &q_{ij}=0 \qquad \text{if } |i-j|>1 \\ &q_{i,i-1}=a_i>0 \qquad q_{i,i+1}=b_i>0 \qquad q_i=-q_{ii}=a_i+b_i \end{aligned} \tag{1}$$

we call the Q processes bilateral birth–death processes.

In this chapter, the Q processes will always refer to bilateral birth–death processes. The bilateral birth–death processes are conservative and, therefore, they surely satisfy the system of backward equations.

In this chapter the construction problem of the bilateral birth–death processes is satisfactorily solved. In other words, all the bilateral birth–death processes are constructed. The bilateral birth–death processes are looked on as diffusion, and the method used is the analytical method. The content of this chapter is derived from Xiang-qun Yang (1964b)

5.2 NATURAL SCALE AND STANDARD MEASURE

For Q of the form (1.1), we call

$$z_i=-b_0\left(1+\frac{b_{-1}}{a_{-1}}+\frac{b_{-1}b_{-2}}{a_{-1}a_{-2}}+\cdots+\frac{b_{-1}b_{-2}\cdots b_{i+1}}{a_{-1}a_{-2}\cdots a_{i+1}}\right) \qquad \text{if } i<-1$$

$$z_{-1}=-b_0 \qquad z_0=0 \qquad z_1=a_0$$

$$z_i=a_0\left(1+\frac{a_1}{b_1}+\frac{a_1a_2}{b_1b_2}+\cdots+\frac{a_1a_2\cdots a_{i-1}}{b_1b_2\cdots b_{i-1}}\right) \qquad \text{if } i>1 \tag{1}$$

the natural scale, we call

$$r_1=\lim_{i\to-\infty} z_i \qquad r_2=\lim_{i\to+\infty} z_i \tag{2}$$

the boundary points and we call

$$\mu_i = \frac{a_{-1}a_{-2}\cdots a_{i+1}}{b_0 b_{-1} b_{-2}\cdots b_{i+1} b_i} \qquad \text{if } i < -1$$

$$\mu_{-1} = \frac{1}{b_0 b_{-1}} \qquad \mu_0 = \frac{1}{a_0 b_0} \qquad \mu_1 = \frac{1}{a_0 a_1}$$

$$\mu_i = \frac{b_1 b_2 \cdots b_{i-1}}{a_0 a_1 a_2 \cdots a_{i-1} a_i} \qquad \text{if } i > 1 \tag{3}$$

the canonical measure.

5.3 CLASSIFICATION OF BOUNDARY POINTS

The boundary points may be classified by means of the natural scale and the canonical measure. The boundary point r_2 is said to be

(a) regular if r_2 is finite and $\sum_{i\geqslant 0}\mu_i$ is finite;
(b) exit if r_2 is irregular but finite, and $\sum_{i\geqslant 0}(r_2 - z_i)\mu_i$ is finite;
(c) entrance if r_2 is irregular, but $\sum_{i\geqslant 0} z_i\mu_i$ is finite; and
(d) natural for all other cases.

Similarly, r_1 may be classified.

If we set

$$R_1 = \sum_{i\leqslant 0}(z_i - r_1)\mu_i = \sum_{i\leqslant 0}(z_i - z_{i-1})\sum_{i\leqslant j\leqslant 0}\mu_j$$

$$S_1 = -\sum_{i\leqslant 0} z_i\mu_i$$

$$R_2 = \sum_{i\geqslant 0}(r_2 - z_i)\mu_i = \sum_{i\geqslant 0}(z_{i+1} - z_i)\sum_{0\leqslant j\leqslant i}\mu_j$$

$$S_2 = \sum_{i\geqslant 0} z_i\mu_i \tag{1}$$

then we find that r_2 is finite when R_2 is finite; and that $\sum_{i\geqslant 0}\mu_i$ is finite when S_2 is finite. Therefore, the above classification of the boundary points is comprehensive and, furthermore, if r_a is entrance then r_a is finite.

Theorem 1. The boundary point r_a is

(a) regular if and only if $R_a < \infty$, $S_a < \infty$;
(b) exit if and only if $R_a < \infty$, $S_a = \infty$;
(c) entrance if and only if $R_a = \infty$, $S_a < \infty$;
(d) natural if and only if $R_a = \infty$, $S_a = \infty$.

Proof. Prove the theorem for $a=2$. Obviously,

$$R_2 \leqslant r_2 \sum_{i\geqslant 0} \mu_i \qquad S_2 \leqslant r_2 \sum_{i\geqslant 0} \mu_i$$

Hence if r_2 is regular, then $R_2<\infty$, $S_2<\infty$. Conversely, if $R_2<\infty$, $S_2<\infty$, it follows that r_2 must be finite by the definition of R_2. Besides

$$\sum_{i\geqslant 0} \mu_i = \frac{1}{r_2}(S_2+R_2)<\infty$$

If r_2 is exit, by the definition we obtain $R_2<\infty$, $\sum_{i\geqslant 0}\mu_i=\infty$. From $S_2 \geqslant z_1 \sum_{i\geqslant 1}\mu_i$ follows $S_2=\infty$. Conversely, if $R_2<\infty$, $S_2=\infty$. By what is proved in the first section it follows obviously that r_2 is irregular. However by $R_2<\infty$ it follows obviously that r_2 is finite. Therefore, r_2 is exit.

If r_2 is entrance, by the definition we obtain $S_2<\infty$; owing to the fact that r_2 is irregular and according to what is proved in the first section, we surely get $R_2=\infty$. Conversely, if $R_2=\infty$, $S_2<\infty$, then by what is verified in the first section it follows that r_2 is irregular. Therefore r_2 is entrance.

By the proof of the above three sections it follows immediately that for r_2 to be natural it is necessary and sufficient that $R_2=\infty$, $S_2=\infty$, and the proof is complete. QED

5.4 SECOND-ORDER DIFFERENCE OPERATOR

Let μ be the column vector on E, and define u^+ and $D_\mu u^+$ as follows:

$$u_i^+ = \frac{u_{i+1}-u_i}{z_{i+1}-z_i} \qquad i\in E$$
$$(D_\mu u^+)_i = \frac{u_i^+ - u_{i-1}^+}{\mu_i} \qquad i\in E \tag{1}$$

Theorem 1. For an arbitrary column vector u,

$$Qu = D_\mu u^+ \tag{2}$$

that is,

$$a_i u_{i-1} - (a_i+b_i)u_i + b_i u_{i+1} = (D_\mu u^+)_i \tag{3}$$

Proof. Since

$$a_i = \frac{1}{(z_i - z_{i-1})u_i}$$
$$b_i = \frac{1}{(z_{i+1}-z_i)u_i} \tag{4}$$

it follows that

$$\begin{aligned}(D_\mu u^+)_i &= \left(\frac{u_{i+1}-u_i}{z_{i+1}-z_i}-\frac{u_i-u_{i-1}}{z_i-z_{i-1}}\right)\Big/u_i\\ &= b_i(u_{i+1}-u_i)-a_i(u_i-u_{i-1})\\ &= a_iu_{i-1}-(a_i+b_i)u_i+b_iu_{i+1}\end{aligned}$$

and the proof is concluded. QED

Let u be a column vector; $u\mu$ represents the row vector with components $u_j\mu_j$. Conversely, if v is a row vector, $v\mu^{-1}$ represents the column vector with components $v_i\mu_i^{-1}$.

Theorem 2. Assume that v is a row vector and $u = v\mu^{-1}$. Then

$$vQ = (Qu)\mu \tag{5}$$

Proof. By observing

$$\mu_{i-1}b_{i-1}\mu_i^{-1} = a_i \qquad a_{i+1}u_{i+1}\mu_i^{-1} = b_i \tag{6}$$

we see that

$$\begin{aligned}(Qu)_i &= a_iv_{i-1}\mu_{i-1}^{-1}-(a_i+b_i)v_i\mu_i^{-1}+b_iv_{i+1}\mu_{i+1}^{-1}\\ &= v_{i-1}b_{i-1}\mu_i^{-1}-(a_i+b_i)v_i\mu_i^{-1}+v_{i+1}a_{i+1}\mu_i^{-1}\\ &= [v_{i-1}b_{i-1}-v_i(a_i+b_i)+v_{i+1}a_{i+1}]\mu_i^{-1}\\ &= (vQ)_i\mu_i^{-1}\end{aligned}$$

QED

Corollary

Suppose that u and f are column vectors, and that $v = u\mu$ and $g = f\mu$. Then u satisfies

$$Qu = f \tag{7}$$

or

$$\lambda u - Qu = f \qquad \lambda > 0 \tag{8}$$

if and only if v satisfies

$$vQ = g \tag{9}$$

or

$$\lambda v - vQ = g \tag{10}$$

Lemma 3. The solution of the system of equations

$$\begin{aligned}&u_i = f_i\\ &a_ku_{k-1}-(a_k+b_k)u_k+b_ku_{k+1} = -f_k \qquad i<k<n\\ &u_n = f_n\end{aligned} \tag{11}$$

is

$$u_k = f_i \frac{z_n - z_k}{z_n - z_i} + f_n \frac{z_k - z_i}{z_n - z_i} + \frac{z_n - z_k}{z_n - z_i} \sum_{j=i+1}^{k-1} (z_i - z_j) f_j u_j + \frac{z_k - z_i}{z_n - z_i} \sum_{j=k}^{n-1} (z_n - z_j) f_j u_j \tag{12}$$

Proof. By Theorem 1, the system (11) of equations becomes

$$\begin{aligned} &u_i = f_i \\ &u_k^+ - u_{k-1}^+ = -f_k \mu_k \qquad i < k < n \\ &u_n = f_n \end{aligned} \tag{13}$$

By the above expression we obtain

$$u_k^+ = u_i^+ + \sum_{l=i+1}^{k} (u_l^+ - u_{l-1}^+) = u_i^+ - \sum_{l=i+1}^{k} f_l \mu_l$$

and from this we get

$$\begin{aligned} u_k &= u_i + \sum_{l=i}^{k-1} (u_{l+1} - u_l) = u_i + \sum_{l=i}^{k-1} u_l^+ (z_{i+1} - z_l) \\ &= u_i + \sum_{l=i}^{k-1} u_i^+ (z_{l+1} - z_l) - \sum_{l=i}^{k-1} \left(\sum_{j=i+1}^{l} f_j u_j \right) (z_{l+1} - z_l) \\ &= f_i + u_i^+ (z_k - z_i) - \sum_{j=i+1}^{k-1} \sum_{l=i}^{k-1} f_j u_j (z_{l+1} - z_l) \end{aligned}$$

Therefore

$$u_k = f_i + u_i^+ (z_k - z_i) - \sum_{j=i+1}^{k-1} (z_k - z_j) f_j \mu_j \tag{14}$$

In particular, if $k = n$, we have

$$f_n = f_i + u_i^+ (z_n - z_i) - \sum_{j=i+1}^{n-1} (z_n - z_j) f_j \mu_j$$

so that

$$u_i^+ = \frac{f_n - f_i}{z_n - z_i} + \frac{1}{z_n - z_i} \sum_{j=i+1}^{n-1} (z_n - z_j) f_j \mu_j \tag{15}$$

Upon substituting (15) into (14) and rearranging it, we obtain (12), and the proof is terminated. QED

5.5 SOLUTION OF THE EQUATION $\lambda u - D_\mu u^+ = 0$

Theorem 1. Let u and v be two solutions of the equation

$$\lambda u - D_\mu u^+ = 0 \qquad \lambda > 0 \tag{1}$$

Then

$$W(u, v) \equiv u^{+}v - uv^{+} = \text{constant} \tag{2}$$

Proof. First note that for arbitrary vectors s, t

$$\begin{aligned} s_i t_i - s_{i-1} t_{i-1} &= s_i(t_i - t_{i-1}) + t_{i-1}(s_i - s_{i-1}) \\ &= s_{i-1}(t_i - t_{i-1}) + t_i(s_i - s_{i-1}) \end{aligned}$$

so that

$$\begin{aligned} [D_\mu(st)]_i &= s_i(D_\mu t)_i + t_{i-1}(D_\mu s)_i \\ &= s_{i-1}(D_\mu t)_i + t_i(D_\mu s)_i \end{aligned} \tag{3}$$

Therefore

$$\begin{aligned} D_\mu W(u, v) &= D_\mu(u^{+}v) - D_\mu(uv^{+}) \\ &= v_i(D_\mu u^{+})_i + u^{+}_{i-1}(D_\mu v)_i \\ &\quad - u_i(D_\mu v^{+})_i - v^{+}_{i-1}(D_\mu u)_i \\ &= \lambda v_i u_i + u^{+}_{i-1}(D_\mu v)_i - \lambda u_i v_i - v^{+}_{i-1}(D_\mu u)_i \\ &= \frac{u_i - u_{i-1}}{z_i - z_{i-1}} \frac{v_j - v_{i-1}}{u_i} - \frac{v_i - v_{i-1}}{z_i - z_{i-1}} \frac{u_i - u_{i-1}}{u_i} \\ &= 0 \end{aligned}$$

QED

Equation (1) can be rewritten as follows:

$$u_i^{+} - u_{i-1}^{+} = \lambda u_i \mu_i \qquad \lambda > 0 \tag{4}$$

Consequently if u is a solution of equation (1), then for $i > 0$

$$\begin{aligned} u_i^{+} &= u_0^{+} + \sum_{k=1}^{i} (u_k^{+} - u_{k-1}^{+}) \\ &= u_0^{+} + \lambda \sum_{k=1}^{i} u_k \mu_k \qquad i > 0 \end{aligned} \tag{5}$$

$$\begin{aligned} u_i &= u_0 + \sum_{k=0}^{i-1} (u_{k+1} - u_k) \\ &= u_0 + \sum_{k=0}^{i-1} u_k^{+}(z_{k+1} - z_k) \qquad i > 0 \end{aligned} \tag{6}$$

Substituting (5) into (6) and rearranging, we obtain

$$u_i = u_0 + u_0^{+}(z_i - z_0) + \lambda \sum_{k=1}^{i-1} u_k(z_i - z_k)\mu_k \qquad i > 0 \tag{7}$$

Similarly, for $i<0$

$$u_i^+ = u_0^+ - \lambda \sum_{i+1 \leqslant k \leqslant 0} u_k \mu_k \qquad i<0 \tag{8}$$

$$u_i = u_0 - u_0^+(z_0 - z_i) + \lambda \sum_{i+1 \leqslant h < 0} u_k(z_k - z_i)\mu_k \qquad i<0 \tag{9}$$

Conversely, arbitrarily fixing the values of u_0, u_0^+, we can determine u_1, u_1^+, u_2, $u_2^+, \ldots,$ by (5) and (6); similarly, we can determine u_{-1}, u_{-1}^+, u_{-2}, $u_{-2}^+, \ldots$. Obviously u determined in this manner is a solution of equation (1).

Given $u_0 = 1$ and $u_0^+ = 0$ the solution of equation (1) is denoted by v. Given $u_0 = 0$, $u_0^+ = 1$ we denote the solution of equation (1) by s. From (7) and (9) we can see that v_i and s_i are positive and strictly increasing as $0 < i \uparrow +\infty$; and that v_i and $-s_i$ are positive and strictly increasing with increase of the absolute value of i, as $0 > i \downarrow -\infty$. Moreover, by Theorem 1,

$$W(s, v) = s_0^+ v_0 - s_0 v_0^+ = 1 \tag{10}$$

Lemma 2. When $i > 0$,

$$v_i/s_i > v_i^+/s_i^+$$

Proof.

$$\frac{v_i}{s_i} - \frac{v_i^+}{s_i^+} = \frac{W(s, v)}{s_i s_i^+} = \frac{1}{s_i s_i^+} > 0 \qquad \text{QED}$$

Lemma 3. v_i/s_i is strictly decreasing as $0 < i \uparrow +\infty$.

Proof.

$$\left(\frac{v}{s}\right)_i^+ = \left(\frac{v_{i+1}}{s_{i+1}} - \frac{v_i}{s_i}\right)\frac{1}{z_{i+1} - z_i}$$

$$= -\frac{W(s, v)}{s_i s_{i+1}} < 0 \qquad \text{QED}$$

Lemma 4. v_i^+/s_i^+ is strictly increasing as $0 < i \uparrow +\infty$.

Proof.

$$\left[D_\mu\left(\frac{v^+}{s^+}\right)\right]_i = \left(\frac{v_i^+}{s_i^+} - \frac{v_{i-1}^+}{s_{i-1}^+}\right)\mu_i^{-1}$$

$$= \frac{s_{i-1}^+ v_i^+ - s_i^+ v_{i-1}^+}{s_i^+ s_{i-1}^+ \mu_i}$$

$$= \frac{s_{i-1}^{+}(D_{\mu}v^{+})_i - v_{i-1}^{+}(D_{\mu}s^{+})_i}{s_i^{+}s_{i-1}^{+}}$$

$$= \frac{\lambda(s_{i-1}^{+}v_i - v_{i-1}^{+}s_i)}{s_i^{+}s_{i-1}^{+}}$$

$$= \frac{\lambda[(s_i^{+} - \lambda s_i)v_i - (v_i^{+} - \lambda v_j)s_i]}{s_i^{+}s_{i-1}^{+}}$$

$$= \frac{\lambda W(s, v)}{s_i^{+}s_{i-1}^{+}} = \frac{\lambda}{s_i^{+}s_{i-1}^{+}} > 0 \qquad \text{QED}$$

Lemma 5. Assume that u is a solution of equation (1), and that u_i is positive and strictly increasing as $0 < i \uparrow +\infty$. Then u_i^{+} is also strictly increasing as $0 < i \uparrow +\infty$. $u(r_2) = \lim_{i\to\infty} u_i < \infty$ if and only if r_2 is regular or exit; $u^{+}(r_2) = \lim_{i\to\infty} u_i^{+} < \infty$ if and only if r_2 is regular or entrance.

Proof. From (5) it can be seen that u_i^{+} $(i > 0)$ is positive and strictly increasing.
Suppose $u(r_2) < \infty$. By (7) we have

$$u_i > \lambda u_0 \sum_{k=1}^{i-1} (z_i - z_k)\mu_k \tag{11}$$

We see $R_2 < \infty$ as $i \to +\infty$, that is, r_2 is regular or exit. Conversely, assume $R_2 < \infty$. By (5)

$$u_{i+1} - u_i < u_0^{+}(z_{i+1} - z_i) + \lambda u_i(z_{i+1} - z_i) \sum_{k=1}^{i} \mu_k$$

$$\frac{u_{i+1}}{u_i} - 1 < \frac{u_0^{+}}{u_0}(z_{i+1} - z_i) + \lambda(z_{i+1} - z_i) \sum_{k=1}^{i} \mu_k$$

The right-hand side of the above expression is the ith term of a convergent series, so that $\sum_{i>0} \log(u_{i+1}/u_i)$ converges. Thus $\lim_{i\to+\infty} u_i < \infty$. Suppose $u^{+}(r_2) < \infty$. By (7) we have $u_i > u_0^{+} z_i$. Therefore, by (5)

$$u_i^{+} > u_0^{+} + u_0^{+}\lambda \sum_{k=1}^{i} z_k\mu_k$$

Therefore $s_2 < \infty$, that is, r_2 is regular or entrance. Conversely if $s_2 < \infty$, by (6)

$$u_i < u_0 + u_{i-1}^{+}(z_i - z_0) = u_0 + u_{i-1}^{+} z_i$$

so that

$$u_i^{+} - u_{i-1}^{+} = \lambda u_i u_i < \lambda u_0 u_i + \lambda u_{i-1}^{+} z_i \mu_i$$

$$\frac{u_i^{+}}{u_{i-1}^{+}} - 1 < \frac{\lambda u_0}{u_0^{+}} u_i + \lambda z_i \mu_i$$

Since $s_2 < \infty$ implies $\sum_{i \geqslant 0} \mu_i < \infty$, it follows from the above expression that $\sum_{i \geqslant 0} \log(u_i^+/u_{i-1}^+)$ converges. Therefore, $\lim_{i \to +\infty} u_i^+ < \infty$, and the proof is terminated. QED

Lemma 6. When $0 < i \uparrow +\infty$

$$\frac{v_i}{s_i} - \frac{v_i^+}{s_i^+} = \frac{1}{s_i s_i^+} \to \begin{cases} 0 & \text{if } r_2 \text{ is not regular;} \\ c > 0 & \text{if } r_2 \text{ is regular.} \end{cases}$$

Proof. It can be deduced from Lemmas 2 and 5.

By Lemmas 2 to 4, we can set

$$\bar{\theta} = \lim_{i \to +\infty} \frac{v_i}{s_i} \qquad \underline{\theta} = \lim_{i \to +\infty} \frac{v_i^+}{s_i^+} \tag{12}$$

Moreover, $\underline{\theta} \leqslant \bar{\theta}$. $\underline{\theta} = \bar{\theta}$ if and only if r_2 is irregular. QED

Theorem 7. For u to be a positive strictly decreasing solution of equation (1), satisfying the condition $u_0 = 1$, it is necessary and sufficient that u has the following form:

$$u = v - \theta s \tag{13}$$

where $\underline{\theta} \leqslant \theta \leqslant \bar{\theta}$. If r_2 is regular, then there exist infinitely many solutions stated above and they are between $\underline{u} = v - \bar{\theta} s$ and $\bar{u} = v - \underline{\theta} s$. If r_2 is irregular, the above mentioned solution u is unique.

Proof. Since v and s are two linearly independent solutions, it follows that every solution u is a linear combination of v and s, so that the solution u satisfying $u_0 = 1$ must have the form (13).

Suppose that u is a strictly decreasing positive solution, then $u = v - \theta s > 0$, $u^+ = v^+ - \theta s^+ > 0$ and, therefore, $\underline{\theta} \leqslant \theta \leqslant \bar{\theta}$. Conversely, if $\underline{\theta} \leqslant \theta \leqslant \bar{\theta}$, then $u_i > 0$ and $u_i^+ < 0$ for $i > 0$. That is, u is positive and strictly decreasing in $i > 0$. In $i < 0$, since v and $(-s)$ are positive and strictly increasing with the increase of absolute value of i as $0 \geqslant i \downarrow -\infty$, it follows that u is positive and strictly decreasing on E, and the proof is concluded. QED

Lemma 8. For $u, \underline{u}$ and $\bar{u}$ in Theorem 7,

$$u(r_2) = \begin{cases} 0 & \text{if } r_2 \text{ is exit or natural; when } r_2 \text{ is regular } \underline{u}(r_2) = 0 \\ 1/s^+(r_2) & \text{if } r_2 \text{ is entrance; when } r_2 \text{ is regular, } \bar{u}(r_2) = 1/s^+(r_2) \end{cases}$$

$$u^+(r_2) = \begin{cases} 0 & \text{if } r_2 \text{ is entrance or natural; when } r_2 \text{ is regular, } \bar{u}^+(r_2) = 0 \\ -1/s(r_2) & \text{if } r_2 \text{ is exit; when } r_2 \text{ is regular, } \underline{u}^+(r_2) = -1/s(r_2) \end{cases}$$

Proof. When r_2 is exit or regular, by Lemma 5 we have $v(r_2) < \infty$, $s(r_2) < \infty$. Therefore, $\underline{u}(r_2) = v(r_2) - \bar{\theta}s(r_2) = 0$.

When r_2 is regular,

$$\bar{u}(r_2) = v(r_2) - \underline{\theta}s(r_2) = \frac{v(r_2)s^+(r_2) - v^+(r_2)s(r_2)}{s^+(r_2)}$$

$$= \frac{1}{s^+(r_2)}$$

When r_2 is entrance or natural, by

$$u_i = \bar{u}_i = v_i - \underline{\theta}s_i \leqslant v_i - \frac{v_i^+}{s_i^+} s_i = \frac{1}{s_i^+}$$

If r_2 is natural, then since $s^+(r_2) = \infty$, it follows that $u(r_2) = \bar{u}(r_2) = 0$. If r_2 is entrance, then

$$u(r_2) = \bar{u}(r_2) \leqslant 1/s^+(r_2)$$

For an arbitrary $\varepsilon > 0$, when i is sufficiently large,

$$u(r_2) + \varepsilon > v_i - \underline{\theta}s_i$$

Upon fixing i, when $j(> i)$ is large enough,

$$u(r_2) + \varepsilon > v_i - \frac{v_j^+}{s_j^+} S_i$$

But when j is fixed,

$$\left(v - \frac{v_j^+}{s_j^+} s\right)_j^+ = v_i^+ - \frac{v_j^+}{s_j^+} s_i^+ = \left(\frac{v_i^+}{s_i^+} - \frac{v_j^+}{s_j^+}\right) s_i^+ < 0$$

so that

$$u(r_2) + \varepsilon > v_j - \frac{v_j^+}{s_j^+} s_j = \frac{1}{s_j^+} \rightarrow \frac{1}{s^+(r_2)}$$

Since ε is arbitrary, $u(r_2) \geqslant 1/s^+(r_2)$. Therefore, when r_2 is entrance,

$$u(r_2) = 1/s^+(r_2).$$

The proof for $u^+(r_2)$ is similar and can be left out. The proof is complete.

QED

Theorem 9. In equation (1) there exist a strictly decreasing positive solution $u_1(\lambda)$ and a strictly increasing positive solution $u_2(\lambda)$ which have the following properties:

(i) $u_1^+(\lambda) < 0$ is strictly increasing; $u_2^+(\lambda) > 0$ is strictly increasing

$$u_2^+(\lambda)u_1(\lambda) - u_2(\lambda)u_1^+(\lambda) = 1 \qquad \lambda > 0 \tag{14}$$

(ii) $u_a(r_a, \lambda) \equiv \lim_{z_i \to r_a} u_{ai}(\lambda)$ is finite if and only if r_a is regular or exit $u_a^+(r_a, \lambda) = \lim_{z_i \to r_a} u_{ai}(\lambda)$ is finite, or equivalently $\sum_i u_{ai}(\lambda)\mu_i < \infty$ if and only if r_a is regular or entrance.

(iii) If r_a is not entrance, then $u_b(r_a, \lambda) = 0$ $(b \neq a)$. If r_a is entrance or natural, then $u_b^+(r_a, \lambda) = 0$ $(b \neq a)$.

Proof. By Theorem 7, it follows that the strictly decreasing positive solution $u_1(\lambda)$ of equation (1) exists; similarly, the strictly increasing positive solution $u_2(\lambda)$ also exists. By (5) and (8) it can be seen that both $u_1^+(\lambda) < 0$ and $u_2^+(\lambda) > 0$ are strictly increasing. Obviously, $W(u_2(\lambda), u_1(\lambda)) > 0$. And (14) is satisfied after properly normalizing.

We shall prove (ii), (iii) for $= 2$. By Lemma 5 we deduce (ii). However,

$$\lambda \sum_i u_{2i}(\lambda)\mu_i = u_2^+(r_2, \lambda) - u_2^+(r_1, \lambda) \tag{15}$$

Obviously, $u_2^+(r_1, \lambda)$ is finite, therefore, that $u_2^+(r_2, \lambda)$ is finite is equivalent to $\sum_i u_{2i}(\lambda)u_i < \infty$. By Lemma 8, $u_1(\lambda)$ and $u_2(\lambda)$ can be selected such that (iii) is satisfied and the proof is complete. QED

Theorem 10. Both $u_1(\lambda)\mu$ and $u_2(\lambda)\mu$ are two linearly independent solutions of the equation

$$(V_\lambda) \qquad \lambda v - vQ = 0 \qquad \lambda > 0 \tag{16}$$

Each solution of the equation (v_λ) is their linear combination.

Proof. This proof follows from the corollary to Theorem 4.2. QED

5.6 MINIMAL SOLUTION

From now on, we shall denote the solutions of Theorem 5.9. by $u_1(\lambda)$ and $u_2(\lambda)$. Set

$$\phi_{ij}(\lambda) = \begin{cases} u_{2i}(\lambda)u_{1j}(\lambda)\mu_j & \text{if } i \leqslant j \\ u_{1i}(\lambda)u_{2j}(\lambda)\mu_j & \text{if } i > j \end{cases} \tag{1}$$

Then

$$\mu_i\phi_{ij}(\lambda) = \mu_j\phi_{ji}(\lambda) \tag{2}$$

Let f be a column vector, and g be a row vector. Then

$$[\phi(\lambda)f]_i = \sum_j \phi_{ij}(\lambda)f_j = u_{1i}(\lambda)\sum_{j \leqslant i} u_{2j}(\lambda)f_j\mu_j + u_{2i}(\lambda)\sum_{j > i} u_{1j}(\lambda)f_j\mu_j \tag{3}$$

$$[g\phi(\lambda)]_j = \sum_i g_i\phi_{ij}(\lambda) = u_{1j}(\lambda)u_j\sum_{i \leqslant j} g_iu_{2i}(\lambda) + u_{2j}(\lambda)\mu_j\sum_{i > j} g_iu_{1i}(\lambda) \tag{4}$$

If $g = f\mu$, then

$$g\phi(\lambda) = [\phi(\lambda)f]\mu \tag{5}$$

Theorem 1. The equation

$$\lambda \sum_i \phi_{ij}(\lambda) = 1 - \frac{u_{1i}(\lambda)}{u_1(r_1, \lambda)} - \frac{u_{2j}(\lambda)}{u_2(r_2, \lambda)} \tag{6}$$

holds. (If $u_a(r_a, \lambda) = \infty$, then the corresponding term in (6) is understood as zero.)

Proof. By (3) and (5.4)

$$\begin{aligned}
\lambda \sum_j \phi_{ij}(\lambda) &= u_{1i}(\lambda) \sum_{j \leqslant i} \lambda u_{2j}(\lambda)\mu_j + u_{2i}(\lambda) \sum_{j > i} \lambda u_{1j}(\lambda)\mu_j \\
&= u_{1i}(\lambda) \sum_{j \leqslant i} [u_{2j}^+(\lambda) - u_{2j-1}^+(\lambda)] + u_{2i}(\lambda) \sum_{j > i} [u_{1j}^+(\lambda) - u_{1,j-1}^+(\lambda)] \\
&= u_{1i}(\lambda)[u_{2i}^+(\lambda) - u_2^+(r_1, \lambda)] + u_{2i}(\lambda)[u_1^+(r_2, \lambda) - u_{1i}^+(\lambda)] \\
&= u_{1i}(\lambda)u_{2i}^+(\lambda) - u_{2i}(\lambda)u_{1i}^+(\lambda) - u_{1i}(\lambda)u_2^+(r_1, \lambda) + u_{2i}(\lambda)u_1^+(r_2, \lambda)
\end{aligned} \tag{7}$$

If we can prove

$$u_1^+(r_2, \lambda) = -\frac{1}{u_2(r_2, \lambda)}, \qquad u_2^+(r_1, \lambda) = \frac{1}{u_1(r_1, \lambda)} \tag{8}$$

then by (5.14), follows (6) from (7).

We prove only the first expression of (8). In fact, if r_2 is entrance or natural, by Theorem 5.9(ii), (iii), we obtain $u_2(r_2, \lambda) = \infty$, $u_1^+(r_2, \lambda) = 0$. Therefore, the first expression of (8) obviously holds. If r_2 is regular or exit, by (5.14), we only need prove

$$\lim_{x_i \to r_2} u_{1i}(\lambda)u_{2i}^+(\lambda) = 0 \tag{9}$$

When r_2 is regular, obviously the above expression holds since $u_2^+(r_2, \lambda) < \infty$, $u_1(r_2, \lambda) = 0$. When r_2 is exit, since $u_1(r_2, \lambda) = 0$ and $u_2^+(\lambda)$ is increasing, $-u_1^+(\lambda)$ is decreasing, it follows that

$$\begin{aligned}
0 \leqslant u_{1i}(\lambda)u_{2i}^+(\lambda) &= u_{2i}^+(\lambda)[u_{1i}(\lambda) - u_1(r_2, \lambda)] \\
&= u_{2i}^+(\lambda) \sum_{j \geqslant i} [-u_{1j}^+(\lambda)(z_{j+1} - z_j] \\
&\leqslant -u_{1i}^+(\lambda) \sum_{j \geqslant i} u_{2j}^+(\lambda)(z_{j+1} - z_j)
\end{aligned}$$

$$= -u_{1i}^{+}(\lambda)[u_2(r_2,\lambda) - u_{2i}(\lambda)]$$
$$\to -u_1^{+}(r_2,\lambda)[u_2(r_2,\lambda) - u_2(r_2,\lambda)]$$
$$= 0 \qquad (z_i \to r_2)$$

and the proof is terminated. QED

Theorem 2. If $f \in m$, $g \in l$ then $\phi(\lambda) f \in m$, $g\phi(\lambda) \in l$ and

$$\lambda\phi(\lambda)f - Q[\phi(\lambda)f] = f \qquad \lambda > 0 \tag{10}$$

$$\lambda g\phi(\lambda) - [g\phi(\lambda)]Q = g \qquad \lambda > 0 \tag{11}$$

Proof. By Theorem 1 we can get that $\phi(\lambda) f \in m$, $g\phi(\lambda) \in l$. On account of (5) and the corollary to Theorem 4.2, we only need to prove (10). In fact, by (3) we have

$$[\phi(\lambda)f]_i^+ = u_{1i}^+(\lambda) \sum_{j \leqslant i} u_{2j}(\lambda) f_j \mu_j + u_{2i}^+(\lambda) \sum_{j > i} u_{1j}(\lambda) f_j \mu_j \tag{12}$$

$$D_\mu[\phi(\lambda)f]_i^+ = D_\mu u_{1i}^+(\lambda) \sum_{j \leqslant i} u_{2j}(\lambda) f_j u_j + D_\mu u_{2i}^+(\lambda) \sum_{j > i} u_{1j}(\lambda) f_j \mu_j$$
$$+ [u_{1,i-1}^+(\lambda) u_{2i}(\lambda) - u_{2,i-1}^+(\lambda) u_{1i}(\lambda)] f_i$$
$$= \lambda u_{1i}(\lambda) \sum_{j \leqslant i} u_{2j}(\lambda) f_j \mu_j + \lambda \mu_{2i}(\lambda) \sum_{j > i} u_{1j}(\lambda) f_j \mu_j$$
$$- [u_{2i}^+(\lambda) u_{1i}(\lambda) - u_{2i}(\lambda) u_{1i}^+(\lambda)] f_i = \lambda[\phi(\lambda)f]_i - f_i$$ QED

Lemma 3. Let $f \in m$ and r_a be regular or exit. Then

$$[\phi(\lambda)f](r_a) = \lim_{z_i \to r_a} [\phi(\lambda)f]_i = 0 \tag{13}$$

Proof. Since $u_a(r_a, \lambda) < \infty$ when r_a is regular or exit and

$$\frac{u_{ai}(\lambda)}{u_a(r_a,\lambda)} \to 1 \qquad u_{bi}(\lambda) \to 0 \qquad (b \neq a) \tag{14}$$

as $z_i \to r_a$, by Theorem 1, $[\phi(\lambda)1](r_a) = 0$. Equation (13) follows from this. QED

Theorem 4. $\phi(\lambda)$ is the minimal Q process. For $\phi(\lambda)$ to be honest it is necessary and sufficient that both r_1 and r_2 are entrance or natural.

Proof. $\phi(\lambda) \geqslant 0$ is trivial. The norm condition follows from (6), and for the equality of (6) to hold it is necessary and sufficient that both r_1 and r_2 are entrance or natural. From (10) and (11) it can be seen that the B condition and F condition for $\phi(\lambda)$ hold.

Let $f \in m$ and set $F(\lambda) = \phi(\lambda)f$. By Theorem 2 it follows that $F(\lambda) - F(\mu) + (\lambda - \mu)\phi(\lambda)F(\mu) \in m$ and is the solution of equation (5.1). Therefore, it is a linear combination of $u_1(\lambda)$ and $u_2(\lambda)$, that is

$$F(\lambda) - F(\mu) + (\lambda - \mu)\phi(\lambda)F(\mu) = c_1 u_1(\lambda) + c_2 u_2(\lambda) \tag{15}$$

where c_1 and c_2 are constants.

If r_2 is regular or exit, by Lemma 3, letting $z_i \to r_2$ in (15), we obtain

$$0 = c_1 u_1(r_2, \lambda) + c_2 u_2(r_2, \lambda) = c_2 u_2(r_2, \lambda)$$

Therefore, $c_2 = 0$. If r_2 is entrance or natural, since $u_2(r_2, \lambda) = \infty$, $u_1(r_2, \lambda) < \infty$ and the left-hand side of (15) is bounded, then $c_2 = 0$ follows, too. Similarly, we can prove $c_1 = 0$. Thus (15) becomes

$$F(\lambda) - F(\mu) + (\lambda - \mu)\phi(\lambda)F(\mu) = 0$$

Letting $f_i = \delta_{ij}$, we find the above expression becomes the resolvent equation for $\phi(\lambda)$.

Thus $\phi(\lambda)$ is a Q process satisfying the system of backward and forward equations. Next we proceed to prove the minimality of $\phi(\lambda)$.

Let $\psi(\lambda)$ be a Q process. Since Q is conservative, and $\psi(\lambda)$ satisfies the B condition, it follows that, for fixed j, $\psi_{ij}(\lambda) - \phi_{ij}(\lambda)$ is a solution of equation (5.1), so that

$$\psi_{ij}(\lambda) - \phi_{ij}(\lambda) = c_1 u_{1i}(\lambda) + c_2 u_{2i}(\lambda) \tag{16}$$

where c_1 and c_2 are constants independent of i.

If r_2 is regular or exit, by $\psi(\lambda) \geqslant 0$ and Lemma 3, after letting $z_i \to r_2$ in (16) it follows that $c_1 u_1(r_2, \lambda) + c_2 u_2(r_2, \lambda) = c_2 u_2(r_2, \lambda) \geqslant 0$. Therefore, $c_2 \geqslant 0$. If r_2 is natural or entrance, then the left-hand side of (16) is bounded; but $u_1(r_2, \lambda) < \infty$, $u_2(r_2, \lambda) = \infty$, and it follows that $c_2 = 0$. Therefore, we always have $c_2 \geqslant 0$. similarly, we can prove $c_1 \geqslant 0$. Hence $\psi(\lambda) \geqslant \phi(\lambda)$, and the proof is complete.

QED

5.7 SEVERAL LEMMAS

From now on, we shall simply write

$$X_i^1(\lambda) = \frac{u_{1i}(\lambda)}{u_1(r_1, \lambda)} \qquad X_i^2(\lambda) = \frac{u_{2i}(\lambda)}{u_2(r_2, \lambda)} \tag{1}$$

$$X_i^1 = \frac{r_2 - z_i}{r_2 - r_1} \qquad X_i^2 = \frac{z_i - r_1}{r_2 - r_1} \tag{2}$$

When r_a is regular or exit, $X^a(\lambda) \neq 0$. When r_a is entrance or natural, $X^a(\lambda) = 0$. If r_1 is finite, and r_2 is infinite, we shall define $X^1 = 1$. If r_1 and r_2 are infinite, we shall define $X^1 = 0$.

We can make similar comments for X^2. X^a is a solution of equation (2.11.18). Equation (6.6) becomes

$$\lambda\phi(\lambda)1 = 1 - X^1(\lambda) - X^2(\lambda) \tag{3}$$

Lemma 1. $X^a(\lambda) \in \mathscr{M}_\lambda^+(1)\,(a = 1, 2)$ are exit families, and

$$\lambda\phi(\lambda)X^a = X^a - X^a(\lambda) \qquad a = 1, 2 \tag{4}$$

Proof. Obviously, $X^a(\lambda) \in \mathscr{M}_\lambda^+(1)$. From (4) and the resolvent equation for $\phi(\lambda)$ we know that $X^a(\lambda)$ is an exit family. Next we prove (4).

Obviously, if both r_1 and r_2 are infinite, then of course (4) holds. If r_a is finite, and $r_b (b \neq a)$ is infinite, then $X^b = X^b(\lambda) = 0$, $X^a = 1$, and (3) becomes

$$\lambda\phi(\lambda)1 = 1 - X^a(\lambda) \tag{5}$$

Therefore, (4) holds.

If r_1 and r_2 are finite, it suffices to prove that (4) holds for $a = 2$; (4) for $a = 1$ can be proved similarly. By (6.3)

$$\lambda \sum_j \phi_{ij}(\lambda)(r_2 - z_j) = u_{1i}(\lambda) \sum_{j \leqslant i} \lambda u_{2j}(\lambda)\mu_j \sum_{k \geqslant j} (z_{k+1} - z_k)$$
$$+ u_{2i}(\lambda) \sum_{j > i} \lambda \mu_{1j}(\lambda)\mu_j \sum_{k \geqslant j} (z_{k+1} - z_k) \tag{6}$$

The first term on the right-hand side is given by

$$u_{1i}(\lambda) \sum_{j \leqslant i} \lambda u_{2j}(\lambda)\mu_j \sum_{k \geqslant j} (z_{k+1} - z_k)$$
$$= u_{1i}(\lambda)\left(\sum_{k < i} (z_{k+1} - z_k) \sum_{j \leqslant k} \lambda u_{2j}(\lambda)\mu_j + \sum_{k \geqslant i} (z_{k+1} - z_k) \sum_{j \leqslant i} \lambda u_{2j}(\lambda)\mu_j \right)$$
$$= u_{1i}(\lambda)\left(\sum_{k < i} (z_{k+1} - z_k)[u_{2k}^+(\lambda) - u_2^+(r_1, \lambda)] \right.$$
$$\left. + \sum_{k \geqslant i} (z_{k+1} - z_k)[u_{2i}^+(\lambda) - u_2^+(r_1, \lambda)] \right)$$
$$= u_{1i}(\lambda)\left(u_{2i}(\lambda) - u_2(r_1, \lambda) + u_{2i}^+(\lambda)(r_2 - z_i) - \sum_k (z_{k+1} - z_k)u_2^+(r_1, \lambda) \right)$$

The second term is given by

$$\begin{aligned}
&u_{2i}(\lambda)\sum_{j>i}\lambda u_{ij}(\lambda)\mu_j\sum_{k\geqslant j}(z_{k+1}-z_k)\\
&=u_{2i}(\lambda)\sum_{k>i}(z_{k+1}-z_k)\sum_{i<j\leqslant k}\lambda u_{1j}(\lambda)\mu_j\\
&=u_{2i}(\lambda)\sum_{k>i}(z_{k-1}-z_k)[u_{1k}^+(\lambda)-u_{1i}^+(\lambda)]\\
&=u_{2i}(\lambda)\sum_{k\geqslant i}(z_{k+1}-z_k)[u_{1k}^+(\lambda)-u_{1i}^+(\lambda)]\\
&=u_{2i}(\lambda)[u_1(r_1,\lambda)-u_{1i}(\lambda)-u_{1i}^+(\lambda)(r_2-z_i)]\\
&=u_{2i}(\lambda)[-u_{1i}(\lambda)-u_{1i}^+(\lambda)(r_2-z_i)]
\end{aligned}$$

Hence substituting them into (6) and noting (5.14), we obtain

$$\lambda\sum_j\phi_{ij}(\lambda)(r_2-z_j)=(r_2-z_i)-u_{1i}(\lambda)u_2(r_1,\lambda)-u_{2i}(\lambda)(r_2-r_1)u_2^+(r_1,\lambda)\qquad(7)$$

Since r_1 being finite implies that r_2 must not be entrance, it follows that $u_2(r_1,\lambda)=0$. Again noting (6.8), and multiplying both sides of (7) by (r_2-r_1), we find that (4) holds for $a=2$, and the proof is finished. QED

Lemma 2. Let r_1 be entrance and r_2 be regular or exit. Set

$$\eta_{1j}=(r_2-z_j)\mu_j\qquad\qquad \eta_{1j}(\lambda)=-\frac{u_{1j}(\lambda)\mu_j}{u_1^+(r_1,\lambda)}\qquad(8)$$

Then $\eta_1(\lambda)\in\mathscr{L}_\lambda^+$ is an entrance family and moreover

$$\lambda\eta_1\phi(\lambda)=\eta_1-\eta_1(\lambda)\qquad(9)$$

Proof. By the corollary to Theorem 4.2 we obtain $\eta_1(\lambda)\in\mathscr{L}_\lambda^+$. If we can prove

$$u_2(r_1,\lambda)=-\frac{1}{u_1^+(r_1,\lambda)}\qquad(10)$$

then by (7) and by noting (6.5), we obtain (9).

To prove (10), by (5.14) it follows that we need only to prove

$$\lim_{z_i\to r_1}[u_{1i}(\lambda)u_{2i}^+(\lambda)]=0\qquad(11)$$

Since $u_2^+(r_1,\lambda)=0$, it follows that

$$\begin{aligned}
0\leqslant u_{1i}(\lambda)u_{2i}^+(\lambda)&=u_{1i}(\lambda)[u_{2i}^+(\lambda)-u_2^+(r_1,\lambda)]\\
&=u_{1i}(\lambda)\sum_{j\leqslant i}\lambda u_{2j}(\lambda)\mu_j
\end{aligned}$$

$$\leqslant u_{2i}(\lambda) \sum_{j \leqslant i} \lambda u_{1j}(\lambda) \mu_i$$

$$= u_{2i}(\lambda)[u_{1i}^{+}(\lambda) - u_1^{+}(r_1, \lambda)] \to 0 \qquad (z_i \to r_1)$$

That is (11). From (9) and the resolvent equation for $\phi(\lambda)$ it follows that $\eta_1(\lambda)$ is an exit family, and the proof is complete. QED

Lemma 3. Let r_2 be regular or exit. For the entrance family $\bar{\eta}(\lambda) \in \mathscr{L}_\lambda^+$ $(\lambda > 0)$ it is necessary and sufficient that $\bar{\eta}(\lambda)$ has the following Riesz representation:

$$\bar{\eta}(\lambda) = p_1 \Phi_1(\lambda) + p_2 X^2(\lambda)\mu \tag{12}$$

where the constant $p_a \geqslant 0 \, (a = 1, 2)$, $p_a = 0$ if r_a is exit or natural, and

$$\Phi_1(\lambda) = \begin{cases} \eta_1(\lambda) & \text{if } r_1 \text{ is entrance} \\ X^1(\lambda)\mu & \text{if } r_2 \text{ is regular} \end{cases} \tag{13}$$

Proof. By Theorem 5.10 it follows that $\bar{\eta}(\lambda) \in \mathscr{L}_\lambda^+$ has the form $\bar{\eta}(\lambda) = c_{1\lambda} u_1(\lambda)\mu + c_{2\lambda} u_2(\lambda)\mu$. Since $\bar{\eta}(\lambda) \in l$ by Theorem 5.9(ii) it follows that $c_{a\lambda} = 0$ if r_a is exist or natural. Therefore, $\bar{\eta}(\lambda) = p_{1\lambda} \Phi_1(\lambda) + p_{2\lambda} X^2(\lambda)\mu$ and, moreover, $p_{a\lambda} = 0$ if r_a is exist or natural. Because $X^a(\lambda)$ is an exit family if r_a is regular, so $X^a(\lambda)\mu$ is an entrance families. Now that $\bar{\eta}(\lambda)$, $\Phi_1(\lambda)$, $X^2(\lambda)\mu$ are all entrance families, it follows that $p_{a\lambda} = p_a$ is independent of λ. Hence $\bar{\eta}(\lambda)$ has the representation (12). Since $\bar{\eta}(\lambda) \geqslant 0$, by Theorem 5.9(iii) we get $p_a \geqslant 0 \, (a = 1, 2)$. Conversely, it is clear that $\bar{\eta}(\lambda)$ given by (12) is an entrance family, and $\bar{\eta}(\lambda) \in \mathscr{L}_\lambda^+$. QED

Lemma 4. Let r_1 and r_2 be regular or exit. If r_a is regular then

$$U_\lambda^{ab} \equiv \lambda[X^a(\lambda)\mu, X^b] \uparrow U^{ab} = \begin{cases} +\infty & \text{if } b = a \\ 1/(r_2 - r_1) & \text{if } b \neq a \end{cases} \tag{14}$$

as $\lambda \uparrow \infty$.

Proof. As in (2.11.46), we can obtain

$$U_\lambda^{ab} - U_v^{ab} = (\lambda - v)[X^a(v)\mu, X^b(\lambda)] \tag{15}$$

Thus the monotonicity of U_λ^{ab} is proved. Next we might as well suppose that r_2 is regular. By (4) and (6.12),

$$[\lambda\phi(\lambda)X^b]_i^+ = \lambda u_{1i}^+(\lambda) \sum_{j \leqslant i} u_{2j}(\lambda) X_j^b \mu_j + \lambda \mu_{2i}^+(\lambda) \sum_{j > i} u_{1j}(\lambda) X_j^b \mu_j$$

$$= \frac{(-1)^b}{r_2 - r_1} - [X_i^b(\lambda)]^+ \tag{16}$$

Letting $z_i \to r_2$ we obtain

$$-U_\lambda^{2b} = \frac{(-1)^b}{r_2 - r_1} - X^{b+}(r_2, \lambda) \tag{17}$$

where

$$X^{b+}(r_2, \lambda) = \lim_{z_i \to r_2} [X_i^b(\lambda)]^+$$

To prove (14), we only need to prove

$$\lim_{\lambda \to \infty} X^{2+}(r_2, \lambda) = +\infty \tag{18}$$

$$\lim_{\lambda \to \infty} X^{1+}(r_2, \lambda) = 0 \tag{19}$$

Since both $X^2(\lambda)$ and $[X^2(\lambda)]^+$ are increasing functions, it follows that

$$\frac{X^2(r_2, \lambda) - X_i^2(\lambda)}{r_2 - z_i} < X^{2+}(r_2, \lambda)$$

By (2.11.29) we have $X^a(\lambda) \downarrow 0 \, (\lambda \uparrow \infty, a = 1, 2)$. From the expression above, we obtain

$$\frac{1}{r_2 - z_i} \leqslant \varliminf_{\lambda \to \infty} X^{2+}(r_2, \lambda)$$

Letting $z_i \to r_2$, (18) follows. Furthermore,

$$0 \leqslant -X^{1+}(r_2, \lambda) < -[X_0^1(\lambda)]^+ = \frac{X_0^1(\lambda) - X_1^1(\lambda)}{z_1 - z_0}$$

Then (19) follows by $X^1(\lambda) \downarrow 0$ again, and the proof is complete. QED

Lemma 5. If r_a is exit or regular, then the standard image of $X^a(\lambda)$ is X^a.

Proof. We shall prove the lemma for $a = 2$. Since $X^2(\lambda) \leqslant X^2$, the standard image of $X^2(\lambda)$, $\bar{X}^2$ is $\leqslant X^2$, so that

$$\lambda \phi(\lambda) \bar{X}^a = \bar{X}^2 - X^2(\lambda) \tag{20}$$

Moreover $u \equiv X^2 - \bar{X}^2$ is a solution of equation (2.11.18). By (4) and (20), we have

$$\lambda \phi(\lambda) u = u \tag{21}$$

Let r_1 be infinite. Since $r_2 - z_i$ and z_i are two linearly independent solutions of $Qu = 0$, it follows that

$$u = c_1(r_2 - z_i) + c_2 z_i = c_1 r_2 + (c_2 - c_1) z_i$$

Since the left-hand side is bounded, surely $c_1 = c_2$. Thus $u = c_1 r_2$ and,

$$\lambda\phi(\lambda)u = c_1 r_2 \lambda\phi(\lambda)\mathbf{1} = c_1 r_2[1 - X^2(\lambda)] = u - c_1 r_2 X^2(\lambda)$$

Comparing (21) it follows that $c_1 = 0$. Consequently $u = 0$, $X^2 = \bar{X}^2$.

Let r_1 be finite. Then $u = c_1 X^1 + c_2 X^2$, since X^1 and X^2 are two linearly independent solutions of (2.11.18). By (4) and (20) we derive

$$\lambda\phi(\lambda)u = u - c_1 X^1(\lambda) - c^2 X^2(\lambda)$$

Comparing (21) we have

$$c_1 X^1(\lambda) + c_2 X^2(\lambda) = 0$$

If r_1 is exit or regular, then $c_1 = c_2 = 0$ and hence $u = 0$; if r_1 is natural (r_1 cannot be entrance because r_1 is finite) then $c_2 = 0$, so that $0 \leqslant c_1 X_i^1 = u_i \leqslant X_i^2$. Letting $i \to -\infty$ we have $c_1 = 0$. Therefore $u = 0$ and the proof is complete.

QED

5.8 ONE OF r_1 AND r_2 IS ENTRANCE OR NATURAL, THE OTHER IS EXIT OR REGULAR

From this section on, we shall construct all Q processes. By Theorem 6.4 it follows that if both r_1 and r_2 are entrance or natural, then the minimal solution $\phi(\lambda)$ is honest, so that the Q process is unique. In this section, we suppose that one boundary point is regular or exit (for example r_2), and that the other boundary point is entrance or natural (say, r_1); whence $u_1(r_1, \lambda) = \infty$, $u_2(r_2, \lambda) < \infty$. Hence equation (5.1) possesses only one linearly independent non-null non-negative bounded solution $\bar{X}(\lambda) = X^2(\lambda)$, that is, the dimension of $\mathscr{M}_\lambda^+$, $m^+ = 1$. In this case, the construction problem is already solved by Theorem 3.2.1. Certainly, we take a relatively special form in the present case.

If we denote $c - [\alpha, X^0] - \bar{\sigma}^0$ of Theorem 3.2.1 by $\bar{c}$, then

$$\begin{aligned} &c + [a, \bar{X} - \bar{X}(\lambda)] + \lambda[\bar{\eta}(\lambda), \bar{X}] \\ &= \bar{c} + [a, X^0 + \bar{X} - \bar{X}(\lambda)] + \lambda[\bar{\eta}(\lambda), X^0 + \bar{X}] \\ &= \bar{c} + [\alpha, 1 - \bar{X}(\lambda)] + \lambda[\bar{\eta}(\lambda), 1] \end{aligned} \tag{1}$$

In addition, if we also rewrite $\bar{c}$ as c, then Theorem 3.2.1 now takes the form below.

Theorem 1. Let r_1 be entrance or natural, and r_2 be exit or regular. For $\psi(\lambda)$ to be a Q process it is necessary and sufficient that either $\psi(\lambda) = \phi(\lambda)$ or $\psi(\lambda)$ can be obtained as follows. Take a row vector $\alpha \geqslant 0$ such that $\alpha\phi(\lambda) \in l$ and select constants $p_a \geqslant 0$ $(a = 1, 2)$ and, moreover, $p_a = 0$ if r_a is exit or natural. According to (7.12) take $\bar{\eta}(\lambda)$ satisfying

$$\eta(\lambda) = \alpha\phi(\lambda) + \bar{\eta}(\lambda) \neq 0 \tag{2}$$

Take a constant $c \geqslant 0$. Finally, set

$$\psi_{ij}(\lambda) = \phi_{ij}(\lambda) + X_i^2(\lambda)\frac{\sum_k \alpha_k \phi_{kj}(\lambda) + \bar{\eta}_j(\lambda)}{c + [\alpha, 1 - X^2(\lambda)] + \lambda[\bar{\eta}(\lambda), 1]} \tag{3}$$

For the process $\psi(\lambda)$ to be honest it is necessary and sufficient that $c = 0$. For the process $\psi(\lambda)$ to satisfy the system of forward equations it is necessary and sufficient that $\alpha = 0$.

5.9 BOTH r_1 AND r_2 ARE REGULAR OR EXIT: LINEARLY DEPENDENT CASE

In this section we suppose that both r_1 and r_2 are regular or exit; whence non-null $X^a(\lambda) \in \mathscr{M}_\lambda^+(1)$ $(a = 1, 2)$ are exit families. Let $\psi(\lambda)$ be a Q process. Since both $\psi(\lambda)$ and $\phi(\lambda)$ satisfy the B condition, thus for fixed j, $\psi_{ij}(\lambda) - \phi_{ij}(\lambda)$ is a solution of equation (5.1), so that

$$\psi_{ij}(\lambda) = \phi_{ij}(\lambda) + X_i^1(\lambda)F_j^1(\lambda) + X_i^2(\lambda)F_j^2(\lambda) \tag{1}$$

Letting $z_i \to r_a$, we obtain $F^a(\lambda) \geqslant 0$.

We shall determine $F^a(\lambda)$ $(a = 1, 2)$ so that $\psi(\lambda)$ of (1) is a Q process. We only need consider the norm conditions and the resolvent equation because $\psi(\lambda)$ in (1) satisfies the B condition.

Because of (7.3) and by observing that $X^a(r_a, \lambda) = 1$, $X^a(r_b, \lambda) = 0\,(b \neq a)$, we can easily see that the norm condition of $\psi(\lambda)$ is equivalent to

$$F^a(\lambda) \geqslant 0 \qquad \lambda[F^a(\lambda), 1] \leqslant 1 \qquad a = 1, 2 \tag{2}$$

Since $\phi(\lambda)$ satisfies the resolvent equation, upon substituting $\psi(\lambda)$ of (1) into the resolvent equation, noticing that $X^a(\lambda)$ $(a = 1, 2)$ are exit families, and noting their linear independence, we find that the resolvent equation for $\psi(\lambda)$ is equivalent to

$$F^a(\lambda)A(\lambda, v) = F^a(v) + (v - \lambda) \sum_{b=1}^{2} [F^a(\lambda), X^b(v)]F^b(v) \qquad a = 1, 2, \lambda, v > 0 \tag{3}$$

To begin with we suppose that $F^1(v)$, $F^2(v)$ are linearly dependent for some $v > 0$, that is,

$$F^a(v) = m_{av}\eta(v) \qquad m_{av} \geqslant 0, \quad \eta(v) \geqslant 0, \quad a = 1, 2 \tag{4}$$

By (2.10.8), multiplying both sides of (3) from the right by $A(v, \lambda)$, we have

$$F^a(\lambda) = \left(m_{av} + (v - \lambda) \sum_{b=1}^{2} [F^a(\lambda), X^b(v)]m_{bv} \right)\eta(v)A(v, \lambda)$$

Therefore, $F^{(a)}_{(\lambda)}$ $(a = 1, 2)$ are linearly dependent for arbitrary $\lambda > 0$. Thus (4) holds for all $v > 0$.

On account of (2.10.8), under (2) and (4), (3) is equivalent to

$$F^a(\lambda) = m_{a\lambda}\eta(\lambda) \qquad \text{quantities } m_{a\lambda} \geqslant 0, \text{ row vector } \eta(\lambda) \geqslant 0 \tag{5}$$

$$\eta(\lambda)\,(\lambda > 0) \qquad \text{are entrance families} \tag{6}$$

$$m_{a\lambda} = m_{av} + (v - \lambda) \sum_{b=1}^{2} m_{av}[\eta(\lambda), X^b(v)]m_{bv}, \qquad a = 1, 2 \tag{7}$$

By (7), if $m_{a\lambda} = 0$ for some a and some $\lambda > 0$, then $m_{a\lambda} = 0$ for all $\lambda > 0$. Hence either $m_{1\lambda} = m_{2\lambda} = 0$ and then $\psi(\lambda) = \psi(\lambda)$, which is trivial; or $\eta(\lambda) \neq 0$, $m_{1\lambda} + m_{2\lambda} > 0$. We discuss the latter case. By (6) and Lemma 7.3.

$$\eta(\lambda) = \alpha\phi(\lambda) + p_1 X^1(\lambda)\mu + p_2 X^2(\lambda)\mu \neq 0 \tag{8}$$

where constant $p_a \geqslant 0$ and $p_a = 0$ if r_a is exist; the row vector $\alpha \geqslant 0$ is such that $\alpha\phi(\lambda) \in l$. Next, by (7) we derive $m_{1\lambda}m_{2\lambda} = m_{1v}m_{2v}$. Therefore, there exist constants $d_a \geqslant 0$, $d_1 + d_2 > 0$ such that

$$d_1 m_{2\lambda} = d_2 m_{1\lambda} \tag{9}$$

We may as well suppose $d_2 > 0$ without loss of generality; hence $m_{2\lambda} > 0$, whereas (7) becomes

$$d_2 m_{2\lambda} = d_2 m_{2v} + (v - \lambda) m_{2\lambda} \sum_{b=1}^{2} d_b[\eta(\lambda), X^b(v)] m_{2v}$$

Dividing the above formula by $m_{2\lambda}m_{2v}$, we obtain

$$d_2 m_{2v}^{-1} = d_2 m_{2\lambda}^{-1} + (v - \lambda) \sum_{b=1}^{2} d_b[\eta(\lambda), X^b(v)] \tag{10}$$

Owing to (2.11.46) we have

$$(v - \lambda)[\eta(\lambda), X^b(v)] = v[\eta(v), X^b] - \lambda[\eta(\lambda), X^b] \tag{11}$$

Therefore

$$d_2 m_{2\lambda}^{-1} - \sum_{b=1}^{2} d_b \lambda[\eta(\lambda), X^b] = c \qquad (c \text{ is constant}) \tag{12}$$

From (9) it follows that

$$m_{a\lambda} = d_a\left(c + \sum_{b=1}^{2} d_b \lambda[\eta(\lambda), X^b]\right)^{-1} \qquad a = 1, 2 \tag{13}$$

By substituting (5) into (2) and noting that $X^1 + X^2 = 1$, the norm condition becomes

$$\begin{aligned} (d_1 - d_2)\lambda[\eta(\lambda), X^2] &\leqslant c \\ (d_2 - d_1)\lambda[\eta(\lambda), X^1] &\leqslant c \end{aligned} \tag{14}$$

The equalities hold if and only if $d_1 = d_2$, $c = 0$. By (8) and (7.4), we have

$$\lambda[\eta(\lambda), X^a] = \lambda[\alpha\phi(\lambda), X^a] + \sum_{b=1}^{2} p_b \lambda[X^b(\lambda)\mu, X^a]$$
$$= [\alpha, X^a - X^a(\lambda)] + p_1 U_\lambda^{1a} + p_2 U_\lambda^{2a}$$

Since $X^a(\lambda)\downarrow 0\,(\lambda\uparrow\infty)$ and on account of (7.14)

$$\lambda[\eta(\lambda), X^a] \uparrow W_a \equiv [\alpha, X^a] + p_1 U^{1a} + p_2 U^{2a} \qquad \lambda\uparrow\infty \tag{15}$$

By (7.14) W_a is finite if and only if

$$[\alpha, X^a] < \infty \qquad p_a = 0 \tag{16}$$

and then

$$W_a = [\alpha, X^a] + \frac{p_b}{r_2 - r_1} \qquad (b \neq a) \tag{17}$$

Therefore, (14) is equivalent to

$$\begin{aligned} c &\geqslant 0 && \text{if } d_1 = d_2 \\ c &\geqslant (d_1 - d_2)W_2 && \text{if } d_1 > d_2 \\ c &\geqslant (d_2 - d_1)W_1 && \text{if } d_1 < d_2 \end{aligned} \tag{18}$$

Theorem 1. Suppose that both r_1 and r_2 are exist or regular and that $F^a(\lambda)$ $(a = 1, 2)$ are linearly dependent for some (hence all) $\lambda > 0$.

For $\psi(\lambda)$ of (1) to be a Q process it is necessary and sufficient that either $\psi(\lambda) = \phi(\lambda)$ or $\psi(\lambda)$ is obtained as follows: Select constants $d_a \geqslant 0$, $d_1 + d_2 > 0$, $p_a \geqslant 0\,(p_a = 0$ if r_a is exit); take a row vector $\alpha \geqslant 0$ such that $\alpha\phi(\lambda) \in l$ and moreover (8) holds. It is also required that (16) holds if $d_a < d_b\,(b \neq a)$. Choose a constant c satisfying (18) (where W_a is the same as in (17)). Finally, set

$$\psi_{ij}(\lambda) = \phi_{ij}(\lambda) + \frac{[d_1 X_i^1(\lambda) + d_2 X_i^2(\lambda)]\,[\sum_k \alpha_k \phi_{kj}(\lambda) + p_1 X_j^1(\lambda)\mu_j + p_2 X_j^2(\lambda)\mu_j]}{c + \sum_{b=1}^{2} d_b\{[\alpha, X^b - X^b(\lambda)] + \lambda[p_1 X^1(\lambda)\mu + p_2 X^2(\lambda)\mu, X^b)\}} \tag{19}$$

For the process $\psi(\lambda)$ to be honest it is necessary and sufficient that $d_1 = d_2$, $c = 0$. For this process to satisfy the system of forward equations it is necessary and sufficient that $\alpha = 0$.

5.10 BOTH r_1 AND r_2 ARE REGULAR OR EXIT: LINEARLY INDEPENDENT CASE

In the previous section, we assumed that $F^a(\lambda)$ $(a = 1, 2)$ were linearly dependent. In this section, we suppose that $F^a(\lambda)$ $(a = 1, 2)$ are linearly independent.

It will be convenient to use matrix notation. We write

$$[y]=\begin{pmatrix} y^1 \\ \hline y^2 \end{pmatrix} \qquad [y]'=(y^1,y^2)$$

where y^1 and y^2 are quantities or vectors. If y^a, v^a $(a=1,2)$ are vectors, the symbol $\{[y,v]\}$ will denote the second-order square matrix whose elements are $[y^a,v^a]$. The second-order identity matrix will be denoted by I.

Using these notations, (9.1), (9.2) and (9.3) may be written as

$$\psi(\lambda)=\phi(\lambda)+[X(\lambda)]'[F(\lambda)] \tag{1}$$

$$[F(\lambda)]\geqslant[0] \qquad \{\lambda[F(\lambda),X]\}[1]\leqslant[1] \tag{2}$$

$$[F(\lambda)A(\lambda,v)]=[F(v)]+(v-\lambda)\{[F(\lambda),X(v)]\}[F(v)] \tag{3}$$

Lemma 1. Let $\psi(\lambda)$ be a Q process having form (1). Then there exist two row vectors $\alpha^a\geqslant 0$ $(a=1,2)$ such that $\alpha^a\phi(\lambda)\in l$, and there exist two second-order square matrices $\mathscr{R}_\lambda=(r_\lambda^{ab})\geqslant 0$ and $\mathscr{M}_\lambda=(M_\lambda^{ab})\geqslant 0$ $(M_\lambda^{1a}=M_\lambda^{2a}=0$ if r_a is exist). Moreover,

$$[F(\lambda)]=\mathscr{R}_\lambda[\alpha\phi(\lambda)]+\mathscr{M}_\lambda[X(\lambda)\mu] \tag{4}$$

Proof. Since $\psi(\lambda)$ is a Q process, (2) and (3) hold.

From (2) and (3) we can see that for arbitrary $\lambda, v>0$, $[F(\lambda)A(\lambda,v)]\geqslant[0]$. Now fix, for the moment, a and $\lambda>0$ and set $\eta(v)=F^a(\lambda)A(\lambda,v)$. Then $\eta(v)$ is an entrance family. Hence, by Lemmas 2.11.3 and 7.3, there exists a row vector $\beta_\lambda^a\geqslant 0$ independent of v (but dependent on a and λ), such that $\beta_\lambda^a\phi(\lambda)\in l$. Moreover,

$$vF^a(\lambda)A(\lambda,v)-F^a(\lambda)A(\lambda,v)Q=\beta_\lambda^a \tag{5}$$

$$F^a(\lambda)A(\lambda,v)=\beta_\lambda^a\phi(v)+M_\lambda^{a1}X^1(\lambda)\mu+M_\lambda^{a2}X^2(\lambda)\mu \tag{6}$$

where $\mathscr{M}_\lambda=(M_\lambda^{ab})\geqslant 0$ is independent of v and $M_\lambda^{1a}=M_\lambda^{2a}=0$ if r_a is exit. In particular, when $v=\lambda$ we have

$$\lambda F^a(\lambda)-F^a(\lambda)Q=\beta_\lambda^a \tag{7}$$

$$[F(\lambda)]=[\beta_\lambda^a\phi(\lambda)]+\mathscr{M}_\lambda[X(\lambda)\mu] \tag{8}$$

Therefore, we should like to prove the following. There exist two row vectors $\alpha^a\geqslant 0$ $(a=1,2)$, independent of λ, such that

$$\beta_\lambda^a=r_\lambda^{a1}\alpha^1+r_\lambda^{a2}\alpha^2 \tag{9}$$

where $\mathscr{R}_\lambda=(r_\lambda^{ab})\geqslant 0$. Then substituting (9) into (8) we obtain (4), and from $\beta_\lambda^a\phi(\lambda)\in l$ we can deduce $\alpha^a\phi(\lambda)\in l$ (if $r_\lambda^{1b}\equiv r_\lambda^{2b}\equiv 0$ for some b, then we can take α^b to be 0 independent of the value of β_λ^a. If we can prove this, the lemma is thus proved.

By (5) and (7), multiplying both sides of (3) from the left by $(\nu I - Q)$ we obtain

$$[\beta_\lambda] = [\beta_\nu] + (\nu - \lambda)\{[F(\lambda), X(\nu)]\}[\beta_\nu] \tag{10}$$

From this we can see $\beta^a_\lambda \downarrow (\lambda \uparrow)$, and if $\beta^a_\lambda = 0$ for some $\lambda > 0$ and some a, then $\beta^1_\nu = \beta^2_\nu = 0$ for $\nu > \lambda$; whence $\beta^1_\lambda = \beta^2_\lambda = 0$ for all $\lambda > 0$. In this case, (9) is trivial, we only need to choose $\alpha^1 = \alpha^2 = 0, \mathscr{R}_\lambda = 0$. We assume $\beta^a_\lambda \neq 0$ $(a = 1, 2)$ for all $\lambda > 0$. Set

$$\mathscr{T}_{\lambda\nu} \equiv I + (\nu - \lambda)\{[F(\lambda), X(\nu)]\}$$

its elements being $t^{ab}_{\lambda\nu}$, and then (10) can be rewritten as

$$\beta^a_\lambda = t^{a1}_{\lambda\nu}\beta^1_\nu + t^{a2}_{\lambda\nu}\beta^2_\nu \qquad a = 1, 2 \tag{11}$$

And when $\nu > \lambda$

$$0 \leqslant t^{ab}_{\lambda\nu}\beta^a_\nu \leqslant \beta^b_\lambda \qquad a, b = 1, 2 \tag{12}$$

If, for arbitrary $\lambda > 0$, when $\nu \to \infty$, we have

$$t^{a1}_{\lambda\nu}\beta^1_\nu \to 0 \qquad a = 1, 2 \tag{13}$$

then, by (11), for arbitrary $\lambda > 0$, we have

$$t^{a2}_{\lambda\nu}\beta^2_\nu \to \beta^a_\lambda \neq 0 \qquad a = 1, 2 \tag{14}$$

as $\nu \to \infty$. Fix an arbitrary $\lambda_0 > 0$, and let $\alpha^2 = \beta^1_{\lambda_0}$. By (11), we have

$$\beta^a_\lambda = t^{a1}_{\lambda\nu}\beta^1_\nu + \frac{t^{a2}_{\lambda\nu}}{t^{12}_{\lambda_0\nu}} t^{12}_{\lambda_0\nu}\beta^2_\nu$$

The left-hand side of the above expression is independent of ν; by (13) it follows that the first term of the right-hand side converges to zero as $\nu \to \infty$. By (14) it follows that the second fraction on the right-hand side of the above formula must converge to some r^{a2}_λ that is finite and non-negative. Therefore, upon taking a limit as $\nu \to \infty$ we obtain

$$\beta^a_\lambda = r^{a2}_\lambda \alpha^2$$

Then we certainly have, the form (9).

If (13) does not hold, then there exist some $\lambda_1 > 0$ and some a_1 (we had better say $a_1 = 1$ without loss of generality) and a subsequence $\nu_n \to \infty$ such that

$$t^{11}_{\lambda_1\nu_n}\beta^1_{\lambda_1} \to \alpha^1 \neq 0 \qquad 0 \leqslant \alpha^1 \leqslant \beta^1_{\lambda_1} \tag{15}$$

If, for every $\lambda > 0$, we have

$$t^{a2}_{\lambda\nu_n}\beta^2_{\nu_n} \to 0 \qquad a = 1, 2$$

as $\nu_n \to \infty$, then in the same way as used in the proof above, we can prove

$$\beta^a_\lambda = r^{a1}_\lambda \alpha^1 \qquad 0 \leqslant \alpha^1 \neq 0$$

Hence we obtain an expression of the form (9) for β_λ^a. Otherwise, there exist some $\lambda_2 > 0$, some a_2 (say, $a_2 = 2$ without loss of generality) and a subsequence $v_n(1) \to \infty$ of v_n such that

$$t^{22}_{\lambda_2 v v_n(1)} \beta^2_{v v_n(1)} \to \alpha^2 \neq 0 \qquad 0 \leqslant \alpha^2 \leqslant \beta^2_{\lambda_1} \tag{16}$$

Now for an arbitrary $\lambda > 0$, (11) can be rewritten as

$$\beta^a_\lambda = \frac{t^{a1}_{\lambda v}}{t^{11}_{\lambda_1 v}} t^{11}_{\lambda_1 v} \beta^1_v + \frac{t^{a2}_{\lambda v}}{t^{22}_{\lambda_2 v}} t^{22}_{\lambda_2 v} \beta^2_v$$

We can also select a subsequence $v_n(2) \to \infty$ of $v_n(1)$ (the subsequence may depend on λ) such that the two fractions in the above formula converge to non-negative numbers r^{a1}_λ and r^{a2}_λ as $v = v_n(2) \to \infty$, respectively. Then, by (15) and (16), letting $v = v_n(2) \to \infty$ in the above formula, we obtain

$$\beta^a_\lambda = r^{a1}_\lambda \alpha^1 + r^{a2}_\lambda \alpha^2$$

Since α^1 and α^2 are not zero, it follows that $\mathscr{R}_\lambda = (r^{ab}_\lambda)$ is finite. Thus the form (9) follows again and the proof is terminated. QED

By Lemma 1, it suffices for us to consider only Q processes $\psi(\lambda)$ that have the form given by (1) and (4).

We introduce the notation:

$$h^{ab}_\lambda = \lambda[\alpha^a \phi(\lambda), X^b] = [\alpha^a, X^b - X^b(\lambda)] \uparrow h^{ab} = [\alpha^a, X^b] \qquad \lambda \uparrow \infty \tag{17}$$

$$\mathscr{H}_\lambda = (h^{ab}_\lambda) \uparrow \mathscr{H} = (h^{ab}) \qquad \lambda \uparrow \infty \tag{18}$$

The above relation follows from $X^a(\lambda) \downarrow 0 \, (\lambda \uparrow \infty)$ and (7.4), By (2.11.40)

$$(v - \lambda)[\alpha^a \phi(\lambda), X^b(v)] = h^{ab}_v - h^{ab}_\lambda \tag{19}$$

We first consider a special case: $\mathscr{M}_\lambda = 0$, $\mathscr{H} < \infty$. That is,

$$[F(\lambda)] = \mathscr{R}_\lambda [\alpha \phi(\lambda)] \tag{20}$$

$$[\alpha^a, 1] < \infty \qquad \alpha = 1, 2 \tag{21}$$

Theorem 2. Assume that $F'(\lambda)$ and $F^2(\lambda)$ are linearly independent, and that $F^1(\lambda)$ and $F^2(\lambda)$ have the forms (20) and (21). Then for $\psi(\lambda)$ given by (1) to be a Q process, it is necessary and sufficient that it can be obtained as follows: Select non-negative row vectors α^1 and α^2, which are linearly independent and satisfy

$$[\alpha^a, 1] \leqslant 1 \qquad \alpha = 1, 2 \tag{22}$$

Then, set

$$\psi(\lambda) = \phi(\lambda) + [X(\lambda)]'(I - \mathscr{T}_\lambda)^{-1} [\alpha \phi(\lambda)] \tag{23}$$

where $\mathscr{T}_\lambda = \{[\alpha, X(\lambda)]\}$.

The process $\psi(\lambda)$ does not satisfy the system of forward equations. The vectors α^1 and α^2 (they satisfy (22)) are uniquely determined by this process. For the process $\psi(\lambda)$ to be honest it is necessary and sufficient that

$$[a^a, 1] = 1 \qquad a = 1, 2 \tag{24}$$

Proof. (a) Let the Q process $\psi(\lambda)$ be of the forms (20) and (21). Substituting (20) into (2), we obtain

$$\mathscr{R}_\lambda \mathscr{H}_\lambda [1] \leqslant [1] \tag{25}$$

Again substituting (20) into (3) and noting (2.10.9) and (19), we obtain

$$\mathscr{R}_\lambda [a\phi(v)] = \mathscr{R}_v[\alpha\phi(v)] + \mathscr{R}_\lambda(\mathscr{H}_v - \mathscr{H}_\lambda)\mathscr{R}_v[\alpha\phi(v)] \tag{26}$$

Multiplying the above formula from the right by $(vI - Q)$, we obtain

$$\mathscr{R}_\lambda[\alpha] = \mathscr{R}_v[\alpha] + \mathscr{R}_\lambda(\mathscr{H}_v - \mathscr{H}_\lambda)\mathscr{R}_v[\alpha]$$

Now that $F^a(\lambda)$ $(a = 1, 2)$ are linearly independent, by (20) it follows that α^1 and α^2 are also linearly independent. Hence from the above formula, we obtain

$$\mathscr{R}_\lambda = [I - \mathscr{R}_\lambda(\mathscr{H}_\lambda - \mathscr{H}_v)]\mathscr{R}_v \tag{27}$$

Select a subsequence $\lambda \to \infty$ such that

$$\mathscr{R}_\lambda \to \mathscr{R} \geqslant 0$$

By (25), (27) and (22) we obtain

$$\mathscr{R}\mathscr{H}[1] \leqslant [1] \tag{28}$$

$$\mathscr{R} = [I - \mathscr{R}(\mathscr{H} - \mathscr{H}_v)]\mathscr{R}_v \tag{29}$$

Set $[\bar{a}] = \mathscr{R}[\alpha] \geqslant [0]$, and then (28) and (29) become

$$[\bar{\alpha}^a, 1] \leqslant 1 \qquad a = 1, 2 \tag{30}$$

$$\mathscr{R} = (I - \bar{\mathscr{T}}_\lambda)\mathscr{R}_\lambda \qquad \bar{\mathscr{T}}_\lambda = \{[\bar{\alpha}, X(\lambda)]\} \tag{31}$$

But by (30), we obtain $[\bar{\alpha}^a, X^1(\lambda) + X^2(\lambda)] < 1$ $(a = 1, 2)$. Therefore the inverse matrix $(I - \mathscr{T}_\lambda)^{-1}$ exists and is non-negative. From (31) it follows that

$$\mathscr{R}_\lambda = (I - \bar{\mathscr{T}}_\lambda)^{-1}\mathscr{R}$$

Substituting the above formula into (20), we obtain

$$[F(\lambda)] = \bar{\mathscr{R}}_\lambda[\bar{\alpha}\phi(\lambda)] \qquad \bar{\mathscr{R}}_\lambda = (I - \bar{\mathscr{T}}_\lambda)^{-1} \tag{32}$$

Thus it remains to prove that $\bar{\alpha}^1$ and $\bar{\alpha}^2$ are linerly independent. This can be seen from the above formula, because if $\bar{\alpha}^1$ and $\bar{\alpha}^2$ are linearly dependent, then $F^1(\lambda)$ and $F^2(\lambda)$ are also linearly dependent.

(b) Let α^1 and α^2 be non-negative and linearly independent, and satisfy (22).

In (a) we have already proved that the inverse matrix $\mathscr{R}_\lambda = (I - \mathscr{T}_\lambda)^{-1}$ exists and is non-negative.

The $F^a(\lambda)$ $(a = 1, 2)$ defined by $[F(\lambda)] = (I - \mathscr{T}_\lambda)^{-1}[\alpha\phi(\lambda)]$ are linearly independent. In fact, let

$$0 = [c]'[F(\lambda)] = [c]'(I - \mathscr{T}_\lambda)^{-1}[\alpha\phi(\lambda)]$$

Multiplying the above formula from the right by $(\lambda I - Q)$, we obtain

$$0 = [c]'(I - \mathscr{T}_\lambda)^{-1}[\alpha]$$

Since α^a $(a = 1, 2)$ are independent, thus $[c]'(I - \mathscr{T}_\lambda)^{-1} = [0]'$; hence

$$[c]' = [0]'.$$

By (22) we obtain $\mathscr{H}[1] \leqslant [1]$, that is, $(I - \mathscr{H})[1] \geqslant [0]$. Since $[I - \mathscr{T}_\lambda)[1] \geqslant \mathscr{H}_\lambda[1]$ and $(I - \mathscr{T}_\lambda)^{-1}$ is non-negative, therefore $[1] \geqslant (I - \mathscr{T}_\lambda)^{-1}\mathscr{H}_\lambda[1]$. Hence (25) is satisfied, so that (2) holds. By a direct verification, we know (27) is satisfied. Therefore (26) holds and, consequently, (3) holds.

(c) Since α^1 and α^2 are linearly independent, and since $u_j \equiv X_i^1(\lambda)F_j^1(\lambda) + X_i^2(\lambda)F_j^2(\lambda)$ satisfies

$$(\lambda u - uQ)_j = [X_i(\lambda)]'[\alpha_j] \neq 0$$

it follows that the Q process $\psi(\lambda)$ does not satisfy the system of forward equations.

For the equality to hold in (2), i.e. in (25), it is necessary and sufficient that $(I - \mathscr{T}_\lambda)^{-1}\mathscr{H}_\lambda[1] = [1]$. That is,

$$(\mathscr{H} - \mathscr{T}_\lambda)[1] = \mathscr{H}_\lambda[1] = (I - \mathscr{T}_\lambda)[1]$$

i.e. $\mathscr{H}[1] = [1]$, and (24) follows.

Assume that both α^a and $\bar{\alpha}^a$ $(a = 1, 2)$ satisfy (22) and, moreover, correspond to the same process, that is,

$$(I - \mathscr{T}_\lambda)^{-1}[\alpha\phi(\lambda)] = (I - \bar{\mathscr{T}}_\lambda)^{-1}[\bar{\alpha}\phi(\lambda)]$$

Multiplying the above formula from the right by $(\lambda I - Q)$, we obtain

$$(I - \mathscr{T}_\lambda)^{-1}[\alpha] = (I - \bar{\mathscr{T}}_\lambda)^{-1}[\alpha]$$

Since $X^a(\lambda) \downarrow 0$ $(\lambda \uparrow \infty)$ and because of (22), we obtain $[\alpha] = [\bar{\alpha}]$ by letting $\lambda \uparrow \infty$ in the above formula.

Thus the proof of the theorem is completed. QED

We now consider the general case.

Theorem 3. Let $F^a(\lambda)$ $(a = 1, 2)$ be linearly independent. For $\psi(\lambda)$ given by (1) and (4) to be a Q process it is necessary and sufficient that it can be obtained as follows: Select two non-negative row vectors $\bar{\alpha}^a$ $(a = 1, 2)$ such that $\bar{\alpha}^a\phi(\lambda) \in l$, and choose two non-negative matrices

$$\varphi = \begin{pmatrix} 0 & \bar{s}^{12} \\ \bar{s}^{21} & 0 \end{pmatrix} \qquad \text{and} \qquad \mathscr{M} = \begin{pmatrix} \bar{M}^{11} & 0 \\ 0 & \bar{M}^{22} \end{pmatrix}$$

Moreover, they possess the following properties:

(i) If r_a is exit, then $\bar{M}^{aa} = 0$.

(ii) $\bar{M}^{aa} > 0$ $(a = 1, 2)$; or $\bar{M}^{aa} = 0$, $\bar{M}^{bb} > 0$ $(b \neq a)$, $\bar{\alpha}^a \neq 0$; or $\bar{M}^{aa} = 0$ $(a = 1, 2)$, α^1 and α^2 are linearly independent.

(iii) $\bar{h}^{ab} < \infty$ $(a \neq b)$.

(iv) $\bar{s}^{12} \leqslant 1$, $\bar{s}^{21} \leqslant 1$, $\bar{s}^{ab} \geqslant \bar{h}^{ab} + \bar{M}^{aa}/(r_2 - r_1)(a \neq b)$.

Set

$$\begin{aligned} \mathscr{R}_\lambda &= (I - \bar{\mathscr{S}} + \bar{\mathscr{H}}_\lambda + \bar{\mathscr{M}}\mathscr{U}_\lambda)^{-1} \\ \mathscr{M}_\lambda &= (I - \bar{\mathscr{S}} + \bar{\mathscr{H}}_\lambda + \bar{\mathscr{M}}\mathscr{U}_\lambda)^{-1}\bar{\mathscr{M}} \end{aligned} \tag{33}$$

where $\bar{\mathscr{H}}_\lambda \uparrow \bar{\mathscr{H}}$ $(\lambda \uparrow \infty)$, which are determined by (17) and (18) for $\bar{\alpha}^a$ $(a = 1, 2)$; by (7.14) we obtain

$$0 < \mathscr{U}_\lambda = (U_\lambda^{ab}) \uparrow \mathscr{U} = \begin{pmatrix} +\infty & 1/(r_2 - r_1) \\ 1/(r_2 - r_1) & +\infty \end{pmatrix} \qquad \lambda \uparrow \infty \tag{34}$$

Finally $\psi(\lambda)$ is determined by (1), (4) and (33).

For the process $\psi(\lambda)$ to be honest it is necessary and sufficient that $\bar{s}^{12} = \bar{s}^{21} = 1$. For this process to satisfy the system of forward equations it is necessary and sufficient that $[\bar{\alpha}] = [0]$. For the process $\psi(\lambda)$ to have the forms (1), (20) and (21) it is necessary and sufficient that $\bar{M}^{11} = \bar{M}^{22} = 0$, $\bar{h}^{aa} < \infty$ $(a = 1, 2)$.

Proof. The proof is to be reached in several steps.

(a) Suppose that the Q process $\psi(\lambda)$ has the forms of (1) and (4) and is such that $F^a(\lambda)$ $(a = 1, 2)$ are linearly independent.

Substituting (4) into (2), we obtain

$$\mathscr{S}_\lambda[1] \leqslant [1] \qquad \mathscr{S}_\lambda \equiv \mathscr{R}_\lambda \mathscr{H}_\lambda + \mathscr{M}_\lambda \mathscr{U}_\lambda \tag{35}$$

Substituting (4) into (3), owing to (2.10.9), (7.15), (19), and to the properties of $X^a(\lambda)$ being an exit family, we obtain

$$\begin{aligned} &\mathscr{R}_\lambda[\alpha\phi(\nu)] + \mathscr{M}_\lambda[X(\nu)\mu] \\ &\quad = \mathscr{R}_\nu[\alpha\phi(\nu)] + \mathscr{M}_\nu[X(\nu)\mu] + (\mathscr{R}_\lambda\mathscr{H}_\nu + \mathscr{M}_\lambda\mathscr{U}_\nu - \mathscr{S}_\lambda)\{\mathscr{R}_\nu[\alpha\phi(\nu)] \\ &\qquad + \mathscr{M}_\nu[X(\nu)\mu]\} \end{aligned} \tag{36}$$

Multiplying the above formula from the right by $(\nu I - Q)$, we have

$$\begin{aligned} \mathscr{R}_\lambda[\alpha] &= \mathscr{R}_\nu[\alpha] + (\mathscr{R}_\lambda\mathscr{H}_\nu + \mathscr{M}_\lambda\mathscr{U}_\nu - \mathscr{S}_\lambda)\mathscr{R}_\nu[\alpha] \\ &= (I - \mathscr{S}_\lambda + \mathscr{R}_\lambda\mathscr{H}_\nu + \mathscr{M}_\lambda\mathscr{U}_\nu)\mathscr{R}_\nu[\alpha] \end{aligned} \tag{37}$$

Substituting it into (36), we obtain

$$\mathscr{M}_\lambda = (I - \mathscr{S}_\lambda + \mathscr{R}_\lambda \mathscr{H}_v + \mathscr{M}_\lambda \mathscr{U}_v)\mathscr{M}_v \tag{38}$$

Set $\delta_\lambda^a = 1 - s_\lambda^{aa}$, then $\delta_\lambda^a > 0$. In fact, $\delta_\lambda^a \geqslant 0$ by (35). If $\delta_\lambda^a = 0$, then $s_\lambda^{aa} = 1$. By (35),

$$s_\lambda^{ab} = \sum_{t=1}^{2} (r_\lambda^{at} h_\lambda^{tb} + M_\lambda^{at} U_\lambda^{tb}) = 0 \qquad (b \neq a)$$

Therefore,

$$r_\lambda^{at} \alpha^t = M_\lambda^{at} = 0 \qquad (t = 1, 2)$$

Hence

$$s_\lambda^{aa} = \sum_{t=1}^{2} (r_\lambda^{at} h_\lambda^{ta} + M_\lambda^{at} U_\lambda^{ta}) = 0$$

This is in contradiction with $s_\lambda^{aa} = 1$.

Dividing the ath of (35), (37) and (38) by δ_λ^a, we obtain

$$\bar{\mathscr{S}}_\lambda[1] \leqslant [1] \qquad \bar{\mathscr{S}}_\lambda \equiv \begin{pmatrix} 0 & \bar{s}_\lambda^{12} \\ \bar{s}_\lambda^{21} & 0 \end{pmatrix} \tag{39}$$

$$\bar{\mathscr{R}}_\lambda[\alpha] = (I - \bar{\mathscr{S}}_\lambda + \bar{\mathscr{R}}_\lambda \mathscr{H}_v + \bar{\mathscr{M}}_\lambda \mathscr{U}_v)\mathscr{R}_v[\alpha] \tag{40}$$

$$\bar{\mathscr{M}}_\lambda = (I - \bar{\mathscr{S}}_\lambda + \bar{\mathscr{R}}_\lambda \mathscr{H}_v + \bar{\mathscr{M}}_\lambda \mathscr{U}_v)\mathscr{M}_v \tag{41}$$

where

$$\bar{s}_\lambda^{aa} = 0 \qquad (a = 1, 2) \tag{42}$$

$$\bar{s}_\lambda^{ab} = \sum_{t=1}^{2} \bar{r}_\lambda^{at} h_\lambda^{tb} + \sum_{t=1}^{2} \bar{M}_\lambda^{at} U_\lambda^{tb} \qquad (a \neq b) \tag{43}$$

$$\bar{r}_\lambda^{ab} = \frac{r_\lambda^{ab}}{\delta_\lambda^a} \qquad \bar{M}_\lambda^{ab} = \frac{M_\lambda^{ab}}{\delta_\lambda^a} \tag{44}$$

Choose a subsequence $\lambda \to \infty$ such that

$$\bar{\mathscr{S}}_\lambda \to \bar{\mathscr{S}} \qquad \bar{\mathscr{R}}_\lambda \to \bar{\mathscr{R}} \qquad \bar{\mathscr{M}}_\lambda \to \bar{\mathscr{M}} \tag{45}$$

Then $\bar{\mathscr{S}} = (\bar{s}^{ab})$, $\bar{\mathscr{R}} = (\bar{r}^{ab})$ and $\bar{\mathscr{M}} = (\bar{M}^{ab})$ are non-negative, and by (39)–(44) we have[1]

$$\bar{s}^{aa} = 0 (a = 1, 2) \qquad \bar{s}^{12} \leqslant 1 \qquad \bar{s}^{21} \leqslant 1 \tag{46}$$

$$\bar{s}^{ab} \geqslant \sum_{t=1}^{2} \bar{r}^{at} h^{tb} + \sum_{t=1}^{2} \bar{M}^{at} U^{tb} \qquad (a \neq b) \tag{47}$$

$$\bar{\mathscr{R}}[\alpha] = (I - \bar{\mathscr{S}} + \bar{\mathscr{R}} \mathscr{H}_v + \bar{\mathscr{M}} \mathscr{U}_v)\mathscr{R}_v[\alpha] \tag{48}$$

$$\bar{\mathscr{M}} = (I - \bar{\mathscr{S}} + \bar{\mathscr{R}} \mathscr{H}_v + \bar{\mathscr{M}} \mathscr{U}_v)\mathscr{M}_v \tag{49}$$

[1] Let $0.\infty = 0$.

By (47) we obtain

$$\bar{M}^{12} = \bar{M}^{21} = 0 \tag{50}$$

If $h^{tb} = \infty$, then

$$\bar{r}^{at} = 0 \qquad (a \neq b) \tag{51}$$

By Lemma 1, it follows that (i) of this theorem holds. Now set $[\bar{\alpha}] = \bar{\mathscr{R}}[\alpha]$; then $\bar{\alpha}^a \geqslant 0$ and $\bar{\alpha}^a \phi(\lambda) \in l$ $(a = 1, 2)$. Equations (50), (46) and (47) become (iv). Hence we obtain (iii). Equations (48) and (49) become

$$[\bar{\alpha}] = (I - \bar{\mathscr{S}} + \bar{\mathscr{H}}_\nu + \bar{\mathscr{M}}\mathscr{U}_\nu)\mathscr{R}_\nu[\alpha] \tag{52}$$

$$\bar{\mathscr{M}} = (I - \bar{\mathscr{S}} + \bar{\mathscr{H}}_\nu + \bar{\mathscr{M}}\mathscr{U}_\nu)\mathscr{M}_\nu \tag{53}$$

Next we prove

$$\bar{\alpha}^a \neq 0 \qquad \text{if } \bar{M}^{aa} = 0 \tag{54}$$

For otherwise, we may as well assume $\bar{M}^{11} = 0$, $\bar{\alpha}^1 = 0$. Then by (52) and (53) we have

$$r_\lambda^{11}\alpha^1 + r_\lambda^{12}\alpha^2 = \bar{s}^{12}(r_\lambda^{21}\alpha^1 + r_\lambda^{22}\alpha^2)$$
$$M_\lambda^{1a} = \bar{s}^{12} M_\lambda^{2a} \qquad (a = 1, 2)$$

Therefore, $F^1(\lambda) = \bar{s}^{12} F^2(\lambda)$. This is in contradiction with the linear independence of $F^1(\lambda)$ and $F^2(\lambda)$.

We proceed to prove that the inverse matrix

$$\mathscr{L}_\nu^{-1} \equiv (I - \bar{\mathscr{S}} + \bar{\mathscr{H}}_\nu + \bar{\mathscr{M}}\mathscr{U}_\nu)^{-1} \tag{55}$$

exists and is non-negative. Hence by (52) and (53) we have

$$\mathscr{R}_\nu[\alpha] = \mathscr{L}_\nu^{-1}[\bar{\alpha}] \qquad \mathscr{M}_\nu = \mathscr{L}_\nu^{-1}\bar{\mathscr{M}}$$

Therefore,

$$\begin{aligned}[F(\lambda)] &= \mathscr{L}_\lambda^{-1}\bar{\mathscr{R}}[\alpha\phi(\lambda)] + \mathscr{L}_\lambda^{-1}\bar{\mathscr{M}}[X(\lambda)\mu] \\ [F(\lambda)] &= \mathscr{L}_\lambda^{-1}[\bar{\alpha}\phi(\lambda)] + \mathscr{L}_\lambda^{-1}\bar{\mathscr{M}}[X(\lambda)\mu]\end{aligned} \tag{56}$$

That is, $\psi(\lambda)$ is determined by (1), (4) and (3).

By (iv), (54) and (34), we obtain

$$1 > \bar{s}^{ab} - (\bar{h}^{ab} - \bar{M}^{aa}U_\nu^{ab}) \geqslant \bar{s}^{ab} - (\bar{h}^{ab} - \bar{M}^{aa}U^{ab}) \geqslant 0 \quad (a \neq b)$$

Hence $I - \bar{\mathscr{S}} + \bar{\mathscr{H}}_\nu + \bar{\mathscr{M}}\mathscr{U}_\nu$ has the form

$$\begin{pmatrix} 1 + l^{11} & -l^{12} \\ -l^{21} & 1 + l^{22} \end{pmatrix} \qquad \begin{aligned} 1 + l^{11} &> l_{12} \geqslant 0 \\ 1 + l^{22} &> l_{21} \geqslant 0 \end{aligned}$$

Consequently, its determinant is $\Delta > 0$, and its inverse matrix $\mathscr{L}_\nu^{-1}$ is

$$\frac{1}{\Delta}\begin{pmatrix} 1 + l^{22} & l^{21} \\ l^{12} & 1 + l^{11} \end{pmatrix} \geqslant 0$$

(b) We proceed to prove that for $F^a(\lambda)$ $(a = 1, 2)$ to be linearly independent the necessary and sufficient condition is (ii).

Obviously, the linear independence of $F^a(\lambda)$ $(a = 1, 2)$ is equivalent to that of $[\bar{\alpha}\phi(\lambda)] + \bar{\mathcal{M}}[X(\lambda)\mu]$. Let

$$[c]'\{[\bar{\alpha}\phi(\lambda)] + \bar{\mathcal{M}}[X(\lambda)\mu]\} = 0 \tag{57}$$

Then multiplying the above expression from the right by $(\lambda I - Q)$, we obtain

$$[c]'[\bar{\alpha}] = 0 \qquad [c]'\bar{\mathcal{M}} = [0]' \tag{58}$$

Conversely, if (58) holds, then (57) certainly holds. In order that (58) is equivalent to $[c] = [0]$, it is necessary and sufficient that (ii) holds.

(c) Suppose that $[\bar{\alpha}], \bar{\mathcal{S}}$ and $\bar{\mathcal{M}}$ satisfy the condition of this theorem.

Because (ii) implies (54), it has been pointed out in (a) that the inverse matrix (55) exists and is non-negative. Therefore, $\psi(\lambda)$ can be determined by (1), (4) (where α^a is replaced by $\bar{\alpha}^a$, $a = 1, 2$) and (33). Now we proceed to prove (2) and (3); that is, (35) and (36) where $[\alpha]$ is replaced by $[\bar{\alpha}]$ hold.

Since

$$\begin{aligned}(\mathcal{R}_\lambda \bar{\mathcal{H}}_\lambda + \mathcal{M}_\lambda \mathcal{U}_\lambda)[1] &= \mathcal{L}_\lambda^{-1}(\bar{\mathcal{H}}_\lambda + \bar{\mathcal{M}}\mathcal{U}_\lambda)[1] = \mathcal{L}_\lambda^{-1}\{\mathcal{L}_\lambda - (I - \bar{\mathcal{S}})\}[1] \\ &= [1] - \mathcal{L}_\lambda^{-1}(I - \bar{\mathcal{S}})[1]\end{aligned}$$

by (iv) and as $\mathcal{L}_\lambda^{-1} \geqslant 0$, it follows that (35) where $[\bar{\alpha}]$ is replaced by $[\alpha]$ holds.

Secondly,

$$\begin{aligned}I - \mathcal{R}_\lambda \bar{\mathcal{H}}_\lambda + \mathcal{M}_\lambda \mathcal{U}_\lambda + \mathcal{R}_\lambda \bar{\mathcal{H}}_\nu + \mathcal{M}_\lambda \mathcal{U}_\nu &= \mathcal{L}_\lambda^{-1}(\mathcal{L}_\lambda - \bar{\mathcal{H}}_\lambda - \bar{\mathcal{M}}\mathcal{U}_\lambda + \bar{\mathcal{H}}_\nu + \bar{\mathcal{M}}\mathcal{U}_\nu) \\ &= \mathcal{L}_\lambda^{-1}(I - \bar{\mathcal{S}} + \bar{\mathcal{H}}_\nu + \bar{\mathcal{M}}\mathcal{U}_\nu) = \mathcal{L}_\lambda^{-1}\mathcal{L}_\nu\end{aligned} \tag{59}$$

According, (37) and (38), where α^a and $\mathcal{H}_\nu$ are replaced by $\bar{\alpha}^a$ and $\bar{\mathcal{H}}_\nu$ respectively, hold, and it follows that (36) where $[\alpha]$ is replaced by $[\bar{\alpha}]$ holds.

(d) Now we proceed to prove that for (56) to have the form given by (21) it is necessary and sufficient that $\bar{M}^{11} = \bar{M}^{22} = 0$, $\bar{h}^{aa} < \infty$ $(a = 1, 2)$.

Suppose that (56) has the form given by (20) and (21). By Theorem 2 that is,

$$\mathcal{L}_\lambda^{-1}\{[\bar{\alpha}\phi(\lambda)] + \bar{\mathcal{M}}[X(\lambda)\mu]\} = (I - \mathcal{T}_\lambda)^{-1}[\alpha\phi(\lambda)] \tag{60}$$

Multiplying the above formula from the right by $(\lambda I - Q)$ we obtain

$$\mathcal{L}_\lambda^{-1}[\bar{\alpha}] = (I - \mathcal{T}_\lambda)^{-1}[\alpha] \tag{61}$$

Hence $\mathcal{L}_\lambda^{-1}\bar{\mathcal{M}} = 0, \bar{\mathcal{M}} = 0$, so that $\bar{M}^{11} = \bar{M}^{22} = 0$. By (61) it follows that $[\bar{\alpha}] = \mathcal{L}_\lambda(I - \mathcal{T}_\lambda)^{-1}[\alpha]$. Since $[\alpha^a, 1] \leqslant 1$, it follows that $\bar{h}^{aa} < \infty$ $(a = 1, 2)$. Conversely, if $\bar{M}^{11} = \bar{M}^{22} = 0$ and $\bar{h}^{aa} < \infty$ $(a = 1, 2)$ then $[F(\lambda)]$ in (56) obviously has the form given by (20) and (21).

(e) The necessary and sufficient condition for the process to be honest or to satisfy the system of forward equations is obvious. Thus the proof is concluded.

QED

5.11 THE CONDITION THAT $\alpha\phi(\lambda)\in l$

Suppose $\alpha \geqslant 0$. By (7.3) it follows that $\alpha\phi(\lambda)\in l$ is equivalent to

$$\sum_i \alpha_i[1 - X_i^1(\lambda) - X_i^2(\lambda)] < \infty \qquad \lambda > 0 \tag{1}$$

In this section, we shall give the necessary and sufficient condition under which we decide whether $\alpha\phi(\lambda)\in l$ directly from Q. Obviously, it suffices to consider $\sum_{i\geqslant 0}$ and $\sum_{i\leqslant 0}$ of (1). We shall only consider $\sum_{i\geqslant 0}$, for the case $\sum_{i\leqslant 0}$ is completely similar.

Lemma 1. We have

$$u_{1i}(\lambda) = u_{2i}(\lambda)\sum_{j\geqslant i}\frac{z_{j+1}-z_j}{u_{2j}(\lambda)u_{2,j+1}(\lambda)} \tag{2}$$

for $u_1(\lambda)$ and $u_2(\lambda)$ in Theorem 5.9.

Proof. Let the quantity determined by the right-hand side of (2) be $v_i(\lambda)$. We prove that $v(\lambda)$ is a decreasing solution of equation (5.1) and (5.14) holds, where $v(\lambda)$ is replaced by $u_1(\lambda)$.

First since $u_2(\lambda)$ and $u_2^+(\lambda)$ are increasing,

$$\begin{aligned}0 < v_i(\lambda) &= u_{2i}(\lambda)\sum_{j\geqslant i}\frac{z_{j+1}-z_j}{u_{2j}(\lambda)u_{2,j+1}(\lambda)} = u_{2i}(\lambda)\sum_{j\geqslant i}\frac{1}{u_{2j}^+(\lambda)}\left(\frac{1}{u_{2j}(\lambda)} - \frac{1}{u_{2,j+1}(\lambda)}\right)\\ &< \frac{u_{2i}(\lambda)}{u_{2i}^+(\lambda)}\sum_{j\geqslant i}\left(\frac{1}{u_{2j}(\lambda)} - \frac{1}{u_{2,j+1}(\lambda)}\right) = \frac{u_{2i}(\lambda)}{u_{2i}^+(\lambda)}\left(\frac{1}{u_{2i}(\lambda)} - \frac{1}{u_2(r_2,\lambda)}\right)\\ &\leqslant \frac{1}{u_{2i}^+(\lambda)} < \infty\end{aligned} \tag{3}$$

Therefore, the series (2) is convergent.

Secondly,

$$v_i^+(\lambda) = u_{2i}^+(\lambda)\sum_{j\geqslant i}\frac{z_{j+1}-z_j}{u_{2j}(\lambda)u_{2.j+1}(\lambda)} - \frac{1}{u_{2i}(\lambda)} \tag{4}$$

From (3) we see that

$$v_i^+(\lambda) < \frac{1}{u_{2i}(\lambda)} - \frac{1}{u_{2i}(\lambda)} = 0$$

Hence $v(\lambda)$ is decreasing. By (2) and (4) it follows that (5.14), where $v(\lambda)$ is replaced by $u_1(\lambda)$, holds. By (4) we obtain

$$D_\mu v_i^+(\lambda) = D_\mu u_{2i}^+(\lambda)\sum_{j\geqslant i}\frac{z_{j+1}-z_j}{u_{2j}(\lambda)u_{2,j+1}(\lambda)} = \lambda u_{2i}(\lambda)\sum_{j\geqslant i}\frac{z_{j+1}-z_j}{u_{2j}(\lambda)u_{2,j+1}(\lambda)} = \lambda v_i(\lambda)$$

that is, $v(\lambda)$ is a strictly decreasing positive solution of equation (5.1).

If r_2 is irregular, by Theorem 5.7, the decreasing solution of equation (5.1) satisfying (5.14) is unique. Hence $v(\lambda)=u_1(\lambda)$.

If r_2 is regular, according to Theorem 5.7

$$v_i(\lambda)=K[v_i-\theta s_i] \qquad (K \text{ is a constant})$$

By (2), we have $0=v(r_2,\lambda)=K[v(r_2)-\theta s(r_2)]$, so that

$$\theta=\frac{v(r_2)}{s(r_2)}=\theta$$

Hence $v(\lambda)=Ku_1(\lambda)$. Therefore, $v^+(\lambda)=Ku_1^+(\lambda)$. By (4) and the first expession of (6.8), we have

$$-\frac{1}{u_2(r_2,\lambda)}=v^+(r_2,\lambda)=Ku_1^+(r_2,\lambda)=-\frac{K}{u_2(r_2,\lambda)}$$

So $K=1$, and thus, $v(\lambda)=u_1(\lambda)$. The proof is completed. QED

Lemma 2. Let r_2 be regular or exit, then

$$\lim_{z_n\to r_2}\frac{u_{1n}(\lambda)}{r_2-z_n}=\frac{1}{u_2(r_2,\lambda)} \tag{5}$$

Proof. By (2)

$$u_{1n}(\lambda)<\frac{u_{2n}(\lambda)}{[u_{2n}(\lambda)]^2}\sum_{j\geqslant n}(z_{j+1}-z_j)=\frac{r_2-z_n}{u_{2n}(\lambda)}$$

$$u_{1n}(\lambda)>\frac{u_{2n}(\lambda)}{[u_2(r_2,\lambda)]^2}\sum_{j\geqslant n}(z_{j+1}-z_j)=\frac{u_{2n}(\lambda)(r_2-z_n)}{[u_2(r_2,\lambda)]^2} \tag{6}$$

That is,

$$\frac{u_{2n}(\lambda)}{[u_2(r_2,\lambda)]^2}<\frac{u_{1n}(\lambda)}{r_2-z_n}<\frac{1}{u_{2n}(\lambda)}$$

From this, we obtain (5), and the proof is completed. QED

Theorem 3. Let r_2 be regular, then

$$\sum_{i\geqslant 0}\alpha_i[1-X_i^1(\lambda)-X_i^2(\lambda)]<\infty \tag{8}$$

if and only if

$$\sum_{i\geqslant 0}\alpha_i(r_2-z_i)<\infty \tag{9}$$

or equivalently

$$\sum_{i\geqslant 0} \alpha_i N_i < \infty \tag{10}$$

where

$$N_i = \sum_{j\geqslant i} (z_{j+1} - z_j) \sum_{k=0}^{j} \mu_k = \left(\sum_{k=0}^{i} \mu_k\right)(r_2 - z_i) + \sum_{k\geqslant i+1} (r_2 - z_k)\mu_k \tag{11}$$

Proof. Since

$$1 - X_i^1(\lambda) - X_i^2(\lambda) = \lambda u_{1i}(\lambda) \sum_{j\leqslant i} u_{2j}(\lambda)\mu_j + \lambda u_{2i}(\lambda) \sum_{j>i} u_{1j}(\lambda)\mu_j \tag{12}$$

and when r_2 is regular we have $\sum_{j\geqslant 0} \mu_j < \infty$, from (5). Letting

$$0 < \frac{u_{2i}(\lambda)}{r_2 - z_i} \sum_{j>i} u_{1j}(\lambda)\mu_j < \frac{u_{2i}(\lambda)u_{1i}(\lambda)}{r_2 - z_i} \sum_{j>i} \mu_j \to 1 \times 0 = 0$$

Therefore

$$\lim_{i\to+\infty} \frac{1 - X_i^1(\lambda) - X_i^2(\lambda)}{r_2 - z_i} = \lim_{i\to+\infty} \frac{\lambda u_{1i}(\lambda)}{r_2 - z_i} \sum_{j\leqslant i} u_{2j}(\lambda)\mu_j = \lambda[X^2(\lambda)\mu, 1]$$

If r_2 is regular, then $0 < \lambda[X^2(\lambda)\mu, 1] < \infty$. From this it follows that (8) and (9) are equivalent. Secondly, (10) clearly implies (9). By (11)

$$N_i < (r_2 - z_i) \sum_{j\geqslant 0} \mu_j$$

Therefore, if r_2 is regular, then (9) also implies (10), and the proof is concluded.

Theorem 4. Suppose that r_2 is exit, then (8) and (10) are equivalent, and (8) or (10) implies (9).

Proof. By (6), (7) and (11), if $j \geqslant i \geqslant 0$

$$u_{1j}(\lambda) < \frac{r_2 - z_j}{u_{2j}(\lambda)} \leqslant \frac{r_2 - z_j}{u_{2i}(\lambda)} \leqslant \frac{r_2 - z_i}{u_{20}(\lambda)} \tag{13}$$

$$u_{1j}(\lambda) > \frac{u_{2j}(\lambda)}{[u_2(r_2, \lambda)]^2}(r_2 - z_j) \geqslant \frac{u_{2i}(\lambda)}{[u_2(r_2, \lambda)]^2}(r_2 - z_j) \tag{14}$$

$$\frac{u_{1j}(\lambda)}{N_j} < \frac{r_2 - z_j}{u_{2j}(\lambda)N_j} \leqslant \frac{\mu_0(r_2 - z_j)}{\mu_0 u_{20}(\lambda)N_j} < \frac{1}{u_{20}(\lambda)\mu_0} \tag{15}$$

If $i \geqslant 0$

$$u_{1i}(\lambda) \sum_{j=0}^{i} u_{2j}(\lambda)\mu_j \leqslant \frac{u_{1i}(\lambda)u_{2i}(\lambda)}{r_2 - z_i}\left(\sum_{j=0}^{i} \mu_j\right)(r_2 - z_i) \tag{16}$$

$$u_{1i}(\lambda) \sum_{j=0}^{i} u_{2j}(\lambda)\mu_j \geqslant \frac{u_{1i}(\lambda)u_{20}(\lambda)}{r_2 - z_i}\left(\sum_{j=0}^{i} \mu_j\right)(r_2 - z_i) \tag{17}$$

Hence by (12), (15), (16) and (13)

$$\frac{1-X_i^1(\lambda)-X_i^2(\lambda)}{N_i}$$

$$\leqslant \frac{\lambda}{u_{20}(\lambda)\mu_0}\sum_{i<0} u_{2j}(\lambda)\mu_j + \lambda\frac{u_{1i}(\lambda)u_{2i}(\lambda)}{r_2-z_i}\frac{\left(\sum_{j=0}^{i}\mu_j\right)(r_2-z_i)}{N_i} + \lambda\frac{u_{2i}(\lambda)}{N_i}\sum_{j>i}\frac{(r_2-z_j)\mu_j}{u_{20}(\lambda)}$$

$$\leqslant \frac{\lambda}{u_{20}(\lambda)\mu_0}\sum_{j<0} u_{2j}(\lambda)\mu_j + \lambda\frac{u_{1i}(\lambda)u_{2i}(\lambda)}{r_2-z_i} + \lambda\frac{u_{2i}(\lambda)}{u_{20}(\lambda)}$$

By (5)

$$\overline{\lim_{i\to+\infty}}\frac{1-X_i^1(\lambda)-X_i^2(\lambda)}{N_i} \leqslant \frac{\lambda}{u_{20}(\lambda)\mu_0}\sum_{j<0} u_{2j}(\lambda)\mu_j + \lambda + \lambda\frac{u_2(r_2,\lambda)}{u_{20}(\lambda)} < \infty \tag{18}$$

Similarly, by (12), (17) and (14)

$$\frac{1-X_i^1(\lambda)-X_i^2(\lambda)}{N_i}$$

$$> \lambda\frac{u_{1i}(\lambda)}{N_i}\sum_{j=0}^{i} u_{2j}(\lambda)\mu_j + \lambda\frac{u_{2i}(\lambda)}{N_i}\sum_{j>i} u_{1j}(\lambda)\mu_j$$

$$> \lambda\frac{u_{i1}(\lambda)u_{20}(\lambda)}{r_2-z_i}\frac{\sum_{j=0}^{i}\mu_j(r_2-z_i)}{N_i} + \frac{\lambda u_{2i}(\lambda)}{N_i}\sum_{j>i}\frac{u_{2i}(\lambda)}{[u_2(r_2,\lambda)]^2}(r_2-z_j)\mu_j$$

$$> \frac{\lambda u_{2i}(\lambda)u_{20}(\lambda)}{[u_2(r_2,\lambda)]^2}\frac{\sum_{j=0}^{i}\mu_j(r_2-z_i)}{N_i} + \frac{\lambda u_{2i}(\lambda)u_{20}(\lambda)}{[u_2(r_2,\lambda)]^2}\frac{\sum_{j>i}(r_2-z_j)\mu_j}{N_j}$$

$$= \lambda X_i^2(\lambda)X_0^2(\lambda)$$

and so

$$\underline{\lim_{i\to+\infty}}\frac{1-X_i^1(\lambda)-X_i^2(\lambda)}{N_i}$$

$$\geqslant \lambda X_0^2(\lambda) > 0 \tag{19}$$

Theorem 4 follows from (18) and (19), and the proof is terminated. QED

Theorem 5. If r_2 is entrance or natural, then (8) holds if and only if

$$\sum_{i\geqslant 0}\alpha_i < \infty \tag{20}$$

Proof. The sufficiency is obvious. The necessity follows from

$$\lim_{i\to+\infty}[1-X_i^1(\lambda)] = 1 - X^1(r_2,\lambda) > 0 \qquad \text{QED}$$

CHAPTER 6

Birth–Death Processes

6.1 INTRODUCTION

Let $E = \{0, 1, 2, 3, \ldots\}$. The matrix Q has the form

$$Q = \begin{pmatrix} -(a_0+b_0) & b_0 & 0 & 0 & \cdots \\ a_1 & -(a_1+b_1) & b_1 & 0 & \cdots \\ 0 & a_2 & -(a_2+b_2) & b_2 & \cdots \\ \vdots & \vdots & \vdots & \vdots & \end{pmatrix} \tag{1}$$

where $a_0 \geqslant 0, b_0 > 0, a_i > 0, b_i > 0\,(i > 0)$. We call the Q process of the form (1) a birth–death process. In this chapter, Q will always have the form (1), and the Q process will always be understood to refer to a birth–death process.

We point out that, for a matrix Q of the form (1), the relation

$$q_i = \sum_{j \neq i} q_{ij} \tag{2}$$

is satisfied when $a_0 = 0$, and is not satisfied for $i = 0$ when $a_0 > 0$. That is, Q is conservative when $a_0 = 0$ and is single non-conservative when $a_0 > 0$. The dimension of the solution space $\mathscr{M}_\lambda$ is $m^+ \leqslant 1$ when $a_0 \geqslant 0$. Therefore construction of the birth–death process in the case $a_0 = 0$ or in the case $a_0 > 0$ and $m^+ = 0$ is solved by Theorem 3.2.1 and Theorem 3.3.3. It is slightly more complex if $a_0 > 0$ and $m^+ = 1$, because in this case it is possible that there exist Q processes that satisfy the system of backward or forward equations, and also Q processes that satisfy neither the system of forward equations nor that of backward equations. Birth–death processes are regarded as diffusions in Feller (1957b, 1971) and all possible birth–death processes that satisfy simultaneously the system of backward and forward equations are found in the case $a_0 \geqslant 0$. The author (Xiang-qun Yang, 1965a) has constructed all possible birth-death processes.

This chapter is concerned with the construction theory of birth–death processes, which is very similar to the construction theory of bilateral birth–death processes dealt with in the previous chapter. For this reason we shall make simple statements or supply brief and simple verifications on some occasions.

6.2 CLASSIFICATION OF BOUNDARY POINT AND SECOND-ORDER DIFFERENCE OPERATOR

For the matrix Q of (1.1), we call

$$\begin{aligned} &z_0 = 1/a_0 \,(\text{if } a_0 > 0) \qquad z_0 = 0 \,(\text{if } a_0 = 0) \\ &z_1 = z_0 + 1/b_0 \\ &\vdots \\ &z_n = z_0 + \frac{1}{b_0} + \cdots + \frac{a_1 a_2 \cdots a_{n-1}}{b_0 b_1 b_2 \cdots b_{n-1}} \qquad (n = 2, 3, \ldots) \end{aligned} \tag{1}$$

the natural scale; we call

$$z = \lim_{n \to +\infty} z_n \tag{2}$$

the boundary point; and we call

$$\mu_0 = 1 \qquad \mu_n = \frac{b_0 b_1 \cdots b_{n-1}}{a_1 \cdots a_{n-1} a_n} \qquad (n > 1) \tag{3}$$

the canonical measure.

The boundary point z may be classified by means of the natural scale and the canonical measure. We say that the boundary point z is

(i) regular if $z < \infty, \sum_i \mu_i < \infty$;
(ii) exit if it is not regular and $\sum_i (z - z_i)\mu_i < \infty$;
(iii) entrance if it is not regular and $\sum_i z_i \mu_i < \infty$;
(iv) natural for all other cases.

We also introduce the characteristic numbers as follows:

$$\begin{aligned} &m_0 = 1/b_0 = (z_1 - z_0)\mu_0 \\ &m_i = \frac{1}{b_i} + \sum_{k=0}^{i-1} \frac{a_i a_{i-1} \cdots a_{i-k}}{b_i b_{i-1} \cdots b_{i-k} b_{i-k-1}} = (z_{i+1} - z_i) \sum_{k=0}^{i} \mu_k \qquad i > 0 \end{aligned} \tag{4}$$

$$\begin{aligned} &e_0 = z_0 \sum_{k=0}^{\infty} \mu_k \\ &e_i = \frac{1}{a_i} + \sum_{k=0}^{\infty} \frac{b_i b_{i+1} \cdots b_{i+k}}{a_i a_{i+1} \cdots a_{i+k} a_{i+k+1}} = (z_i - z_{i-1}) \sum_{k=i}^{\infty} \mu_k \qquad i > 0 \end{aligned} \tag{5}$$

$$N_i = \sum_{j=i}^{\infty} m_j = (z - z_i) \sum_{j=0}^{i} \mu_j + \sum_{j=i+1}^{\infty} (z - z_j)\mu_j \tag{6}$$

$$R = \sum_{j=0}^{\infty} m_j = \sum_{j=0}^{\infty} (z - z_j)\mu_j$$

$$S = \sum_{j=0}^{\infty} e_j = \sum_{j=0}^{\infty} z_j\mu_j \tag{7}$$

Theorem 1. The boundary point z is

(i) regular if and only if $R < \infty, S < \infty$;
(ii) exit if and only if $R < \infty, S = \infty$;
(iii) entrance if and only if $R = \infty, S < \infty$;
(iv) natural if and only if $R = \infty, S = \infty$.

Proof. It is similar to Theorem 5.3.1.

We can introduce the second-order difference operator by (5.4.1), but in the case $i = 0$. We shall make appropriate modifications.

Suppose that u is a column vector on E. We define u^+ as follows:

$$u_i^+ = \frac{u_{i+1} - u_i}{z_{i+1} - z_i} \qquad (i \geqslant 0) \tag{8}$$

For convenience, we will adopt the following convention from now on:

$$u_{-1}^+ = a_0 u_0 \qquad u_{-1} = 0 \tag{9}$$

Let u be a column vector on $\{-1\} \cup E$. We define $D_\mu u$ as follows:

$$D_\mu u_i = \frac{u_i - u_{i-1}}{\mu_i} \qquad (i \geqslant 0) \tag{10}$$

Theorem 2. If u is a column vector on E, then

$$Qu = D_\mu u^+ \tag{11}$$

That is,

$$a_i u_{i-1} - (a_i + b_i)u_i + b_i u_{i+1} = D_\mu u_i^+ \qquad i \geqslant 0 \tag{12}$$

Proof. It is similar to Theorem 5.4.1, but we must notice convention (9).

Theorem 3. Theorem 5.4.2, together with its corollary, and Lemma 5.4.3 are also true, and even their notation need not be changed, if only we understand E as referring to the set of non-negative integers and Q is the matrix of (1.1).

Lemma 4. The solutions of equations

$$\begin{aligned} &-(a_0 + b_0)u_0 + b_0 u_1 = -f_0 \\ &a_i u_{i-1} - (a_i + b_i)u_i + b_i u_{i+1} = -f_i \qquad 0 < i < n \\ &u_n = f_n \end{aligned} \tag{13}$$

are

$$u_i = \frac{z_n - z_i}{a_0(z_n - z_0) + 1} f_0 + \frac{a_0(z_i - z_0) + 1}{a_0(z_n - z_0) + 1} f_n$$

$$+ \frac{z_n - z_i}{a_0(z_n - z_0) + 1} \sum_{j=1}^{i-1} [a_0(z_j - z_0) + 1] f_j \mu_j$$

$$+ \frac{a_0(z_i - z_0) + 1}{a_0(z_n - z_0) + 1} \sum_{j=i}^{n-1} (z_n - z_j) f_j \mu_j \tag{14}$$

Proof. By (5.4.12)

$$u_i = \frac{z_n - z_i}{z_n - z_0} u_0 + \frac{z_i - z_0}{z_n - z_0} f_n + \frac{z_n - z_i}{z_n - z_0} \sum_{j=1}^{i-1} (z_j - z_0) f_j \mu_j$$

$$+ \frac{z_i - z_0}{z_n - z_0} \sum_{j=i}^{n-1} (z_n - z_j) f_j \mu_j \qquad 0 < i < n \tag{15}$$

In particular,

$$u_1 = \frac{z_n - z_1}{z_n - z_0} u_0 + \frac{z_1 - z_0}{z_n - z_0} f_n + \frac{z_1 - z_0}{z_n - z_0} \sum_{j=1}^{n-1} (z_n - z_j) f_j \mu_j \tag{16}$$

So by (13)

$$u_0 = \frac{b_0}{a_0 + b_0} u_1 + \frac{f_0}{a_0 + b_0} = \frac{u_1}{a_0(z_1 - z_0) + 1} + \frac{(z_1 - z_0) f_0}{a_0(z_1 - z_0) + 1} \tag{17}$$

By (16) and (17) we derive

$$u_0 = \frac{z_n - z_0}{a_0(z_n - z_0) + 1} f_0 + \frac{1}{a_0(z_n - z_0) + 1} f_n + \frac{1}{a_0(z_n - z_0) + 1} \sum_{j=1}^{n-1} (z_n - z_j) f_j \mu_j$$

Substituting the above expression into (15) we obtain (14), and the proof is completed. QED

Lemma 5. The solution of the equation

$$D_\mu u^+ = f \tag{18}$$

namely the system of equations

$$\begin{aligned} -(a_0 + b_0) u_0 + b_0 u_1 &= f_0 \\ a_i u_{i-1} - (a_i + b_i) u_i + b_i u_{i+1} &= f_i \qquad (i > 0) \end{aligned} \tag{19}$$

is

$$u_i = [a_0(z_i - z_0) + 1] u_0 + \sum_{j=0}^{i-1} (z_i - z_j) f_j u_j \tag{20}$$

Proof. By (19) we have

$$
\begin{aligned}
u_0^+ &= a_0 u_0 + f_0\mu_0 \\
u_i^+ - u_{i-1}^+ &= f_i\mu_i \qquad i>0
\end{aligned}
\tag{21}
$$

Therefore

$$u_i^+ = u_0^+ + \sum_{j=1}^{i}(u_j^+ - u_{j-1}^+) = a_0u_0 + f_0\mu_0 + \sum_{j=1}^{i} f_j\mu_j = a_0u_0 + \sum_{j=0}^{i} f_j\mu_j$$

$$
\begin{aligned}
u_i &= u_0 + \sum_{j=0}^{i-1}(u_{j+1}-u_j) = u_0 + \sum_{j=0}^{i-1} u_j^+(z_{j+1}-z_j) \\
&= u_0 + \sum_{j=0}^{i-1}\left(a_0u_0 + \sum_{k=0}^{i} f_k\mu_k\right)(z_{j+1}-z_j) \\
&= u_0 + a_0(z_i - z_0)u_0 + \sum_{j=0}^{i-1}(z_{j+1}-z_j)\sum_{k=0}^{j} f_k\mu_k \\
&= [a_0(z_i-z_0)+1]u_0 + \sum_{j=0}^{i-1}(z_i - z_j)f_j\mu_j
\end{aligned}
$$

QED

Corollary

The solution of the equation

$$Qu = 0 \tag{22}$$

is

$$u_i = [a_0(z_i - z_0)+1]u_0 \qquad i \geqslant 0 \tag{23}$$

6.3 SOLUTION OF THE EQUATION $\lambda u - D_\mu u^+ = 0$

Theorem 1. For every $\lambda > 0$, the solution $u(\lambda)$ of the equation

$$\lambda u - D_\mu u^+ = 0 \tag{1}$$

with condition $u_0 = 1$ exists and is unique, and has the following properties:

(i) Both $u(\lambda)$ and $u^+(\lambda)$ are strictly increasing;
(ii) $u(z,\lambda) \equiv \lim_{i\to\infty} u_i(\lambda) < \infty$ if and only if z is exit or regular;
(iii) $u(\lambda)\mu \in l$, i.e. $u^+(z,\lambda) \equiv \lim_{i\to\infty} u_i^+(\lambda) < \infty$ if and only if z is regular or entrance.

Proof. By (2.20) it follows that $u_i(\lambda)$ necessarily satisfies

$$u_i(\lambda) = 1 + a_0(z_i - z_0) + \sum_{j=0}^{i-1}(z_i - z_j)u_j(\lambda)\mu_j \tag{2}$$

From this we can determine $u_1(\lambda), u_2(\lambda), \ldots,$ and hence $u(\lambda)$ exists and is unique. By (2) it follows that $u(\lambda)$ is strictly increasing.

Secondly,

$$u_i^+(\lambda) = a_0 + \sum_{j=0}^{i} D_\mu u_j^+(\lambda)\mu_j = a_0 + \lambda \sum_{j=0}^{i} u_j(\lambda)\mu_j \tag{3}$$

So $u^+(\lambda)$ is strictly increasing.

The proof of (ii) and (iii) is similar to Lemma 5.4.5, and the proof is completed.

Theorem 2. Suppose that $u(\lambda)$ is the solution of (1) in Theorem 1; set

$$v_i(\lambda) = u_i(\lambda) \sum_{j=i}^{\infty} \frac{z_{j+} - z_j}{u_j(\lambda)u_{j+1}(\lambda)} \tag{4}$$

Then $v(\lambda)$ is strictly decreasing while $v^+(\lambda)$ is strictly increasing, and

$$\lambda v_i(\lambda) - D_\mu v_i^+(\lambda) = \begin{cases} 0 & \text{if } i > 0 \\ 1 & \text{if } i = 0 \end{cases} \tag{5}$$

$$u_i^+(\lambda)v_i(\lambda) - v_i^+(\lambda)u_i(\lambda) = 1 \qquad i \geqslant 0 \tag{6}$$

$$v^+(z, \lambda) \equiv \lim_{i\to\infty} v_i^+(\lambda) = -\frac{1}{u(z, \lambda)} \tag{7}$$

(The right-hand side is zero when $u(z, \lambda) = \infty$.)

Proof. As we did in Lemma 5.11.1, we can obtain the convergence of the series in (4) in a way similar to (5.11.3). Similarly to (5.11.4) we have

$$v_i^+(\lambda) = u_i^+(\lambda) \sum_{j\geq i} \frac{z_{j+1} - z_j}{u_j(\lambda)u_{j+1}(\lambda)} - \frac{1}{u_i(\lambda)} \tag{8}$$

$$v_i^+(\lambda) < \frac{1}{u_i(\lambda)} - \frac{1}{u_i(\lambda)} = 0 \tag{9}$$

Hence $v(\lambda)$ is strictly decreasing. From (8) we obtain (6). And again by (8)

$$\begin{aligned} -\frac{1}{u_i(\lambda)} \leqslant v_i^+(\lambda) &\leqslant \sum_{j=i}^{\infty} u_j^+(\lambda)\frac{z_{j+1} - z_j}{u_j(\lambda)u_{j+1}(\lambda)} - \frac{1}{u_i(\lambda)} \\ &= \sum_{j=i}^{\infty} \left(\frac{1}{u_j(\lambda)} - \frac{1}{u_{j+1}(\lambda)}\right) - \frac{1}{u_i(\lambda)} \\ &= -\frac{1}{u(z, \lambda)} \end{aligned} \tag{10}$$

From this follows (7). By (8) and (2.10) we obtain

$$D_\mu v_i^+(\lambda) = D_\mu u_i^+(\lambda) \sum_{j\geq i} \frac{z_{j+1} - z_j}{u_j(\lambda)u_{j+1}(\lambda)} = \lambda v_i(\lambda)$$

when $i > 0$, and we have proved the first line of (5).

To prove the second line of (5) note that by definition (4) of $v(\lambda)$ we have

$$v_1(\lambda) = v_0(\lambda)u_1(\lambda) - \frac{z_1 - z_0}{u_0(\lambda)} = v_0(\lambda)u_1(\lambda) - b_0^{-1} \tag{11}$$

As $u(\lambda)$ is the solution of (1), so

$$b_0 u_1(\lambda) = (\lambda + a_0 + b_0)u_0(\lambda) = \lambda + a_0 + b_0 \tag{12}$$

Substituting the above expression into (11) we get the second line of (5), and the proof is terminated. QED

6.4 CONSTRUCTION OF THE MINIMAL SOLUTION

For the solution $u(\lambda)$ of (3.1) and $v(\lambda)$ determined by (3.4),

$$\phi_{ij}(\lambda) = \begin{cases} u_i(\lambda)v_j(\lambda)\mu_j & \text{if } j \geqslant i \\ v_i(\lambda)u_j(\lambda)\mu_j & \text{if } j \leqslant i \end{cases} \tag{1}$$

Similarly to (5.6.1)–(5.6.5) we have

$$\mu_i \phi_{ij}(\lambda) = \mu_j \phi_{ji}(\lambda) \tag{2}$$

If f is a column vector and g is a row vector, then

$$[\phi(\lambda)f]_i = \sum_j \phi_{ij}(\lambda)f_j = v_i(\lambda)\sum_{j=0}^{i} u_j(\lambda)f_j u_j + u_i(\lambda)\sum_{j=i+1}^{\infty} v_j(\lambda)f_j\mu_j \tag{3}$$

$$[g\phi(\lambda)]_i = \sum_i g_i\phi_{ij}(\lambda) = v_j(\lambda)\mu_j \sum_{i=0}^{j} g_i u_i(\lambda) + u_j(\lambda)\mu_j \sum_{i=j+1}^{\infty} g_i v_i(\lambda) \tag{4}$$

If $g = f\mu$ then

$$g\phi(\lambda) = (\phi(\lambda)f)\mu \tag{5}$$

Theorem 1. The following holds[1]:

$$\lambda \sum_j \phi_{ij}(\lambda) = 1 - a_0 v_i(\lambda) - \frac{u_i(\lambda)}{u(z, \lambda)} \tag{6}$$

Proof. By (3) and recalling convention (2.9),

$$\lambda \sum_j \phi_{ij}(\lambda) = v_i(\lambda)\sum_{j=0}^{i} [u_j^+(\lambda) - u_{j-1}^+(\lambda)] + u_i(\lambda)\sum_{j=i+1}^{\infty} [v_j^+(\lambda) - v_{j-1}^+\lambda)]$$

$$= v_i(\lambda)u_i^+(\lambda) - u_i(\lambda)v_i^+(\lambda) - a_0 v_i(\lambda) + u_i(\lambda)v^+(z, \lambda)$$

On account of (3.6) and (3.7) the above expression is precisely (6). QED

[1] Suppose 0/00 = 0.

Theorem 2. If $f \in m, g \in l$ then $\phi(\lambda)f \in m, g\phi(\lambda) \in l$ and

$$\lambda[\phi(\lambda)f] - Q[\phi(\lambda)f] = f \qquad \lambda > 0 \tag{7}$$

$$\lambda[g\phi(\lambda)] - [g\phi(\lambda)]Q = g \qquad \lambda > 0 \tag{8}$$

Proof. We only prove (7). In the case $i > 0$ the proof is similar to Theorem 5.6.2. But in the case $i = 0$ we must take care. By (3) we obtain

$$[\phi(\lambda)f]_i^+ = v_i^+(\lambda) \sum_{j=0}^{i} u_j(\lambda) f_j \mu_j + u_i^+(\lambda) \sum_{j=i+1}^{\infty} v_j(\lambda) f_j \mu_j \tag{9}$$

$$[\phi(\lambda)f]_0^+ = v_0^+(\lambda) f_0 + u_0^+(\lambda) \sum_{j=1}^{\infty} v_j(\lambda) f_j \mu_j \tag{10}$$

Recall the convention (2.9),

$$[\phi(\lambda)f]_{-1}^+ = a_0[\phi(\lambda)f]_0 = a_0 v_0(\lambda) f_0 + a_0 u_0(\lambda) \sum_{j=1}^{\infty} v_i(\lambda) f_j \mu_j$$

$$= v_{-1}^+(\lambda) f_0 + u_{-1}^+(\lambda) \sum_{j=1}^{\infty} v_j(\lambda) f_j \mu_j \tag{11}$$

By (10), (11) and (3.5)

$$D_\mu[\phi(\lambda)f]_0^+ = D_\mu v_0^+(\lambda) f_0 + D_\mu u_0^+(\lambda) \sum_{j=1}^{\infty} v_j(\lambda) f_j \mu_j$$

$$= [\lambda v_0(\lambda) - 1] f_0 + \lambda u_0(\lambda) \sum_{j=1}^{\infty} v_j(\lambda) f_j \mu_j$$

$$= [\phi(\lambda)f]_0 - f_0$$

QED

Lemma 3. Suppose that $f \in m$, and z is regular or exit. Then

$$[\phi(\lambda)f](z) = \lim_{i \to \infty} [\phi(\lambda)f]_i = 0 \tag{12}$$

Proof. Because if z is regular or exit, then $z < \infty$ and $u(z, \lambda) < \infty$. By the definition (3.4) of $v(\lambda)$ we have $\lim_{i \to \infty} v_i(\lambda) = v(z, \lambda) = 0$. By (6) we obtain $[\lambda\phi(\lambda)\mathbf{1}](z) = 0$. From this follows (12). QED

Theorem 4. $\phi(\lambda)$ is the minimal Q process. For $\phi(\lambda)$ to be honest it is necessary and sufficient that $a_0 = 0$ and z is entrance or natural.

If z is natural, then $\phi(\lambda)$ is the unique Q process satisfying the system of either forward or backward equations.

If z is entrance, then $\phi(\lambda)$ is the unique Q process satisfying the system of backward equations.

If z is exit, then $\phi(\lambda)$ is the unique Q process satisfying the system of forward equations.

Proof. From Theorem 1 the norm condition of $\phi(\lambda)$ follows. By Theorem 2 we obtain the B condition and F condition for $\phi(\lambda)$. To prove the resolvent equation for $\phi(\lambda)$ let $f \in m$ and $F(\lambda) = \phi(\lambda)f$. Then $F(\lambda) - F(v) + (\lambda - v)\phi(\lambda)F(v) \in m$ is the solution of equation (3.1). Consequently,

$$F(\lambda) - F(v) + (\lambda - v)\phi(\lambda)F(v) = cu(\lambda) \tag{13}$$

where c is a constant. If z is regular or exit, by Lemma 3 we get $cu(z, \lambda) = 0$ and $c = 0$ as $i \to \infty$ in the above expression. If z is entrance or natural, then because the left-hand side of (13) is bounded and $u(\lambda)$ is unbounded, we get $c = 0$. Hence we always have $c = 0$. Take $f_i = \delta_{ij}$ and obtain the resolvent equation of $\phi(\lambda)$. Thus, $\phi(\lambda)$ is the Q process satisfying the system of backward and forward equations.

Suppose that $\psi(\lambda)$ is an arbitrary Q process. By the backward inequality (2.8.10) and Theorem 2.7.3 it follows that for fixed $j, u_i \equiv \psi_{ij}(\lambda) - \phi_{ij}(\lambda)$ satisfy the following equation:

$$\lambda u_i - \sum_k q_{ik} u_k = \begin{cases} c_1 \geqslant 0 & \text{if } i = 0 \\ 0 & \text{if } i > 0 \end{cases} \tag{14}$$

Hence $u_i - c_1 v_i(\lambda)$ is a solution of equation (3.1), so that $u_i - c_1 v_i(\lambda) = c_2 u_i(\lambda)$; that is,

$$\psi_{ij}(\lambda) = \phi_{ij}(\lambda) + c_1 v_i(\lambda) + c_2 u_i(\lambda) \tag{15}$$

where c_1 and c_2 are constants which are independent of i, and $c_1 \geqslant 0$.

If z is exit or regular, by Lemma 3 and $v(z, \lambda) = 0$, which is proved in Lemma 3, we obtain $c_2 u(z, \lambda) \geqslant 0$ as $i \to \infty$ in (15) and, hence, $c_2 \geqslant 0$. If z is entrance or natural, as the left-hand side of (15) is bounded, and only $u(\lambda)$ is unbounded in the right-hand side of (15) it follows that $c_2 = 0$. Thus, we always have $c_2 \geqslant 0$, that is $\psi(\lambda) \geqslant \phi(\lambda)$. We have proved the minimality of $\phi(\lambda)$.

If $\psi(\lambda)$ satisfies the system of backward equations, then $u_1 = \psi_{ij}(\lambda) - \phi_{ij}(\lambda)$ is a solution of equation (3.1), so that

$$\psi_{ij}(\lambda) = \phi_{ij}(\lambda) + cu_i(\lambda) \tag{16}$$

where the constant $c \geqslant 0$ is independent of i. If z is entrance or natural, we can prove $c = 0$ in the same way as was used in the proof of the minimality of $\phi(\lambda)$; hence $\phi(\lambda)$ is the unique Q process satisfying the system of backward equations.

If $\psi(\lambda)$ satisfies the system of forward equations, then $v_j = \psi_{ij}(\lambda) - \phi_{ij}(\lambda)$ is a solution of the equation

$$\lambda v - vQ = 0 \tag{17}$$

whereas $u(\lambda)\mu$ is the unique linearly independent solution of (17), so that

$$\psi_{ij}(\lambda) = \phi_{ij}(\lambda) + cu_j(\lambda)\mu_j \tag{18}$$

where c is a constant independent of j. If z is exit or natural, then by the norm condition of $\psi(\lambda)$ and $\phi(\lambda)$ we obtain $c = 0$. Hence $\phi(\lambda)$ is the unique Q process satisfying the system of forward equations, and the proof is concluded.

Remark

If $a_0 = 0$ then $c_1 = 0$ in (14). Therefore by the proof of this theorem it can be seen that any Q process $\psi(\lambda)$ has the following form:

$$\psi_{ij}(\lambda) = \phi_{ij}(\lambda) + a_0 v_i(\lambda) F_j^1(\lambda) + \frac{u_i(\lambda)}{u(z,\lambda)} F_j^2(\lambda) \tag{19}$$

where $F^a(\lambda) \geqslant 0$. Here $\psi(\lambda)$ satisfies the system of backward equations if and only if $F^1(\lambda) = 0$; and $\psi(\lambda)$ satisfies the system of forward equations if and only if $\psi(\lambda)$ has the form (18).

6.5 SEVERAL LEMMAS

From now on, we will simply write

$$X_i^1(\lambda) = a_0 v_i(\lambda) \qquad X_i^2(\lambda) = \frac{u_i(\lambda)}{u(z,\lambda)} \tag{1}$$

$$X_i^1 = \frac{a_0(z - z_i)}{a_0(z - z_0) + 1} \qquad X_i^2 = \frac{a_0(z_i - z_0) + 1}{a_0(z - z_0) + 1} \tag{2}$$

Obviously, $X^1 + X^2 = 1$, and (4.6) becomes

$$\lambda\phi(\lambda)\mathbf{1} = 1 - X^1(\lambda) - X^2(\lambda) \tag{3}$$

Lemma 1. $X^a(\lambda)\,(a = 1, 2)$ are two exit families and

$$X^1(\lambda)\downarrow 0 \qquad \lambda X_i^1(\lambda) \to \begin{cases} 0 & \text{if } i > 0 \\ a_0 & \text{if } i = 0 \end{cases} \quad (\lambda \uparrow \infty) \tag{4}$$

$$X_2(\lambda)\downarrow 0 \qquad \lambda X_2(\lambda) \to 0 \qquad (\lambda \uparrow \infty) \tag{5}$$

$$\lambda\phi(\lambda)X^a = X^a - X^a(\lambda) \qquad a = 1, 2 \tag{6}$$

Proof. It suffices to prove (6). When $a_0 = 0$, (6) is obvious for $a = 1$. For $a = 2$, (6) follows from (3). In what follows, we suppose $a_0 > 0$.

If $z = \infty$, then $X^2 = X^2(\lambda) = 0, X^1 = 1$, by (3) we also get (6).

If $z < \infty$, then

$$\begin{aligned}
\lambda \sum_j \phi_{ij}(\lambda)(z - z_j) \\
&= v_i(\lambda) \sum_{j=0}^{i} \lambda u_j(\lambda)\mu_j \sum_{k=j}^{\infty} (z_{k+1} - z_k) \\
&\quad + u_i(\lambda) \sum_{j=i+1}^{\infty} \lambda v_j(\lambda)\mu_j \sum_{k=j}^{\infty} (z_{k+1} - z_k) \\
&= v_i(\lambda)\left(\sum_{k=0}^{i} (z_{k+1} - z_k) \sum_{j=0}^{k} \lambda u_j(\lambda)\mu_j \right. \\
&\quad \left. + \sum_{k=i+1}^{\infty} (z_{k+1} - z_k) \sum_{j=0}^{i} \lambda u_j(\lambda)\mu_j \right) \\
&\quad + u_i(\lambda) \sum_{k=i+1}^{\infty} (z_{k+1} - z_k) \sum_{j=i+1}^{k} \lambda v_j(\lambda)\mu_j \\
&= v_i(\lambda)\left(\sum_{k=0}^{i} (z_{k+1} - z_k)[u_k^+(\lambda) - u_{-1}^+(\lambda)] \right. \\
&\quad \left. + \sum_{k=i+1}^{\infty} (z_{k+1} - z_k)[u_i^+(\lambda) - u_{-1}^+(\lambda)] \right) \\
&\quad + u_i(\lambda) \sum_{k=i+1}^{\infty} (z_{k+1} - z_k)[v_k^+(\lambda) - v_i^+(\lambda)] \\
&= v_i(\lambda)\left(\sum_{k=0}^{i} [u_{k+1}(\lambda) - u_k(\lambda)] + u_i^+(\lambda)(z - z_{i+1}) \right. \\
&\quad \left. - u_{-1}^+(\lambda)(z - z_0) \right) \\
&\quad + u_i(\lambda)\left(\sum_{k=i+1}^{\infty} [v_{k+1}(\lambda) - v_k(\lambda)] - v_i^+(\lambda)(z - z_{i+1}) \right) \\
&= v_i(\lambda)[u_i(\lambda) - u_0(\lambda) + u_i^+(\lambda)(z - z_i) - a_0 u_0(\lambda)(z - z_0)] \\
&\quad + u_i(\lambda)[v(z, \lambda) - v_i(\lambda) - v_i^+(\lambda)(z - z_i)] \\
&= [u_i^+(\lambda)v_i(\lambda) - u_i(\lambda)v_i^+(\lambda)](z - z_i) - [a_0(z - z_0) + 1]v_i(\lambda) \\
&= (z - z_i) - [a_0(z - z_0) + 1]v_i(\lambda) \qquad (7)
\end{aligned}$$

Multiplying both sides by $a_0/[a_0(z - z_0) + 1]$ we obtain (6) for $a = 1$. From this and (3), (6) follows for $a = 2$ and this completes the proof. QED

Lemma 2. (i) Let X^0 and $\bar{X}$ be the maximal passive solution and the maximal exit solution in Definition 2.11.3, respectively. Then when z is regular or exit, $X^0=0, \bar{X}=X^2$; when z is entrance or natural, $X^0=X^2, \bar{X}=0$.

(ii) The standard image of $X^1(\lambda)$ is X^1. If z is regular or exit, then the standard image of $X^2(\lambda)$ is X^2.

Proof. By the corollary to Lemma 2.5 it follows that the unique linearly independent solution of equation (2.11.18) is X^2. But both X^0 and $\bar{X}$ are solutions of (2.11.18); hence $X^0=cX^2, \bar{X}=\bar{c}X^2$ (c and $\bar{c}$ are constants).

Suppose z is exit or regular, then $X^2 \neq 0$. By (6)

$$\begin{aligned} \lambda\phi(\lambda)X^0 &= c[X^2-X^2(\lambda)] = X^0 - cX^2(\lambda) \\ \lambda\phi(\lambda)\bar{X} &= \bar{c}[X^2-X^2(\lambda)] = \bar{X} - \bar{c}X^2(\lambda) \end{aligned} \tag{8}$$

Comparing (2.11.17) with (2.11.19), we get $c=0$, and $\bar{c}=1$.

Suppose z is entrance or natural. If $a_0 z=\infty$, then $X^2=0$ and hence $X^0=X^2, \bar{X}=0$. If $a_0 z<\infty$ then $X^2 \neq 0$. And because $X^2(\lambda)=0$ the solution space $\mathscr{M}^+_\lambda(1)$ is a null space, so the standard image of $\bar{X}(\lambda)=0$ is $\bar{X}=0$ while (6) becomes

$$\lambda\phi(\lambda)X^2 = X^2$$

By the maximality of X^0 in Lemma 2.11.2 we obtain $X^2 \leqslant X^0 = cX^2$, so that $c \geqslant 1$. But obviously, $X^0_i = cX^2_i \leqslant 1$, and we obtain $c \leqslant 1$ by setting $r \to \infty$. Hence $c=1, X^0=X^2$.

By (6) for the standard image $\bar{X}^2$ of $X^2(\lambda)$ we have $\bar{X}^2 \leqslant X^2$. As is the case with Lemma 5.7.5, $u=X^2-\bar{X}^2$ is the solution of equation (2.11.18) satisfying (5.7.21). By the maximality of X^0 we obtain $X^2-\bar{X}^2 \leqslant X^0$. When z is regular or exit $X^0=0, X^2=\bar{X}^2$. That is, the standard image of $X^2(\lambda)$ is X^2; hence the standard image of $X^1(\lambda)$ is X^1.

If z is entrance or natural, from (2.11.13) follows that the standard image of $X^1(\lambda)$ is $\bar{X}^1 = 1 - X^0 - \bar{X} = X^1$ and the proof is concluded. QED

Lemma 3. Set

$$\bar{\eta}_i = \begin{cases} X^2_i \mu_i & \text{if } z \text{ is regular} \\ [a_0(z_i - z_0)+1]\mu_i & \text{if } z \text{ is entrance} \end{cases} \tag{9}$$

$$\bar{\eta}_j(\lambda) = \begin{cases} X^2_i(\lambda)\mu_i & \text{if } z \text{ is regular} \\ a_0 u_i(\lambda)\mu_i/u^+(z,\lambda) & \text{if } z \text{ is entrance} \end{cases}$$

Then $\bar{\eta}(\lambda) \in \mathscr{L}^+_\lambda$ $(\lambda>0)$ is an entrance family, and

$$\lambda\bar{\eta}\phi(\lambda) = \bar{\eta} - \bar{\eta}(\lambda) \tag{10}$$

Proof. We need only prove (10). By (6) and (4.5) we derive (10) when z is regular.

If z is entrance, then $u(z,\lambda)=\infty, u^+(z,\lambda)<\infty$. By (3.7)

$$v^+(z,\lambda)=0 \tag{11}$$

Furthermore

$$v(z,\lambda)=\frac{1}{u^+(z,\lambda)} \tag{12}$$

In fact, notice that $u(\lambda)$ is increasing and $v(\lambda)$ is decreasing, so

$$0\leqslant -u_i(\lambda)v_i^+(\lambda)=u_i(\lambda)[v^+(z,\lambda)-v_i^+(\lambda)]=u_i(\lambda)\sum_{k=i+1}^{\infty}\lambda v_k(\lambda)\mu_k$$

$$\leqslant v_0(\lambda)\sum_{k=i+1}^{\infty}\lambda u_k(\lambda)\mu_k=v_0(\lambda)[u^+(z,\lambda)-u_i^+(\lambda)]\to 0 \qquad (i\to\infty)$$

From this and (3.6) follows (12).

Secondly, by noting (11) and (12),

$$\begin{aligned}
&\lambda\sum_k \phi_{ik}(\lambda)(z_k-z_0)\\
&=v_i(\lambda)\sum_{k=0}^{i}\lambda\mu_k(\lambda)\mu_k\sum_{j=0}^{k-1}(z_{j+1}-z_j)+u_i(\lambda)\sum_{k=i+1}^{\infty}\lambda v_k(\lambda)\mu_k\sum_{j=0}^{k-1}(z_{j+1}-z_j)\\
&=v_i(\lambda)\sum_{j=0}^{i-1}(z_{j+1}-z_j)\sum_{k=i+1}^{\infty}\lambda u_k(\lambda)\mu_k\\
&\quad+u_i(\lambda)\left(\sum_{j=0}^{i}(z_{j+1}-z_j)\sum_{k=i+1}^{\infty}\lambda v_k(\lambda)\mu_k+\sum_{i=j+1}^{\infty}(z_{j+1}-z_i)\sum_{k=i+1}^{\infty}\lambda v_k(\lambda)\mu_k\right)\\
&=v_i(\lambda)\sum_{j=0}^{i-1}(z_{j+1}-z_j)[u_i^+(\lambda)-u_j^+(\lambda)]\\
&\quad+u_i(\lambda)\left(\sum_{j=0}^{i}(z_{j+1}-z_j)[v^+(z,\lambda)-v_i^+(\lambda)]\right.\\
&\quad\left.+\sum_{j=i+1}^{\infty}(z_{j+1}-z_j)[v^+(z,\lambda)-v_i^+(\lambda)]\right)\\
&=v_i(\lambda)\{u_i^+(\lambda)(z_i-z_0)-[u_i(\lambda)-u_0(\lambda)]\}\\
&\quad+u_i(\lambda)\{-v_i^+(\lambda)(z_{j+1}-z_0)-[v(z,\lambda)-v_{i+1}(\lambda)]\}\\
&=v_i(\lambda)[u_i^+(\lambda)(z_i-z_0)-u_i(\lambda)+1]\\
&\quad+u_i(\lambda)\left(-v_i^+(\lambda)(z_i-z_0)-\frac{1}{u^+(z,\lambda)}+v_i(\lambda)\right)
\end{aligned}$$

$$= [u_i^+(\lambda)v_i(\lambda) - u_i(\lambda)v_i^+(\lambda)](z_i - z_0) + v_i(\lambda) - \frac{u_i(\lambda)}{u^+(z,\lambda)}$$

$$= (z_i - z_0) - v_i(\lambda) - \frac{u_i(\lambda)}{u^+(z,\lambda)}$$

Hence by (4.5) and (6) follows (10), and the proof is terminated. QED

Lemma 4. For a family $\eta(\lambda)\,(\lambda > 0)$ to be an entrance family it is necessary and sufficient that it has the following Riesz expression:

$$\eta(\lambda) = \alpha\phi(\lambda) + d\bar{\eta}(\lambda) \tag{13}$$

where row vector $\alpha \geqslant 0$ is such that $\alpha\phi(\lambda) \in l$, the constant $d \geqslant 0$, and $d = 0$ if z is entrance or natural. Also

$$\bar{\eta}(\lambda) = \begin{cases} X^2(\lambda)\mu & \text{if } z \text{ is regular} \\ a_0 u(\lambda)\mu/u^+(z,\lambda) & \text{if } z \text{ is entrance} \end{cases} \tag{14}$$

Proof. Since $\mathscr{L}_\lambda^+$ contains a non-null entrance family $\bar{\eta}(\lambda)$ if and only if z is regular or entrance, this lemma is a particular case of Lemma 2.11.3. QED

Lemma 5. If z is regular, then

$$U_\lambda^a = \lambda[X^2(\lambda)\mu, X^a] \uparrow U^a \qquad \lambda \uparrow \infty \tag{15}$$

where

$$U^1 = \frac{a_0}{a_0(z - z_0) + 1} \qquad U^2 = +\infty \tag{16}$$

Proof. Now (2.11.40) becomes

$$\begin{aligned}\lambda[X^2(\lambda)\mu, X^a] - \nu[X^2(\nu)\mu, X^a] &= (\lambda - \nu)[X^2(\lambda)\mu, X^a(\nu)] \\ &= (\lambda - \nu)[X^2(\nu)\mu, X^a(\lambda)] \qquad \lambda, \nu > 0\end{aligned} \tag{17}$$

From this it follows that U_λ^a is monotone.

Next, by (6) and (4.9) we have

$$\begin{aligned}[\lambda\phi(\lambda)X^a]_i^+ &= \lambda v_i^+(\lambda) \sum_{j=0}^{i} u_j(\lambda)X_j^a\mu_j + \lambda u_i^+(\lambda) \sum_{j=i+1}^{\infty} v_j(\lambda)X_j^a\mu_j \\ &= [X^a]_i^+ - [X^a(\lambda)]_i^+\end{aligned}$$

Noticing (3.7) and letting $i \to \infty$ in the above expression we obtain

$$-U_\lambda^a = [X^a]^+(z) - [X^a(\lambda)]^+(z)$$

From (2) we have

$$[X^1]^+(z) = -\frac{a_0}{a_0(z-z_0)+1}$$

$$[X^2](z) = \frac{a_0}{a_0(z-z_0)+1}$$

Therefore to prove this lemma, we need only prove

$$\lim_{\lambda\to\infty} [X^2(\lambda)]^+(z) = \infty \qquad \lim_{\lambda\to\infty} (X^1(\lambda)]^+(z) = 0 \tag{18}$$

This is similar to the proof of (5.7.18) and (5.7.19). Thus we complete the proof of this lemma. QED

6.6 CONSTRUCTION OF THE Q PROCESSES SATISFYING THE BACKWARD EQUATIONS

By Theorem 4.4 it follows that the Q process satisfying the system of backward equations is unique if z is entrance or natural. Hence we will suppose that z is exit or regular. In that case $z < \infty, u(z, \lambda) < \infty$.

By the remark of Theorem 4.4 it follows that the Q process satisfies the system of backward equations if and only if $\psi(\lambda)$ has the form

$$\psi_{ij}(\lambda) = \phi_{ij}(\lambda) + X_i^2(\lambda)F_j(\lambda) \tag{1}$$

where the row vector $F(\lambda) \geqslant 0$. Hence the construction problem of the Q processes satisfying the system of backward equations has already been contained in Theorem 3.2.1. Certainly, in the case of birth–death processes we will obtain a more specific and simpler form.

In this case, in Theorem 3.2.1, $H = \{0\}, \bar{\eta}(\lambda)$ becomes $dX^2(\lambda)\mu$, while the constant $d \geqslant 0$, and $d = 0$ if z is exit, where

$$\bar{V}^1 = dU^1 = \frac{da_0}{a_0(z-z_0)+1}$$

Note that $X^0 = 0, \bar{X} = X^2$ follows from Lemma 5.2. Therefore Theorem 3.2.1 takes the following form.

Theorem 1. Let z be regular or exit. For $\psi(\lambda)$ to be a Q process satisfying the system of backward equations it is necessary and sufficient that either $\psi(\lambda) = \phi(\lambda)$ or $\psi(\lambda)$ can be obtained as follows: Select a row vector $a \geqslant 0$ such that $\alpha\phi(\lambda) \in l$, take a constant $d \geqslant 0$, and $d = 0$ if z is exit. Moreover

$$\eta(\lambda) = \alpha\phi(\lambda) + dX^2(\lambda)\mu \neq 0 \tag{2}$$

Take a constant c so that

$$c \geqslant [\alpha, X^1] + \frac{da_0}{a_0(z - z_0) + 1} \tag{3}$$

is satisfied, where X^a $(a = 1, 2)$ are determined by (5.2). Finally set

$$\psi_{ij}(\lambda) = \phi_{ij}(\lambda) + X_i^2(\lambda) \frac{\sum_k \alpha_k \phi_{kj}(\lambda) + dX_j^2(\lambda)\mu_j}{c + [\alpha, X^2 - X^2(\lambda)] + d\lambda[X^2(\lambda)\mu, X^2]} \tag{4}$$

For the process $\psi(\lambda)$ to be honest it is necessary and sufficient that $a_0 = 0$, and $c = 0$. For the process $\psi(\lambda)$ to satisfy the system of forward equations it is necessary and sufficient that $\alpha = 0$.

6.7 CONSTRUCTION OF THE Q PROCESSES SATISFYING THE FORWARD EQUATIONS

Because of Theorem 4.4, we need only consider the case that z is regular or entrance. Since the Q process $\psi(\lambda)$ satisfies the system of forward equations if and only if the $\psi(\lambda)$ has the form (4.18), that is

$$\psi_{ij}(\lambda) = \phi_{ij}(\lambda) + F_i(\lambda)\bar{\eta}_j(\lambda) \tag{1}$$

where $\bar{\eta}(\lambda)$ is determined by (5.14), the problem of construction of the Q processes satisfying the system of forward equations is solved in Theorem 3.4.1.

First let us suppose that z is regular. In this case, $\bar{\eta}(\lambda) = X^2(\lambda)\mu$. According to the notation in Theorem 3.4.1, either $\bar{\xi}(\lambda) = 0$ or $\bar{\xi}(\lambda) = X^2(\lambda)$. If $\bar{\xi}(\lambda) = 0$, then

$$\bar{W}_\lambda = \lambda[\bar{\eta}(\lambda), \bar{X}] = U_\lambda^2 \uparrow U^2 = +\infty$$

This is in contradiction with (3.4.3). Therefore it necessarily follows that $\bar{\xi}(\lambda) = X^2(\lambda)$, and hence $\delta > 0$. According to the same consideration, the k in (3.4.3) must be δ^{-1}. By (3.4.4) we obtain $\beta^0 \leqslant \delta$. In addition,

$$\bar{V}^0 = U^1 = \frac{a_0}{a_0(z - z_0) + 1} \qquad X^0 = 0$$

Therefore if we replace β^0 by the constant β, then Theorem 3.4.1 takes the following form.

Theorem 1. Let z regular. For $\psi(\lambda)$ to be a Q process satisfying the system of forward equations it is necessary and sufficient that either $\psi(\lambda) = \phi(\lambda)$ or $\psi(\lambda)$ is obtained as follows: Choose a non-negative constant β and a positive constant $\delta, \beta \leqslant \delta$, and a constant c that satisfies

$$\frac{(\delta - \beta)a_0}{a_0(z - z_0) + 1} \leqslant c \tag{2}$$

Set

$$\psi_{ij}(\lambda)=\phi_{ij}(\lambda)+\frac{[\beta X_i^1(\lambda)+\delta X_i^2(\lambda)]X_j^2(\lambda)\mu_j}{c+\lambda[X^2(\lambda)\mu,\beta X^1+\delta X^2]} \tag{3}$$

For the process $\psi(\lambda)$ to be honest it is necessary and sufficient that $a_0>0,\beta=\delta,c=0$ or $a_0=0,c=0$. For the process $\psi(\lambda)$ to satisfy the system of backward equations it is necessary and sufficient and $a_0=0$ or $a_0>0,\beta=0$.

When z is entrance, the construction of the Q processes satisfying the system of forward equations, of course, can be deduced from Theorem 3.4.1. We shall, however, proceed as follows.

We assume that z is entrance or natural, whence $u(z,\lambda)=\infty$. If $a_0=0$, then the minimal solution is honest, and the Q process is unique. Consequently we further assume that $a_0>0$. Thus, (5.3) becomes

$$\lambda\phi(\lambda)\mathbf{1}=\mathbf{1}-X^1(\lambda)$$

By (4.19), an arbitrary Q process has the form

$$\psi_{ij}(\lambda)=\phi_{ij}(\lambda)+X_i^1(\lambda)F_j(\lambda) \tag{4}$$

Hence, in this case, the construction of the Q process is a special case of Theorem 3.3.3.

On account of the decreasing property of $v(\lambda)$ it follows that

$$\sup_i X_i^1(\lambda)=a_0v_0(\lambda)<1 \tag{5}$$

Therefore, if z is natural, then is null entrance, by Theorem 3.3.3 the Q process is unique.

If z is entrance, by Theorem 3.3.3 it follows that every non-minimal Q process is obtained by Theorem 3.3.1. According to the notation in Theorem 3.3.1, in this case

$$\bar{n}(\lambda)=d\frac{u(\lambda)\mu}{u^+(z,\lambda)}$$

and constant $d\geqslant 0$, whereas

$$Y_\lambda=\lambda[\bar{n}(\lambda),\mathbf{1}]=\frac{d}{u^+(z,\lambda)}\sum_i\lambda u_i(\lambda)\mu_i$$

$$=\frac{d}{u^+(z,\lambda)}[u^+(z,\lambda)-u^+_{-1}(\lambda)]=d-\frac{da_0}{u^+(z,\lambda)}$$

Therefore $Y\leqslant d<\infty$. By (3.3.5), if $\alpha\neq 0$ then it must follow that $[\alpha,1]=\infty$. By (5) and Lemma 3.3.2 there does not exit such a row vector α. Therefore it is certain that $\alpha=0$. Hence $d>0$. Thus the Q process of Theorem 3.3.2 satisfies the system of forward equations. If we write $\bar{c}$ for $(c-\bar{\sigma}^0/d)$ in Theorem 3.3.1,

$$c+d\lambda[\bar{n}(\lambda),\mathbf{1}-X^0]=d(\bar{c}+\lambda[\bar{n}(\lambda),\mathbf{1}])$$

If we still write also c for $\bar{c}$, then we obtain the form of Theorem 3.3.3 as follows.

Theorem 2. If z is natural, $a_0\geqslant 0$, then the Q process is unique. Assume that z is entrance. Then the Q process is unique if $a_0=0$; if $a_0>0$ every Q process $\psi(\lambda)$ satisfies the system of forward equations and, moreover, either $\psi(\lambda)=\phi(\lambda)$ or

$$\psi_{ij}(\lambda)=\phi_{ij}(\lambda)+\frac{X_i^1(\lambda)\bar{n}_j(\lambda)}{c+\lambda[\bar{n}(\lambda),\mathbf{1}]} \tag{6}$$

where constant $c\geqslant 0, \bar{n}(\lambda)=u(\lambda)\mu/u^+(z,\lambda)$. The process $\psi(\lambda)$ is honest if and only if $c=0$.

6.8 CONSTRUCTION OF THE Q PROCESSES SATISFYING NEITHER BACKWARD NOR FORWARD EQUATIONS

When z is entrance or natural it has already been investigated in the preceding section. We assume that z is regular or exit in this section. Since when $a_0=0$ any Q process satisfies the system of backward equations, we assume further that $a_0>0$.

By (4.19), each Q process $\psi(\lambda)$ has the form

$$\psi_{ij}(\lambda)=\phi_{ij}(\lambda)+X_i^1(\lambda)F_j^1(\lambda)+X_i^2(\lambda)F_j^2(\lambda) \tag{1}$$

where $F^a(\lambda)\geqslant 0$. By Theorem 2.12.1 we have $\lambda[F^1(\lambda),1]\leqslant 1$.

We shall determine $F^a(\lambda)\,(a=1,2)$ such that $\psi(\lambda)$ determined by (1) satisfies the norm condition, the resolvent equation are the Q condition.

If the norm condition holds for $\psi(\lambda)$, by (53), that is

$$X^1(\lambda)\lambda[F^1(\lambda),\mathbf{1}]+X^2(\lambda)\lambda[F^2(\lambda),\mathbf{1}]\leqslant X^1(\lambda)+X^2(\lambda)$$

holds. Letting $i\to\infty$ and noting that $X^1(z,\lambda)=0,\ X^2(z,\lambda)=1$ we obtain $\lambda[F^2(\lambda),\mathbf{1}]\leqslant 1$. Therefore, the norm condition is equivalent to

$$F^a(\lambda)\geqslant 0 \qquad \lambda[F^a(\lambda),\mathbf{1}]\leqslant 1 \qquad a=1,2 \tag{2}$$

As in bilateral birth–death processes, the resolvent equation for $\psi(\lambda)$ is also equivalent to

$$F^a(\lambda)A(\lambda,v)=F^a(v)+(v-\lambda)\sum_{b=1}^{a}[F^a(\lambda),X^b(v)]F^b(v) \qquad (a=1,2,\lambda,v>0) \tag{3}$$

(a) Now suppose that $F'(v),F^2(v)$ are linearly dependent for some $v>0$. Therefore provided we make some proper modifications the discussion in section 5.9 is still valid. We shall state them in the form of the following lemma.

Lemma 1. Suppose that z is regular or exit, and $F^a(\lambda)$ $(a=1,2)$ are linearly dependent for some (hence for all) $\lambda>0$.

For $\psi(\lambda)$ of (1) to satisfy the norm condition and the resolvent equation it is necessary and sufficient that either $\psi(\lambda)=\phi(\lambda)$ or $\psi(\lambda)$ is obtained as follows: Choose constants $d_a\geqslant 0$ $(a=1,2)$, $d_1+d_2>0$, $p\geqslant 0$ ($p=0$ if z is exit). Select the row vector $\alpha\geqslant 0$ so that $\alpha\phi(\lambda)\in l$ and

$$\eta(\lambda)=\alpha\phi(\lambda)+pX^2(\lambda)\mu\neq 0 \tag{4}$$

If $d_1>d_2$ it is also required that

$$[\alpha, X^2]<\infty \qquad p=0 \tag{5}$$

Choose a constant c satisfying

$$\begin{aligned} &c\geqslant 0 && \text{if } d_1=d_2\\ &c\geqslant(d_1-d_2)W_2 && \text{if } d_1>d_2\\ &c\geqslant(d_2-d_1)W_1 && \text{if } d_1<d_2 \end{aligned} \tag{6}$$

where

$$\begin{aligned} W_1&=[\alpha, X^1]+\frac{pa_0}{a_0(z-z_0)+1}\\ W_2&=[\alpha, X^2]+pU^2. \end{aligned} \tag{7}$$

Finally, set

$$\psi_{ij}(\lambda)=\phi_{ij}(\lambda)+\frac{[d_1X_i^1(\lambda)+d_2X_i^2(\lambda)]\eta_j(\lambda)}{c+\lambda[\eta(\lambda), d_1X^1+d_2X^2]} \tag{8}$$

Remark

Upon comparison with Theorem 5.9.1, this lemma does not require that

$$W_1=\lim_{\lambda\to\infty}\lambda[\eta(\lambda), X^1]<\infty$$

since by Lemma 2.11.4 it follows that W_1 is finite.

To prove that $\psi(\lambda)$ of Lemma 1 is a Q process, we must also verify the Q condition, that is

$$\lim_{\lambda\to\infty}\lambda X_i^q(\lambda)\lambda F_j^a(\lambda)=0 \qquad a=1,2 \tag{9}$$

By (8) and (2.11.26), we have

$$\lim_{\lambda\to\infty}\lambda F_j^a(\lambda)=\frac{d_a\alpha_j}{c+d_1W_1+d_2W_2} \tag{10}$$

Therefore, by (5.4) and (5.5) it follows that (9) holds except for the case $a=1$,

$i=0$. For the case $a=1$ and $i=0$, (9) becomes

$$\frac{a_0 d_1 \alpha}{c + d_1 W_1 + d_2 W_2} = 0 \tag{11}$$

Now assume $d_1 > d_2 \geqslant 0$. By (5) and (6) we obtain $W_2 = [\alpha, X^2] < \infty$, $p=0$. By (4) we derive $\alpha \neq 0$. The above remark has already pointed out that $W_1 < \infty$. Therefore, (11) cannot possibly hold. Therefore, it must be that $d_2 \geqslant d_1 \geqslant 0$. Thus (6) becomes

$$\begin{aligned} &c \geqslant 0 && \text{if } d_1 = d_2 \\ &c \geqslant (d_2 - d_1)\left([\alpha, X^1] + \frac{p a_0}{a_0(z - z_0) + 1}\right) && \text{if } d_1 < d_2 \end{aligned} \tag{12}$$

For (11) to hold it is necessary and sufficient that either $d_1\alpha = 0$ or $W_2 = \infty$. Since $[\alpha, X^1] \leqslant W_1 < \infty$ this necessary and sufficient condition becomes

$$d_1\alpha = 0, \text{ or } p > 0, \text{ or } p = 0 \text{ and } [\alpha, 1] = \infty \tag{13}$$

We have thus obtained the following theorem.

Theorem 2. Let z be regular or exit, $a_0 > 0$. Let $F^a(\lambda)$ $(a = 1, 2)$ be linearly dependent. For $\psi(\lambda)$ given by (1) to be a Q process it is necessary and sufficient that either $\psi(\lambda) = \phi(\lambda)$ or $\psi(\lambda)$ is obtained as follows: Take a row vector $\alpha \geqslant 0$ so that $\alpha\phi(\lambda) \in l$ and select constants $d_2 \geqslant d_1 \geqslant 0$, $d_2 > 0$, $p \geqslant 0$ (with $p = 0$ if z is exit) and c satisfying (4), (12) and (13). Set

$$\psi_{ij}(\lambda) = \phi_{ij}(\lambda) + \frac{[d_1 X_i^1(\lambda) + d_2 X_i^2(\lambda)][\sum_k \alpha_k \phi_{kj}(\lambda) + p X_j^2(\lambda)\mu_j]}{c + \sum_{b=1}^{2} d_b\{[\alpha, X^b - X^b(\lambda)] + p\lambda[X^2(\lambda)\mu, X^b]\}} \tag{14}$$

This process is honest if and only if $d_1 = d_2$, $c = 0$. A necessary and sufficient condition for this process to satisfy neither forward nor backward equations is $d_1 > 0$, $\alpha \neq 0$.

The proof of the last conclusion is as follows: Multiplying both sides of (1) by $(\lambda I - Q)$ from the left-hand side, we find that the B condition is equivalent to

$$0 = (\lambda I - Q)[X_i^1(\lambda)F_j^1(\lambda) + X_i^2(\lambda)F_j^2(\lambda)] = \begin{cases} 0 & \text{if } i > 0 \\ a_0 F_j^1(\lambda) & \text{if } i = 0 \end{cases}$$

Multiplying both sides of (1) by $(\lambda I - Q)$ from the right-hand side it follows that the F condition is equivalent to

$$\begin{aligned} 0 &= [X_i^1(\lambda)F_j^1(\lambda) + X_i^2(\lambda)F_j^2(\lambda)](\lambda I - Q) \\ &= \frac{[d_1 X_i^1(\lambda) + d_2 X_i^2(\lambda)]a_j}{c + \lambda[\eta(\lambda), d_1 X^1 + d_2 X^2]} \end{aligned}$$

But $X^1(\lambda), X^2(\lambda)$ are linearly independent and, moreover, $d_1 + d_2 > 0$. Therefore, $d_1 X^1(\lambda) + d_2 X^2(\lambda) \neq 0$.

(b) We assume in (a) that $F^a(\lambda)$ $(a = 1, 2)$ are linearly dependent. We now assume that $F^a(\lambda)$ $(a = 1, 2)$ are linearly independent for some (hence for all) $\lambda > 0$.

In this case we can follow the notation and discussion in section 5.10. In particular, Lemma 5.10.1 is still valid after some obvious modifications.

Lemma 3. Let $\psi(\lambda)$ be a Q process possessing form (1). Then there exist row vectors $\alpha^a \geqslant 0$ $(a = 1, 2)$ such that $\alpha^a \phi(\lambda) \in l$ and a second-order square matrix $\mathscr{R}_\lambda = (r_\lambda^{ab}) \geqslant 0$ and quantities $p_\lambda^a \geqslant 0$ $(a = 1, 2, \lambda > 0)$ ($[p_\lambda] = [0]$ if z is exit) such that

$$[F(\lambda)] = \mathscr{R}_\lambda [\alpha \phi(\lambda)] + [p_\lambda] X^2(\lambda) \mu \tag{15}$$

We introduce the notations

$$h_\lambda^{ab} = \lambda[\alpha^a \phi(\lambda), X^b] = [\alpha^a, X^b - X^b(\lambda)] \uparrow h^{ab} = [\alpha^a, X^b] \qquad \lambda \uparrow \tag{16}$$

$$\mathscr{H}_\lambda = (h_\lambda^{ab}) \uparrow \mathscr{H} = (h^{ab}) \qquad \lambda \uparrow \infty \tag{17}$$

Consider a special case: $[p_\lambda] = [0], \mathscr{H} < \infty$, that is

$$[F(\lambda)] = \mathscr{R}_\lambda [\alpha \phi(\lambda)]$$

$$[\alpha^a, 1] < \infty \qquad a = 1, 2 \tag{19}$$

Theorem 4. Let z be regular or exit, $a_0 > 0$. It is impossible that a Q process $\psi(\lambda)$ having the form (1) satisfies conditions (18) and (19), and moreover $F^a(\lambda)$ $(a = 1, 2)$ are linearly independent.

Proof. Suppose that $\psi(\lambda)$ is a Q process having the form (1) given by (18) and (19), and moreover that $F^a(\lambda)$ $(a = 1, 2)$ are linearly independent.

As we have done in the proof of Theorem 5.10.1, by the norm condition and the resolvent equation for $\psi(\lambda)$ it follows that $[F(\lambda)]$ has the form (5.10.33), that is,

$$[F(\lambda)] = (I - \bar{\mathscr{T}}_\lambda)^{-1} [\bar{\alpha} \phi(\lambda)] \tag{20}$$

where row vectors $\bar{\alpha}^a \geqslant 0$ so that $[\bar{\alpha}^a, 1] \leqslant 1, \bar{\mathscr{T}}_\lambda = \{[\bar{\alpha}, X(\lambda)]\}$.

The $\psi(\lambda)$ must also satisfy the Q condition, that is

$$0 = \lim_{\lambda \to \infty} \lambda [X(\lambda)]' \lambda [F(\lambda)] = \lim_{\lambda \to \infty} \lambda [X(\lambda)]' (I - \bar{\mathscr{T}}_\lambda)^{-1} [\lambda \bar{\alpha} \phi(\lambda)] \tag{21}$$

Since by (5.4), (5.5) and the dominated convergence theorem

$$\lim_{\lambda \to \infty} (I - \bar{\mathscr{T}}^\lambda)^{-1} = \lim_{\lambda \to \infty} \sum_{n=0}^{\infty} \bar{\mathscr{T}}_\lambda^n = I$$

Therefore, (21) holds for $i>0$. For $i=0$, (21) becomes

$$0=\begin{pmatrix} a_0 \\ 0 \end{pmatrix}[\bar{\alpha}]=a_0\bar{\alpha}^1$$

Thus $\bar{\alpha}^1=0$. By (20) it follows that $F^a(\lambda)$ $(a=1,2)$ are linearly dependent, and the proof is completed.

Remark

From this proof it can be seen that $\psi(\lambda)$ determined by (20) and (1) is a $\bar{Q}$ process although it is not a Q process, where $\bar{q}_{ij}=q_{ij}$ $(i>0, j\in E)$, $\bar{q}_{0j}=q_{0j}+a_0\bar{\alpha}_l^1$.

We now consider the general case of (15).

Lemma 5. Let z be exit or regular, $a_0>0$. Suppose that $F^a(\lambda)$ $(a=1,2)$ are linearly independent. For $\psi(\lambda)$ of the form (1) to satisfy the norm conditions and the resolvent equation it is necessary and sufficient that $\psi(\lambda)$ can be obtained as follows: Select two non-negative row vectors $\bar{\alpha}^a$ $(a=1,2)$ such that $\bar{\alpha}^a\phi(\lambda)\in l$. Take a non-negative matrix

$$\bar{\mathscr{S}}=\begin{pmatrix} 0 & \bar{s}^{12} \\ \bar{s}^{21} & 0 \end{pmatrix}$$

and take a constant $\bar{p}^2\geqslant 0$ (with $\bar{p}^2=0$ if z is exit), and moreover, the following properties are satisfied:

(i) either $\bar{p}^2>0$, or $\bar{p}^2=0$ and $\bar{\alpha}^1$ and $\bar{\alpha}^2$ furthermore are linearly independent.

(ii) $\bar{h}^{ab}<\infty$ $(a\neq b)$.

(iii) $\bar{s}^{12}\leqslant 1, \bar{s}^{21}\leqslant 1, \bar{s}^{ab}\geqslant \bar{h}^{ab}+\dfrac{\bar{p}^2 a_0}{a_0(z-z_0)+1}$ $(a\neq b)$.

Finally, set

$$[\bar{p}]=\begin{pmatrix} 0 \\ \bar{p}^2 \end{pmatrix}$$

and

$$\begin{aligned} \mathscr{R}_\lambda &= (I-\bar{\mathscr{S}}+\bar{\mathscr{H}}_\lambda+[\bar{p}][U_\lambda]')^{-1} \\ [p_\lambda] &= (I-\bar{\mathscr{S}}+\bar{\mathscr{H}}_\lambda+[\bar{p}][U_\lambda]')^{-1}[\bar{p}] \end{aligned} \tag{22}$$

whereas $\psi(\lambda)$ is determined by (1), (15) and (22). Here $\bar{\mathscr{H}}_\lambda=(\bar{h}_\lambda^{ab})\uparrow\bar{\mathscr{H}}=(\bar{h}^{ab})$ $(\lambda\uparrow\infty)$ are determined by (16) and (17) for $\bar{\alpha}^a$. By (5.15) and (5.16)

$$[U_\lambda]'=(U_\lambda^1, U_\lambda^2)\uparrow[U]'=\left(\frac{a_0}{a_0(z-z_0)+1}, +\infty\right). \tag{23}$$

For $\psi(\lambda)$ to have the forms (1), (18) and (19) it is necessary and sufficient that $\bar{p}^2=0$, $\bar{h}^{22}<\infty$.

Proof. To repeat the proof of Theorem 5.10.3, it suffices to make evident modifications. We must also notice that by Lemma 2.11.4, if $\bar{\alpha}^1\geqslant 0$ so that $\bar{\alpha}^1\phi(\lambda)\in l$, then $\bar{h}^{11}=\lim_{\lambda\to\infty}\bar{h}_1^\lambda<\infty$ and the proof is completed. QED

We discuss the condition that $\psi(\lambda)$ of Lemma 5 is a Q process. By Theorem 4, certainly

$$\bar{p}^2>0 \qquad \text{or} \qquad \bar{h}^{22}=\infty \tag{24}$$

To prove that $\psi(\lambda)$ is a Q process, we must also verify the Q condition, that is

$$\lim_{\lambda\to\infty}[\lambda X(\lambda)]'\mathscr{L}_\lambda^{-1}\lambda([\bar{\alpha}\phi(\lambda)]+[\bar{p}]X^2(\lambda)\mu)=0 \tag{25}$$

where

$$\mathscr{L}_\lambda=(I-\bar{\mathscr{S}}+\bar{\mathscr{H}}_\lambda+[\bar{p}][U_\lambda]')=\begin{pmatrix}1+\bar{h}_\lambda^{11} & -\bar{s}^{12}+\bar{h}_\lambda^{12}\\ -\bar{s}^{21}+\bar{h}_\lambda^{21}+\bar{p}^2U_\lambda^1 & 1+\bar{h}_\lambda^{22}+\bar{p}^2U_\lambda^2\end{pmatrix}$$

By (24) and Lemma 5(iii), we obtain $\lim_{\lambda\to\infty}\det\mathscr{L}_\lambda=+\infty$. Notice that $\bar{h}^{11}<\infty, \bar{h}^{ab}<\infty$ $(a\neq b)$, and it follows that

$$\begin{aligned}\lim_{\lambda\to\infty}\mathscr{L}_\lambda^{-1}&=\lim_{\lambda\to\infty}\frac{1}{\det\mathscr{L}_\lambda}\begin{pmatrix}1+\bar{h}_\lambda^{22}+\bar{p}^2U_\lambda^2 & \bar{s}^{21}-\bar{h}^{21}-\bar{p}^2U_\lambda^1\\ \bar{s}^{12}-\bar{h}_\lambda^{12} & 1+\bar{h}_\lambda^{11}\end{pmatrix}\\ &=\begin{pmatrix}(1+\bar{h}^{11})^{-1} & 0\\ 0 & 0\end{pmatrix}\end{aligned}$$

Owing to (5.4) and (5.5), (25) holds for $i>0$. For $i=0$ the limit on the left-hand side of (25) is

$$\begin{pmatrix}a_0\\0\end{pmatrix}'\begin{pmatrix}(1+\bar{h}^{11})^{-1} & 0\\ 0 & 0\end{pmatrix}[\bar{\alpha}]=\frac{a_0\bar{\alpha}^1}{1+\bar{h}^{11}}$$

Therefore, (25) holds if and only if $\bar{\alpha}^1=0$. Thus by (22) and (15)

$$[F(\lambda)]=\mathscr{L}_\lambda^{-1}\begin{pmatrix}0\\ \bar{\alpha}^2\phi(\lambda)+\bar{p}^2X^2(\lambda)\mu\end{pmatrix}$$

Therefore $F^a(\lambda)$ $(a=1,2)$ are linearly dependent. This is contradictory to the linear independence of $F^a(\lambda)$ $(a=1,2)$; thus the Q condition (25) cannot possibly hold. QED

Combining this with Theorem 4, we have obtained the following theorem.

Theorem 6. If z is regular or exit, $a_0>0$, then there does not exist a Q process $\psi(\lambda)$ which has the form (1) so that $F^1(\lambda)$ and $F^2(\lambda)$ are linearly independent.

Remark

From the preceding part of the proof of Theorem 6 it can be seen that $\psi(\lambda)$ which satisfies the condition (24) in Lemma 5 is a $\bar{Q}$ process although it is not a Q process, where

$$\bar{q}_{ij}=q_{ij} \qquad (i>0, j\in E) \qquad \bar{q}_{0j}=q_{0j}+\frac{\alpha_0\bar{\alpha}_j^1}{1+\bar{h}^{11}}$$

6.9 THE CONDITION THAT $\alpha\phi(\lambda)\in l$

Suppose $\alpha\geqslant 0$. By (5.3) it follows that $\alpha\phi(\lambda)\in l$ is equivalent to

$$\sum_{i=0}^{\infty}\alpha_i[1-X_i^1(\lambda)-X_i^2(\lambda)]<\infty \tag{1}$$

In this section, we shall give the condition under which $\alpha\phi(\lambda)\in l$ is directly determined from Q

Lemma 1. Assume that $a_0\geqslant 0$, z is regular or exit. Then

$$\lim_{i\to\infty}\frac{v_i(\lambda)}{z-z_i}=\frac{1}{u(z,\lambda)} \tag{2}$$

Proof. Using (3.4), we can give the proof by following that of Lemma 5.11.2. QED

Lemma 2. Suppose $a_0\geqslant 0$, z is regular. Then

$$\lim_{i\to\infty}\frac{1-X_i^1(\lambda)-X_i^2(\lambda)}{z-z_i}=\lambda[X^2(\lambda)\mu,1] \tag{3}$$

Proof. The proof is similar to Theorem 5.11.3. QED

Theorem 3. If $a_0\geqslant 0$ and z is regular, then the necessary and sufficient condition under which $\alpha\phi(\lambda)\in l$, that is, (1) holds, is that

$$\sum_{i=0}^{\infty}\alpha_i(z-z_i)<\infty \tag{4}$$

which is equivalent to

$$\sum_{i=0}^{\infty}\alpha_i N_i<\infty \tag{5}$$

where N_i is determined by (2.6).

Proof. Applying Lemma 2, this proof can be provided by following that of Theorem 5.11.3. QED

Theorem 4. If $a_0 \geqslant 0$ and z is exit, then the necessary and sufficient condition for $\alpha\phi(\lambda) \in l$ is that (5) holds.

Proof. This proof is similar to Theorem 5.11.4. QED

Theorem 5. If $a_0 \geqslant 0$ and z is entrance or natural, then the necessary and sufficient condition for $\alpha\phi(\lambda) \in l$ is that

$$\sum_{i=0}^{\infty} \alpha_i < \infty \tag{6}$$

Proof. The necessity follows from

$$1 - X_i^1(\lambda) \geqslant 1 - X_0^1(\lambda) > 0 \qquad (i \geqslant 0)$$

The sufficiency is obvious. QED

PART III MARTIN BOUNDARY AND ITS APPLICATION IN THE CONSTRUCTION THEORY

CHAPTER 7

Martin Boundary and Q Processes

7.1 INTRODUCTION

When the Q process is not unique, we must compactify the state space E so as to solve the problem of construction of the Q process. That is, we must adjoin some 'boundary points' to E so that certain requirements can be satisfied. In other words, we need to introduce and study the boundary of the Q process. The simplest compactification of E is one-point compactification. For instance, the unilateral birth–death process is done in just this manner, and moreover the one-point compactification is enough. However, in general it is not sufficient to perform the one-point compactification. For example, for the bilateral birth–death process, it is obvious that we must compactify E by two 'points'.

In Feller (1956) a kind of boundary is introduced, which is called a Feller boundary. In his article, Feller (1957a) introduces the exit and passive boundaries for conservative Q, and then by applying these boundaries, he constructs all the Q processess that satisfy the systems of backward and forward equations simultaneously in the case that both the exit and entrance boundaries are finite. The method used by Feller is a purely analytical one.

In Doob (1959) the Martin boundary of the Markov chain is introduced, its characteristic being that the Martin boundary is closely connected with the path of the Markov chain (henceforth the Markov chain will be abbreviated as MC). Doob proves the convergence of the MC's path in the Martin boundary. In Dynkin (1969), Hunt (1960) and Watanabe (1960a, b), the theory of the Martin boundary of MC is further investigated and developed. In Kunita (1962), the Martin boundary is applied to a study of the instantaneous return process. But all the discussions in the articles mentioned above are conducted so that some restrictions are imposed on the Markov chain (e.g. if the MC is required to be non-recurrent, there must be a 'centre' at least). Though these restrictions are inessential, they need to be removed. In Hunt (1960) the introduction of standard measure can exclude the restriction that a 'centre' is needed. The following formula in Dynkin (1969, section 9)

$$K(i,j) = \frac{G(i,j)}{\sum_s r_s G(s,j)} = \frac{f_{ij}}{\sum_s r_s f_{sj}}$$

provides the way to rule out the restriction of the non-recurrence of the MC. Zhen-ting Hou (1974) derives the Martin boundary without imposing any restriction on the MC.

In this chapter we first introduce the Martin boundary of general MC's, whose detailed description basically follows Dynkin (1969). Then we shall conduct extensive discussions in close association with the Q process. We apply the standard image introduced in Feller (1957a). It is similar to Definition 2.11.2, but narrower. We shall present and discuss in detail the λ-image. Following the method in Kunita (1962), we derive the Martin exit and passive boundaries of the Q matrix without any conditions attached, and moreover, by applying these boundaries, we further describe the general analytical expressions of the Q processes treated in section 2.12.

The content of this chapter is absolutely necessary and fundamental either to the construction of the Q process by means of probability methods or to the probabilistic interpretation of the analytical method.

7.2 MARKOV CHAINS

We are given that $(\Omega, \mathscr{F}, P)$ is a complete probability space[1], on which β is defined to be a random variable with values of non-negative integers and '∞'. We say that $X_T(\omega) = \{x_n(\omega)^2, n \leqslant \beta\}$ ($\omega \in \Omega$, n is a non-negative integer) is a homogeneous MC or simply MC defined on the probability space $(\Omega, \mathscr{F}, P)^3$ and taking values in the denumerable index set E. If for any integer $n \geqslant 2$, the non-negative integers $0 \leqslant t_1 < t_2 < \cdots < t_{n+1}$ and any $i_1, i_2, \ldots, i_{n+1} \in E$, provided $P\{x(t_a) = i_a, 1 \leqslant a \leqslant n\} > 0$, then we have

$$P\{x(t_{n+1}) = i_{n+1} | x(t_a) = i_a, 1 \leqslant a \leqslant n\} = P\{x(t_{n+1}) = i_{n+1} | x(t_n) = i_n\} \quad (1)$$

and moreover the value on the right-hand side is dependent only on $t_{n+1} - t_n$, and not on t_n.

Write

$$p_{ij}^n = P\{x(n) = j | x(0) = i\} \quad (2)$$

We call matrix (p_{ij}^n) the n-step transition probability matrix of the chain and

[1] From now on when considering the sub-Borel field of $\mathscr{F}$ we always mean it to be completed according to P and will not make any more statement about it.
[2] We take $x_n(\omega) \equiv x(n, \omega)$, and shall leave out ω if it is not emphasized hereafter.
[3] If $(\Omega, \mathscr{F}, P)$ in the definition of the MC is relaxed into a finite measure space, X_T is likewise called a Markov chain.

simply write $p_{ij} = p'_{ij}$. The matrix, $P = (p_{ij})$ is non-negative and its row sums are not more than one. Every such matrix P may become a one-step transition probability matrix for some Markov chain X_T. Therefore we also say the matrix P is a Markov chain. Note that

$$p^n_{ij} = (P^n)_{ij} \tag{3}$$

We call

$$d_i = 1 - \sum_j p_{ij} \qquad i \in E \tag{4}$$

the stopping quantity of the chain. We call $H = \{i | d_i > 0\}$ the set of stopping states. If H is an empty set, we say that the chain is honest; otherwise it is stopping. Evidently, for the chain to be honest, the necessary and sufficient condition is

$$P_i\{\beta = \infty\} = 1 \qquad i \in E \tag{5}$$

Here and afterwards we write $P_i\{\cdot\} = P\{\cdot | x_0 = i\}, E_i\{\cdot\}$ represents the mathematical expectation taken with respect to P_i. For instance,

$$E_i\{f, \Lambda\} = E_i f I_\Lambda = \int_\Lambda f \, \mathrm{d}p_i \tag{6}$$

Here I_Λ denotes the indicator function of the set Λ.

Set

$$\eta^*_i = \inf\{n | 1 \leqslant n \leqslant \beta, x_n = i\} \tag{7}$$

$$\eta_i = \inf\{n | 0 \leqslant n \leqslant \beta, x_n = i\} \tag{8}$$

We set $\inf \varnothing = \infty$, where $\varnothing$ is the empty set. We write

$$\begin{aligned} f^{n*}_{ij} &= P_i\{\eta^*_j = n\} \qquad n \geqslant 1 \\ f^n_{ij} &= P_i\{\eta_j = n\} \qquad n \geqslant 0 \\ f^*_{ij} &= \sum_{n=1}^{\infty} f^{n*}_{ij} \qquad f_{ij} = \sum_{n=0}^{\infty} f^n_{ij} \end{aligned} \tag{9}$$

We call f^{n*}_{ij} (or f^n_{ij}) the probability that the MC first visits the state j from the state i at the nth step, starting from the first (or zeroth) step.

And we call f^*_{ij} (or f_{ij}) the probability that the MC reaches j from i at the nth step for some $n \geqslant 1$ (some $n \geqslant 0$). Obviously

$$\begin{aligned} &f^0_{ij} = 0 \qquad f^{n*}_{ij} = f^n_{ij} \, (n \geqslant 1) \qquad f^*_{ij} = f_{ij} \qquad \text{if } i \neq j \\ &f_{ii} = f^0_{ii} = 1 \qquad f^n_{ii} = 0 \qquad (n \geqslant 1) \\ &f^*_{ij} = \sum_k p_{ik} f_{kj} \qquad f_{ik} f_{kj} \leqslant f_{ij} \end{aligned} \tag{10}$$

If $f^*_{ij} > 0$ $(i \neq j)$, we say that the chain can lead to j from i, and write $i \Rightarrow j$. Set $i \Rightarrow i$. If $i \Rightarrow j, j \Rightarrow i$, we say that i and j communicate and write $i \Leftrightarrow j$. Assume $C \subset E$; if any two states in C communicate, and furthermore $p_{ik} = 0$ $(i \in C)$ for any $k \in E - C$, then we call C an irreducible class. If $f^*_i \equiv f^*_{ii} = 1$, then we say that the state i is recurrent.

Theorem 1. Write

$$G(i,j) = \sum_{n=0}^{\infty} p^n_{ij} \tag{11}$$

Then

$$G(i,j) = f_{ij} G(j,j) \qquad G(i,i) = \frac{1}{1 - f^*_i} \tag{12}$$

Especially, for i to be recurrent, the necessary and sufficient condition is $G(i,i) = \infty$.

Proof. Write

$$\delta_j(i) = \delta_{ij} \tag{13}$$

Then

$$p^n_{ij} = E_i \delta_j(x_n) \qquad G(i,j) = E_i \sum_{n=0}^{\beta} \delta_j(x_n) \tag{14}$$

Write

$$A_m = (x_0 \neq j, x_1 \neq j, \ldots, x_{m-1} \neq j, x_m = j)$$

Then

$$P_i(A_m) = f^m_{ij} \qquad P_i(A_m, x_{m+k} = j) = f^m_{ij} p^k_{jj}$$

Therefore

$$\begin{aligned}
G(i,j) &= E_i \sum_{n=0}^{\beta} \delta_j(x_n) \\
&= E_i \sum_{m=0}^{\infty} I_{A_m} \sum_{n=m}^{\beta} \delta_j(x_n) \\
&= \sum_{m=0}^{\infty} \sum_{n=m}^{\infty} E_i I_{A_m} \delta_j(x_n) \\
&= \sum_{m=0}^{\infty} \sum_{k=0}^{\infty} P_i(A_m, x_{m+k} = j) \\
&= \sum_{m=0}^{\infty} \sum_{k=0}^{\infty} f^m_{ij} p^k_{jj} = f_{ij} G(j,j)
\end{aligned}$$

The proof of the second formula in (12) can be found in Zi-kun Wang and Xiang-qun Yang (1988, section 2.2, Theorem 1). QED

Remark

If j is non-recurrent, then $G(i,j) < \infty$; thus $\sum_{n=0}^{\beta} \delta_j(x_n) < \infty$ P_i almost surely; that is, with probability 1, the number of times that the chain visits j is finite.

Theorem 2 (Decomposition theorem of state space). The state space E of the MC has the following unique decomposition:

$$E = E_0 \cup \left(\bigcup_{a \in \mathscr{A}} E_a \right) \tag{15}$$

where E_0 is composed of all the non-recurrent states. $\mathscr{A}$ is an empty set or finite set or denumerably infinite set, and moreover $0 \notin \mathscr{A}$. Every E_a $(a \in \mathscr{A})$ is an irreducible recurrent class.

The verification can be seen in Zi-kun Wang and Xiang-qun Yang (1988, section 2.3, Theorem 1). We point out that every stopping state must be non-recurrent, that is,

$$H \equiv \left\{ i : \sum_j p_{ij} < 1 \right\} \subset E_0 \tag{16}$$

The chain $X_r = \{x_n, n \leqslant \beta\}$ moves as follows: If x_0 is in some E_a $(a \in \mathscr{A})$, then $\beta = \infty$, and the chain will move in E_a forever, and moreover visits every state in E_a infinitely many times. If $x_0 \in E_0$, then either $\beta < \infty$, and at this moment there must exist $x_\beta \in H$; or $\beta = \infty$, and in this case the chain either moves in E_0 forever, or gets into some recurrent class E_a at some step, and then moves in E_a forever. As for further consideration of E_0, we have the Blackwell decomposition theorem below.

Let

$$\Omega_\infty = (\beta = \infty) \qquad \Omega_F = (\beta < \infty) \tag{17}$$

Provided $A \subset E$, set

$$\begin{aligned} \bar{\mathscr{L}}(A) &= \Omega_\infty \cap \limsup_{n \to \infty} \{x_n \in A\} \\ \underline{\mathscr{L}}(A) &= \Omega_\infty \cap \liminf_{n \to \infty} \{x_n \in A\} \end{aligned} \tag{18}$$

If $P_i\{\bar{\mathscr{L}}(A)\} = 0$ $(i \in E)$, then we call A a transient set; if $P_i\{\underline{\mathscr{L}}(A)\} > 0$ for some

$i \in E$, then we say that A is a sojourn set. Suppose that A is a sojourn set, and moreover, $P_i\{\bar{\mathscr{L}}(A)\} = P_i\{\underline{\mathscr{L}}(A)\}$ $(i \in E)$, then we say that A is an almost closed set. When A is a transient set or an almost closed set, we write $\mathscr{L}(A)$ for any one set that is equal to $\underline{\mathscr{L}}(A)$ with the exception of a P_i-null probability set (for all i).

Assume A to be an almost closed set. If for any $B \subset A$, either B or $A - B$ is a transient set, then we call A an atomic almost closed set. If for any $B \subset A, B$ is not an atomic almost closed set, then we call A a completely non-atomic almost closed set.

The following is the Blackwell decomposition theorem.

Theorem 3 (Chung, 1967, I, 17, Theorem 4). The state space E has the following decomposition:

$$E = A_0 + \left(\bigcup_{a \in \mathscr{A}} A_a \right) \tag{19}$$

where A_0 is a completely non-atomic almost closed set, and so can be dropped. $\mathscr{A}$ is an empty set or a finite set or a denumerably infinite set, and moreover $0 \notin \mathscr{A}$. A_a $(a \in \mathscr{A})$ is an atomic almost closed set. Furthermore

$$P\{\mathscr{L}(A_0)\} + \sum_{a \in \mathscr{A}} P\{\mathscr{L}(A_a)\} = P(\Omega_\infty) \tag{20}$$

If we rule out the difference of a transient set, the decomposition (19) is unique.

Note that in the decomposition (15) every irreducible recurrent class is an atomic almost closed set.

7.3 MARTIN BOUNDARY THEORY

The study of the ultimate behaviour of the path of the MC, $X_T = \{x_n, n \leqslant \beta\}$, that is the behaviour of x_n, when $n \uparrow \beta$, gives rise to the Martin boundary theory of the MC. Applying the boundary theory, we can also describe all the excessive and harmonic functions relative to the chain. Assume that the one-step transition matrix of X_T is $P = (p_{ij})$. If the stopping state set H is not empty, we can select $\Delta \notin E$, and let

$$p_{\Delta\Delta} = 1 \qquad p_{\Delta i} = 0 \qquad p_{i\Delta} = 1 - \sum_j p_{ij} \qquad i \in E \tag{1}$$

$$x_n = \Delta \qquad \text{if } \beta < \infty, n > \beta \tag{2}$$

Then $\{x_n, n \geqslant 0\}$ is an MC, whose one-step transition matrix is $(p_{ij}, i, j \in E \cup \{\Delta\})$. From now on it is to be understood always like this.

7.3.1 Excessive function and excessive measure

Assume u to be a function on E[1]. The function Pu is defined as follows:

$$(Pu)_i = \sum_j p_{ij} u_j$$

Assume that v is a measure on E. The measure vP is defined like this:

$$(vP)_j = \sum_i v_i p_{ij}$$

Definition 1. A non-negative (including $+\infty$) function u is called P-excessive, if $Pu \leqslant u$; it is said to be P-harmonic if $Pu = u$. We write the class composed of bounded harmonic functions as $\mathscr{M}^+$, and denote the class of harmonic functions bounded by k by $\mathscr{M}^+(k)$.

Evidently, if u is excessive, then $P^n u \leqslant u$ for all n.

Definition 2. A non-zero excessive (or harmonic) function u is said to be minimal if $u = u^1 + u^2$ and moreover u^1 and u^2 are excessive (or harmonic) functions. Then $u = c^a u^a$ $(a = 1, 2)$, c^a is a constant. The non-zero harmonic function u is called completely non-minimal if, for any non-zero harmonic function $v \leqslant u$, v is not minimal.

Definition 3. A (non-negative) measure μ is called P-excessive, if $\mu P \leqslant \mu$; it is said to be harmonic if $\mu P = \mu$.

Assume that u is a P-excessive function on E and furthermore let $u(\Delta) = 0$. Then $(u_i, i \in E \cup \{\Delta\})$ is likewise excessive with respect to $(p_{ij}, i, j \in E \cup \{\Delta\})$. Applying the Markov property of X_T and the excessive property of u:

$$E[u(x_{n+1}) | x_0, x_1, \ldots, x_n] = E[u(x_{n+1}) | x_n] = \sum_j P_{x_n j} u_j \leqslant u(x_n) \tag{3}$$

Hence $\{u(x_n), n \geqslant 0\}$ is a supermartingale while $\{-u(x_n), n \geqslant 0\}$ is a submartingale, and thus

$$Eu(x_0) \geqslant Eu(x_1) \geqslant Eu(x_2) \geqslant \cdots \tag{4}$$

Assume $0 < a < b$. Let U_N be the number of times that the sequence $u(x_0), \ldots,$ $u(x_N)$ goes down across the interval $[a, b]$, that is the number of times that the sequence passes from the right side of the interval $[a, b]$ to its left side, and hence the number of times that $-u(x_0), \ldots, -u(x_N)$ passes from the left of the interval $[-b, -a]$ to its right. According to Zi-kun Wang (1965a, section 1.4,

[1]Hereafter u_i will also be written as $u(i)$; similarly p_{ij} will be denoted by $p(i, j)$ and so on.

Lemma 3), or Doob (1953, Ch. VII, Theorem 3.3),

$$EU_N \leqslant \frac{E[-u(x_N)-(-b)]^+}{(-a)-(-b)}$$
$$= \frac{E[u(x_N)-b]^-}{b-a}$$
$$\leqslant \frac{Eu(x_N)}{b-a}$$
$$\leqslant \frac{Eu(x_0)}{b-a} \tag{5}$$

Let v be the number of times that $u(x_0), u(x_1), \ldots, u(x_n), \ldots$ moves down across the interval $[a, b]$. Then $v_N \uparrow v$. Accordingly

$$Ev \leqslant \frac{Eu(x_0)}{b-a} \tag{6}$$

In particular,

$$E_i v \leqslant \frac{u(i)}{b-a} \tag{7}$$

Hence if $u(i) < \infty$, then $E_i v < \infty$, and therefore P_i almost surely[1] $v < \infty$; that is, the number of times that $\{u(x_n), n \geqslant 0\}$ goes down across $[a, b]$ is finite. Use the following fact: if the number of times that a sequence goes down across any interval $[a, b]$ with rational ends is finite, then the sequence must have a finite or infinite limit. Therefore, if $u(i) < \infty$, then there is a limit

$$\xi = \lim_{n\to\infty} u(x_n) \qquad P_i \text{ almost surely} \tag{8}$$

But

$$E_i u(x_n) = \sum_j p_{ij}^* u(j) \leqslant u(i)$$

so that according to Fatou's lemma,

$$E_i \xi \leqslant u(i)$$

Consequently we have P_i almost surely $\xi < \infty$. So we have proved the theorem below.

Theorem 1. Assume u to be an excessive function, $u(i) < \infty$. Then there almost surely exists a finite limit $\lim_{n\to\infty} u(x_n)$ on Ω_∞.

[1] If P is a probability, 'P almost surely' means 'with P probability one'.

Theorem 2. Assume that the states of the MC communicate. Then the chain is recurrent if and only if all its excessive functions are constants.

Proof. Suppose that u is excessive. If for some j we have $u(j)=\infty$, then for any i, according to the fact that i and j communicate, there must exist n such that $p_{ij}^n > 0$, thus

$$u_i \geqslant \sum_k p_{ik}^n u(k) \geqslant p_{ij}^n u(j) = \infty$$

Hence $u \equiv \infty$. And therefore we only need to consider the finite excessive function u.

Let us assume X_T to be recurrent. By Theorem 1 there exists a finite limit

$$\lim_{n\to\infty} u(x_n) = \xi \qquad P_i \text{ almost surely}$$

According to the recurrence property of X_T and the fact that the states communicate, for an arbitrary j we have

$$P_i(x_n = j \text{ for infinitely many } n) = 1$$

so that

$$P_i(\xi = u(j)) = 1$$

Since j is arbitrary,

$$P_i(\xi = u(k)) = 1$$

for the state k. It follows that $u(j)=u(k)$, that is, $u \equiv$ constant.

Suppose that all the excessive functions of X_T are constants. Fix k and write $u_i = f_{ik}$. Then in accordance with the notation in (2.7),

$$\sum_j p_{ij} u_j = p_i(\eta_k^* < \infty) = f_{ik}^* \leqslant f_{ik} = u_i \tag{9}$$

This means that u is excessive, therefore $u =$ constant. But it is obvious that $u_k = f_{kk} = 1$; hence $u \equiv 1$, that is, $f_{ik} \equiv 1$ for all i and k. Taking $i \neq j$ arbitrarily, by (2.10),

$$1 = f_{ij} = \sum_m p_{im} f_{mj} = \sum_m p_{im}$$

$$f_{ii}^* = \sum_j p_{ij} f_{ji} = \sum_j p_{ij} = 1$$

that is, i is recurrent, so the chain is recurrent, and the proof is completed. QED

The following corollary does not require the condition that the states communicate.

Corollary

When j is fixed, f_{ij} as a function of i is an excessive function.

Proof. From (9) the corollary follows. QED

Corollary

Assume u to be an excessive function; then u takes the constant value in each recurrent class.

Theorem 3. For any measure $\gamma, \gamma G$ is an excessive measure. Especially when i is fixed, $G(i,j)$ is an excessive measure.

Proof. Since

$$G=\sum_{n=0}^{\infty} p^n$$

it follows that

$$GP=\sum_{n=1}^{\infty} p^n \leqslant G \qquad (\gamma G)P \leqslant \gamma G$$

QED

Corollary

Suppose that there is a measure γ such that $\sum_i \gamma_i f_{ij} > 0$ $(j \in E)$, and h is an excessive function. If h is γ-integrable, then h is a finite function; if $\sum_i \gamma_i h_i = 0$, then $h = 0$.

Proof. Since $P^n h \leqslant h$, it follows that

$$\gamma_i p_{ij}^n h \leqslant \sum_i \gamma_i \sum_j p_{ij}^n h_j \leqslant \sum_i \gamma_i h_i$$

Because of the assumption with respect to γ there exist i and n such that $\gamma_i p_{ij}^n > 0$ for every j. Therefore if $\sum_i \gamma_i h_i < \infty$, then $h_j < \infty$; if $\sum_i \gamma_i h_i = 0$, then $h_j = 0$. The proof is concluded. QED

7.3.2 Density function of an excessive measure

Suppose that the MC X_T has an initial distribution γ[1]. In order to emphasize the dependence on γ we had better write the measure P as P_γ and the mathematical expectation E as E_γ. So the mathematical expectation of the

[1]The total measure of γ may not necessarily be 1.

number of times that the chain X_T is in the state j is

$$\eta_j = \sum_i \gamma_i G(i, j)$$

The measure

$$\eta = \gamma G \tag{10}$$

is a P-excessive measure.

Suppose D is a finite subset of E; let τ_D (or simply τ) be the last exit time from D of the chain X_T, that is,

$$\tau = \sup\{n: 0 \leqslant n \leqslant \beta, x_n \in D\} \tag{11}$$

If for all n $(0 \leqslant n < \beta)$, we have $x_n \notin D$, then we leave τ undefined.

Write

$$L_D(i) = P_i(\tau = 0) = P_i(x_0 \in D, x_n \notin D \text{ for } 1 \leqslant n \leqslant \beta) \tag{12}$$

so that $L_D(i) = 0$ for $i \notin D$; when $i \in D$, $L_D(i)$ is the probability that the chain leaves D at the first step and will never come back to D. Thus when $i \in D$ and moreover when i is recurrent, $L_D(i) = 0$.

Evidently, for any $j \in E$

$$P_i(x_\tau = j) = \sum_{m=0}^{\infty} P_i(\tau = m, x_m = j)$$

$$= \sum_{m=0}^{\infty} p_{ij}^m L_D(j) = G(i, j) L_D(j) \tag{13}$$

$$\sum_j G(i, j) L_D(j) = \sum_j P_i(x_\tau = j) \leqslant 1 \tag{14}$$

From (13) it follows that

$$P_\gamma(x_\tau = j) = \eta(j) L_D(j) \qquad \text{where } \eta = \gamma G \tag{15}$$

Suppose n is a non-negative integer. When $\tau < n$, or $\tau = \infty$, or τ is undefined, we shall set $x_\tau = \Delta$. Assume $i_0, i_1, \ldots, i_n \in E$; moreover write

$$R(i_n, \ldots, i_0) = P(i_n, i_{n-1}) \cdots P(i_1, i_0)$$

Then it follows that

$$P_i(x_\tau = i_0, x_{\tau-1} = i_1, \ldots, x_{\tau-n} = i_n) = \sum_{m=n}^{\infty} P_i(\tau = m, x_m = i_0, x_{m-1} = i_1, \ldots, x_{m-n} = i_n)$$

$$= \sum_{m=n}^{\infty} P^{m-n}(i, i_n) R(i_n, \ldots, i_0) L_D(i_0)$$

$$= G(i, i_n) R(i_n, \ldots, i_0) L_D(i_0)$$

Multiplying both sides of the above formula by γ_i and summing over i we obtain

$$P_\gamma(x_\tau = i_0, \ldots, x_{\tau-n} = i_n) = \eta(i_n)R(i_n, \ldots, i_0)L_D(i_0) \tag{16}$$

When $n = 0$, the formula above becomes (15).

Lemma 4. Suppose every state of the chain X_T is non-recurrent and let u be a non-negative function on E; set $u(\Delta) = 0$ also. If $u\eta = (u_j\eta_j, j \in E)$ is an excessive measure, then with respect to the measure P_γ, $\{u(x_{\tau-n}), n \geqslant 0\}$ is a supermartingale relative to $\{x_{\tau-n}, n \geqslant 0\}$. That is,

$$E_\gamma[u(x_{\tau-n}) | x_\tau, x_{\tau-1}, \ldots, x_{\tau-(n-1)}] \leqslant u(x_{\tau-(n-1)}) \tag{17}$$

Proof. Since $u\eta$ is an excessive measure it follows that

$$\sum_{i_n} u(i_n)\eta(i_n)P(i_n, i_{n-1}) \leqslant u(i_{n-1})\eta(i_{n-1})$$

Thus

$$\sum_{i_n} \eta(i_n)R(i_n, \ldots, i_0)L_D(i_0)u(i_n) \leqslant \eta(i_{n-1})R(i_{n-1}, \ldots, i_0)L_D(i_0)u(i_{n-1})$$

Noting (16) we obtain (17) from the formula above. The proof is terminated.

QED

The function u in Lemma 4 is the density of the excessive measure $\mu \equiv u\eta$ with regard to η.

Let K_γ denote all non-negative functions u satisfying the following condition:

$$S \equiv \sup\{E_\gamma u(x_{\tau_D}) : \text{finite } D \subset E\} < \infty \tag{18}$$

Note that when $\tau_D = \infty$ or τ_D is undefined, $x_{\tau_D} = \Delta$, and furthermore $u(\Delta) = 0$.

Theorem 5. Suppose that every state of the chain X_T is non-recurrent, and assume that $u \in K_\gamma$ and that $u\eta$ is an excessive measure. Then on Ω_∞, there exists P_γ almost surely a finite limit

$$\lim_{n\to\infty} u(x_n)$$

Proof. Let v_N be the number of times that $u(x_\tau), \ldots, u(x_{\tau-N})$ passes down across $[a, b]$, that is, the number of times that $u(x_{\tau-N}), \ldots, u(x_\tau)$ moves up across $[a, b]$. By (5)

$$E_\gamma v_N \leqslant \frac{E_\gamma u(x_\tau)}{b-a}$$

Let v_D denote the number of times that $u(x_0), \ldots, u(x_\tau)$ goes up across $[a, b]$. By the non-recurrence property when D is finite, for P_γ almost certainly $\tau < \infty$; hence $v_N \uparrow v_D$ $(N \to \infty)$. Therefore

$$E_\gamma v_D \leqslant \frac{E_\gamma u(x_{\tau_D})}{b - a}$$

Since $u \in K_\gamma$ it follows that

$$E_\gamma v_D \leqslant S/(b - a)$$

Let v represent the number of times that $u(x_0), u(x_1), \ldots, u(x_n), \ldots$ passes up across $[a, b]$, then $v_D \uparrow v$ $(D \uparrow E)$, and therefore

$$E_\gamma v \leqslant S/(b - a) \tag{19}$$

Consequently $P_\gamma(v < \infty) = 1$, thereby there is P_γ almost certainly a finite or infinite limit $\xi = \lim_{n \to \infty} u(x_n)$. It remains only to prove that P_γ almost surely holds $\xi < \infty$.

Suppose $v(c)$ is the number of times that $u(x_0), \ldots, u(x_n)$ moves up across $[c, 2c]$. Let

$$\varliminf_{c \to \infty} v(c) = \bar{v}$$

Evidently

$$(\xi = \infty) \subseteq (\bar{v} \geqslant 1)$$

so that

$$P_\gamma(\xi = \infty) \leqslant P_\gamma(\bar{v} \geqslant 1) \leqslant E_\gamma \bar{v}$$

By (19) and Fatou's lemma we obtain

$$P_\gamma(\xi = \infty) \leqslant E_\gamma \varliminf_{c \to \infty} v(c) \leqslant \varliminf_{c \to \infty} E_\gamma v(c)$$

$$\leqslant \varliminf_{c \to \infty} \frac{S}{c} = 0$$

and the theorem is proved.

7.3.3 Martin kernel

Definition 4. We call the measure $\gamma = (\gamma_i, i \in E)$ the standard measure of the chain X_T or P if

$$\sum_i \gamma_i < \infty \qquad 0 < \sum_i \gamma_i f_{ij} \tag{20}$$

From now on when considering a Martin boundary we always fix a standard measure γ. By Section 1.4, the measure generated by the initial distribution γ and the matrix $P=(P_{ij})$ is represented by P_γ. Since $P_\gamma=\sum_i \gamma_i P_i$, it follows that if $\gamma_i>0$ $(i\in E)$, then $P_\gamma(\Lambda)=0$ if and only if $P_i(\Lambda)=0$ for all i. Obviously,

$$0<A_j\equiv\sum_i \gamma_i f_{ij}<\infty \tag{21}$$

Definition 5. We call

$$K(i,j)=\frac{f_{ij}}{A_j}=\frac{f_{ij}}{\sum_s \gamma_s f_{sj}} \tag{22}$$

the Martin kernel of the chain

By Theorem 1 we have

$$A_i G(j,j)=(\gamma G)_j \qquad (j\in E_0)$$

Therefore

$$K(i,j)=\frac{G(i,j)}{\eta(j)} \qquad \text{if } j\in E_0 \tag{23}$$

where $\eta=\gamma G$ and E_0 is the set of all non-recurrent states.

By the first corollary to Theorem 2 and Theorem 3, when j is given, $K(\cdot,j)$ is an excessive function.

According to the definition and the last formula of (2.10) we have

$$K(i,j)\leqslant\frac{1}{A_i} \qquad K(i,j)\leqslant\frac{1}{A_j} \quad \sum_i \gamma_i K(i,j)=1 \tag{24}$$

Theorem 6. Suppose μ is a totally finite measure. Then on Ω_∞ there P_γ-almost surely exists a finite limit

$$\lim_{n\to\infty}\sum_i \mu(i)K(i,x_n) \tag{25}$$

In particular, for any i, on Ω_∞, there P_γ-almost surely exists a finite limit

$$\lim_{n\to\infty} K(i,x_n) \tag{26}$$

Proof. By (24) we have

$$\sum_i \mu(i)K(i,j)\leqslant\sum_i \mu(i)A_j^{-1}<\infty \tag{27}$$

Suppose that E_a is an irreducible recurrent class. When the class property of

E_a is taken into account, the states in E_a may be considered to be equal. For instance, every state in E_a is equal to one state ξ_a. When $j \in E_a$, f_{ij} is independent of j and is denoted by $f_i(\xi_a)$. Thus $K(i, j)$ is also independent of $j \in E_a$, and is denoted by $K(i,\xi_a)$. Therefore when the chain gets into some irreducibly recurrent class E_a at a certain step, for all sufficiently large n,

$$\sum_i \mu(i)K(i, x_n) = \sum_i \mu(i)K(i, \xi_a) < \infty$$

Consequently the limit (25) exists and is finite.

What remains to be proved is: when $x_n \in E_0$ for all $n < \beta = \infty$, the limit (25) exists and is finite. In this case we only need to consider the non-recurrent chain on E_0, $\tilde{X}_T = \{x_n, n \leqslant \tilde{\beta}\}$, where $\tilde{\beta} = \sup\{n: x_n \in E_0\}$. The initial distribution of $\tilde{X}_T$ is $(\gamma_i, i \in E_0)$; the one-step transition matrix is $\tilde{P} = (p_{ij}, i, j \in E_0)$; and as for $\tilde{P}$, we have the corresponding $\tilde{f}_{ij} = f_{ij}$, $\tilde{K}(i, j) = K(i, j)$ $(i, j \in E_0)$. Moreover,

$$\sum_i \mu(i)K(i, j) = \sum_{i \in E_0} \mu(i)\tilde{K}(i, j) \qquad j \in E_0$$

Accordingly we might as well suppose that every state of the chain $P = (p_{ij}, i, j \in E)$ is non-recurrent; then we shall prove that the limit (25) exists and is finite.

Letting $u(j) = \sum_i \mu(i)K(i, j)$, and noting (23), we have $u\eta = \mu G$, while μG is an excessive measure. Now we proceed to prove $u \in K\gamma$. Actually by (15),

$$\begin{aligned} E_\gamma u(x_\tau) &= \sum_j P_\gamma(x_\tau = j)u(j) \\ &= \sum_j u(j)\eta(j)L_D(j) \\ &= \sum_{i,j} \mu(i)G(i, j)L_D(j) \end{aligned}$$

By (14)

$$E_\gamma u(x_\tau) \leqslant \sum_i \mu(i) < \infty$$

Thus

$$\sup\{E_\gamma u(x_{\tau_D}): \text{finite } D \subset E\} \leqslant \sum_i \mu(i) < \infty$$

That is $u \in K_\gamma$. The conclusion of the theorem is immediately reached by an application of Theorem 5, and the proof is completed. QED

7.3.4 Martin boundary

In the theory of the Martin boundary we usually regard every state in the irreducibly recurrent class E_a as one and the same state ξ_a, and we still

write

$$E = E_0 \cup \{\xi_a, a \in \mathscr{A}\} \tag{28}$$

We arrange arbitrarily the order of the E in (28): $E = \{e_1, e_2, \dots\}$, and write $N(e_m) = m$. Let

$$d(i,j) = |2^{-N(i)} - 2^{-N(j)}| + \sum_s |K(s,i) - K(s,j)| A_s 2^{-N(s)} \tag{29}$$

where A_j is defined by (21). Then d is a metric in E and induces a discrete topology from E; moreover $d(i,j) \leqslant 3$ $(i, j \in E)$. Completing E according to the metric d we obtain the complete metric space E^*.

Definition 6. We call $\partial E = E^* - E_0$ the Martin boundary of the chain P. The Borel field generated by the open sets in E^* is written as $\mathscr{E}^*$. An element in $\mathscr{E}^*$ is called a Borel set in E^*. And an $\mathscr{E}^*$-measurable function defined on a Borel set Γ is called a measurable function on Γ.

Evidently,

$$\partial E = (\partial E)_1 \cup (\partial E)_2 \qquad (\partial E)_1 = \{\xi_0, a \in \mathscr{A}\} \qquad (\partial E)_2 = \partial E - (\partial E)_1 \tag{29a}$$

Following the definition of the metric d, we evidently have the following.

Theorem 7. An infinite sequence $\{j_n\}$ in E is a fundamental sequence in the metric space E if and only if

(i) $\lim_{n\to\infty} N(j_n)$ exists (finite or infinite);

(ii) for every $i \in E$, $\{K(i, j_n)\}$ is a real Cauchy's fundamental sequence.

By Theorem 7 we can extend the domain E of definition for $N(i)$ to E^*; therefore

$$\begin{aligned} N(\xi) &< \infty \qquad (\xi \in (\partial E)_1) \\ N(\xi) &= \infty \qquad (\xi \in (\partial E)_2) \end{aligned} \tag{30}$$

For every i, $K(i,j)$ as a function of j can be continuously extended to E^*, that is

$$K(i,\xi) = \lim_{j \xrightarrow{d} \xi} K(i,j)$$

By (24) and Fatou's lemma, $K(\cdot, \xi)$ is an excessive function, and

$$K(i,\xi) \leqslant \frac{1}{A_i} \qquad \sum_i \gamma_i K(i,\xi) \leqslant 1 \qquad \xi \in E^* \tag{31}$$

Thus (24) is a special case of (31), and the second formula above is an equality for $\xi \in E_0 \cup (\partial E)_1$.

Thus (29) can also be extended to $i \in E^*$ and $j \in E^*$; hence Theorem 7 is still correct if E in Theorem 7 is changed to E^* and $\{j_n\}$ is an infinite sequence in E^*.

Theorem 8. E^* is a sequentially compact space.

Proof. By Theorem 2, §3, Chapter 1 in Guan (1988), for the metric space the concept of sequential compactness and that of compactness are identical. Therefore it suffices to prove the sequential compactness.

Suppose $\{\xi_n\}$ is an infinite sequence; let $\xi_n \in E^*$. Then we can surely select a subsequence of $\{\xi_n\}$, which is also denoted by $\{\xi_n\}$, such that $\lim_{n\to\infty} N(\xi_n)$ exists (finite or infinite). By (31) for every $i \in E$ we have

$$K(i, \xi_n) \leqslant 1/A_i$$

Applying the diagonal process, we can select a subsequence $\{\xi_{nn}\}$ of $\{\xi_n\}$ such that for every $i \in E$, $\{K(i, \xi_{nn})\}$ are all fundamental sequences of real numbers. According to the paragraph preceding Theorem 8, the sequence $\{\xi_{nn}\}$ must converge in E^* to some point $\xi \in E^*$. This is just the sequential compactness. The proof is terminated. QED

Theorem 9. For P_γ-almost all $\omega \in \Omega$ either

$$x_\beta \in H \qquad \text{if } \beta < \infty \tag{32}$$

or there exists a limit

$$d - \lim_{n\to\infty} x_n = x_\infty \in \partial E \qquad \text{if } \beta = \infty \tag{33}$$

Here

$$H = \left\{ i : \sum_i p_{ij} < 1 \right\}$$

is the stopping state set.

Proof. On Ω_F, $x_\beta \in H$ is obvious. On Ω_∞, as Theorem 6 has already pointed out, there P_γ almost surely exists a finite limit

$$\lim_{n\to\infty} K(i, x_n) \qquad \text{for all } i \in E$$

Thus there exists a limit

$$d - \lim_{n\to\infty} x_n = x_\infty \in E^*$$

If the chain gets into some recurrent class E_a, then $x_\infty = \xi_a \in \partial E$. Otherwise the chain will move in E_0 forever. As the number of times that the chain stays at any non-recurrent state $i\in E_0$ is always finite, it follows that $x_\infty \notin E_0$, and hence $x_\infty \in E^* - E_0 = \partial E$. The proof is concluded. QED

7.3.5 Distribution of ultimate states

By Theorem 9 the ultimate state x_β is defined P_γ-almost surely. We write the distribution of x_β as follows:

$$\mu(\Gamma) \equiv \mu_1(\Gamma) = P_\gamma(x_\beta \in \Gamma) \qquad \Gamma \in \mathscr{E}^* \tag{34}$$

Then the mass of μ is distributed on $H \cup \partial E$.

Theorem 10. Suppose that u is a continuous function or a non-negative Borel function on E^* Then

$$E_i u(x_\beta) = \int_{H\cup\partial E} K(i, \xi) u(\xi) \mu(\mathrm{d}\xi) \tag{35}$$

$$E_i u(x_\beta) 1_{\beta<\infty} = \sum_j G(i, j) u_j \left(1 - \sum_s p_{js}\right) \tag{36}$$

$$E_i u(x_\beta) 1_{\beta=\infty} = \int_{\partial E} K(i, \xi) u(\xi) \mu(\mathrm{d}\xi) \tag{37}$$

In particular,

$$P_i(x_\beta \in \Gamma) = \int_\Gamma K(i, \xi) \mu(\mathrm{d}\xi) \qquad \Gamma \subset H \cup \partial E \tag{38}$$

$$\begin{aligned} P_i(x_\beta = j) &= P_i(x_\beta = j, \beta < \infty) \\ &= G(i, j)\left(1 - \sum_s p_{js}\right) \qquad j \in E_0 \end{aligned} \tag{39}$$

$$\begin{aligned} P_i(x_\beta \in \Gamma) &= P_i(x_\beta \in \Gamma, \beta = \infty) \\ &= \int_\Gamma K(i, \xi) \mu(\mathrm{d}\xi) \qquad \Gamma \subset \partial E \end{aligned} \tag{40}$$

$$\mu(j) \equiv P_\gamma(x_\beta = j) = \eta(j)\left(1 - \sum_s p_{js}\right) \qquad j \in E_0 \tag{41}$$

Proof. When $\xi_0 \in (\partial E)_1$

$$P_i(x_\beta = \xi_a) = f_i(\xi_a)$$

$$\mu(\xi_a) \equiv P_\gamma(x_\beta = \xi_a) = \sum_s \gamma_s f_s(\xi_a)$$

So

$$P_i(x_\beta = \xi) = K(i, \xi)\mu(\xi) \qquad \xi \in (\partial E)_1$$

Hence

$$E_i u(x_\beta) 1_{x_\beta \in (\partial E)_1} = \int_{(\partial E)_1} K(i, \xi) u(\xi) \mu(\mathrm{d}\xi) \tag{42}$$

Secondly, when $j \in E_0$,

$$P_i(x_\beta = j) = P_i(x_\beta = j, \beta < \infty)$$
$$= \sum_{n=0}^{\infty} P_i(x_n = j, n = \beta) = \sum_{n=0}^{\infty} p_{ij}^n \left(1 - \sum_s p_{js}\right)$$

From this (39) follows. Then (36) and (41) follow from (39), and furthermore from (39) and (41) we obtain

$$E_i u(x_\beta) 1_{\beta < \infty} = \int_H K(i, \xi) u(\xi) \mu(\mathrm{d}\xi) \tag{43}$$

Thirdly, comparing (13), (15) and (22) we have

$$P_i(x_\tau = j) = K(i, j) P_\gamma(x_\tau = j) \qquad j \in E \tag{44}$$

Hence

$$E_i u(x_\tau) = \sum_j u(j) P_i(x_\tau = j) = \sum_j u(j) K(i, j) P_\gamma(x_\tau = j)$$
$$= E_\gamma K(i, x_\tau) u(x_\tau) \tag{45}$$

Let $D \uparrow E$, then $\tau_D \uparrow \beta$, $x_{\tau_D} \to x_\beta$ when $x_\beta \in H \cup (\partial E)_2$; when $x_\beta \in (\partial E)_1$, if only D is sufficiently large, that is, when D and $(\partial E)_1$ have some non-empty intersection, $\tau_D = \infty$. According to the supposition $x_{\tau_D} = \Delta$, $u(\Delta) = 0$. Thus when u is a continuous function on E^*, noting the first formula of (31) we get

$$E_i u(x_\beta) 1_{x_\beta \in H \cup (\partial E)_1} = E_r K(i, x_\beta) u(x_\beta) 1_{x_\beta \in H \cup (\partial E)_1} = \int_{H \cup (\partial E)_2} K(i, \xi) u(\xi) \mu(\mathrm{d}\xi) \tag{46}$$

From (42) and (46) it follows that (35) is valid for the continuous function u on $H \cup \partial E$ and therefore it also holds for the non-negative Borel function u. By (35) and (43) we derive (37), and the proof is completed. QED

Theorem 11. Assuming that u is a continuous function on E^*, then

$$\int_{E^*} u(\xi) \mu(\mathrm{d}\xi) = E_r u(x_\beta) = \sum_j u(j) \eta(j) \left(1 - \sum_s p_{js}\right) + \lim_{n \to \infty} \sum_{i,j} r(i) p_{ij}^n u(j) \tag{47}$$

Proof. When u is a continuous function on E^*,

$$E_i u(x_\infty) = \lim_{n\to\infty} E_i u(x_n) = \lim_{n\to\infty} (p^n u)_i$$

By this and (39) we obtain (47). QED

7.3.6 *h*-Chain and the Martin representation of an excessive function

Suppose that h is a P-excessive function, and moreover is γ-integrable, that is, $\sum_i \gamma_i h_i < \infty$. On account of the corollary to Theorem 3, h is finite-valued. We write

$$E^h = \{i: h_i > 0\} \qquad p^h_{ij} = \frac{p_{ij} h_j}{h_i} \qquad i, j \in E^h \tag{48}$$

By the excessive property of h we can easily get

$$\begin{aligned} p_{ij} &= 0 \qquad i \notin E^h, j \in E^h \\ f_{ij} &= 0 \qquad i \notin E^h, j \in E^h \end{aligned} \tag{49}$$

whereas $P^h = (p^h_{ij}, i, j \in E^h)$ is a matrix that is non-negative and whose row sum is not more than 1. The MC $X_T = \{x_n, n \leqslant \beta\}$, whose one-step transition matrix is P^h, is called an h-chain. When $h \equiv 1$ or constant, the h-chain becomes an MC whose one-step transition matrix is P. Henceforth, all the characteristics of the h-chain will be preceded by h. For example,

$$G^h = \sum_{n=0}^{\infty} (P^h)^n$$

measure P^h_i and so on.

Clearly, the measure $\gamma^h_i = \gamma_i h_i$ ($i \in E^h$) is a standard measure for the h-chain, and furthermore

$$P^{hn}_{ij} = \frac{P^n_{ij} h_j}{h_i} \qquad i, j \in E^h \tag{50}$$

Hence

$$G^h(i, j) = \frac{G(i, j) h_j}{h_i} \qquad i, j \in E^h \tag{51a}$$

$$\eta^h(j) = \eta(j) h_j \qquad j \in E^h \tag{51b}$$

where $\eta^h = \gamma^h G^h$, $\eta = \gamma G$.

By (51a) it can be seen that for an h-chain the state decomposition theorem in Theorem 2.2 takes the following form:

$$E^h = E^h_0 \cup \left(\bigcup_{a \in \mathscr{A}} E^h_a \right) \tag{52a}$$

where

$$E_0^h = E^h \cap E_0 \qquad E_a^h = E^h \cap E_a \qquad a \in \mathscr{A} \tag{52b}$$

Let the stopping state set for the h-chain be denoted by H^h, that is,

$$H^h = \left\{ i : i \in E^h, \sum_{j \in E^h} p_{ij}^h < 1 \right\} = \left\{ i : i \in E^h, \sum_{j \in E} p_{ij} h_j < h_i \right\} \tag{52c}$$

Then we have

$$H^h \subset E_0^h \subset E_0 \tag{52d}$$

Suppose $i, j \in E^h$. By (49)

$$f_{ij}^{hn} = \sum{}' p_{ij_1}^h p_{j_1 j_2}^h \cdots p_{j_{n-1} j}^h = \frac{1}{h_i} \sum{}'' p_{ij_1} p_{j_1 j_2} \cdots p_{j_{n-1} j} h_j = \frac{f_{ij}^n h_j}{h_i} \tag{53}$$

where $\sum'$ represents summing $j_l \neq j$, $j_1 \in E^h$, $1 \leqslant l \leqslant n-1$ while $\sum''$ indicates summing $j_l \neq j$, $j_l \in E$, $1 \leqslant l \leqslant n-1$. By (51b) we obtain

$$f_{ij}^h = \frac{f_{ij} h_j}{h_i} \qquad i, j \in E^h \tag{54}$$

Thus

$$A_j^h = \sum_{i \in E^h} \gamma_i^h f_{ij}^h = \sum_{i \in E} \gamma_i f_{ij} h_j = A_j h_j \qquad j \in E^h \tag{55}$$

Consequently the Martin kernel of P^h corresponding to γ^h is

$$K^h(i, j) = \frac{f_{ij}^h}{A_j^h} = \frac{K(i, j)}{h_i} \qquad i, j \in E^h \tag{56}$$

From the foregoing formula we can see that if $\{K^h(i, j_n)\}$ is a Cauchy sequence of real numbers, then so is $\{K(i, j_n)\}$. Therefore by Theorem 7 we can find that the Martin topology of the h-chain is identical with that of the 1-chain. Hence the Martin sequentially compact space E^{h*} of the h-chain can be looked on as a closed subspace of E^*, that is E^{h*} is the closure of E^h in E^*. And the Martin boundary ∂E^h of the h-chain is precisely the boundary of E^h *in* E^*, i.e.

$$\partial E^h = \partial E \cap E^{h*}$$

Suppose that $\{x_n, n \leqslant \beta\}$ is an h-chain; let the initial distribution be γ^h. We write μ_h for the distribution of the ultimate state x_β. The support of μ_h is contained in $H^h \cup \partial E^h$, but μ_h can be looked on as a measure on E^{h*} and more reasonably as a measure on E^*, that is

$$\mu_h(\Gamma) = p_{\gamma^h}^h(x_\beta \in \Gamma) \qquad \Gamma \in \mathscr{E}^* \tag{57}$$

And it follows that

$$\mu_h(E^*) = \sum_{i\in E^h} \gamma_i^h P_i^h(x_\beta \in E^*) = \sum_{j\in E^h} \gamma_i^h = \sum_i \gamma_i h_i \tag{58}$$

$$\mu_h(E^* - E^{h*}) = 0 \tag{59}$$

Theorem 12. Assume u to be a continuous function or a non-negative Borel function on E^*, then when $i\in E^h$,

$$E_i^h u(x_\beta) = \frac{1}{h_i}\int_{E^*} K(i,\xi)u(\xi)\mu_h(\mathrm{d}\xi) \tag{60}$$

$$E_i^h u(x_\beta) L_{\beta<\infty} = \frac{1}{h_i}\sum_j G(i,j)u(j)\left(h_j - \sum_s p_{js}h_s\right) \tag{61}$$

$$E_i^h u(x_\beta) L_{\beta=\infty} = \frac{1}{h_i}\int_{\partial E} K(i,\xi)u(\xi)\mu_h(\mathrm{d}\xi) \tag{62}$$

Especially, when $i\in E^h$,

$$P_i^h(x_\beta\in\Gamma) = \frac{1}{h_i}\int_\Gamma K(i,\xi)\mu_h(\mathrm{d}\xi) \tag{63}$$

$$\begin{aligned} P_i^h(x_\beta = j) &= P_i^h(x_\beta = j, \beta<\infty) \\ &= \frac{1}{h_i} G(i,j)\left(h_i - \sum_s p_{js}h_s\right) \qquad j\in E_0 \end{aligned} \tag{64}$$

$$\begin{aligned} P_i^h(x_\beta\in\Gamma) &= P_i(x_\beta\in\Gamma, \beta=\infty) \\ &= \frac{1}{h_i}\int_\Gamma K(i,\xi)\mu_h(\mathrm{d}\xi) \qquad \Gamma\subset\partial E \end{aligned} \tag{65}$$

$$\mu_h(j) = \eta(j)\left(h_j - \sum_s p_{js}h_s\right) \qquad j\in E_0 \tag{66}$$

If u is a continuous function on E^*, then

$$\int_{E^*} u(\xi)\mu_h(\mathrm{d}\xi) = \sum_j u_j\eta_j\left(h_j - \sum_s p_{js}h_s\right) + \lim_{n\to\infty}\sum_{i,j}\gamma_i p_{ij}^n h_j u_j \tag{67}$$

Proof. Applying the conclusions of Theorems 10 and 11 to the h-chain and noting (56) as well as $E_0^h\subset E_0$, we can prove the theorem. QED

Theorem 13. Suppose that h is a γ-integrable excessive function. Then

$$h_i = \int_{E^*} K(i, \xi)\mu_h \, d\xi \tag{68a}$$

$$= \sum_j G(i, j)\left(h_j - \sum_s p_{js} h_s \right) + \int_{\partial E} K(i, \xi)\mu_h(d\xi) \tag{68b}$$

Proof. Letting $u \equiv 1$ in (60) it follows that (68a) is valid for $i \in E^h$. Suppose $i \notin E^h$, that is $h_i = 0$. By (49) we have $K(i, j) = 0$ $(j \in E^h)$, so that $K(i, \xi) = 0$ $(\xi \in E^{h*})$. Noticing (59),

$$\int_{E^*} K(i, \xi)\mu_h(d\xi) = \int_{E^{h*}} K(i, \xi)\mu_h(d\xi) = 0$$

Therefore (68a) is also true for $i \notin E^h$.

Using (66) we obtain (68b) from (68a). The proof is terminated. QED

Remark

The measure μ_h is called the spectral measure of h whereas (68a) is said to be the Martin representation of the excessive function h.

7.3.7 Essential Martin boundary

Suppose $j \in E$, then $K(\cdot, j)$ is an excessive measure; hence the measure $\mu_{K(\cdot, j)}$ can be well defined. By (58) and (24),

$$\mu_{K(\cdot, j)}(E^*) = \sum_i r_i K(i, j) = 1 \tag{69}$$

Theorem 14. Suppose δ_j is the unit measure concentrated at j, then

$$\mu_{K(\cdot, j)} = \delta_j \qquad j \in E_0 \cup (\partial E)_1 \tag{70}$$

Proof. Write $h = K(\cdot, j)$. When $j \in E_0$, then

$$(ph)_i = \sum_s p_{is} K(s, j) = \frac{G(i, j) - \delta_j(i)}{\eta(j)} = h_i - \frac{\delta_j(i)}{\eta(j)}$$

and so

$$h_j - (ph)_j = 1/\eta(j)$$

Thereupon by (66), $\mu_{K(\cdot, j)}(j) = 1$. Observing (69) we have (70).

When $j = \xi_a$ $(a \in \mathscr{A})$, we have

$$K(i, \xi_a) = f_i(\xi_a)/A(\xi_a)$$

and

$$\mu_h(\xi_a) = \sum{}' \gamma_i^h p_i^h(x_\beta = \xi_a) = \sum{}' \gamma_i^h f_i^h(\xi_a)$$

where $\sum'$ represents summation over E^h. Consider (54) and $h(\xi_a) = K(\xi_a, \xi_a) = 1/A(\xi_a)$ and then

$$\mu_h(\xi_a) = \sum' \gamma_i h_i \frac{f_i(\xi_a)h(\xi_a)}{h_i}$$

$$= \sum_i \gamma_i f_i(\xi_a) \frac{1}{A(\xi_a)} = 1 \tag{71}$$

Again applying (69) we obtain $\mu_{K(\cdot,\xi_a)} = \delta_{\xi_a}$. The proof is concluded. QED

Definition 7. Let

$$B = \{\xi : \xi \in \partial E, \mu_{K(\cdot,\xi)} = \delta_\xi\} \tag{72}$$

B is called the essential Martin boundary. If $\xi \in B$ and $\mu(\xi) > 0$, then ξ is said to be an atomic boundary point. The set B_1 composed of all the atomic boundary points is called the atomic boundary of the chain; $B_2 = B - B_1$ is called the non-atomic boundary of the chain. $B = B_1 \cup B_2$.

By Theorem 14, $\xi_a \in B$. Again since

$$\mu(\xi_a) = p_\gamma(x_\beta = \xi_a) = \sum_i \gamma_i p_i(x_\beta = \xi_a)$$

$$= \sum_i \gamma_i f_i(\xi_a) = A(\xi_a) > 0$$

it follows that every recurrent boundary point ξ_a is an atomic boundary point, that is,

$$(\partial E)_1 \subset B_1 \tag{73}$$

Theorem 15. Suppose $\xi \in B$, then $K(\cdot, \xi)$ is a harmonic function; moreover

$$\sum_i \gamma_i K(i, \xi) = 1 \qquad \xi \in B \tag{74}$$

Proof. Write $h = K(\cdot, \xi)$; since $\mu_h = \delta_\xi$ from (68b) we get $Gf = 0$, where $f = h - ph$. Accordingly,

$$\sum_j \eta(j) f(j) = \sum_j \gamma_j (Gf)_j = 0$$

and thus $f = 0$. That is, $K(\cdot, \xi)$ $(\xi \in B)$ is harmonic. Again by (58) and (3) when $\xi \in B$,

$$1 = \mu_{K(\cdot,\xi)} = \sum_i \gamma_i K(i, \xi) \leqslant 1$$

From what precedes, (74) follows. The proof is completed. QED

Theorem 16. B is a Borel subset of ∂E. For any γ-integrable excessive function h, we have

$$\mu_h(\partial E - B) = 0$$

Proof. As $\xi \in \partial E$, by (58) and (31)

$$\mu_{K(\cdot,\xi)}(E^*) = \sum_i \gamma_i K(i,\xi) \leqslant 1$$

so that

$$B = \{\xi : \xi \in \partial E, \mu_{K(\cdot,\xi)}(\xi) = 1\} \tag{75a}$$

For $\xi \in \partial E$, by (67),

$$\begin{aligned}\mu_{K(\cdot,\xi)}(\xi) &= \lim_{m\to\infty} \int_{E^*} \mathrm{e}^{-md(\zeta,\xi)} \mu_{K(\cdot,\xi)}(\mathrm{d}\zeta) \\ &= \lim_{m\to\infty} \sum_j \mathrm{e}^{md(j,\xi)} \eta_i \left(K(j,\xi) - \sum_s p_{js} K(s,\xi) \right) \\ &\quad + \lim_{m\to\infty} \lim_{n\to\infty} \sum_{i,j} \gamma_i p_{ij}^n K(j,\xi) \mathrm{e}^{-md(i,j)} \end{aligned} \tag{75b}$$

From this we know B is a Borel set.

Suppose $\xi \in \partial E$; simply write $K = K(\cdot, \xi)$. Assume φ and ψ are both continuous functions on E^*, $m \geqslant 0$, $n \geqslant 0$. Then

$$\begin{aligned} E_{rh}^h \varphi(x_n)\psi(x_{n+m}) &= \sum_{i,j,s} \gamma_i h_i p_{ij}^{hn} \varphi(j) p_{js}^{hm} \psi(s) \\ &= \sum_{i,j} \gamma_i p_{ij}^n h_j \varphi(j) E_j^h \psi(x_m) \end{aligned}$$

Let $m \to \infty$; by an application of (62) we have

$$\begin{aligned} E_{rh}^h \varphi(x_n)\psi(x_\infty) &= \sum_{i,j} \gamma_i p_{ij}^n h_j \varphi(j) \int_{\partial E} K^h(j,\xi)\psi(\xi)\mu_h(\mathrm{d}\xi) \\ &= \int_{\partial E} \sum_{i,j} \gamma_i K(i,\xi) p_{ij}^K(n) \varphi(j)\psi(\xi)\mu_h(\mathrm{d}\xi) \\ &= \int_{\partial E} E_{rk}^K \varphi(x_n)\psi(\xi)\mu_h(\mathrm{d}\xi) \end{aligned} \tag{76}$$

Considering

$$E_{rh}^K \varphi(x_\infty) = \int_{\partial E} \varphi(\zeta)\mu_K(\mathrm{d}\zeta)$$

and letting $n\to\infty$ in (76), we obtain

$$\int_{\partial E}\varphi(\xi)\psi(\xi)\mu_h\,\mathrm{d}\xi=\int_{\partial E}\left(\int_{\partial E}\varphi(\zeta)\mu_K\,\mathrm{d}\zeta\right)\psi(\xi)\mu_h(d\xi)$$

Since the continuous function ψ can be arbitrarily selected, from the formula above it follows that

$$\varphi(\xi)=\int_{\partial E}\varphi(\zeta)\mu_{K(\cdot,\xi)}(\mathrm{d}\zeta)$$

holds for μ_h-almost all $\xi\in\partial E$. In particular, taking φ as

$$\varphi_m(\zeta)=\mathrm{e}^{-md(\zeta,\xi)}\qquad m=1,2,3,\ldots$$

we get that

$$1=\int_{\partial E}\mathrm{e}^{-md(\zeta,\xi)}\mu_{K(\cdot,\xi)}(\mathrm{d}\zeta)\qquad m=1,2,\ldots$$

holds for μ_h-almost all $\xi\in E$. Letting $m\to\infty$ in the foregoing formulae we have that

$$1=\mu_{K(\cdot,\xi)}(\xi)$$

holds for μ_h almost all $\xi\in\partial E$. From (75) it follows immediately that $\mu_h(\partial E-B)=0$. The proof is completed. QED

Corollary

Write the essential Martin boundary of an h-chain as B^h, and then

$$B^h=B\cap E^{h*}$$

Proof. Similar to (75),

$$B^h=\{\xi:\xi\in\partial E^h,\mu_{K^h(\cdot,\xi)}(\xi)=1\}$$

However $\partial E^h=\partial E\cap E^{h*}$. When $\xi\in\partial E^h$ according to (75b) we may calculate $\mu_{K^h(\cdot,\xi)}(\xi)$:

$$\begin{aligned}\mu_{K^h(\cdot,\xi)}(\xi)=&\lim_{m\to\infty}\sum_{j\in E^h}e^{md(j,\xi)}\eta_j^h\left(K^h(j,\xi)-\sum_{s\in E^h}p_{js}^hK^h(s,\xi)\right)\\&+\lim_{m\to\infty}\lim_{n\to\infty}\sum_{i,j\in E^h}\gamma_i^hp_{ij}^{hn}K^h(j,\xi)\mathrm{e}^{-md(i,j)}\end{aligned}$$

Noting (49)–(52) and (56) we can see that the right-hand side of the formula above becomes the right-hand side of (75b); thus $\mu_{K(\cdot,\xi)}(\xi)=\mu_{K^h(\cdot,\xi)}(\xi)$. From this we can prove $B^h=B\cap E^{h*}$. The proof is completed. QED

Because of Theorem 16, Theorem 9 can be strengthened as follows:

Theorem 17. Suppose $X_T = \{x_n, n \leqslant \beta\}$ is a 1-chain. Then for P_γ-almost all $\omega \in \Omega$, either

$$x_\beta \in H \qquad \text{if } \beta < \infty \tag{77}$$

or there exists a limit

$$d - \lim_{n \to \infty} x_n = x_\infty \in B \qquad \text{if } \beta = \infty \tag{78}$$

Equations (60) and (62) can be strengthened like this:

$$E_i^h u(x_\beta) = \frac{1}{h_i} \int_{E_0 \cup B} K(i, \xi) u(\xi) \mu_h(\mathrm{d}\xi) \tag{79}$$

$$E_i^h u(x_\beta) 1_{\beta = \infty} = \frac{1}{h_i} \int_B K(i, \xi) u(\xi) \mu_h(\mathrm{d}\xi) \tag{80}$$

Especially

$$E_i u(x_\beta) = \int_{E_0 \cup B} K(i, \xi) u(\xi) \mu(\mathrm{d}\xi) \tag{81}$$

$$E_i u(x_\beta) 1_{\beta = \infty} = \int_B K(i, \xi) u(\xi) \mu(\mathrm{d}\xi) \tag{82}$$

7.3.8 Uniqueness of Martin representation

Theorem 18. Every γ-integrable excessive function h can be uniquely represented by

$$h_i = \int_{E_0 \cup B} K(i, \xi) \lambda(\mathrm{d}\xi) \tag{83}$$

where λ is a totally finite measure on the Borel set $E_0 \cup B$; therefore $\lambda = \mu_h$. Conversely, given any arbitrary totally finite measure λ on $E_0 \cup B$, (83) defines a γ-integrable excessive function. This function is harmonic if and only if $\lambda(E_0) = 0$.

Proof. Let $u = 1$ in (79). It follows that h has the representation (83), where $\lambda = \mu_h$ is a measure on $E_0 \cup B$, and the total finiteness of μ_h can be derived from (58).

Now we assume that h has the representation (83). We proceed to prove $\lambda = \mu_h$. For the continuous function u on E^*, applying h and $K(\cdot, \xi)$ to (67) respectively and noting (83), we have

$$\int_{E^*} u(\xi)\mu_h(\mathrm{d}\xi) = \int_{E_0 \cup B} \left(\int_{E^*} u(\xi)\mu_{K(\cdot,\zeta)}(\mathrm{d}\xi) \right) \lambda(\mathrm{d}\zeta)$$

But $\mu_{K(\cdot,\zeta)} = \delta_\zeta$ ($\zeta \in E_0 \cup B$); hence

$$\int_{E^*} u(\xi)\mu_h(\mathrm{d}\xi) = \int_{E_0 \cup B} u(\xi)\lambda(\mathrm{d}\xi)$$

Because the mass of μ_h all concentrates on $E_0 \cup B$, on account of the arbitrariness of u we obtain $\lambda = \mu_h$ from the formula above.

Since $K(\cdot, \xi)$ is an excessive function, and moreover (31) is valid, it follows that (83) defines a γ-integrable excessive function. By Theorem 15, $K(\cdot, \xi)$ ($\xi \in B$) is harmonic; therefore if $\lambda(E_0) = 0$, then the h in (83) is harmonic. Conversely, assume that h in (83) is harmonic. By (66) we obtain $\mu_h(j) = 0$ for all $j \in E_0$, that is, $\lambda(E_0) = \mu_h(E_0) = 0$. The proof is completed. QED

7.3.9 Minimal excessive function

Recall Definition 2 in 7.3.1 of the minimal excessive function. We have:

Theorem 19. The general form of the γ-integrable minimal excessive function is $CK(\cdot, \xi)$, $\xi \in E_0 \cup B$, where C is a constant.

Proof. It can be seen from (68) that $\mu_{h_1+h_2} \simeq \mu_{h_1} + \mu_{h_2}$. Suppose $\xi \in E_0 \cup B$ and $K(\cdot, \xi) = h_1 + h_2$, and furthermore both h_1 and h_2 are excessive functions. Then $\mu_{h_1+h_2} = \mu_k(\cdot, \xi) = \delta_\xi$. Accordingly,

$$\mu_{h_1}(E^* - \xi) + \mu_{h_2}(E^* - \xi) = \delta_\xi(E^* - \xi) = 0$$

and it follows that

$$\mu_{h_e}(E^* - \xi) = 0\,(e = 1, 2)$$

According to the spectral representation (68) of h_e,

$$h_e(i) = \int_{E^*} K(i, \zeta)\mu_{h_e}(\mathrm{d}\zeta)$$

$$= K(i, \xi)\mu_{h_e}(\xi) = C_e K(i, \xi)$$

where $C_e = \mu_{h_e}(\xi)$ is a constant. Therefore $K(\cdot, \xi)$ ($\xi \in E_0 \cup B$) is a minimal excessive function.

Secondly, suppose h is a γ-integrable minimal excessive function, and let μ_h be its spectral measure. Also we may as well suppose $h \neq 0$. By (58) and the corollary to Theorem 3, $\mu_h(E_0 \cup B) > 0$. Hence there exists $\xi \in E_0 \cup B$ such that μ_h has a positive measure in any neighbourhood of ξ. Let $U_n = \{\zeta : d(\zeta, \xi) < 1/n\}$,

$h_n = \int_{U_n} K(\cdot, \zeta)\mu_h \mathrm{d}\zeta$. Then both h_n and

$$h - h_n = \int_{(E_0 \cup B) - U_n} K(\cdot, \zeta)\mu_h(\mathrm{d}\zeta)$$

are excessive functions. Owing to the minimality of h, $h_n = C_n h$ (C_n is a constant), so that

$$\sum_i \gamma_i h_n(i) = C_n \sum_i \gamma_i h(i)$$

Again on account of the uniqueness of the representation of h_n, we get

$$\mu_{hn}(\Gamma) = \mu_h(U_n \cap \Gamma)$$

Considering (58),

$$\mu_h(U_n) = \mu_{hn}(E^*) = \sum_i \gamma_i h_n(i)$$

$$\mu_h(E^*) = \sum_i \gamma_i h(i)$$

Thus

$$h_i = \frac{\mu_h(E^*)}{\mu_h(U_n)} \int_{U_n} K(i, \zeta)\mu_h(\delta\zeta)$$

Letting $n \to \infty$ we obtain $h_i = CK(i, \xi)$, where $C = \mu_h(E^*)$. The proof is terminated.
QED

Theorem 20. The essential Martin boundary B is

$$B = \left\{ \xi : \xi \in \partial E, K(\cdot, \xi) \text{ is a minimal harmonic function}, \sum_i \gamma_i K(i, \xi) = 1 \right\} \tag{84}$$

Proof. Write the set on the right-hand side of (84) as C. By Theorems 15 and 19, $B \subset C$. Suppose $\xi \in C$. Then $h \equiv K(\cdot, \xi)$ is γ-integrable. Write μ_h for its spectral measure. Evidently h has the representation

$$h_i = \int_{E_0 \cup B} K(i, \zeta)\delta_\xi(\mathrm{d}\zeta)$$

By the uniqueness theorem, Theorem 18, $\mu_h = \delta_\xi$, that is $\xi \in B$, $C \subset B$. The proof is concluded.
QED

7.3.10 Ultimate field and ultimate random variables

The infinite-dimensional function $f(j_0, j_1, j_2, \ldots)$ ($j_k \in E, k = 0, 1, 2, \ldots$) is said to be invariant if for any $j_k \in E$, $k = 0, 1, 2, \ldots$, $f(j_0, j_1, j_2, \ldots) = f(j_1, j_2, j_3, \ldots)$ holds.

Definition 8. A function Φ defined on Ω_∞ is called an ultimate random variable of the chain if there exists an invariant function f such that

$$\Phi(\omega) = f\{x_n(\omega), x_{n+1}(\omega), \ldots\} \qquad \omega \in \Omega_\infty$$

holds for $n = 0$ and therefore for all $n \geqslant 0$. We define an ultimate random variable Φ to take the value 0 on $\Omega_F = \Omega - \Omega_\infty$. We call the set $\Lambda \subset \Omega_\infty$ an ultimate set if the indicator I_Λ is an ultimate random variable. The Borel field on Ω_∞ composed of ultimate sets is denoted by $\mathscr{B}_\infty$ and is called the ultimate field.

Suppose $\Lambda_1, \Lambda_2 \in \mathscr{F}$. If $P_\gamma\{(\Lambda_1 - \Lambda_2) + (\Lambda_2 - \Lambda_1)\} = 0$, then we write $\Lambda_1 \simeq \Lambda_2$. If Φ_1 and Φ_2 are random variables and $P_\gamma(\Phi_1 \neq \Phi_2) = 0$, then we write $\Phi_1 \simeq \Phi_2$.

Theorem 21. There exists a one-to-one correspondence between the non-null ultimate set and the almost closed set A (modulo transient sets) under the following condition:

$$\Lambda \simeq \mathscr{L}(A) \tag{85}$$

A can be taken as $A = \{i : P_i(\Lambda) > \varepsilon\}$, $0 < \varepsilon < 1$.

Proof. See Chung (1967, I.17, Theorem 1) QED

Theorem 22. There exists a one-to-one correspondence between the non-negative bounded P-harmonic function u and the non-negative bounded ultimate random variable Φ under the following condition:

$$u_i = E_i\Phi \qquad \Phi = \lim_{n\to\infty} u(x_n) \tag{86}$$

Proof. See Chung (1967, I.17, Theorem 5). QED

Definition 9. The non-null ultimate set $\Lambda \in \mathscr{B}_\infty$ is said to be atomic if Λ cannot be decomposed into the union of two non-null ultimate sets; the non-null ultimate set $\Lambda \in B_\infty$ is said to be completely non-atomic if for any $\Lambda_1 \subset \Lambda$, $\Lambda_1 \in \mathscr{B}_\infty$, then Λ_1 is not atomic.

Theorem 23. Ω_∞ has the following decomposition:

$$\Omega_\infty = \Lambda_0 + \left(\bigcup_{a \in \mathscr{A}} \Lambda_a\right) \tag{87}$$

where Λ_0 is a completely non-atomic ultimate set and can be dropped. $\mathscr{A}$ is an empty set or a finite set or a denumerable infinite set, and moreover $0 \notin \mathscr{A}$. Also, Λ_a $(a \in \mathscr{A})$ is an atomic ultimate set. The decomposition is unique.

Proof. Making use of Theorem 2.3 and Theorem 5.6, which is to be proved latter, we obtain the proof of this theorem. QED

7.3.11 Martin entrance boundary

The Martin boundary discussed previously is actually the Martin exit boundary, that is, to depict how the process 'goes to infinity' from a finite state. As for the description of how the process comes to a finite state from infinity, we need the Martin entrance boundary.

Suppose the decomposition of the state space E for the Markov chain $P = (p_{ij}, i, j \in E)$ is (2.15) and E_0 is the set of non-recurrent states, we write

$$E_{00} = \{i: i \in E_0, \text{ there exists } j \in E - E_0 \text{ such that } i \Rightarrow j\}$$

$$E_1 = E - E_{00} \qquad P_1 = (p_{ij}, i, j \in E_1)$$

Zhen-ting Hou (1974, Chapter 8) points out: any finite excessive measure $(\alpha_j, j \in E)$ of P must be zero on E_{00}, that is, $\alpha_j = 0 (j \in E_{00})$, and moreover, there exist excessive measures that are positive on E_1. We fix such a P-excessive measure $(\alpha_j, j \in E)$, then the measure $(V_j, j \in E)$ is a finite excessive (finite harmonic) measure of P if and only if $V_j = 0$ $(j \in E_{00})$ and $(V_j/\alpha_j, j \in E_1)$ is the finite excessive (finite harmonic) function of $\tilde{P}$. Here

$$\tilde{P} = (\tilde{p}_{ij}, i, j \in E_1) \qquad \tilde{p}_{ij} = \alpha_j p_{ji}/\alpha_i$$

Select a standard measure $\tilde{\gamma}$ of $\tilde{P}$. According to $\tilde{\gamma}$ and $\tilde{P}$, we can well define the Martin boundary $\partial\tilde{E}_1$ of $\tilde{P}$, the essential Martin boundary $\tilde{B}$, the atomic boundary $\tilde{B}_1$ and the non-atomic boundary $\tilde{B}_2$. We call $\partial\tilde{E}_1$, $\tilde{B}, \tilde{B}_1$ and $\tilde{B}_2$ respectively the Martin entrance boundary of P, the essential Martin entrance boundary, the atomic entrance boundary and the non-atomic entrance boundary. Of course the boundaries mentioned above are relative to α and $\tilde{\gamma}$. Therefore the study of the entrance boundary and excessive measure can be obtained with the help of the study of the exit boundary and excessive function. We are not going to discuss it in detail.

7.4 PROBABILITY REPRESENTATION OF A STOPPING POTENTIAL

Definition 1. We call a non-negative (including $+\infty$) function u the potential of the chain P if there exists a non-negative function v such that

$$u = Gv = \sum_{n=0}^{\infty} P^n v$$

that is,

$$u_i = \sum_j G(i, j) v_j \qquad i \in E \tag{1}$$

The non-zero potential u is said to be minimal if, when $u = u^1 + u^2$, where u^1 and u^2 are potentials, then $u = c^a u^a$ $(a = 1, 2,)$, where c^a is a constant.

If $u = Gv$, then

$$Pu = PGv = (G - I)v = u - v$$

Hence if $u_i < \infty$, then $v_i = u_i - (Pu)_i$ is uniquely determined by u.

Definition 2. We call u the stopping potential of P if $u = Gv$, and furthermore $v_i = 0$ $(i \in E - H)$, where H is the stopping state set.

Theorem 1. u is a stopping potential of P if and only if there exists a non-negative function f_a $(a \in H)$ such that

$$u_i = E_i\{f_{x(\beta)}, \beta < \infty\} = \sum_{a \in H} f_a u_i^a \qquad i \in E \tag{2}$$

where

$$u_i^a = P_i\{x_\beta = a\} = G(i, a) d_a \qquad a \in H \tag{3}$$

is a minimal potential whereas d_a is the stopping quantity of the state a.

If u is finite then there exists a one-to-to correspondence between u and f, and moreover f_a $(a \in H)$ is also finite

Proof. Sufficiency: By (3.77),

$$\begin{aligned} u_i &= E_i\{f_{x(\beta)}, \beta < \infty\} \\ &= E_i\{f_{x(\beta)}, \beta < \infty, x(\beta) \in H\} \\ &= \sum_{n=0}^{\infty} E_i\{f_{x(n)}, \beta = n, x(n) \in H\} \\ &= \sum_{n=0}^{\infty} \sum_{a \in H} p_{ia}^n d_a f_a \\ &= \sum_{a \in H} G(i, a) d_a f_a \end{aligned} \tag{4}$$

Hence the u determined by (2) is a stopping potential. In particular, (3) is valid.

Now we are going to prove that u^a is a minimal potential. Suppose

$$u^a = u_1 + u_2 \qquad u_b = Gv_b \qquad (b = 1, 2)$$

are potentials. Then

$$d_a \delta_a(i) = (u^a - pu^a)_i = \sum_{b=1}^{2} (u_b - pu_b)_i = v_1(i) + v_2(i)$$

Thereby we obtain $v_b = c^b(d_a\delta_a)(b=1,2, c^b \geqslant 0$ are constants); consequently $u_b = c^b u^a (b=1,2)$.

Necessity: Suppose $u = Gv$ is a stopping potential. Take $f_a = v_a/d_a(a \in H)$. By (4) we know that (2) holds.

The one-to-one correspondence follows from the paragraph preceding Definition 2. The proof is terminated. QED

7.5 SOJOURN SOLUTION, ULTIMATE SET, ALMOST CLOSED SET AND BOUNDARY

Theorem 1. The (non-negative) bounded harmonic function h corresponds one-to-one to the non-negative bounded Borel function f on B according to the following relation:

$$h_i = \int_B K(i,\xi) f(\xi)\mu(\mathrm{d}\xi) \tag{1}$$

$$\lim_{n\to\infty} h(x_n) = f(x_\infty) \qquad E_i f(x_\infty) = h_i \tag{2}$$

where the measure μ is the distribution of the ultimate state, determined by (3.34).

Proof. According to Theorem 3.18, the function h defined by (1) is bounded and harmonic. Conversely, suppose h is a bounded harmonic function. We shall assume its bound to be 1. Then both h and $g = 1 - h$ are excessive. By Theorem 3.18, $\mu_h(E_0) = 0$, and moreover

$$1 = \int_{E_0 \cup B} K(i,\xi)u(\mathrm{d}\xi) = \int_{E_0 \cup B} K(i,\xi)(\mu_h + \mu_g)(\mathrm{d}\xi)$$

Therefore $\mu = \mu_h + \mu_g$, and μ_h has a density function f bounded by 1 relative to $\mu = \mu_h(\mathrm{d}\xi) = f(\xi)\mu(\mathrm{d}\xi)$. Hence (3.83) becomes (1).

Rewrite (1) as $h_i = E_i\{f(x_\beta), \Omega_\infty\}$. Consequently the formula (2) follows from Theorem 3.22. The proof is completed. QED

Theorem 2. Write $\mathscr{B}_{x(\infty)}$ for the Borel field on Ω_∞ produced by sets $\{x(\infty) \in \Gamma\}$ (Borel sets $\Gamma \subset B$), then $\mathscr{B}_\infty = \mathscr{B}_{x(\infty)}$.

Proof. Since the non-negative bounded $\mathscr{B}_{x(\infty)}$-measurable function Φ can be written in the form $f(x_\infty)$, where f is a non-negative bounded Borel function on B, it follows that by (2) we know $\Phi = f(x_\infty)$ is $\mathscr{B}_\infty$-measurable. That is, $\mathscr{B}_{x(\infty)} \subset \mathscr{B}_\infty$.

Suppose that Φ is a non-negative bounded $\mathscr{B}_\infty$-measurable function. According to Theorem 3.22, $h_i = E_i\Phi$ is bounded harmonic and $\Phi = \lim_{n\to\infty} h_{x(n)}$.

By Theorem 1, there exists a bounded Borel-measurable function f on B such that (2) holds. As a result, $\Phi = f(x_\infty)$. Consequently Φ is $\mathscr{B}_{x(\infty)}$-measurable, that is, $\mathscr{B}_\infty \subset \mathscr{B}_{x(\infty)}$, and the proof is completed. QED

Theorem 3. For any almost closed set A, there exists a Borel set $\Gamma \subset B$ such that

$$\mathscr{L}(A) \simeq \{x_\infty \in \Gamma\} \tag{3}$$

Apart from μ-null sets, Γ is uniquely defined.

Proof. As $\mathscr{L}(A) \in \mathscr{B}_\infty$, making use of Theorem 2 we can prove the theorem. QED

Definition 1. Call non-zero $u \in \mathscr{M}^+(1)$ a sojourn solution, if for any $v \in \mathscr{M}^+(1)$: $v \leqslant u, v \leqslant \bar{u} - u$ we have $v = 0$. Here

$$\bar{u}_i = P_i(\Omega_\infty) = P_i(\mathscr{L}(E)) \in \mathscr{M}^+(1) \tag{4}$$

Theorem 4. There exists a one-to-one correspondence between sojourn solutions u and the closed sets A (modulo transient sets) under the following condition:

$$u_i = P_i\{\mathscr{L}(A)\} \tag{5}$$

Proof. Suppose that u is a sojourn solution so that u is harmonic; according to Theorem 3.22, there exists a non-zero ultimate random variable $\Phi: 0 \leqslant \Phi \leqslant 1$ such that $u_i = E_i\Phi$. Evidently

$$\psi = \min\{\Phi, (1 - \Phi)I_{(\Phi < 1)}\}$$

is an ultimate random variable. Hence $v_i = E_i\psi \in \mathscr{M}^+(1)$, $v \leqslant u, v \leqslant \bar{u} - u$. So $v = 0$, and therefore $\psi = 0$. From this we obtain $\Phi = I_{(\Phi = 1)}$. Because $(\Phi = 1)$ is a non-negative ultimate set, according to Theorem 3.21, there exists an almost closed set A such that $(\Phi = 1) \simeq \mathscr{L}(A)$. Thus

$$u_i = E_i\Phi = E_i I_{(\Phi = 1)} = E_i I_{\mathscr{L}(A)} = P_i(\mathscr{L}(A))$$

Now suppose that u is defined by (5) and assume $v \in \mathscr{M}^+(1), v \leqslant u, v \leqslant \bar{u} - u$. According to Theorems 3.21 and 3.22,

$$I_{\mathscr{L}(E-A)} = \lim_{n\to\infty} u_{x(n)} \qquad I_{\mathscr{L}(E-A)} = \lim_{n\to\infty} (\bar{u}_{x(n)} - u_{x(n)})$$

and there exists an ultimate random variable $\Phi: 0 \leqslant \Phi \leqslant 1$, such that

$$v_i = E_i\Phi \qquad \Phi = \lim_{n\to\infty} v_{x(n)}$$

Since $v \leqslant \min(u, \bar{u} - u)$, it follows that $\Phi \leqslant \min(I_{\mathscr{L}(A)}, I_{\mathscr{L}(E-A)}) = 0$.

Accordingly $v_i = E_i\Phi = 0$, and the proof is concluded. QED

Theorem 5. The sojourn solution u, the non-null ultimate set Λ, the almost closed set A (modulo transient sets) and the μ-non-null Borel set Γ ($\subset B$ modulo μ-null sets) are one-to-one correspondent under the following condition:

$$u_i = P_i(\mathscr{L}(A)) \qquad \Lambda \simeq \mathscr{L}(A) \simeq \{x_\infty \in \Gamma\} \tag{6}$$

Proof. It suffices to summarize Theorems 3 and 4 and Theorem 3.21. QED

Definition 2. We say a sojourn solution u is minimal if u as a harmonic function is minimal; a sojourn solution u is said to be completely non-minimal if for any sojourn solution $v < u, v$ is not minimal.

Evidently, if the completely non-atomic A_0 appears in the Blackwell decomposition, then $u_i^0 = P(\mathscr{L}(A_0))$ is the greatest completely non-minimal sojourn solution.

As a special case of Theorem 5, we have the following theorem.

Theorem 6. The completely non-minimal sojourn solution $u(\leqslant u^0)$ and the completely non-atomic ultimate set $\Lambda(\subset \Lambda_0$, see (3.87)), the completely non-atomic almost closed set A ($\subset A_0$, Modulo transient sets) and the μ-non-null Borel set $\Gamma(\subset B_2$, modulo μ-null sets) are one-to-one correspondent according to (6). And, the minimal sojourn solution u, the atomic ultimate set Λ, the atomic almost closed set A (modulo transient sets) and the atomic boundary point $\xi \in B_1$ are one-to-one correspondent according to the following relation:

$$\Lambda \simeq \mathscr{L}(A) \simeq \{x_\infty = \xi\} \qquad u_i = P_i\{\mathscr{L}(A)\} \tag{7}$$

Theorem 7. The bounded harmonic function u, the non-negative bounded ultimate random variable Φ and the non-negative bounded Borel function f (defined on B, apart from the difference of the function values on μ-null sets) are one-to-one correspondent according to the following relation:

$$u_i = E_i\Phi \qquad \Phi \simeq \lim_{n\to\infty} u_{x(n)} \qquad f(x_\infty) \simeq \Phi \tag{8}$$

Proof. Summing up Theorem 3.22 and Theorem 1, the conclusions of this theorem follow. QED

7.6 CANONICAL PROCESS

Suppose that E has a discrete topology. Compactifying the denumerable set E by one point '∞' ($\infty \notin E$), we get $\bar{E} = E \cup \{\infty\}$. Assume σ to be a non-negative (including ∞) random variable defined on the complete probability space

$(\Omega, \mathscr{F}, P)$, and call $X(\omega)=\{x_t(\omega), t<\sigma(\omega)\}$[1] $(\omega\in\Omega)$ a homogeneous Markov process, or simply process, if for any $\omega\in\Omega$ and $t<\sigma(\omega)$ we have $x(t,\omega)\in\bar{E}$, but

$$P\{x(t)=\infty\}=0 \qquad t\geqslant 0 \tag{1}$$

While for any $l\geqslant 2, 0\leqslant t_1<t_2<\cdots<t_{l+1}, i_1, i_2, \ldots, i_{l+1}\in E$, provided that $P\{x(t_a)=i_a, 1\leqslant a\leqslant l\}>0$, we have

$$P\{x(t_{\mu+1})=i_{l+1}|x(t_a)=i_a, 1\leqslant a\leqslant l\}=P\{x(t_{l+1})=i_{l+1}|x(t_l)=i_l\} \tag{2}$$

And moreover the value of the right-hand side of the formula above is independent of t_1 and only depends on $t_{l+1}-t_l$. We call

$$p_{ij}(t)=P\{x(s+t)=j|x(s)=i\} \tag{3}$$

the transition probability of the process, whose transition probability matrix $P(t)=\{p_{ij}(t)\}$ must satisfy conditions (2.2.A) and (2.2.B) and is supposed to satisfy (2.2.C) (see section 2.2). According to the notation of section 2.1, that is, $P(t)\in\mathscr{P}$. If two processes X and $\bar{X}$ have the same transition probability matrix $P(t)$, we say that X and $\bar{X}$ are the same process.

Conversely, for every $P(t)\in\mathscr{P}$, there exists a process $X=\{x(t), t<\sigma\}$ with $P(t)$ as its transition probability matrix. According to Zi-kun Wang and Xiang-qun Yang (1988) we can also suppose X is well separable, Borel-measurable, and right-lower semicontinuous. That is, for every $\omega\in\Omega$.

$$\underline{\lim}_{s\downarrow t}\, x(s,\omega)=x(t,\omega)\in\bar{E} \qquad \text{for all } t<\sigma(\omega) \tag{4}$$

Such a process is said to be a canonical one. If for process X, (4) is valid for almost all $\omega\in\Omega$, and for exceptional $\omega, x(\omega)$ may even have no definition. In this case for the exceptional ω, we can revise or add to the definition as follows:

$$x(t,\omega)=i_0\in E \qquad t<\sigma(\omega) \tag{5}$$

Denote the class composed of all canonical processes X by $\mathscr{H}$. Evidently, $\mathscr{H}$ is one-to-one corresponding to $\mathscr{P}$. According to the notation in section 2.1, the class composed of processes $X\in\mathscr{H}$ corresponding to $P(t)\in\mathscr{P}_s$ is denoted by $\mathscr{H}_s$. The class composed of processes $X\in\mathscr{H}$ corresponding to $P(t)\in\mathscr{F}_s(Q)$ is denoted by $\mathscr{H}_s(Q)$ and X in $\mathscr{H}_s(Q)$ is known as a Q process.

When $P_i(\sigma=\infty)=1\,(i\in E)$, the process X is said to be honest. If and only if condition (2.2.D) is valid, X is honest.

For a stopping process $X=\{x(t), t<\sigma\}$, we can select $\Delta\notin\bar{E}$ arbitrarily, and then let

$$\tilde{x}(t)=\begin{cases} x(t) & \text{if } t<\sigma \\ \Delta & \text{if } t\geqslant\sigma \end{cases} \tag{6}$$

[1] Take $x_t(\omega)\equiv x(t,\omega)$ and we frequently omit ω.

Then $\tilde{X} = \{\tilde{x}(t), t < \infty\}$ is an honest process, whose transition probability matrix $\tilde{P}(t) = \{\tilde{p}_{ij}(t)\}$ $(i, j \in E \cup \{\Delta\})$ is well defined by (2.2.3).

Let $X \in \mathscr{H}$. Write $\mathscr{F}_t^0$ for the Borel field generated by the sets $\{x(s) = i\}$ $(s \leqslant t, i \in E)$. Write

$$\mathscr{F}_{t+0}^0 = \bigcap_{s>t} \mathscr{F}_s^0$$

Theorem 1.

$$\mathscr{F}_{t+0}^0 = \mathscr{F}_t^0$$

The proof is seen in Chung (1967, II.8, Theorem 1).

We call non-negative random variable β a Markov time for the process X, if $\beta \leqslant \sigma$, and for any $t \geqslant 0$,

$$(\beta \leqslant t < \sigma) \in \mathscr{F}_t^0 \tag{7}$$

or equivalently,

$$(\beta < t < \sigma) \in \mathscr{F}_t^0 \tag{8}$$

For a Markov time β, let $\mathscr{F}_\beta$ be the class of sets Λ that satisfy $\Lambda \subset \Omega_\beta \equiv (\beta < \sigma)$ and $\Lambda \cap (\beta \leqslant t < \sigma) \in \mathscr{F}_t^0, t \geqslant 0$; $\mathscr{F}_\beta$ is called the pre-β field. Let $\mathscr{F}'_\beta$ be the Borel field on Ω_β generated by the sets $\{x(\beta + t) = i\}$ $(i \in E, t \geqslant 0)$, it is called the post-β field.

For the Markov time β of X, we can define a set translation operator θ_β from $\mathscr{F}_\infty^0$ to $\mathscr{F}_\infty^0$ such that the operations union, intersection and complement are invariant under action of the operator θ_β; for instance, $\theta_\beta(A \cup B) = (\theta_\beta A) \cup (\theta_\beta B)$ and so on. Moreover

$$\theta_\beta\{x(t) = i\} = \{x(\beta + t) = i\} \tag{9}$$

Hence for a $\mathscr{F}_\infty^0$-measurable function ξ, we can define a function $\theta_\beta \xi$: $(\theta_\beta \xi)(\omega) = a$ if $\omega \in \theta_\beta(\xi = a)$, especially

$$\theta_\beta x(t) = x(\beta + t) \qquad (t \geqslant 0) \tag{10}$$

The details are found in Dynkin (1963, pp. 121–44).

Theorem 2. $X \in \mathscr{H}$ has the strong Markov property, that is, it possesses the following property. For any Markov time β, provided $P(\Omega_\beta) > 0$,

(i) suppose $M \in \mathscr{F}_\infty^0, \Lambda \in \mathscr{F}_\beta$, then

$$P\{\theta_\beta M \mid \Lambda, x(\beta) = i\} = P_i\{M\} \qquad i \in E \tag{11}$$

$$P\{x(\beta + t) = \infty\} = 0 \qquad t > 0 \tag{12}$$

(ii) $\{X(\beta + t), 0 < t < \sigma - \beta\}$ is an open Markov chain on $(\Omega_\beta, \Omega_\beta \mathscr{F}, P(\cdot \mid \Omega_\beta))$ whose transition probability is $(p_{ij}(t))$.

(iii) Let $\Delta' = \{x(\beta) \neq \infty\}, \sigma = \sigma - \beta, x'(t) = x(\beta + t)$. Then $X'(\omega) = \{x'(t, \omega), t < \sigma'(\omega)\}(\omega \in \Delta')$ is a canonical process defined on the probability space $(\Delta', \Delta'\mathscr{F}, P(\cdot|\Delta'))$; the state space E' of X' is included in E, and its transition probability matrix is the restriction to $E' \times E'$ of the transition probability matrix of X.

We call X' the post-β process.

The proof is seen in Zi-kun Wang and Xiang-qun Yang (1988, section 3.3) or Chung (1967, II.9).

Theorem 3. Suppose that $X \in \mathscr{H}$ and that β is a Markov time of X, and assume that α is a Markov time of the post-β process X', then $\beta + \alpha$ is a Markov time of X.

The proof is encountered in Chung (1967, II.15, Theorem 1).

7.7 PROBABILISTIC Q PROCESS

The probabilistic Q process $X \in \mathscr{H}_s(Q)$ and the analytical Q process $P(t) \in \mathscr{P}_s(Q)$ are identical. We say that a Q process X satisfies the Kolmogorov equations: certainly by saying so we mean that $P(t) \in \mathscr{P}_s(Q)$ corresponding to X satisfies the system of Kolomogorov equations, and so on.

For a Q process $X \in \mathscr{H}_s(Q)$, the right-lower semicontinuity property (6.4) becomes right-continuity in $\bar{E}$:

$$\lim_{s \downarrow t} x(s, \omega) = x(t, \omega) \in \bar{E} \qquad \text{for all } t < \sigma(\omega) \tag{1}$$

If Q process $X \in \mathscr{H}_s(Q)$ satisfies the stronger condition:

$$\lim_{s \downarrow t} x(s, \omega) = x(t, \omega) \in E \qquad \text{for all } t < \sigma(\omega) \tag{2}$$

then we call X the D-type process. The class composed of the D-type processes is written as $\mathscr{H}_D$. When Q is fixed, the class composed of the D-type Q processes X is written as $\mathscr{H}_D(Q)$.

Definition 1. Suppose $X \in \mathscr{H}$, $q_i < \infty$. We call $[a, b)$ an i-interval of $X(\omega)$, if $x(t, \omega) = i$ for all $t \in [a, b)$, but $x(t, \omega) \not\equiv i$ in any interval $[c, d)$ containing $[a, b)$ as a proper subset. When i is used in general an i-interval is called a constancy interval.

Theorem 1. Suppose $X \in \mathscr{H}, q_i < \infty$. Then for almost all $\omega \in \Omega, X(\omega)$ has only finitely many i-intervals in any finite interval.

The proof is seen in Zi-kun Wang and Xiang-qun Yang (1988, section 3.1, Theorem 2) or Chung (1967, II.5, Theorem 7).

Let $X \in \mathscr{H}_s(Q)$. We call

$$\tau_1 = \begin{cases} \inf\{t \mid 0 < t < \sigma, x(t) \neq x(0)\} \\ \sigma \qquad \text{if the set above is empty} \end{cases} \tag{3}$$

the first discontinuity of X.

Theorem 2. Let $X \in \mathscr{H}_s(Q)$. Then

$$P_i\{\tau_1 > t\} = e^{-q_i t} \tag{4}$$

when $q_i > 0$,

$$P_i\{x(\tau_1) = j\} = \Pi_{ij} \qquad j \in E \tag{5}$$

$$P_i\{x(\tau_1) = \infty\} = P_i\{\tau_1 < \sigma\} - \sum_j \Pi_{ij} \tag{6}$$

where $\Pi = (\Pi_{ij})$ is determined by (2.9.7).

For the proof see Zi-kun Wang and Xiang-qun Yang (1988, section 2.2, Theorems 5 and 6).

Theorem 3. Suppose that $X \in \mathscr{H}, 0 < q_i < \infty, \beta$ is a Markov time of X, $P\{x(\beta) \neq i\} = 0, \alpha$ is the first discontinuity after β, and let $\rho = \alpha - \beta$. Then for $\Lambda \in \mathscr{F}_\beta, M \in \mathscr{F}'_\alpha$,

$$P\{\Lambda, \rho > t, M\} = P\{\Lambda\} e^{-q_i t} P\{M \mid \Omega_\beta\}$$

In particular, $\mathscr{F}_\alpha$ and $\mathscr{F}'_\alpha$ are conditionally independent with regard to Ω_β. More particularly, $x(\tau_1)$ and τ_1 are conditionally independent with respect to the measure P_i.

The proof is to be seen in Zi-kun Wang and Xiang-qun Yang (1988, section 3.2, corollary 2 to Theorem 1 or Chung (1967, II.15, Theorem 2).

Suppose $X \in \mathscr{H}_s(Q)$. We call

$$\eta_i \simeq \begin{cases} \inf\{t \mid \tau_1 < t < \sigma, x(t) = i\} \\ \sigma \qquad \text{if the set above is empty} \end{cases} \tag{7}$$

the first time of returning to i after first discontinuity. If $P_i\{\eta_i^* < \sigma\} = 1$, we say that i is recurrent. If i is recurrent, and moreover

$$m_{ii} = E_i \eta_i^* < \infty \tag{8}$$

then we say i is ergodic. If all i are recurrent or ergodic, then we say process X is recurrent or ergodic.

Theorem 4.

(i) i is recurrent iff for almost all $\omega\in\{x(0)=i\}, X(\omega)$ has infinitely many i-intervals, or equivalently

$$\int_0^\infty p_{ii}(t)\,dt=\infty$$

(ii) Suppose i is recurrent, then $P_i\{\sigma=\infty\}=1$, and i is ergodic iff

$$\lim_{t\to\infty} p_{ii}(t)=\pi_i>0 \tag{9}$$

If i is ergodic, C is the irreducible recurrent class including i, then

$$\lim_{t\to\infty} p_{ji}(t)=\frac{1}{q_i m_{ii}} \qquad j\in C \tag{10}$$

The proof is found in Zi-kun Wang and Xiang-qun Yang (1988, section 4.2, Theorem 1) or Chung (1967, II. 10, Theorem 4 and its corollary in II. 12 formula (9)).

7.8 PROBABILISTIC MINIMAL PROCESS

Definition 1. Suppose $X\in\mathscr{H}_s$. We call $t\in(0,\sigma(\omega)]$ a jumping point of $X(\omega)$, if $t=\sigma(\omega)<\infty$ and there exist $i\in E$ and $\varepsilon>0$ such that $x(u,\omega)=i$ for $u\in(t-\varepsilon,t)$ or $t<\sigma(\omega)$ and there exist different $i,j\in E$ and $\varepsilon>0$ such that we have $x(u,\omega)=i$ for $u\in(t-\varepsilon,t)$, and $x(u,\omega)=j$ for $u\in(t,t+\varepsilon)$. We call $t\in(0,\sigma(\omega)]$ a leaping point of $X(\omega)$, if $t=\sigma(\omega)<\infty$, or $t<\sigma(\omega)$ and for any $\varepsilon>0, X(\omega)$ has infinitely many jumping points in $(t-\varepsilon,t+\varepsilon)$. We agree that $t=0$ is a jumping point and leaping point; it is called the zeroth jumping point and the zeroth leaping point.

Every Q process $X\in\mathscr{H}_s(Q)$ has first leaping point τ:

$$\tau=\begin{cases}\inf\{t\,|\,0<t<\sigma(\omega), \lim_{s\to t} x(s,\omega)=\infty\} \\ \sigma(\omega) \qquad \text{if the set above is empty}\end{cases} \tag{1}$$

The first discontinuity τ_1 of X may not necessarily be a jumping point. But when Q is conservative, τ_1 is a jumping point. If $\tau_1\,(<\sigma)$ is not a jumping point, according to Theorem 7.2, then τ_1 is the first leaping point.

Theorem 1. Suppose $X\in\mathscr{H}_s(Q), q_i>0$. Then

$$P_i\{\tau_1=\tau\}=d_i/q_i \tag{2}$$

where d is the non-conservative quantity of Q, determined by (2.2.6).

Proof. By (7.6)

$$P_i\{\tau_1=\tau\}=P_i\{x(\tau_1)=\infty\}+P_i\{\tau_1=\sigma\}$$

$$=1-\sum_j \Pi_{ij}=\frac{d_i}{q_i} \qquad \text{QED}$$

Theorem 2. Suppose $X\in\mathscr{H}_s(Q)$, the set Λ and the non-negative random variable ξ are all $\mathscr{F}^0_\infty$-measurable.

(i) If

$$P\{\Lambda=\theta_{\tau_1}\Lambda\}=P_i\{\xi=\tau_1+\theta_{\tau_1}\xi\}=1 \tag{3}$$

then $u_j=E_j\{\xi,\Lambda\}$ satisfies the equation

$$\sum_j q_{ij}u_j=-P_i\{\Lambda\} \tag{4}$$

and $u_j(\lambda)=E_j\{e^{-\lambda\xi},\Lambda\}(\lambda>0)$ satisfies the equation

$$\lambda u_i-\sum_j q_{ij}u_j=0 \tag{5}$$

(ii) If

$$P_i\{\tau_1=\tau,\Lambda\}=P_i\{\tau_1=\tau\} \tag{6}$$

$$P_i\{\tau_1<\tau,\theta_{\tau_1}\Lambda\}=P_i\{\tau_1<\tau,\Lambda\} \tag{7}$$

then $u_j=P_j\{\Lambda\}$ satisfies the equation

$$\sum_j q_{ij}u_j=d_i \tag{8}$$

Proof. It is trivial when $q_i=0$. Suppose $q_i>0$.

By the strong Markov property, Theorem 7.2 and the independence between τ_1 and $x(\tau_1)$,

$$\begin{aligned}E_i\{\xi,\Lambda\}&=E_i\{\tau_1+\theta_{\tau_1}\xi,\theta_{\tau_1}\Lambda\}\\&=\sum_j \Pi_{ij}E_i\{\tau_1+\theta_{\tau_1}\xi,\theta_{\tau_1}\Lambda\,|\,x(\tau_1)=j\}\\&=\sum_j \Pi_{ij}[E_i\{\tau_1,\theta_{\tau_1}\Lambda\,|\,x(\tau_1)=j\}+E_i\{\theta_{\tau_1}\xi,\theta_{\tau_1}\Lambda\,|\,x(\tau_1)=j\}]\\&=\sum_j \Pi_{ij}(E_i\{\tau_1\}P_j\{\Lambda\}+E_j\{\xi,\Lambda\})\\&=\frac{1}{q_i}P_i\{\Lambda\}+\sum_j \Pi_{ij}E_j\{\xi,\Lambda\}\end{aligned}$$

So we get (4). Secondly

$$E_i\{e^{-\lambda\xi},\Lambda\}=\sum_j\Pi_{ij}E_i\{e^{-\lambda\tau_1}\theta_{\tau_1}e^{-\lambda\xi},\theta_{\tau_1}\Lambda|x(\tau_1)=j\}$$
$$=\sum_j\Pi_{ij}E_i\{e^{-\lambda\tau_1}|x(\tau_1)=j\}E_i\{\theta_{\tau_1}e^{-\lambda\xi},\theta_{\tau_1}\Lambda|x(\tau_1)=j\}$$
$$=\sum_j\Pi_{ij}E_i\{e^{-\lambda\tau_1}\}E_j\{e^{-\lambda\xi},\Lambda\}$$
$$=\frac{q_i}{\lambda+q_i}\sum_j\Pi_{ij}E_j\{e^{-\lambda\xi},\Lambda\}$$

(5) follows. Thirdly,

$$P_i\{\Lambda\}=P_i\{\Lambda,\tau=\tau_1\}+P_i\{\theta_{\tau_1}\Lambda,\tau_1<\tau\}$$
$$=P_i\{\tau=\tau_1\}+\sum_j\Pi_{ij}P_j\{\Lambda\}$$

Noting (2), we obtain (8). The proof is completed. QED

Theorem 3. Suppose $X\in\mathscr{H}_s(Q)$. X satisfies the backward equations iff for almost all $\omega\in\Omega$, $X(\omega)$ the first discontinuity is a jumping point. X satisfies the forward equations iff for arbitrarily given $t>0$, for almost all $\omega\in(t<\sigma)$, $X(\omega)$ has a last discontinuity in $[0,t]$ and moreover, it is a jumping point.

For the proof see Zi-kun Wang and Xiang-qun Yang (1988, section 2.3, Theorems 1 and 2).

Theorem 4. Suppose $X\in\mathscr{H}_s(Q)$, τ is the first leaping point, then $\bar{X}=\{x(t),t<\tau\}\in\mathscr{H}_s(Q)$, whose transition probability matrix is the minimal solution $f(t)=\{f_{ij}(t)\}$ in section 2.9.

The proof is in Zi-kun Wang and Xiang-qun Yang (1988, section 2.3, Theorem 5).

Definition 2. We call $X=\{x(t),t<\sigma\}\in\mathscr{H}_s(Q)$ the minimal Q process, if $P\{\tau=\sigma\}=1$, where τ is the first leaping point.

Suppose $X\in\mathscr{H}_s(Q)$, τ_1 is the first discontinuity, $\tau_1\leqslant\tau$. If $\tau_1=\infty$, then $\tau=\infty$, and we put $\tau_n=\infty\,(n>1)$. If $\tau_1<\infty$, then either $\tau_1=\tau<\infty$ and in this case $\tau_n(n>2)$ is undefined, or $\tau_1<\tau$ and in this case τ_1 is a jumping point. Hence after τ_1 there exists a first discontinuity $\tau_2\leqslant\tau$. If $\tau_2=\infty$, then $\tau=\infty$, and we put $\tau_n=\infty\,(n>2)$. If $\tau_2<\infty$, then either $\tau_2=\tau<\infty$ and in this case $\tau_n(n>2)$ is undefined, or $\tau_2<\tau$, in which case τ_2 is a jumping point. Therefore after τ_2 there exists a first discontinuity $\tau_3\leqslant\tau$. And it continues like this.

Write $\tau_0 = 0, \Omega = \Omega_F \cup \Omega_\infty$

$$\begin{aligned} \Omega_F &= \bigcup_{n=1}^{\infty} (\tau_n = \tau < \infty) \\ \Omega_\infty &= \bigcap_{n=1}^{\infty} (\tau_n < \tau) + \bigcup_{n=1}^{\infty} (\tau_n = \tau = \infty) \end{aligned} \tag{9}$$

$$\beta = \begin{cases} +\infty & \text{if } \omega \in \Omega_\infty \\ \sup\{n: n \geqslant 0, \tau_n < \tau\} & \text{if } \omega \in \Omega_F \end{cases} \tag{10}$$

$$\tau_\beta = \lim_{n \to \beta} \tau_n \tag{11}$$

Then on Ω_F we have $\tau_\beta < \tau$, on Ω_∞ we have $\tau_\beta = \tau$.

When $\tau_n = \infty$, we set $x(\tau_n) = x(\tau_K)$, where $K = \max\{m: \tau_m < \infty\}$. Thus, when $n \leqslant \beta$, that is $\tau_n \leqslant \tau_\beta$, $x(\tau_n)$ is well defined.

By the strong Markov property and Theorem 7.2 we obtain the following.

Theorem 5. $X_T = \{X(\tau_n), n \leqslant \beta\}$ or write $X_T = \{X(\tau_n), \tau_n \leqslant \tau_\beta\}$ is a Markov chain. Its one-step transition probability matrix $\Pi = (\Pi_{ij})$ is defined by (2.9.7).

We call the Markov chain X_T the embedded chain of the process $\{X(t), t < \tau\}$, and call the matrix Π the embedding matrix of the matrix Q.

7.9 RESOLVENT PROCESS AND INDUCED PROCESS

Suppose $X = \{x(t), t < \sigma\} \in \mathscr{H}$, and ρ is a random variable independent of X, its distribution being the exponential distribution with parameter 1:

$$P\{\rho > t\} = e^{-t} \qquad t \geqslant 0 \tag{1}$$

For $\lambda > 0$ let

$$\begin{aligned} \rho^\lambda &= \rho/\lambda & \sigma^\lambda &= \min(\sigma, \rho^\lambda) \\ x^\lambda(t) &= x(t) & t &< \sigma^\lambda \end{aligned} \tag{2}$$

Theorem 1. The process $X^\lambda = \{x^\lambda(t), t < \sigma^\lambda\} \in \mathscr{H}$. Has transition probabilities

$$p_{ij}^\lambda(t) = e^{-\lambda t} p_{ij}(t) \tag{3}$$

where $p_{ij}(t)$ are the transition probabilities of X.

Proof. Since the path of X^λ is the front section of the path of X, we need only prove the homogeneous Markov property of X^λ.

Suppose $0 \leqslant t_1 < t_2 < \cdots < t_{n+1}, i_1, i_2, \ldots, i_{n+1} \in E$. By the homogeneous

Markov property of X and the independence between ρ and X,

$$P\{x^\lambda(t_a)=i_a,\ 1\leqslant a\leqslant n+1\}=P\{x(t_a)=i_a,\ 1\leqslant a\leqslant n+1,\ t_{n+1}<\rho^\lambda\}$$

$$=P\{x(t_1)=i_1\}\prod_{a=1}^{n} P_{i_a i_{a+1}}(t_{a+1}-t_a)\mathrm{e}^{-\lambda t_{n+1}}$$

$$=P\{x(t_1)=i_1\}P\{t_1<\rho^\lambda\}\prod_{a=1}^{n} p^\lambda_{i_a i_{a+1}}(t_{a+1}-t_a)$$

$$=P\{x^\lambda(t_1)=i_1\}\prod_{a=1}^{n} p^\lambda_{i_a i_{a+1}}(t_{a+1}-t_a)$$ QED

We call X^λ the resolvent process of X.

Theorem 2. Suppose that $\bar{X}=\{x(t), t<\tau\}$ is the minimal Q process, Π is the embedding matrix. Then the resolvent process X^λ is the minimal Q^λ process, where

$$q^\lambda_{ij}=q_{ij}(i\neq j) \qquad q^\lambda_i=\lambda+q_i \tag{4}$$

The embedding matrix $\Pi(\lambda)$ is determined by (2.9.8).

Proof. It suffices to show that Q^λ has the same form as (4), the rest being obvious. Suppose that τ^λ_1 is the first discontinuity of X^λ, then $\tau^\lambda_1=\min(\tau_1,\rho^\lambda)$, so

$$\exp(-q^\lambda_i t)=P_i\{\tau^\lambda_1>t\}=P_i\{\tau_1>t,\ \rho^\lambda>t\}$$
$$=P_i\{\tau_1>t\}P_i\{\rho^\lambda>t\}=\mathrm{e}^{-q_i t}\mathrm{e}^{-\lambda t}=\exp[-(\lambda+q_i)t]$$

so that $q^\lambda_i=\lambda+q_i$. Secondly, for $j\neq i$,

$$\frac{q^\lambda_{ij}}{q^\lambda_i}=P_i\{x^\lambda(\tau^\lambda_1)=j\}=P_i\{x(\tau_1)=j,\ \tau_1<\rho^\lambda\}$$

$$=P_i\{x(\tau_1)=j\}P\{\tau_1<\rho^\lambda\}=\frac{q_i}{\lambda+q_i}\Pi_{ij}$$

hence $q^\lambda_{ij}=q_{ij}$ and the proof is completed. QED

Theorem 3. Suppose $\bar{X}=\{x(t), t<\tau\}$ is the minimal Q process, Λ is the ultimate set of the embedding chain X_T, $P(\Lambda)>0$. Then the process $X^\lambda(\omega)=\{x(t,\omega), t<\tau(\omega)\}$ $(\omega\in\Lambda)$ confined to Λ is the minimal Q^λ process on the probability space $(\Lambda, \Lambda\mathscr{F}, P\{\cdot|\Lambda\})$, and its state space is

$$E^\Lambda=\{i|P_i(\Lambda)>0\} \tag{5}$$

Its transition probability is

$$P_{ij}^{\Lambda}(t)=\frac{f_{ij}(t)P_j(\Lambda)}{P_i(\Lambda)} \tag{6}$$

The Q matrix $Q^{\Lambda}=(q_{ij}^{\Lambda})$ $(i,j\in E^{\Lambda})$ is conservative, and furthermore

$$q_{ij}^{\Lambda}=\frac{q_{ij}D_j(\Lambda)}{P_j(\Lambda)} \tag{7}$$

Proof. First notice: for any $t\geqslant 0$, we have $\Lambda=\theta_t\Lambda$ on $(t<\tau)$ where θ_t is the translation operator. Actually, because Λ is a ultimate set, there exists an infinite-dimensional invariant function g such that

$$1_{\Lambda}=g(x(\tau_n),x(\tau_{n+1}),\ldots) \qquad \text{for all } n \tag{8}$$

So by Dynkin (1963, p. 122),

$$\begin{aligned}1_{\theta_t\Lambda}&=g(\theta_t x(\tau_n),\theta_t x(\tau_{n+1}),\ldots)\\&=g(x(\theta_t\tau_n),x(\theta_t\tau_{n+1}),\ldots)\end{aligned} \tag{9}$$

On $(t<\tau)$ there must exist l such that $\tau_1\leqslant t<\tau_{l+1}$; therefore $\theta_t\tau_m=\tau_{m+l}$. Accordingly by (8) and (9),

$$1_{\theta_t\Lambda}=g(x(\tau_{n+l}),x(\tau_{n+l+1}),\ldots)=1_{\Lambda}$$

that is, we have $\Lambda=\theta_j\Lambda$ on $(t<\tau)$.

Suppose $0\leqslant t_1<t_2<\cdots<t_{n+1},i_1,i_2,\ldots,i_{n+1}\in E^{\Lambda}$. By the homogeneous Markov property of X

$$\begin{aligned}P\{x^{\Lambda}(t_{n+1})&=i_{n+1}\mid\Lambda,\ x^{\Lambda}(t_a)=i_a,\ 1\leqslant a\leqslant n\}\\&=P\{x(t_{n+1})=i_{n+1}\mid\Lambda,\ x(t_a)=i_a,\ 1\leqslant a\leqslant n\}\\&=\frac{P\{x(t_a)=i_a,\ 1\leqslant a\leqslant n+1,\Lambda\}}{P\{x(t_a)=i_a,\ 1\leqslant a\leqslant n,\Lambda\}}\\&=\frac{P\{x(t_a)=i_a,\ 1\leqslant a\leqslant n+1,\theta_{t_{n+1}}\Lambda\}}{P\{x(t_a)=i_a,\ 1\leqslant a\leqslant n,\theta_{t_n}\Lambda\}}\\&=\frac{P\{x(t_a)=i_a,\ 1\leqslant a\leqslant n\}f_{i_ni_{n+1}}(t_{n+1}-t_n)P_{i_{n+1}}(\Lambda)}{P\{x(t_a)=i_a,\ 1\leqslant a\leqslant n\}P_{i_n}(\Lambda)}\\&=P_{i_ni_{n+1}}^{\Lambda}(t_{n+1}-t_n)\end{aligned}$$

It remains to be proved that Q^{λ} is conservative, and this follows from

$$P_i(\Lambda)=P_i(\theta_{\tau_1}\Lambda)=\sum_j \Pi_{ij}P_j(\Lambda)$$

The proof is over. QED

Corollary

Suppose that X is the minimal Q process, $P(\Omega_\infty)>0$. Then X^{Ω_∞} is the minimal Q^{Ω_∞} process on $(\Omega_\infty, \Omega_\infty \mathscr{F}, P(\cdot|\Omega_\infty))$ and moreover is conservative.

Definition 1. Call process X^Λ the induced process of the minimal Q process on the ultimate set Λ.

7.10 PROBABILITY REPRESENTATION OF THE $\Pi(\lambda)$ POTENTIAL

Suppose that $X=\{x(t), t<\tau\}$ is the minimal Q process, H the set of non-conservative states of Q, d the non-conservative quantity of Q, and $\Pi(\lambda)$ the embedding chain of the resolvent process X^λ of X. For convenience, we write $f_i \equiv f(i)$.

Theorem 1. For the chain $\Pi(\lambda), u(\lambda)$ is the potential of a non-negative function v equal to zero on $E-H$ iff $u(\lambda)$ has the representation:

$$u_i(\lambda)=E_i\{\mathrm{e}^{-\lambda\tau}f(x_{\tau-0}), \Omega_F\}=[\phi(\lambda)df]_i \tag{1}$$

where $f(a)\geqslant 0$ $(a\in H)$.

Proof. Sufficiency: By (1)

$$u_i(\lambda)=\sum_{n=1}^{\infty} E_i\{\mathrm{e}^{-\lambda\tau_n}f(x_{\tau_n-0}), \tau_n=\tau\}=\sum_{n=1}^{\infty} T_i^n$$

where

$$T_i^1=E_i\{\mathrm{e}^{-\lambda\tau_1}f(x_{\tau_1-0}), \tau_1=\tau\}=f_i\frac{d_i}{q_i}\frac{q_i}{\lambda+q_i}=\frac{1}{\lambda+q_i}d_if_i$$

$$T_i^n=E_i\{\mathrm{e}^{-\lambda\tau_n}f(x_{\tau_n-0}), \tau_n=\tau\}$$

$$=\sum_j \Pi_{ij}\frac{q_i}{\lambda+q_i}T_j^{n-1}=\sum_j \Pi_{ij}(\lambda)T_j^{n-1}$$

$$T^n=\Pi(\lambda)T^{n-1}=\cdots=\Pi^{n-1}(\lambda)(\lambda+q)^{-1}df$$

$$u(\lambda)=\sum_{n=1}^{\infty}\Pi^{n-1}(\lambda)(\lambda+q)^{-1}df=\phi(\lambda)df$$

So $u(\lambda)$ is the potential of the function $v_j=d_jf_j/(\lambda+q_j)$ while v is zero on $E-H$.

Necessity: Take

$$f_a=\frac{(\lambda+q_a)v_a}{d_a} \qquad (a\in H)$$

By the proof of the sufficiency, we know $u(\lambda)$ has the representation (1), and the proof is terminated. QED

Theorem 2. Suppose

$$G(\lambda)=\sum_{n=0}^{\infty}\Pi^n(\lambda)=\phi(\lambda)(\lambda+q)$$

and column vector $v\geqslant 0$. Then

$$[\Pi^n(\lambda)v]_i=E_i\{e^{-\lambda\tau_n}v(x_{\tau_n})\} \tag{2}$$

$$[G(\lambda)v]_i=\sum_{n=0}^{\infty}E_i\{e^{-\lambda\tau_n}v(x_{\tau_n})\} \tag{3}$$

Proof. Following Theorem 1 we can prove (2), and hence (3) follows. QED

Corollary

$$\Pi^n_{ij}(\lambda)=E_i\{e^{-\lambda\tau_n}, x(\tau_n)=j\}.$$

7.11 λ-IMAGE AND STANDARD IMAGE

Recall that a harmonic function is non-negative. We write the class of finite Π- (or Π(λ)-) harmonic functions as $\hat{\mathcal{M}}^+$ (or $\hat{\mathcal{M}}^+_\lambda$), denote the class of bounded Π- (or Π(λ)-) harmonic functions by $\mathcal{M}^+$ (or $\mathcal{M}^+_\lambda$), and denote the class of Π- (or Π(λ)-) harmonic functions having the upper bound K by $\mathcal{M}^+(K)$ (or $\mathcal{M}^+_\lambda(K)$).

Suppose $u\in\hat{\mathcal{M}}^+$. Then $\Pi(\lambda)u\leqslant\Pi u=u$, $\Pi^{n+1}(\lambda)u\leqslant\Pi^n(\lambda)u\leqslant u$, so $\Pi^n(\lambda)u\downarrow u(\lambda)\leqslant u$, and moreover $u(\lambda)\in\hat{\mathcal{M}}^+_\lambda$. If $u\in\mathcal{M}^+(K)$, then $u(\lambda)\in\mathcal{M}^+_\lambda(K)$.

Definition 1. Suppose $u\in\hat{\mathcal{M}}^+$, λ is fixed. We call $u(\lambda)=\lim_{n\to\infty}\Pi^n(\lambda)u\in\hat{\mathcal{M}}^+_\lambda$ the λ-image of u.

Theorem 1. $u\in\hat{\mathcal{M}}^+$ and its λ-image $u(\lambda)\in\hat{\mathcal{M}}^+_\lambda$ have the following relation:

$$\lambda\phi(\lambda)u=u-u(\lambda) \tag{1}$$

Proof. Obviously

$$\sum_{a=0}\Pi^a(\lambda)=\sum_{a=0}\Pi^a(\lambda)(\lambda I+q)^{-1}q\Pi+I \tag{2}$$

Consequently if we let

$$\phi^n(\lambda)=\sum_{a=0}^{n-1}\Pi_a(\lambda)(\lambda I+q)^{-1}$$

then (2) becomes

$$\phi^{n+1}(\lambda)(\lambda+q)=\phi^n(\lambda)q\Pi+I$$
$$\lambda\phi^{n+1}(\lambda)+\phi^{n+1}(\lambda)q=\phi^n(\lambda)q\Pi+I \tag{3}$$

Multiplying from the right by u and noticing that $\Pi u=u$, we obtain

$$\lambda\phi^{n+1}(\lambda)u+\phi^{n+1}(\lambda)q\Pi u=\phi^n(\lambda)q\Pi u+u$$
$$\lambda\phi^{n+1}(\lambda)u+\Pi^n(\lambda)(\lambda I+q)^{-1}q\Pi u=u \tag{4}$$
$$\lambda\phi^{n+1}(\lambda)u+\Pi^{n+1}(\lambda)u=u$$

Setting $n\to\infty$ we obtain (1). The proof is concluded. QED

Suppose that $u(\lambda)$ is a $\Pi(\lambda)$-harminoc function but not necessarily finite. Then $\Pi u(\lambda)\geqslant\Pi(\lambda)u(\lambda)=u(\lambda)$, $\Pi^{n+1}u(\lambda)\geqslant\Pi^n u(\lambda)$. Therefore $\Pi^n u(\lambda)\uparrow u\geqslant u(\lambda)$, and moreover $\Pi u=u$; that is, u is a $\Pi(\lambda)$-harmonic function, but not necessarily finite.

Definition 2. Let $u(\lambda)$ be a $\Pi(\lambda)$-harmonic function. We call the Π-harmonic function $u=\lim_{n\to\infty}\Pi^n u(\lambda)$ the standard image of $u(\lambda)$.

Theorem 2. Assume that $u(\lambda)\in\hat{\mathscr{M}}^{+}_{\lambda}$. Then its standard image is[1]

$$u=u(\lambda)+\lambda\Gamma u(\lambda) \tag{5}$$

where

$$\Gamma=\lim_{\lambda\downarrow 0}\phi(\lambda)=\sum_{n=0}\Pi^n\frac{1}{q}$$

Proof. Rewrite $\Pi(\lambda)u(\lambda)=u(\lambda)$ as

$$u(\lambda)+\lambda q^{-1}u(\lambda)=\Pi u(\lambda)$$

Multiplying from the left by Π^n and summing over $0\leqslant n\leqslant a-1$, we obtain

$$u(\lambda)+\lambda\sum_{n=0}^{a-1}\Pi^n q^{-1}u(\lambda)=\Pi^a u(\lambda)$$

Letting $n\to\infty$, (5) follows, and the proof is completed. QED

Remark

From (5) it can be seen that the standard image of the harmonic exit family $u(\lambda)$ ($\lambda>0$) agrees with Definition 2.10.2. But the λ in Definition 2 may be a fixed non-negative number.

[1] When $q_j=0$, set $1/q_j=\infty$. But from $\Pi(\lambda)u(\lambda)=u(\lambda)$ we know $u_j(\lambda)=0$, and q_j; hence $(1/q_j)u_j(\lambda)=\infty.$ $0=0$.

Theorem 3. Suppose $u(\lambda)\in\hat{\mathscr{M}}_\lambda^+$. The standard image of $u(\lambda)$ is $u\in\hat{\mathscr{M}}^+$. Then (1) is true.

Proof. Equation (3) is already known to be true. Multiplying (3) from the right by $u=u(\lambda)+\lambda\Gamma u(\lambda)\in\hat{\mathscr{M}}^+$ we have

$$\lambda\phi^{n+1}(\lambda)u+\phi^{n+1}(\lambda)q\Pi u=\phi^n(\lambda)q\Pi u+u$$

$$\lambda\phi^{n+1}(\lambda)u+\sum_{a=0}^{n}\Pi^a(\lambda)(\lambda+q)^{-1}q\Pi[u(\lambda)+\lambda\Gamma u(\lambda)]$$

$$=\sum_{a=0}^{n-1}\Pi^a(\lambda)(\lambda+q)^{-1}q\Pi[u(\lambda)+\lambda\Gamma u(\lambda)]+u$$

$$\lambda\phi^{n+1}(\lambda)u+\Pi^{n+1}(\lambda)[u(\lambda)+\lambda\Gamma u(\lambda)]=u$$

Noticing that $u(\lambda)\in\hat{\mathscr{M}}_\lambda^+$, that is,

$$\lambda\phi^{n+1}(\lambda)u+u(\lambda)+\Pi^{n+1}(\lambda)\lambda\Gamma u(\lambda)=u \tag{6}$$

However as $u\in\hat{\mathscr{M}}_\lambda^+$,

$$\Pi^{n+1}(\lambda)\lambda\Gamma u(\lambda)\leqslant\Pi^{n+1}\lambda\Gamma u(\lambda)=\Pi^{n+1}[u-u(\lambda)]=u-\Pi^{n+1}u(\lambda)\downarrow u-u=0$$

Letting $n\to\infty$ in (6), (1) follows. The proof is terminated. QED

Corollary

Suppose $u(\lambda)\in\hat{\mathscr{M}}_\lambda^+$, and its standard image is $u\in\hat{\mathscr{M}}^+$. Then

$$\phi(\lambda)u=\Gamma u(\lambda) \tag{7}$$

Proof. The assertion follows from Theorems 2 and 3. QED

Let

$$\bar{\mathscr{M}}_\lambda^+=\{u(\lambda)\in\hat{\mathscr{M}}_\lambda^+\,|\,\text{the standard image of } u(\lambda) \text{ is } u\in\hat{\mathscr{M}}^+\} \tag{8}$$

$$\bar{\mathscr{M}}^+=\{u\in\bar{\mathscr{M}}^+\,|\,u \text{ is the standard image of some } u(\lambda)\in\bar{\mathscr{M}}_\lambda^+\}. \tag{9}$$

Theorem 4. Between $\bar{\mathscr{M}}_\lambda^+$ and $\bar{\mathscr{M}}^+$ we have established a one-to-one correspondence between the standard image and the λ-image. Furthermore, the standard map and the λ-map are mutually inverse maps.

Proof. Clearly, the standard map maps $\bar{\mathscr{M}}_\lambda^+$ onto $\bar{\mathscr{M}}^+$ by Theorem 3. The standard map is one-to-one. Now we suppose the standard image of $u(\lambda)\in\bar{\mathscr{M}}_\lambda^+$ is $u\in\bar{\mathscr{M}}^+$, and the λ-image of u is $v(\lambda)$. By Theorems 1 and 3, $v(\lambda)=u(\lambda)$, which means that $u(\lambda)$ is the λ-image of u. The proof is finished. QED

Theorem 5. Under the standard map or the λ-map, there exists a one-to-one correspondence between the $\Gamma(\lambda)$-minimal $u(\lambda)\in\bar{\mathscr{M}}_{\lambda}^{+}$ and the Π-minimal $u\in\bar{\mathscr{M}}$.

Proof. According to the definition of minimality, and by using Theorem 4, the assertion follows. QED

7.12 BOUNDARY OF MINIMAL Q PROCESS

Suppose that $X=\{x(t), t<\tau\}$ is the minimal Q process and that $X_T=\{x(\tau_n), n\leqslant\beta\}$ is its embedding chain with its embedding matrix denoted by Π. According to section 7.3, for the chain Π and X_T we can introduce the concepts of its Martin boundary ∂E, essential Martin boundary B, atomic boundary B_1, non-atomic boundary B_2, ultimate Borel field $\mathscr{B}_\infty$ and so on. We still use the notation employed in sections 7.3 to 7.5. But in the present case, Ω_F and Ω_∞ in (2.17) should be understood according to (8.9) and (8.10); H should be the set of non-conservative states of Q; (3.11) becomes

$$x(\tau-0)=\lim_{\tau_n\uparrow\tau} x(\tau_n)\in H\cup B \tag{1}$$

or more precisely, $x(\tau-0)\in H$ on Ω_F, $x(\tau-0)\in B$ on Ω_∞; whereas the measure μ in (3.34) becomes

$$\mu(\Gamma)=P_\gamma\{x(\tau-0)\in\Gamma\} \qquad \Gamma\subset H\cup B \tag{2}$$

Obviously, $K(\cdot,\xi)\in\hat{\mathscr{M}}^{+}$ when $\xi\in B$. Suppose the λ-image of $K(\cdot,\xi)$ is $K_\lambda(\cdot,\xi)$. By Theorem 11.1 we have

$$\lambda\phi(\lambda)K(\cdot,\xi)=K(\cdot,\xi)-K_\lambda(\cdot,\xi) \qquad \xi\in B \tag{3}$$

According to the resolvent equation of $\phi(\lambda)$, and from the formula above, we obtain

$$K_\lambda(\cdot,\xi)-K_\nu(\cdot,\xi)+(\lambda-\nu)\phi(\lambda)K_\nu(\cdot,\xi)=0 \qquad (\xi\in B,\ \lambda,\nu>0) \tag{4}$$

That is, $K_\lambda(\cdot,\xi)$ is a harmonic exit family. Thus either for all $\lambda>0$, $K_\lambda(\cdot,\xi)=0$ or $\neq 0$. And so we can let

$$B_{\mathrm{e}}=\{\xi\in B\,|\,K_\lambda(\cdot,\xi)\neq 0\} \tag{5}$$

$$B_{\mathrm{p}}=\{\xi\in B\,|\,K_\lambda(\cdot,\xi)=0\} \tag{6}$$

Definition 1. We call B_{e} and B_{p} respectively the Martin exit boundary and the passive boundary of the minimal Q process X or of the matrix Q.

Theorem 1. B_{e} and B_{p} are Borel sets.

Proof. Since B is a Borel set and moreover $K(i,\xi)$ $(\xi\in B)$ is a Borel-measurable function for every $i \in E$, it follows that $K_\lambda(i, \xi)$ $(\xi \in B)$ is also a Borel-measurable function. Accordingly

$$B_e = \bigcup_{i \in E} \{\xi | K_\lambda(i, \xi) \neq 0\}$$

is a Borel set, and so is $B - B_e = B_p$. The proof is completed. QED

Theorem 2. Suppose $\xi \in B_e$, then $K(\cdot, \xi) \in \bar{\mathscr{M}}^+, K_\lambda(\cdot, \xi) \in \bar{\mathscr{M}}_\lambda^+$. Moreover the standard image of $K_\lambda(\cdot, \xi)$ is $K(\cdot, \xi)$, and the λ-image of $K(\cdot, \xi)$ is $K_\lambda(\cdot, \xi)$.

Proof. By (3), $K_\lambda(\cdot, \xi) \leqslant K(\cdot, \xi)$, $\Pi^n K_\lambda(\cdot, \xi) \leqslant \Pi^n K(\cdot, \xi) = K(\cdot, \xi)$. So the standard image of $K_\lambda(\cdot, \xi)$ is $u = \lim_{n \to \infty} \Pi^n K_\lambda(\cdot, \xi) \leqslant K(\cdot, \xi)$, that is $K_\lambda(\cdot, \xi) \in \bar{\mathscr{M}}_\lambda^+, u \in \bar{\mathscr{M}}^+$. According to Theorem 11.4, $u = K(\cdot, \xi)$. QED

Corollary

$\{u(\lambda) \in \bar{\mathscr{M}}_\lambda^+ | u(\lambda)$ is $\Pi(\lambda)$-minimal, the standard image of $u(\lambda)$ is γ-integrable$\} = \{cK_\lambda(\cdot, \xi) | \xi \in B_e$, constant $c > 0\}$.

Proof. If follows from Theorem 2 and Theorem 8.19. QED

For any Borel set $\Gamma \subset B$, let

$$X_i^\Gamma \equiv P_i\{x(\tau - 0) \in \Gamma\} = \int_\Gamma K(i, \xi) \mu(\mathrm{d}\xi) \tag{7}$$

$$X_i^\Gamma(\lambda) \equiv E_i\{\mathrm{e}^{-\lambda\tau}, x(\tau - 0) \in \Gamma\} \tag{8}$$

Theorem 3. For any Borel set $\Gamma \subset B$,

$$\lambda \phi(\lambda) X^\Gamma = X^\Gamma - X^\Gamma(\lambda) \tag{9}$$

Proof. By the homogeneous Markov property of X and $\Lambda = \{x(\tau - 0) \in \Gamma\}$ being the ultimate set,

$$\begin{aligned} X_i^\Gamma(\lambda) &= \int_0^\infty \mathrm{e}^{-\lambda t} \mathrm{d}P_i\{\Lambda, \tau \leqslant t\} \\ &= \int_0^\infty \mathrm{e}^{-\lambda t} \frac{\mathrm{d}}{\mathrm{d}t} P_i\{\Lambda, \tau \leqslant t\} \, \mathrm{d}t \\ &= \lambda \int_0^\infty \mathrm{e}^{-\lambda t} P_i\{\Lambda, \tau \leqslant t\} \, \mathrm{d}t \\ &= \lambda \int_0^\infty \mathrm{e}^{-\lambda t} [P_i\{\Lambda\} - P_i\{\Lambda, \tau > t\}] \, \mathrm{d}t \end{aligned}$$

$$= X_i^\Gamma - \lambda \sum_j \int_0^\infty e^{-\lambda t} P_i\{t < \tau, x(t) = j, \theta_t \Lambda\}\, dt$$

$$= X_i^\Gamma - \lambda \sum_j \int_0^\infty e^{-\lambda t} f_{ij}(t) P_j\{\Lambda\}\, dt$$

$$= X_i^\Gamma - \lambda \sum_j \phi_{ij}(\lambda) X_j^\Gamma$$

QED

Theorem 4. For any Borel set $\Gamma \subset B$,

$$X_i^\Gamma(\lambda) = \int_\Gamma K_\lambda(i, \xi)\mu(d\xi) = \int_{\Gamma \cap B_e} K_\lambda(i, \xi)\mu(d\xi) = X_i^{\Gamma \cap B_e}(\lambda) \tag{10}$$

The standard image of $X^\Gamma(\lambda)$ is $X^{\Gamma \cap B_e} \in \bar{\mathscr{M}}^+$. And the λ-image of X^Γ is $X^\Gamma(\lambda)$.

Proof. By the definition of B_p, we have the second equality in (10). Since

$$\Pi^n(\lambda) X^\Gamma = \int_\Gamma \Pi^n(\lambda) K(\cdot, \xi)\mu(d\xi)$$

setting $n \to \infty$, it follows that the λ-image of X^Γ is $\int_\Gamma K_\lambda(\cdot, \xi)\mu(d\xi)$. By Theorem 11.1 and Theorem 3 we obtain the first equality in (10). Hence by (10),

$$\Pi^n X^\Gamma(\lambda) = \int_{\Gamma \cap B_e} \Pi^n K_\lambda(\cdot, \xi)\mu(d\xi)$$

Taking the limit, we know that the standard image of $X^\Gamma(\lambda)$ is

$$\int_{\Gamma \cap B_e} K(\cdot, \xi)\mu(d\xi) = X^{\Gamma \cap B_e}$$

and the proof is completed.

QED

Theorem 5. For all $i \in E$,

$$P_i\{\tau < \infty \mid x(\tau - 0) \in B_e\} = 1 \tag{11}$$

$$P_i\{\tau = \infty \mid x(\tau - 0) \in B_p\} = 1 \tag{12}$$

Proof. By (10)

$$X_i^{B_p}(\lambda) = E_i\{e^{-\lambda\tau}, x(\tau - 0) \in B_p\} = E_i\{e^{-\lambda\tau}, \tau < \infty, x(\tau - 0) \in B_p\}$$

$$= \int_{B_p} K_\lambda(i, \xi)\mu(d\xi) = 0$$

From what precedes we get (12). Secondly

$$\begin{aligned}\Pi^n X_i^\Gamma(\lambda) &= E_i\{E_{x(\tau_n)}[\mathrm{e}^{-\lambda\tau}, \tau<\infty, x(\tau-0)\in\Gamma]\}\\ &= E_i\{\theta_{\tau_n}[\mathrm{e}^{-\lambda\tau}, \tau<\infty, x(\tau-0)\in\Gamma]\}\\ &= E_i\{\mathrm{e}^{-\lambda(\tau-\tau_n)}, \tau<\infty, x(\tau-0)\in\Gamma\}\end{aligned}$$

Taking the limit we know that the standard image of $X^\Gamma(\lambda)$ is $P_i\{\tau<\infty, x(\tau-0)\in\Gamma\}$. Especially, the standard image of $X^{B_e}(\lambda)$ is $P_i\{\tau<\infty, x(\tau-0)\in B_e\}$. But according to Theorem 4, the standard image of $X^{B_e}(\lambda)$ is X^{B_e}. And so $P_i\{\tau<\infty, x(\tau-0)\in B_e\}=P_i\{x(\tau-0)\in B_e\}$. From this we get (11), and the proof is terminated. QED

Theorem 6. Suppose f is a non-negative Borel function on B_e. Then the standard image of

$$u_i(\lambda)=E_i\{\mathrm{e}^{-\lambda\tau}f[x(\tau-0)], \Omega_\infty\}=\int_{B_e} K_\lambda(i,\xi)f(\xi)\mu(\mathrm{d}\xi) \tag{13}$$

is

$$u_i=E_i\{f[x(\tau-0)], x(\tau-0)\in B_e\}=\int_{B_e} K(i,\xi)f(\xi)\mu(\mathrm{d}\xi) \tag{14}$$

Conversely, if $u_i<\infty$ $(i\in E)$ determined by the formula above or f is a non-negative Borel function on B, and furthermore

$$\bar{u}_i=E_i\{f[x(\tau-0)], \Omega_\infty\}=\int_B K(i,\xi)f(\xi)\mu(\mathrm{d}\xi)<\infty. \tag{15}$$

then the λ-image of u or $\bar{u}$ is $u(\lambda)$.

Proof. Similar to the proof of Theorem 4. QED

Theorem 7. $(\partial E)_1=\{\xi_a, a\in\mathscr{A}\}\subset B_\mathrm{p}$.

Proof. Fix a $j\in E_a$. Suppose the time that the minimal Q process X stays at j after reaching j for the nth time is ρ_j^n. Then $\rho_j^n (n\geqslant 1)$ are mutually independent with respect to P_j, $\tau\geqslant\sum_{n=1}^\infty \rho_j^n=\infty$. That is $P_i\{\tau=\infty\}=1$. Hence for any i, $P_i\{\tau=\infty \mid x(\tau-0)=\xi_a\}=1$. According to Theorem 5, $\xi_a\in B_\mathrm{p}$. QED

Theorem 8. For any $i\in E$,

$$P_i\{\tau<\infty \mid x(\tau-0)\in H\}=1 \tag{16}$$

Proof. Note that $q_j>0$ when $j\in H$, that is, j is non-absorbing. If $x(\tau-0)=j$ and $\tau=\infty$, then there exists t_0 such that for all $t\geqslant t_0$ we have $x(t,\omega)=j$. Since j is non-absorbing, except for the ω sets whose probability is 0, this is impossible. The proof is completed. QED

Combining Theorems 5 and 8, we have P_γ almost surely QED

$$x(\tau-0)\in H\cup B_{\mathrm{e}} \qquad \text{if } \tau<\infty \tag{17}$$

$$x(\tau-0)\in B_{\mathrm{p}} \qquad \text{if } \tau=\infty \tag{18}$$

7.13 PROBABILITY REPRESENTATION OF $\mathscr{M}_\lambda^+$

Theorem 1. Suppose that f is a non-negative, bounded Borel function on $B, u(\lambda)$ is defined by (12.13). Then we have P_γ-almost surely

$$\lim_{n\to\infty} u_{x(\tau_n)}(\lambda)=\begin{cases} f[x(\tau-0)] & \text{when } x(\tau-0)\in B_{\mathrm{e}}\\ 0 & \text{when } x(\tau-0)\in B_{\mathrm{p}}\end{cases} \tag{1}$$

Proof. Because on Ω_∞,

$$\begin{aligned} u_{x(\tau_n)}(\lambda)&=E_i\{\theta_{\tau_n}(\mathrm{e}^{-\lambda\tau}f[x(\tau-0)],x(\tau-0)\in B_{\mathrm{e}})|\mathscr{F}_{\tau_n}\}\\ &=E_i\{\mathrm{e}^{-\lambda(\tau-\tau_n)}f[x(\tau-0)],x(\tau-0)\in B_{\mathrm{e}}|\mathscr{F}_{\tau_n}\}\\ &=\mathrm{e}^{-\lambda\tau_n}E_i\{\mathrm{e}^{-\lambda\tau}f[x(\tau-0)],x(\tau-0)\in B_{\mathrm{e}}|\mathscr{F}_{\tau_n}\}\end{aligned}$$

When $n\to\infty$, making use of the convergence theorem for martingales, we get

$$\lim_{n\to\infty} u_{x(\tau_n)}(\lambda)=\mathrm{e}^{-\lambda\tau}\mathrm{e}^{-\lambda\tau}f[x(\tau-0)]1_{(x(\tau-0)\in B)}$$

By this formula and using Theorem 12.5 we obtain (1), and the proof is concluded. QED

Theorem 2. Suppose that for a Borel set $\Gamma\subset B, X^\Gamma$ is defined by (12.7). Then we have P_γ-almost surely

$$\lim_{n\to\infty} X^\Gamma_{x(\tau_n)}=\begin{cases}1 & \text{if } x(\tau-0)\in\Gamma\\ 0 & \text{if } x(\tau-0)\in B-\Gamma\end{cases} \tag{2}$$

Proof. This is a special case of Theorem 3.22. QED

Suppose that f is a bounded Borel function which is zero on B_{p} and is non-negative on B_{e}. The class composed of such functions f is written as $\mathscr{F}_{\mathrm{e}}^+$. If $f_1, f_2\in\mathscr{F}_{\mathrm{e}}^+$ and $\mu\{f_1\neq f_2\}=0$, then regard f_1 and f_2 as the same function.

Theorem 3. The element $u(\lambda)$ in $\mathscr{M}_\lambda^+$ has the following general form:

$$u_i(\lambda)=E_i\{\mathrm{e}^{-\lambda\tau}f[x(\tau-0)]\}=\int_{B_{\mathrm{e}}}K_\lambda(i,\xi)f(\xi)\mu(\mathrm{d}\xi) \qquad f\in\mathscr{F}_{\mathrm{e}}^+. \tag{3}$$

The element u of $\mathscr{M}^+\cap\bar{\mathscr{M}}^+$ has the following general form:

$$u_i = E_i\{f[x(\tau-0)]\} = \int_{B_e} K(i,\xi)f(\xi)\mu(d\xi) \qquad f\in\mathscr{F}_e^+ \tag{4}$$

$u(\lambda)\in\mathscr{M}_\lambda^+, u\in\mathscr{M}^+\cap\bar{\mathscr{M}}^+$, $f\in\mathscr{F}_e^+$ are one-to-one correspondent under the following conditions:

$$u_i(\lambda) = E_i\{e^{-\lambda\tau}f[x(\tau-0)]\} = E_i\left\{e^{-\lambda\tau}\lim_{n\to\infty} u_{x(\tau_n)}\right\} \tag{5}$$

$$u_i = E_i f[x(\tau_\infty-0)] = E_i\left\{\lim_{n\to\infty} u_{x(\tau_n)}(\lambda)\right\} \tag{6}$$

$$f[x(\tau-0)] \simeq \lim_{n\to\infty} u_{x(\tau_n)}(\lambda) \simeq \lim_{n\to\infty} u_{x(\tau_n)} \tag{7}$$

Proof. By Theorem 12.6, if $f\in\mathscr{F}_e^+$, then for $u(\lambda)$ and u, determined by (5) and (6), $u(\lambda)\in\mathscr{M}_\lambda^+, u\in\mathscr{M}^+\cap\bar{\mathscr{M}}^+$ hold.

Suppose $u(\lambda)\in\mathscr{M}_\lambda^+$, whose standard image is $u\in\mathscr{M}^+\cap\bar{\mathscr{M}}^+$. According to Theorem 5.7, there exists a non-negative bounded Borel function g on B such that $u_i = E_i g[x(\tau-0)]$. By Theorem 12.6, the λ-image of u is $u_i(\lambda) = E_i\{e^{-\lambda\tau}g[x(\tau-0)]\}$. Taking $f = g1_{B_e}\in\mathscr{F}_e^+$ we know $u(\lambda)$ has the form (3).

By Theorem 1 we derive the first formula in (7). Again by (4) and Theorem 5.7 we obtain $f[x(\tau-0)] = \lim_{n\to\infty} u_{x(\tau_n)}$, hence we get the second formula in (7). The proof is terminated. QED

7.14 ATOMIC AND NON-ATOMIC EXIT BOUNDARY OF THE MINIMAL Q PROCESS

Definition 1. We call $B_{ea} = B_e\cap B_a\,(a=1,2)$ the atomic exit boundary and the non-atomic exit boundary of the minimal Q process (or matrix Q) respectively. Similarly we can define the atomic passive boundary and the non-atomic passive boundary.

Theorem 1. When $\xi\in B_{e1}, K_\lambda(\cdot,\xi)$ is bounded; when $\xi\in B_{e2}, K_\lambda(\cdot,\xi)$ is unbounded.

Proof. By the definition of standard image and λ-image and by Theorem 12.2, when $\xi\in B_e$, both $K(\cdot,\xi)$ and $K_\lambda(\cdot,\xi)$ are bounded, or unbounded.

Suppose $\xi\in B_{e1}$, then $\mu(\xi)>0$. By (3.40),

$$1 \geqslant p_i(x(\tau-0)=\xi) = K(i,\xi)\mu(d\xi)$$
$$K(i,\xi) \leqslant 1/\mu(\xi) < \infty$$

That is, $K(\cdot,\xi)$ is bounded.

Suppose $\xi\in B_e, K(\cdot,\xi)$ is bounded. So $K(\cdot,\xi)$ is the standard image of $K_\lambda(\cdot,\xi)$

and moreover is a bounded Π-harmonic function, that is $K(\cdot,\xi)\in\mathscr{M}^+\cap\bar{\mathscr{M}}^+$. According to Theorem 13.3, there exists a non-negative bounded Borel function f on B_e such that

$$K(\cdot,\xi)=\int_{B_e} K(\cdot,\zeta)f(\zeta)\mu(d\zeta)$$

By the uniqueness Theorem 3.18, surely $f(\zeta)\mu(d\zeta)=\delta_\xi d\zeta$, so that $f(\xi)\mu(\xi)=1$. But f is bounded, therefore $\mu(\xi)>0$, that is $\xi\in B_{e1}$. The proof is finished. QED

Definition 2. We call the elements of $\hat{\mathscr{M}}_\lambda^+$ the non-negative solutions of the equation

$$\lambda u - Qu = 0 \qquad \lambda>0 \tag{1}$$

We call $u(\lambda)\in\hat{\mathscr{M}}_\lambda^+$ the minimal solution or the completely non-minimal solution of the equation (1), if $u(\lambda)$ as a $\Pi(\lambda)$-harmonic function is minimal or completely non-minimal.

Theorem 2. $B_{e1}=\phi$ iff there does not exist any bounded non-negative minimal solution to equation (1); $B_{e2}=\phi$ iff there does not exist any bounded non-negative completely non-minimal solution to equation (1), or equivalently, there does not exist a non-negative minimal solution $u(\lambda)$ to equation (1) such that $u(\lambda)$ is unbounded and the standard image of $u(\lambda)$ is γ-integrable.

Proof. It follows from Theorem 1, the corollary to Theorem 12.2 and Theorem 13.3. QED

7.15 EXITING ALMOST CLOSED SET AND THE BLACKWELL DECOMPOSITION OF THE MINIMAL Q PROCESS

Assume that X is the minimal Q process. According to Theorem 5.6, and in accordance with the Blackwell decomposition (2.2.19) of the embedding matrix Π, the potency of the index set $\mathscr{A}$ is equal to that of the atomic boundary B_1, so we can assume $\mathscr{A}=B_1$. Hence we have

$$\mathscr{L}(A_a)=\{x(\tau-0)=a\} \qquad a\in B_1 \tag{1}$$

$$\mathscr{L}(A_0)=\{x(\tau-0)\in B_2\} \tag{2}$$

Definition 1. Suppose that A is an almost closed set. We say that A is exit if

$$P\{\tau<\infty\,|\,\mathscr{L}(A)\}=1 \tag{3}$$

We say that A is passive if

$$P\{\tau=\infty\,|\,\mathscr{L}(A)\}=1 \tag{4}$$

Obviously, if A is an atomic almost closed set, and not exit, then it must be passive. But if A is a non-atomic almost closed set, the conclusion above is not true.

As for the completely non-atomic almost closed set A_0, according to Theorems 5.6 and 12.5, we can decompose it into $A_0 = A_{0e} \cup A_{0p}$, where both A_{0e} and A_{0p} are almost closed sets, so that

$$\mathscr{L}(A_{0e}) = \{x(\tau - 0) \in B_{e2}\} = \{x(\tau - 0) \in B_2, \tau < \infty\}$$
$$\mathscr{L}(A_{0p}) = \{x(\tau - 0) \in B_{p2}\} = \{x(\tau - 0) \in B_2, \tau = \infty\}$$

Accordingly A_{0e} is exit while A_{0p} is passive. Thus we follow the Blackwell decomposition of the minimal Q process.

Theorem 1. Suppose that $X = \{x(t), t < \tau\}$ is the minimal Q process. Then its state space E has the following decomposition:

$$E = A_{0e} \cup \left(\bigcup_{a \in B_{e_1}} A_a \right) \cup A_{0p} \cup \left(\bigcup_{a \in B_{p_1}} A_a \right) \tag{5}$$

where A_{0e} and A_{0p} are completely non-atomic almost closed sets, exit and passive respectively, and therefore may be not present. $A_a (a \in B_{e1})$ are exiting atomic almost closed sets while $A_a (a \in B_{p1})$ are passive atomic almost closed sets. The decomposition is unique modulo transient sets.

7.16 THE CONDITION FOR FINITE EXIT

Theorem 1. Suppose that finitely or denumerable infinitely many exit almost closed sets A_a are mutually disjoint. Write

$$X_i^a(\lambda) = E_i\{e^{-\lambda\tau}, \mathscr{L}(A_a)\} \tag{1}$$

If $\sum_a |c_a| < \infty, \sum_a c_a X^a(\lambda) = 0$, then $c_a = 0$.

Proof. By Theorem 5.6, there exist mutually disjoint Borel sets Γ_a such that $\mathscr{L}(A_a) = \{x(\tau - 0) \in \Gamma_a\}$ and $\mu(\Gamma_a) > 0$. According to Theorem 12.4, the standard image of $X^a(\lambda)$ is $X_i^a = P_i\{x(\tau - 0) \in \Gamma_a\}$. Consequently by $\sum_a c_a X^a(\lambda) = 0$ we get $\sum_a c_a X^a = 0$. According to Theorem 3.22 $\sum_a c_a 1_{(x(\tau-0)\in\Gamma_a)} = 0$, that is $\sum_a c_a 1_{\Gamma_a} = 0$ (μ, a.s.). Thus $c_a = 0$. The proof is concluded. QED

Theorem 2. Suppose $a, b \in B_{e1}$. Then

$$X_i^a(\lambda) = E_i\{e^{-\lambda\tau}, x(\tau - 0) = a\} \to \begin{cases} 1 & \text{when } i \to a \\ 0 & \text{when } i \to b \neq a \end{cases} \tag{2}$$

Proof. According to Zi-kun Wang and Xiang-qun Yang (1988, section 0.2,

Theorem 4), for any $b \in B_{e1}$, the limit $\lim_{i\to b} X_i^a(\lambda)$ exists, and so we only need to calculate the limit, which follows from Theorem 13.2. QED

Theorem 3. Suppose $0 \leqslant n < \infty$. The following conditions are equivalent

(i) In the decomposition (15.5), A_{0e} vanishes, the potency of the set B_{e1} is n.

(ii) The dimension of $\mathscr{M}_\lambda^+$ is n.

(iii) B_e is composed of only n atomic boundary points.

Proof. It is obvious that (i) and (ii) are equivalent.

(i)$\Rightarrow$(ii). Suppose that $A_1, \ldots, A_n$ are exit atomic. By Theorem 1, $X^a(\lambda)(1 \leqslant a \leqslant n)$ are mutually independent. Secondly, by Theorem 13.3, for $u(\lambda) \in \mathscr{M}_\lambda^+$, there exists f such that

$$u_i(\lambda) = E_i\{e^{-\lambda\tau} f[x(\tau-0)]\} = \sum_{a=1}^{n} E_i\{e^{-\lambda\tau} f[x(\tau-0)], \mathscr{L}(A_a)\}$$

$$= \sum_{a=1}^{n} f(a) X_i^a(\lambda)$$

That is, the dimension of $\mathscr{M}_\lambda^+$ is n.

(ii)$\Rightarrow$(i). First we show $P_i\{\mathscr{L}(A_{0e})\} = 0 (i \in E)$. Otherwise, there exists i such that $P_i(\mathscr{L}(A_{0e})\} > 0$. Since A_{0e} is completely non-atomic, it follows that there exist infinitely many disjoint almost closed sets. Thereby by Theorem 1, to $\mathscr{M}_\lambda^+$ there are more than n linear independent solutions. This is in contradiction with (ii). So A_{0e} vanishes. By the process of proof of (i)$\Rightarrow$(ii) we know the number of exit atoms is identical to the dimension of $\mathscr{M}_\lambda^+$, and the proof is completed. QED

7.17 A CONDITIONAL INDEPENDENCE THEOREM

Suppose that $X = \{x(t), t < \sigma\} \in \mathscr{H}_s(Q)$ is the non-minimal Q process, and that α is a Markov time of $X, P\{x(\alpha) = \infty\} = 0, P\{\alpha < \sigma\} < 0$. We write the first leaping point after α as τ_α; on$(\alpha < \sigma), \theta_\alpha \tau = \tau_\alpha$; let $\tau_\alpha^n = \theta_\alpha \tau_n$. We write $\mathscr{F}_{\tau_\alpha^n}$ for the pre-τ_α^n field, and $\mathscr{F}_{\tau_\alpha^n}$ for the minimal Borel field on Ω_{τ_α} containing all $\mathscr{F}_{\tau_\alpha^n}(n \geqslant 1)$.

Theorem 1. Suppose $A \in \mathscr{F}_\infty^0, \Lambda \in \mathscr{F}_{\tau_\alpha^-}$. There exists a Borel function f which is defined on $H \cup B_e$ and is independent of the conditional distribution of $X(\alpha)$, relative to $(\alpha < \sigma)$, such that on $\{x(\tau_\alpha - 0) \in H \cup B_e\}, P_\gamma$-almost surely holds

$$P\{\theta_{\tau_\alpha} A | \Lambda, x(\tau - 0)\} = f\{x(\tau_\alpha - 0)\} \tag{1}$$

The condition on the left side of the above formula should be understood as $|_A, x(\tau_\alpha - 0)$.

In order to prove Theorem 1, we need to quote a theorem in Dynkin (1963, p. 782), which is stated as follows.

Theorem 2. Suppose that $(\Omega, \mathscr{F}, P)$ is a probability space, $E|\xi| < \infty, \Omega_n \in \mathscr{F}$, $\Omega_n \downarrow \Omega'(n \uparrow \infty), \mathscr{A}_n \subseteq \mathscr{F}, \mathscr{A}_n$ is a Borel field on Ω_n and moreover $\mathscr{A}_m \cap \Omega_n \subseteq \mathscr{A}_n$ $(m < n)$. Then on Ω' we have almost surely

$$\lim_{n \to \infty} E(\xi | \mathscr{A}_n) = E(\xi | \mathscr{A})$$

where $\mathscr{A}$ is the minimal Borel field on Ω including all $\mathscr{A}_n$.

Proof of Theorem 1. First we prove (1) is almost surely true on $M \equiv \{X(\tau_a - 0) \in H\}$. For this, let

$$\beta = \begin{cases} \tau_a^{n-1} & \text{if } \omega \in M, \tau_\alpha = \tau_\alpha^n, \tau_\alpha^{n-1} < \tau_\alpha^n \\ \infty & \text{if } \omega \notin M \end{cases}$$

Note that when $\Lambda \in \mathscr{F}_{\tau_\alpha^-}$, $\Lambda(X(\beta) = i) \in \mathscr{F}_\beta$. What is more, $\{X(\tau_\alpha - 0) = i\} = \{X(\beta) = i\}$. According to Theorem 6.2(i), when $X(\tau_\alpha - 0) = i \in H$, (1) is true, and furthermore the restriction of f on H, $f(a), a \in H$, is independent of the conditional distribution of $X(\alpha)$, relative to $(\alpha < \sigma)$. So (1) is almost everywhere true on M.

Secondly, we verify that (1) is almost surely true on $N \equiv \{X(\tau_\alpha - 0) \in B_e\}$. That is, we must prove that

$$1_N P\{\theta_{\tau_\alpha} A | \Lambda, X(\tau_\alpha - 0\} = 1_N f\{X(\tau_\alpha - 0)\} \tag{2}$$

According to the definition of conditional probability, on N we have almost surely

$$P\{\theta_{\tau_\alpha} A, M | \Lambda, X(\tau_\alpha - 0)\} = 0$$

that is

$$1_N P\{\theta_{\tau_\alpha} A, M | \Lambda, X(\tau_\alpha - 0)\} = 0$$

so that the left side of (2) is equal to

$$1_N P\{\theta_{\tau_\alpha} A, M | \Lambda, X(\tau_\alpha - 0)\} \tag{3}$$

Now let $\Omega_n = (\tau_\alpha^n < \tau_\alpha), \mathscr{A}_n = \mathscr{F}_{\tau_\alpha^n}, \xi = 1_{N \cap \theta_{\tau_\alpha A}}$. Then the conditions in Theorem 2 are satisfied. Obviously, $N \subset \Omega' = \bigcap_n \Omega_n$; on Ω_n we have $\xi = \theta_{\tau_\alpha^n} \eta$, where $\eta = 1_{(\theta_\tau A) \cap (X(\tau - 0) \in B_e)}$. And it is easy to see

$$u(i) \equiv E_i \eta \in \mathscr{M}^+$$

Thus by the strong Markov property and Theorem 2, on Ω' we have almost surely

$$E(\xi|\mathscr{A}) = \lim_{n\to\infty} E(\xi|\mathscr{A}_n) = \lim_{n\to\infty} E_{X(\tau_\alpha^n)}\eta$$

$$= \lim_{n\to\infty} u\{X(\tau_\alpha^n)\}$$

Thereby we have

$$1_N E(\xi|\mathscr{A}) = \lim_{n\to\infty} 1_N u\{X(\tau_\alpha^n)\}$$

We can easily see that $\phi = \lim_{n\to\infty} 1_N u\{X(\tau_\alpha^n)\}$ is a non-negative ultimate random variable; so there exists a non-negative Borel-measurable function f on B such that $\phi = f\{x(\tau_\alpha - 0)\} = 1_N f\{x(\tau_\alpha - 0)\}$.

Consequently

$$1_N E(\xi|\mathscr{A}) = 1_N f\{x(\tau_\alpha - 0)\}$$

Because $\Lambda \in \mathscr{F}_{\tau_\alpha^-} \subset \mathscr{A}$, $x(\tau_\alpha - 0)$ is $\mathscr{A}$-measurable, and hence the left side of (2), that is, expression (3), equals

$$\begin{aligned} 1_N E\{\xi|\Lambda, x(\tau_\alpha - 0)\} &= 1_N E\{E(\xi|\mathscr{A})|\Lambda, x(\tau_\alpha - 0)\} \\ &= E\{1_N E(\xi|\mathscr{A})|\Lambda, x(\tau_\alpha - 0)\} \\ &= E\{1_N f[x(\tau_\alpha - 0)]|\Lambda, x(\tau_\alpha - 0)\} \\ &= 1_N f\{x(\tau_\alpha - 0)\} \end{aligned}$$

Thus we have proved (2).

We are going to prove that the restriction of f to B_e, $f(a), a \in B_e$, is independent of the conditional distribution of $x(\alpha)$ with respect to $(\alpha < \sigma)$, but we should think of the function and a function which is equal to f μ-almost surely as the same function, μ being the ultimate state distribution. For this, it suffices to apply the following fact: Suppose (x_n) is a Markov chain, and $u(i) (i \in E)$ is a real-valued function. If for any initial distribution, with probability 1 limit, the $\lim_{n\to\infty} u(x_n)$, exists, then this limit is independent of the initial distribution. In the foregoing fact, taking the Markov chain $x_n = x(\tau_\alpha^n)$ defined on the probability space $((\alpha < \sigma), (\alpha < \sigma)\mathscr{F}, P(\cdot|\alpha < \sigma))$ we get what we intend to prove.

The proof of the above fact is as follows. Suppose the distribution $v = v(i)$ satisfies $v_i > 0$, for all i. By $P_v(u(x_n) \to A(v)) = 1$ we derive $P_i(u(x_n) \to A(v)) = 1$ for all i. On the other hand under the supposition that $P_i(u(x_n) \to A(\delta_i)) = 1$, δ_i represents the unit distribution concentrated at i. So $P_i(A(v) = A(\delta_i)) = 1$. Now suppose v' is any distribution. When $v_i' > 0$, repeating the argument above we obtain

$$P_i(A(v) = A(\delta_i) = A(v')) = 1 \qquad P_i(A(v) = A(v')) = 1$$

Thus we have $P_{v'}(A(v) = A(v')) = 1$. By symmetry, $P_v(A(v) = A(v')) = 1$. If $\tilde{v}$ is another distribution, from the above $P_v(A(v)=A(\tilde{v}))=1$. It follows that

$$P_v(A(v')=A(\tilde{v}))=1 \qquad P_i(A(v')=A(\tilde{v}))=1 \qquad \text{(for all } i)$$
$$P_\gamma(A(v')=A(\tilde{v}))=1$$

According to Theorem 5.7, there exist non-negative Borel functions $f_{v'}$ and $f_{\tilde{v}}$, such that $A(v')=f_{v'}(x_\infty), A(\tilde{v})=f_{\tilde{v}}(x_\infty)$. Therefore $P_\gamma(f_{v'}(x_\infty)=f_{\tilde{v}}(x_\infty))=1$. That is, $f_{v'}=f_{\tilde{v}}$ is true μ-almost surely. The proof is over. QED

Corollary

Suppose $\Lambda\in\mathscr{F}_{\tau^-}$, $A\in\mathscr{F}^0_\infty$, then under $\{\tau<\infty, x(\tau-0)\}$, Λ and $\theta_\tau A$ are conditionally independent. That is, on $\{x(\tau-0)\in H\cup B_e\}$, almost surely holds

$$P\{\Lambda\theta_\tau A|x(\tau-0)\}=P\{\Lambda|x(\tau-0)\}P\{\theta_\tau A|x(\tau-0)\} \tag{4}$$

7.18 FURTHER DESCRIPTION OF GENERAL Q PROCESSES

Suppose that $X=\{x(t), t<\sigma\}\in\mathscr{H}_s(Q)$ is a non-minimal Q process and its resolvent operator is $\psi(\lambda)$.

Theorem 1. $\psi(\lambda)$ has the following representation:

$$\psi_{ij}(\lambda)=\phi_{ij}(\lambda)+\sum_{\alpha\in H\cup B_{e1}} X_i^a(\lambda)F_j^a(\lambda)+\int_{B_{e2}} X_i(\lambda, d_a)F_j(\lambda, a) \tag{1}$$

where

$$X_i^a(\lambda)=E_i\{e^{-\lambda\tau}, x(\tau-0)=a\}=\phi_{ia}(\lambda)d_a \qquad a\in H \tag{2}$$
$$X_i^a(\lambda)=E_i\{e^{-\lambda\tau}, x(\tau-0)=a\} \qquad a\in B_{e1} \tag{3}$$
$$X_i(\lambda,\Gamma)=E_i\{e^{-\lambda\tau}, x(\tau-0)\in\Gamma\} \qquad \Gamma\subset B_{e2} \tag{4}$$

and

$$F^a(\lambda)\geqslant 0 \qquad \lambda[F^a(\lambda),1]\leqslant 1 \qquad a\in H\cup B_{e1} \tag{5}$$
$$F(\lambda,a)\geqslant 0 \qquad \lambda[F(\lambda,a),1]\leqslant 1 \qquad \text{for } \mu\text{-almost all } a\in B_{e2} \tag{6}$$

Proof.

$$\begin{aligned}\psi_{ij}(\lambda)&=E_i\int_0^\sigma e^{-\lambda t}1_j[x(t)]\,dt\\ &=E_i\int_0^\tau e^{-\lambda t}1_j[x(t)]\,dt+E_i\int_\tau^{\sigma-\tau} e^{-\lambda t}1_j[x(t)]\,dt\\ &=\phi_{ij}(\lambda)+E_i e^{-\lambda\tau}\int_\tau^{\sigma-\tau} e^{-\lambda(t-\tau)}1_j[x(t)]\,dt\end{aligned}$$

By the corollary to Theorem 17.1, the second term of the above formula is equal to

$$E_i\left\{e^{-\lambda\tau}1_{(\tau<\infty)}\theta_\tau\int_0^\sigma e^{-\lambda\tau}1_j[x(t)]\,dt\right\}$$

$$=\int_{H\cup B_e} E_i\{e^{-\lambda\tau}, x(\tau-0)\in d_a\}E_i\left\{\theta_\tau\int_0^\sigma e^{-\lambda t}1_j[x(t)]\,dt\,|\,x(\tau-0)=a\right\}$$

$$=\int_{H\cup B_e} X_i(\lambda, d_a)F_j(\lambda, a)$$

Parts (1), (5) and (6) follow from this. As for the second equality in (2), it is derived from Theorem 10.1, and the proof is concluded. QED

7.19 INSTANTANEOUS RETURN PROCESS AND ITS BOUNDARY

Recall the class $\mathscr{H}_D$ of D-type processes, defined in section 7.7, and the definition 8.1 of the jumping and leaping points. Declare the minimal process to be a zero-order instantaneous return process.

Definition 1. Suppose $X=\{x(t), t<\sigma\}\in\mathscr{H}_D$ is a non-minimal process. If for almost all ω and any $t<\sigma(\omega)$, $X(\omega)$ has at most only finitely many leaping points in $[0, t)$, then we call X a first-order instantaneous return process, or simply a first-order process. The class composed of first-order processes (or first-order Q processes) is written as $\mathscr{H}_1$ (or $\mathscr{H}_1(Q)$).

Let $X\in\mathscr{H}_1$. Then its discontinuities can be arranged in an order:

$$0=\tau_0, \tau_1, \tau_2, \ldots; \tau_\omega, \tau_{\omega+1}, \ldots; \tau_{\omega^2}, \tau_{\omega^2+1}, \ldots; \tau_{\omega^n}, \tau_{\omega^n+1}, \ldots \tag{1}$$

where either for some ordinal number $\alpha<\omega^2$ we have $\tau_\alpha=\sigma$, or for all ordinal numbers $\alpha<\omega^2$ we have $\tau_\alpha<\alpha$, and $\lim_{n\to\infty}\tau_{\omega^n}=\sigma$.

If $X=\{x(t), t<\sigma\}\in\mathscr{H}_D$, then its discontinuouities can also be arranged in this order:

$$0=\tau_0, \tau_1, \tau_2, \ldots; \tau_\omega, \tau_{\omega+1}, \ldots; \tau_{\omega^2}, \ldots, \tau_{\omega^n}, \ldots, \tau_{\sum_{i=0}^{l}\omega^{l-i}k_i}, \ldots \tag{2}$$

where the k_i are non-negative integers and $l>n$. We shall define the kth-order instantaneous return process by induction.

Definition 2. We say $X=\{x(t), t<\sigma\}\in\mathscr{H}_D$ is a kth-order instantaneous return process, if X is not an nth-order $(0\leqslant n\leqslant k-1)$ instantaneous return process, and either there exists some ordinal number $\alpha<\omega^{k+1}$ such that $\tau_\alpha=\sigma$, or for all ordinal numbers $\alpha<\omega^{k+1}$ it is true that $\tau_\alpha<\sigma$, and furthermore $\tau_{\omega^{k+1}}=\sigma$. The class composed of kth-order instantaneous return processes is written as

$\mathscr{H}_k$. The class composed of kth-order instantaneous return Q processes is denoted by $\mathscr{H}_k(Q)$.

Suppose $X \in \mathscr{H}_k(Q)$. We call the embedding chain $X_T^0 = \{x(\tau_n), \tau_n \leqslant \tau_\beta\}$ the zero-order embedding chain of X, where τ_β is defined by (8.9)–(8.11), the one-step transition probability of X_T^0 is

$$_0\Pi_{ij} = P_i\{x(\tau_1) = j\} \tag{3}$$

We call $X_T^1 = \{x(\tau_{\omega n}), \tau_{\omega n} \leqslant \tau_{\omega\beta}\}$ the first-order embedding chain, where $\tau_{\omega\beta}$ can be defined by following (8.9)–(8.11). The one-step transition probability of X_T^1 is

$$_1\Pi_{ij} = P_i\{x(\tau_\omega) = j\} \tag{4}$$

$x^l = \{x(\tau_{\omega^l n}), \tau_{\omega^l n} \leqslant \tau_{\omega^l \beta}\}$ is the lth-order embedding chain, where $\tau_{\omega^l\beta}$ can be defined by following (8.9)–(8.11). The one-step transition probability of X_T^l is

$$_l\Pi_{ij} = P_i\{x(\tau_{\omega^l}) = j\} \tag{5}$$

Similarly, we can define

$$_0\Pi_{ij}(\lambda) = E_i\{e^{-\lambda\tau_1}, x(\tau_1) = j) \qquad \lambda > 0 \tag{6}$$

$$_l\Pi_{ij}(\lambda) = E_i\{e^{-\lambda\tau_{\omega^l}}, x(\tau_{\omega^l}) = j\} \qquad \lambda > 0 \tag{7}$$

Just as in sections 7.12 and 7.14 according to $_0\Pi, {}_0\Pi(\lambda)$ we could determine Martin boundary ∂E, essential Martin boundary B, Martin exit boundary B_e, Martin passive boundary B_p, atomic exit boundary B_{e1}, non-atomic exit boundary B_{e2}, atomic passive boundary B_{p1}, non-atomic passive boundary B_{p2}, and so on, so according to $_l\Pi$ and $_l\Pi(\lambda)$ we can determine lth-order Martin boundary $_l(\partial E)$, l-order essential Martin boundary $_lB$, lth-order exit boundary $_lB_e$, lth-order passive boundary $_lB_p$, lth-order atomic exit boundary $_lB_{e1}$, lth-order non-atomic exit boundary $_lB_{e2}$, lth-order atomic passive boundary $_lB_{p1}$, lth-order non-atomic passive boundary $_lB_{p2}$, and so on. Much of the boundary theory of the minimal process, i.e. zero-order process, can be transplanted into kth-order instantaneous return processes. For instance, for a kth-order instantaneous return process $X = \{x(t), t < \sigma\} \in \mathscr{H}_k$, if we let

$$_k\Omega_F = \{\text{there exists an ordinal number } \alpha < \omega^{k+1} \text{ such that } \tau_\alpha = \sigma < \infty\}$$

and

$$_k\Omega_\infty = \{\text{for all ordinal numbers } \alpha < \omega^{k+1} \text{ it is true that } \tau_\alpha < \sigma\}$$
$$+ \{\text{there exists an ordinal number } \alpha < \omega^{k+1} \text{ such that } \tau_a = \sigma = \infty\}$$

$$H = ({}_0H) \cup ({}_0B_e) \cup ({}_1B_e) \cup \cdots \cup ({}_{k-1}B_e) \qquad {}_0H = H \qquad {}_0B_e = B_e$$

Then on $_k\Omega_F$ we have $x(\sigma - 0) \in {}_kH$, on $_k\Omega_\infty$ we have $x(\sigma - 0) \in {}_kB$. Similarly to Theorem 12.5 and Theorem 12.8 we have the following.

Theorem 1. Suppose $X = \{x(t), t < \sigma\} \in \mathscr{H}_k$. Then for all $i \in E$ we have

$$P_i\{\sigma < \infty \mid x(\sigma - 0) \in_k B_e\} = 1 \tag{8}$$

$$P_i\{\sigma = \infty \mid x(\sigma - 0) \in_k B_p\} = 1 \tag{9}$$

$$P_i\{\sigma < \infty \mid x(\sigma - 0) \in_k H\} = 1 \tag{10}$$

CHAPTER 8

Construction of Q Processes with Finite Non-conservative States and Finite Exit Boundary

8.1 INTRODUCTION

For construction of general Q processes we often suppose that Q is conservative. Under this supposition, Feller (1957a) for the case of finite exit boundary and finite entrance boundary has constructed all Q processes satisfying the system of forward equations. Under the same supposition, Xiang-qun Yang (1966a) has constructed all Q processes. Williams (1964, 1966) and Chung (1963, 1966) in the case that Q is conservative and has a finite exit boundary found all the Q processes. Xiang-qun Yang (1982, 1983a), for the case of Q having a finite set of non-conservative states and a finite exit boundary (simply called bifinite), has constructed all the Q processes, but the results obtained are not very obviously related with those in Williams (1964, 1966). For this reason, by means of methods similar to those used in Williams (1964, 1966), Da-guo Xiong (1980, 1981) has constructed all the bifinite Q processes, but his results are still not very obviously associated with those in Feller (1957a) and Xiang-qun Yang (1966a). In this chapter, we shell, under the bifinite condition and according to the methods in Feller (1957a) and Xiang-qun Yang (1966a), construct all the Q processes, and point out the relation with the results obtained in Williams (1964, 1966) and Da-guo Xiong (1980, 1981). The content of this chapter is taken from Xiang-qun Yang (1983b).

8.2 BASIC HYPOTHESIS AND THE CONDITION SATISFIED BY $F^a(\lambda)$

Given a matrix $Q=(q_{ij})$ satisfying (2.2.6) and with $\boldsymbol{d}=Q1$ as the non-conservative quantity, let $H_e=\{i:d_i>0\}$ be the set of non-conservative states and B_e be the Martin exit boundary induced by Q.

Basic Hypothesis

Suppose that $A_e = H_e \cup B_e$ is a finite set, H_e or B_e may be empty. But when B_e is empty, either $\mathscr{L}_\lambda^+$ is not empty, or

$$\inf_\lambda \lambda \sum_j \phi_{ij}(\lambda) = 0 \qquad \lambda > 0 \tag{1}$$

Under the basic hypothesis the Q process is not unique. When $a \in H_e$, $X_i^a(\lambda) = \phi_{ia}(\lambda) d_a$ is an exit family, and

$$X_i^a(\lambda) \uparrow X_i^a = \Gamma_{ia} d_a \qquad \lambda \downarrow 0 \tag{2}$$

$$\lambda X_i^a(\lambda) \to \delta_{ia} d_a \qquad \lambda \to \infty \tag{3}$$

Write $\bar{X}(\lambda)$ for the maximal solution of $\mathscr{M}_\lambda^+(1)$, $\bar{X}(\lambda) \uparrow \bar{X}(\lambda \downarrow 0)$. If B_e is non-empty and finite, according to the discussion of sections 7.13 to 7.16, we can select an exit family $X^a(\lambda)$ $(a \in B_e)$ such that $\bar{X}(\lambda) = \sum_{a \in B_e} X^a(\lambda)$. Moreover,

$$X_i^a(\lambda) \to \delta_{ab} \qquad a, b \in B_e, i \to b \tag{4}$$

$$X^a(\lambda) \uparrow X^a \qquad \lambda \downarrow 0, a \in B_e \tag{5}$$

$$\lambda X^a(\lambda) \to 0 \qquad a \in B_e, \lambda \to \infty \tag{6}$$

$$X_i^a(\lambda) \to 0 \qquad a \in H_e, i \to b \in B_e \tag{7}$$

Thus under the basic hypothesis, we can select a harmonic exit family $X^a(\lambda)$ $(a \in A_e)$ whose standard image is X^a $(a \in A_e)$, such that

$$\lambda \phi(\lambda) 1 = 1 - Z(\lambda) \qquad Z(\lambda) = \sum_{a \in A_e} X^a(\lambda) \tag{8}$$

$$Z(\lambda) \uparrow Z = \sum_{a \in A_e} X^a \quad \lambda \downarrow 0 \tag{9}$$

$\bar{X} = \sum_{a \in B_e} X^a$ and $X^0 = 1 - Z$ are the maximal exit solution and the maximal passive solution of Q, respectively. The $X^a(\lambda)(a \in A_e)$ are linearly independent. In fact, suppose $\sum_{a \in A_e} c^a X^e(\lambda) = 0$. By (7) we get $\sum_{a \in B_e} c^a X^a(\lambda) = 0$ and so by (4) we obtain $c^a = 0$ $(a \in B_e)$. Hence $\sum_{a \in H_e} c^a X^a(\lambda) = 0$. By Lemma 2.11.6 $c^a = 0$ $(a \in H_e)$.

According to Theorem 7.18.1, and Q process $\psi(\lambda)$ has the following form:

$$\psi_{ij}(\lambda) = \phi_{ij}(\lambda) + \sum_{a \in A_e} X_i^a(\lambda) F_j^a(\lambda) \tag{10}$$

where

$$F^a(\lambda) \geqslant 0 \qquad \lambda[F^a(\lambda), 1] \leqslant 1 \tag{11}$$

Lemma 1. For $\psi_{(\lambda)}$ in (10), the norm condition is equivalent to that (11) holds for every $a \in A_e$. The resolvent equation is equivalent to that for every $a \in A_e$ the following holds:

$$F^a(\lambda)A(\lambda,\mu) = F^a(\mu) + (\mu-\lambda)\sum_{b\in A_e}[F^a(\lambda), X^b(\mu)]F^b(\mu) \tag{12}$$

where $A(\lambda,\mu)$ is as in (2.10.7); (12) is also equivalent to

$$F^a(\lambda) - F^a(\mu) = (\mu-\lambda)F^a(\lambda)\psi(\mu) \tag{13}$$

The Q condition is equivalent to that, for every $a\in H_e$, the following holds:

$$\lim_{\lambda\to\infty} \lambda F^a(\lambda) = 0 \tag{14}$$

Proof. The necessity of condition (11) for the norm condition has already been pointed out previously. The sufficiency follows from

$$\lambda\psi(\lambda)1 = 1 - \sum_{a\in A_e} X^a(\lambda) + \sum_{a\in A_e} X^a(\lambda)\lambda[F^a(\lambda), 1]$$

and (11).

Substituting $\psi_{(\lambda)}$ in (10) into the resolvent equation, because $\phi(\lambda)$ satisfies the resolvent equation and $X^a(\lambda)(a\in A_e)$ is a harmonic exit family, and moreover, is linearly independent, we know that the resolvent equation for $\psi(\lambda)$ is equivalent to (12). Substituting $\psi(\lambda)$ (10) into (13), we find (13) is equivalent to (12).

Noting that $\phi(\lambda)$ satisfies the Q condition and observing the finiteness of A_e, we find that the Q condition for $\psi(\lambda)$ is equivalent to

$$\lim_{\lambda\to\infty} \lambda X_i^a(\lambda)\lambda F_j^a(\lambda) = 0 \qquad a\in A_e$$

By (3) and (6), the formula above is equivalent to that for every $a\in H_e$ (14) holds.

QED

8.3 SIMPLIFICATION OF THE PROBLEM

Definition 1. Suppose a and $b\in A_e$. We call a and b indistinguishable if for all $\lambda>0$, $F^a(\lambda)=F^b(\lambda)$ holds.

Obviously, the indistinguishable relation is an equivalence relation. According to the indistinguishable relation, A_e can be decomposed into disjoint equivalence classes $a_1, a_2, \ldots,$. We can think of the equivalence class a_n as a new boundary point, and write

$$A = \{a_1, a_2, \ldots\} \qquad A = H\cup B \tag{1}$$

$$H = \{a: a\in A, a\cap H_e \neq \varnothing\} \qquad B = \{a: a\in A, a\cap H_e = \phi\} \tag{2}$$

Write

$$Y^a(\lambda) = \sum_{b \in a} X^b(\lambda) \uparrow Y^a = \sum_{b \in a} X^b \qquad a \in A, \lambda \downarrow 0 \tag{3}$$

Then (2.10) becomes

$$\psi_{ij}(\lambda) = \phi_{ij}(\lambda) + \sum_{a \in A} Y^a(\lambda) F^a(\lambda) \tag{4}$$

All the boundary points in A are distinguishable.

Lemma 1. There exists a subset $J \subset A$ and an $A \times J$ matrix $G = (G^{ab})$, whose columns are non-zero and whose rows satisfy

$$G^{ab} \geqslant 0 \qquad \sum_{b \in J} G^{ab} \leqslant 1 \, (a \in A) \qquad G^{aa} = 1 \, (a \in J) \tag{5}$$

such that

$$F^g(\lambda) = \sum_{b \in J} G^{ab} F^b(\lambda) \qquad a \in A, \lambda > 0 \tag{6}$$

And for each $a \in J$, $F^a(\lambda)$ cannot be represented by the following form:

$$F^a(\lambda) = \sum_{b \in J} h^{ab} F^b(\lambda) \qquad \lambda > 0 \tag{7}$$

where

$$h^{aa} = 0 \qquad h^{ab} \geqslant 0 \qquad \sum_{b \in J} h^{ab} \geqslant 1 \tag{8}$$

Especially, we have $F^a(\lambda) \neq 0 \, (\alpha \in J)$.

Proof. It is obvious that there exist a subset $J \subset A$ and an $A \times J$ matrix G such that (5) and (6) hold. For instance, it suffices to take A to be J, and to take G to be the unit matrix. But then (7) and (8) may hold for some $a \in J$.

Suppose there exist $J \subset A$ and an $A \times J$ matrix G such that (5) and (6) hold, and there exist $a_0 \in J$ such that (7) and (8) hold for $a = a_0$. Then by (6) and (7) we obtain

$$F^a(\lambda) = \sum_{b \in J_0} G_0^{ab} F^b(\lambda) \qquad a \in A, \lambda > 0$$

where $J_0 = J - \{a_0\}$, and the element of the $A \times J_0$ matrix $G_0 = (G_0^{ab})$ are $G_0^{ab} = G^{ab} + G_0^{aa} h^{a_0 b}, a \in A, b \in J_0$.

By (5) and (8), G satisfies

$$G_0^{ab} \geqslant 0 \qquad \sum_{b \in J_0} G_0^{ab} \leqslant 1 \, (a \in A) \qquad G_0^{aa} = 1 \, (a \in J_0)$$

Thus, after substituting J_0 and $A \times J_0$ matrix G_0 for J and G, (5) and (6) still hold. Going on like this, we can finally get a subset J and an $A \times J$ matrix G such that (5) and (6) hold, while for every $a \in J$, (7) and (8) cannot hold simultaneously. Finally if some columns in G are zero columns, then substituting $J - \{b \in J$: the bth column of G is a zero column$\}$ for J we can meet the needs of the lemma. The proof is completed. QED

Lemma 2. Every Q process $\psi(\lambda)$ has the form

$$\psi_{ij}(\lambda) = \phi_{ij}(\lambda) + \sum_{a \in J} Z_i^a(\lambda) F_j^a(\lambda) \tag{9}$$

where the subset $J \subset A$ and all the boundary points in J are distinguishable, and

$$Z^b(\lambda) = \sum_{a \in A} Y^a(\lambda) G^{ab} \qquad b \in J \tag{10}$$

is a linerly independent harmonic exit family. Each column of the $A \times J$ matrix $G = (G^{ab})$ is non-zero and satisfies (5), and is such that (6) holds; for every $a \in J$, (7) and (8) cannot hold simultaneously. Especially, $F^a(\lambda) \neq 0 (a \in J)$.

Proof. Substituting into (4) we get (9). Since $X^a(\lambda)$, $a \in A_e$, is a linearly independent harmonic exit family, it follows that $Z^b(\lambda)(b \in J)$ in (10) is a linearly independent harmonic exit family. The remainder of the conclusion follows from Lemma 1. The proof is completed. QED

Lemma 3. Suppose the Q process $\psi(\lambda)$ has the form (9) in Lemma 2. Then the norm condition of $\psi(\lambda)$ is equivalent to

$$F^a(\lambda) \geqslant 0 \qquad \lambda[F^a(\lambda), 1] \leqslant 1 \qquad a \in J \tag{11}$$

The resolvent equation is equivalent to

$$F^a(\lambda) A(\lambda, \mu) = F^a(\mu) + (\mu - \lambda) \sum_{b \in J} [F^a(\lambda), Z^b(\mu)] F^b(\mu) \qquad a \in J \tag{12}$$

The Q condition is equivalent to

$$\lim_{\lambda \to \infty} \lambda F^a(\lambda) = 0 \qquad a \in J_H \tag{13}$$

where

$$J_H = J \cap H = \{a \in J : a \cap H_e \neq \varnothing\} \tag{14}$$

$$J_B = J \cap B = \{a \in J : a \cap H_e = \varnothing\} \tag{15}$$

$$J = J_H \cup J_B \tag{16}$$

Proof. It follows by using Lemma 2.1 and Lemma 3.2. QED

Write

$$Z^b = \sum_{a\in A} Y^a G^{ab} \qquad b\in J \tag{17}$$

$$Z^*(\lambda) = Z(\lambda) - \sum_{b\in J} Z^b(\lambda) = \sum_{a\in A} Y^a(\lambda)\left(1 - \sum_{b\in J} G^{ab}\right) \tag{18}$$

$$Z^* = Z - \sum_{b\in J} Z^b = \sum_{a\in A} Y^a\left(1 - \sum_{b\in J} G^{ab}\right) \tag{19}$$

Then obviously

$$Z^b(\lambda)\uparrow Z^b (b\in J) \qquad Z^*(\lambda)\uparrow Z^* \qquad \lambda\downarrow 0 \tag{20}$$

$$X^0 + Z^* + \sum_{a\in J} Z^a = 1 \tag{21}$$

8.4 GENERAL FORM OF $F^a(\lambda)$

Lemma 1. Suppos $g\in l$, then

$$C_{\nu\lambda}^{-1}\|g\| \leqslant \|gA(\lambda,\nu)\| \leqslant C_{\lambda\nu}\|g\| \tag{1}$$

Here $\|g\|$ represents the norm in the Banach space l, and

$$C_{\lambda\nu} = 1 + \frac{|\lambda-\nu|}{\nu} \tag{2}$$

Proof. From the definition of $A(\lambda,\nu)$,

$$\|gA(\lambda,\nu)\| \leqslant \|g\| + \frac{|\lambda-\nu|}{\nu}[|g|, \nu\phi(\nu)1]$$

$$\leqslant \|g\| + \frac{|\lambda-\nu|}{\nu}[|g|, 1] = C_{\lambda\nu}\|g\|$$

The right-hand inequality in (1) follows. The left-hand inequality follows from

$$\|g\| = \|gA(\lambda,\nu)A(\nu,\lambda)\| \leqslant C_{\nu\lambda}\|gA(\lambda,\nu)\| \qquad \text{QED}$$

Lemma 2. Suppose the non-zero $F^a(\lambda)\,(a\in J)$ satisfies (3.11) and (3.12). Then there exist non-negative numbers $M^{ab}(\lambda)\,(a,b\in J)$ and a harmonic entrance family $\eta^a(\lambda)\,(a\in J)$ such that

$$F^a(\lambda) = \sum_{b\in J} M^{ab}(\lambda)\eta^b(\lambda) \qquad a\in J \tag{3}$$

Furthermore we can also select numbers $M^{ab}(\lambda)$ and $\eta^b(\lambda)$ having the following properties:

(i) $\eta^b(\lambda)=0$ if and only if $M^{ab}(\lambda)=0\,(a\in J)$,

(ii) there exists a non-negative $J\times J$ matrix $H=(H^{ab})$ such that

$$\eta^a(\lambda)=\sum_{b\in J}H^{ab}\eta^b(\lambda)\qquad a\in J \tag{4}$$

(iii) there exist positive numbers $v_a(a\in J)$ such that when μ moves along some subsequence $\mu_n\to\infty$ then

$$H^{ab}(\mu)=\frac{M^{ab}(\mu)}{\|F^a(\mu)A(\mu,v_a)\|}\to H^{ab}\qquad a,b\in J \tag{5}$$

If $\eta^a(\lambda)\neq 0$ then in the sense of strong convergence, when $\mu=\mu_n\to\infty$,

$$\eta^a(\lambda,\mu)=\frac{F^a(\mu)A(\mu,\lambda)}{\|F^a(\mu)A(\mu,v_a)\|}\to\eta^a(\lambda) \tag{6}$$

Proof. First note that for all $\lambda,\mu>0$, $F^a(\lambda)A(\lambda,\mu)$ is non-negative. In fact, by the definition of $A(\lambda,\mu)$ when $\lambda\geqslant\mu$ the conclusion is obvious. When $\lambda<\mu$, since the right-hand side of (3.12) is non-negative, so is the left-hand side. Because $F^a(\lambda)$ is non-zero we know $\|F^a(\lambda)A(\lambda,\mu)\|>0$ by (1). Multiplying (3.12) by $A(\mu,\lambda)$ from the right we obtain

$$F^a(\lambda)=\sum_{b\in J}M^{ab}(v_b,\lambda,\mu)\eta^b(v_b,\lambda,\mu) \tag{7}$$

where $v_b>0$ is arbitrary, and

$$M^{ab}(v,\lambda,\mu)=\sigma^{ab}(\lambda,\mu)\|F^b(\mu)A(\mu,v)\| \tag{8}$$

$$\sigma^{ab}(\lambda,\mu)=\delta_{ab}+(\mu-\lambda)[F^a(\lambda),Z^b(\mu)] \tag{9}$$

$$\eta^b(v,\lambda,\mu)=\frac{F^b(\mu)A(\mu,\lambda)}{\|F^b(\mu)A(\mu,v)\|} \tag{10}$$

Obviously

$$\|\eta^b(v,v,\mu)\|=1 \tag{11}$$

On account of (2.10.9) and Lemma 1

$$C_{\lambda v}^{-1}\leqslant\|\eta^b(v,\lambda,\mu)\|\leqslant C_{v\lambda} \tag{12}$$

$$C_{v\lambda}^{-1}M^{ab}(\lambda,\lambda,\mu)\leqslant M^{ab}(v,\lambda,\mu)\leqslant C_{\lambda v}M^{ab}(\lambda,\lambda,\mu) \tag{13}$$

Taking $v_b=v>0$ in (7) we have

$$F^a(\lambda)=\sum_{b\in J}M^{ab}(v,\lambda,\mu)\eta^b(v,\lambda,\mu) \tag{14}$$

From (12) we derive

$$1/\lambda \geqslant \| F^a(\lambda) \| \geqslant \sum_{b \in J} M^{ab}(v, \lambda, \mu) C_{\lambda v}^{-1} \tag{15}$$

$$C_{\lambda v}/\lambda \geqslant \sum_{a \in J} M^{ab}(v, \lambda, \mu) \tag{16}$$

Especially, taking $\lambda = v$ in (15) we get

$$\| F^a(v) \| \geqslant \sum_{b \in J} M^{ab}(v, v, \mu) \tag{17}$$

Consequently for every $b \in J$, either

$$\overline{\lim_{\mu \to \infty}} M^{ab}(v, \lambda, \mu) = 0 \qquad \text{for all } v > 0, \lambda > 0, a \in J \tag{18}$$

or

$$\overline{\lim_{\mu \to \infty}} M^{ab}(v, \lambda, \mu) > 0 \qquad \text{for some } \lambda > 0, \text{ some } v > 0 \text{ and some } a \in J \tag{19}$$

By (13) the formula above is equivalent to

$$\overline{\lim_{\mu \to \infty}} M^{ab}(v, v, \mu) > 0 \qquad \text{for some } v > 0 \text{ and some } a \in J \tag{20}$$

Assume the whole of $b \in J$ that makes (18) hold to be G_0; then $J - G_0 \neq \varnothing$. Because, otherwise, by (14), we derive $F^a(\lambda) = 0$ which contradicts the hypothesis that $F^a(\lambda)$ is non-zero. Therefore there exists $b_1 \in J - G_0$, $v(b_1) > 0$, $a(b_1) \in J$ and a subsequence $\mu_n(1) \to \infty$ such that

$$M^{a(b_i)b_i}(v(b_i), v(b_i), \mu_n(i)) \to M^{a(b_c)b_c}(v(b_i)) > 0 \tag{21}$$

hold for $i = 1$. Assume the whole of $b \in J$ that makes (18) hold to be G_1 when $\mu = \mu_n(1) \to \infty$. Then $G_0 \subset G_1$, $b_1 \in J - G_1$. If $G_1 \cup \{b_1\} = J$, then we take a subsequence $\mu_n = \mu_n(1)$. Otherwise, there exist $b_2 \in J - (G_1 \cup \{b_1\})$, $v(b_2) > 0$, $a(b_2) \in J$ and a subsequence $\mu_n(2)$ of $\mu_n(1)$ such that (21) is true for $i = 2$. Because J is finite, there exist $\Delta = \{b_1, b_2, \ldots, b_k\} \subset J$ and $a(b_1), a(b_2), \ldots, a(b_k) \in J$, $v(b_1)$, $v(b_2), \ldots, v(b_k) > 0$, and a subsequence $\mu_n \to \infty$, such that

$$\lim_{n \to \infty} M^{ab}(v, \lambda, \mu_n) = 0 \tag{22}$$

For $b \in J - \Delta$ and all $a \in J$, $v > 0$, $\lambda > 0$,

$$\lim_{n \to \infty} M^{a(c)c}(v(c), v(c), \mu_n) = M^{a(c)c}(v(c)) > 0 \qquad c \in \Delta \tag{23}$$

On account of (11) and (17), and by applying the diagonal method, we can select a subsequence of μ_n, again to be denoted by μ_n, such that for every $c \in \Delta$ we have

$$\lim_{n\to\infty} M^{ab}(v(c), v(c), \mu_n) = M^{ab}(v(c)) \qquad a, b \in J \tag{24}$$

$$\lim_{n\to\infty} \eta_j^b(v(c), v(c), \mu_n = \eta_j^b(v(c)) \qquad b \in J, j \in E \tag{25}$$

and moreover

$$M^{a(c)c}(v(c)) > 0\,(c \in \Delta) \qquad M^{ab}(v(c)) = 0\,(c \in \Delta, a \in J, b \in J - \Delta) \tag{26}$$

Letting $\lambda = v = v(c)$ in (14) we obtain

$$F^a(v(c)) = \sum_{b \in J} M^{ab}(v(c), v(c), \mu)\eta^b(v(c), v(c), \mu) \qquad c \in \Delta \tag{27}$$

Letting $\mu = \mu_n \to \infty$, we get

$$F^a(v(c)) = \sum_{b \in \Delta} M^{ab}(v(c))\eta^b(v(c)) \tag{28}$$

Thus

$$\| F^a(v(c)) \| = \sum_{b \in \Delta} M^{ab}(v(c)) \| \eta^b(v(c)) \| \tag{29}$$

Letting $v = v(c)$ in (17) and taking the limit we derive

$$\| F^a(v(c)) \| \geqslant \sum_{b \in \Delta} M^{ab}(v(c)) \tag{30}$$

By (11) and Fatou's lemma we have $\| \eta^b(v(c)) \| \leqslant 1$ ($c \in \Delta$, $b \in J$). Comparing (29) and (30) we obtain

$$\| \eta^b(v(b)) \| = 1 \qquad b \in \Delta \tag{31}$$

Thus when $b \in \Delta$, $\eta^b(v(b), v(b), \mu_n)$ converges to $\eta^b(v(b))$ in coordinates, and the sequence of norms $\| \eta^b(v(b), v(b), \mu_n) \| = 1$ obviously converges to the norm $\| \eta^b(v(b)) \| = 1$. Thus the convergence is a strong one in the Banach space l, that is

$$\| \eta^b(v(b), v(b), \mu_n) - \eta^b(v(b)) \| \to 0 \qquad b \in \Delta \tag{32}$$

Write

$$\eta^b(\lambda) = \eta^b(v(b))A(v(b), \lambda)$$

Then $\eta^b(\lambda)\,(b \in \Delta)$ is obviously a non-zero harmonic entrance family. By Lemma 1

$$\| \eta^b v(b), \lambda, \mu_n) - \eta^b(\lambda) \| \leqslant C_{v(b)\lambda} \| \eta^b(v(b), v(b), \mu_n) - \eta^b(v(b)) \| \to 0 \tag{33}$$

Now for every $a \in J$ and $\lambda > 0$, we can select a subsequence of μ_n, and still write it as μ_n, such that

$$M^{ab}(v(b), \lambda, \mu_n) \to M^{ab}(\lambda) \qquad b \in \Delta \tag{34}$$

Letting $\mu = \mu_n \to \infty$ in (7) and noting (34) and (32) we find

$$F^a(\lambda) = \sum_{b \in \Delta} M^{ab}(\lambda) \eta^b(\lambda)$$

For $b \in J - \Delta$, we add the definition that $\eta^b(\lambda) = 0$, $M^{ab}(\lambda) = 0\,(a \in J)$. Thus we have proved (3), and (i) in Lemma 2, and (6).

By (3) we have

$$\eta^a(\lambda, \mu) = \frac{F^a(\mu)A(\mu, \lambda)}{\| F^a(\mu)A(\mu, v_a) \|} = \sum_{b \in J} H^{ab}(\mu) \eta^b(\lambda) \tag{35}$$

where $v_a = v(a)$, and

$$H^{ab}(\mu) = \frac{M^{ab}(\mu)}{\| F^a(\mu)A(\mu, v_a) \|} = \frac{M^{ab}(\mu)}{\sum_{b \in J} M^{ab}(\mu) \| \eta^b(v_a) \|} \tag{36}$$

Because

$$\sum_{b \in J} H^{ab}(\mu) \| \eta^b(v_a) \| = 1 \tag{37}$$

and $\| \eta^b(v_a) \| > 0$ when $b \notin \Delta$, then $H^{ab}(\mu) = 0$ when $b \in J - \Delta$. So that when μ tends to infinity along some subsequence of μ_n the limit in (5) exists and is finite.

When $\eta^a(\lambda) \neq 0$, that is, $a \in \Delta$, by (33), $\eta^a(\lambda, \mu_n) = \eta^a(v_a, \lambda, \mu_n)$ converges strongly to $\eta^a(\lambda)$. Hence when $\eta^a(\lambda) \neq 0$, (4) follows from (35). When $\eta^a(\lambda) = 0$, taking $H^{ab} = 0\,(b \in J)$ (4) obviously holds. The lemma is proved. QED

8.5 NON-STICKY CASE

Definition 1. A harmonic exit family $(\eta(\lambda),\ \lambda > 0)$ is said to be sticky if $\lim_{\lambda \to \infty} \lambda \| \eta(\lambda) \| = \infty$, otherwise it is said to be non-sticky. If in the representation (3.9) and (4.3) of a Q process $\psi(\lambda)$ every $\eta^a(\lambda)$ is non-sticky, then we say that the process $\psi(\lambda)$ is non-sticky.

Theorem 1. (i) Decompose $A_e = H_e \cup B_e$ into $A_e = a_1 \cup a_2 \cup \ldots$, where every a_m is non-empty and they are mutually disjoint. Write $A = \{a_1, a_2, \ldots\}$. Take a non-empty subset J of A, and $J = J_H \cup J_B$, where

$$J_H = \{a \in J : a \cap H_e \neq \varnothing\} \qquad J_B = \{a \in J : a \cap H_e = \varnothing\}$$

(ii) Take an $A \times J$ matrix $G = (G^{ab})$ satisfying (3.5), where each column of G

is non-zero. Write

$$Z^b(\lambda) = \sum_{a\in A}\sum_{c\in a} X^c(\lambda)G^{ab} \qquad Z^b = \sum_{a\in A}\sum_{c\in a} X^c G^{ab} \qquad b\in J \tag{1}$$

$$Z^*(\lambda) = Z(\lambda) - \sum_{a\in J} Z^a(\lambda) = \sum_{a\in A}\left(\sum_{c\in a} X^c(\lambda)\right)\left(1 - \sum_{b\in J} G^{ab}\right) \tag{2}$$

$$Z^* = Z - \sum_{a\in J} Z^a = \sum_{a\in A}\left(\sum_{c\in a} X^c\right)\left(1 - \sum_{b\in J} G^{ab}\right)$$

(iii) Take a non-zero and non-sticky harmonic exit family $(\eta^a(\lambda), \lambda > 0)(a\in J)$. Write

$$\bar{W}^{ab}(\lambda) = \lambda[\bar{\eta}^a(\lambda), Z^b] \uparrow \bar{W}^{ab} \qquad \lambda\uparrow\infty \tag{3}$$

$$\bar{W}^{a*}(\lambda) = \lambda[\bar{\eta}^a(\lambda), Z^*] \uparrow \bar{W}^{a*} \qquad \lambda\uparrow\infty \tag{4}$$

$$\bar{\sigma}^a = \lambda[\bar{\eta}^a(\lambda), X^0] \qquad (\bar{\sigma}^a \text{ is independent of } \lambda) \tag{5}$$

$$\bar{\alpha}^a = \lim_{\lambda\to\infty} \lambda\bar{\eta}^a(\lambda) \tag{6}$$

such that

$$\bar{\sigma}^a + \bar{W}^a + \sum_{b\in J} \bar{W}^{ab} \leqslant 1 \tag{7}$$

$$\bar{\alpha}^a = 0 \qquad \text{if } a\in J_H \tag{8}$$

(iv) Let

$$\psi_{ij}(\lambda) = \phi_{ij}(\lambda) + \sum_{a\in J}\sum_{b\in J} Z_i^a(\lambda)K^{ab}(\lambda)\bar{\eta}^b(\lambda) \tag{9}$$

where the $J \times J$ matrix $K(\lambda)$ is given by

$$\begin{aligned} K(\lambda) &= (K^{ab}(\lambda)) = [I - \bar{W} + \bar{W}(\lambda)]^{-1} \\ \bar{W}(\lambda) &= (\bar{W}^{ab}(\lambda)) \uparrow \bar{W} = (\bar{W}^{ab}) \qquad \lambda\uparrow\infty \end{aligned} \tag{10}$$

and $I = (\delta_{ab})$ is the $J \times J$ unit matrix.

Then $\psi(\lambda)$ is a non-sticky Q process, which is honest if and only if

$$\bar{W}^{a*} = 0 \qquad \bar{\sigma}^a + \sum_{b\in J} \bar{W}^{ab} = 1 \tag{11}$$

The $\psi(\lambda)$ is of B-type if and only if $J_H = \varnothing$. The $\psi(\lambda)$ is of F-type if and only if

$$\bar{\alpha}^a = 0 \qquad a\in J_B \tag{12}$$

Every non-minimal and non-sticky Q process $\psi(\lambda)$ can be derived in the way above.

Proof. (a) Suppose that the Q process $\psi(\lambda)$ is non-minimal and non-sticky. Substituting (4.3) into (3.11) and (3.12) we obtain

$$\sum_{b\in J} M^{ab}(\lambda)\sigma^b + S^{a*}(\lambda) + \sum_{b\in J} S^{ab}(\lambda) \leqslant 1 \tag{13}$$

$$\sum_{b\in J} M^{ab}(\lambda)\eta^b(\mu) = \sum_{b\in J} [I - S(\lambda) + M(\lambda)W(\mu)]^{ab} F^b(\mu) \tag{14}$$

where $W(\lambda) = (W^{ab}(\lambda), S(\lambda) = (S^{ab}(\lambda))$ are $J \times J$ matrices,

$$W^{ab}(\lambda) = \lambda[\eta^a(\lambda), Z^b] \uparrow W^{ab} \qquad \lambda \uparrow \infty \tag{15}$$

$$W^{a*}(\lambda) = \lambda[\eta^a(\lambda), Z^*] \uparrow W^{a*} \qquad \lambda \uparrow \infty \tag{16}$$

$$\bar{\sigma}^a = \lambda[\bar{\eta}^a(\lambda), X^0] \qquad (\sigma^a \text{ being independent of } \lambda) \tag{17}$$

$$S^{ab}(\lambda) = \sum_{c\in J} M^{ac}(\lambda) W^{cb}(\lambda) \tag{18}$$

$$S^{a*}(\lambda) = \sum_{c\in J} M^{ac}(\lambda) W^{c*}(\lambda) \tag{19}$$

As $\psi(\lambda)$ is non-sticky, $W = (W^{ab})$ and W^{a*} are all finite.

Select a subsequence $\lambda \to \infty$ such that

$$M^{ab}(\lambda) \to M^{ab} \qquad S^{ab}(\lambda) \to S^{ab} \qquad S^{a*}(\lambda) \to S^{a*} \tag{20}$$

It is plain that S^{ab}, S^{a*} are all finite. We are going to prove M^{ab} finite. When $\eta^b(\lambda) = 0$ obviously $M^{ab}(\lambda) = 0 \to M^{ab} = 0$. Suppose $\Delta = \{b \in J : \eta^b(\lambda) \neq 0\}$. By (13), (18), (19) and Fatou's lemma we have

$$\sum_{c\in J} M^{ac}\sigma^c + S^{a*} + \sum_{b\in J} S^{ab} \leqslant 1 \tag{21}$$

$$\sum_{c\in J} M^{ac} W^{c*} \leqslant S^{a*} \tag{22}$$

$$\sum_{c\in J} M^{ac} W^{cb} \leqslant S^{ab} \tag{23}$$

(setting $\infty \cdot 0 = 0$). So by (21)–(23),

$$\begin{aligned} \sum_{c\in\Delta} M^{ac}\lambda[\eta^c(\lambda), 1] &= \sum_{c\in J} M^{ac}\lambda[\eta^c(\lambda), 1] \\ &= \sum_{c\in J} M^{ac}\left(\sigma^c + W^{c*}(\lambda) + \sum_{b\in J} W^{cb}(\lambda)\right) \\ &\leqslant \sum_{c\in J} M^{ac}\sigma^c + \sum_{c\in J} M^{ac} W^{c*} + \sum_{b\in J}\left(\sum_{c\in J} M^{ac} W^{cb}\right) \\ &\leqslant 1 \end{aligned}$$

and therefore when $b\in\Delta$, $M^{ab}<\infty$. Thus the $J\times J$ matrix $M=(M^{ab})$ is finite. Hence the equalities hold in (22) and (23).

Taking the limit in (14) we get

$$\sum_{b\in J} M^{ab}\eta^b(\mu)=\sum_{b\in J}[I-MW+MW(\mu)]^{ab}F^b(\mu) \tag{24}$$

Write

$$\bar{\eta}^a(\lambda)=\sum_{c\in J} M^{ac}\eta^c(\lambda) \tag{25}$$

Then (24) becomes

$$\bar{\eta}^a(\mu)=\sum_{b\in J}[I-\bar{W}+\bar{W}(\mu)]^{ab}F^b(\mu) \tag{26}$$

Note that $\bar{\eta}^a(\mu)$ is a non-zero and harmonic entrance family. Actually, if $\bar{\eta}^a(\mu)=0$, then $\bar{W}^{ab}(\mu)=0$, $\bar{W}^{ab}=0$, and hence by (26) we have $F^a(\mu)=0$, which contradicts the non-zero property of $F^a(\mu)$. Thus $\mu[\bar{\eta}^a(\mu),1]>0$.

The matrix $\bar{W}-\bar{W}(\mu)$ is non-negative, and the row sum satisfies

$$\begin{aligned}\sum_{c\in J}[\bar{W}^{ac}-\bar{W}^{ac}(\mu)]&=\sum_{c\in J}\bar{W}^{ac}-\mu[\bar{\eta}^a(\mu),\mathbf{1}-X^0-Z^*]\\&=\bar{\sigma}^a+\bar{W}^{a*}+\sum_{c\in J}\bar{W}^{ac}-\mu[\bar{\eta}^a(\mu),\mathbf{1}]\\&\leqslant 1-\mu[\bar{\eta}^a(\mu),\mathbf{1}]<1\end{aligned}$$

Accordingly the inverse matrix in (10) exists and is non-negative, and furthermore

$$K(\lambda)=\sum_{n=0}^{\infty}[\bar{W}-\bar{W}(\lambda)]^n \tag{27}$$

By (26) we have

$$F^a(\mu)=\sum_{b\in J}K^{ab}(\mu)\bar{\eta}^b(\mu) \tag{28}$$

An so we obtain (9).

Noting that

$$\lim_{\lambda\to\infty}K(\lambda)=\sum_{n=0}^{\infty}(\bar{W}-\bar{W})^n=I \tag{29}$$

by (28) we know that the Q condition (3.13) is just (8).

(b) Suppose that $\psi(\lambda)$ is determined according to (i)–(iv) in Theorem 1. Because

$$\bar{W}(\lambda)=K(\lambda)^{-1}-I+\bar{W}$$

by (28) and after simple computations we derive

$$\lambda[F^a(\lambda), \mathbf{1}] = 1 - \sum_{b\in J} K^{ab}(\lambda)\left(1 - \bar{\sigma}^b - \bar{W}^{b*} - \sum_{c\in J} \bar{W}^{bc}\right)$$

$$- \sum_{b\in J} K^{ab}(\lambda)[\bar{W}^{b*} - \bar{W}^{b*}(\lambda)] \leqslant 1 \tag{30}$$

The equality in the formula above holds if and only if the equality in (7) holds and, moreover, $\bar{W}^{b*}(\lambda) = \bar{W}^{b*}$ $(b\in J)$. Considering (2.11.40) we have

$$(\mu - \lambda)[\bar{\eta}^b(\mu), Z^*(\lambda)] = \bar{W}^{b*}(\mu) - \bar{W}^{b*}(\lambda)$$

Therefore when $\mu \neq \lambda$, $[\bar{\eta}^b(\mu), Z^*(\lambda)] = 0$. Whence $[\bar{\eta}^b(\mu), Z^*] = 0$, $\bar{W}^{b*}(\mu) = 0$. Thus $\bar{W}^{b*}(\lambda) = \bar{W}^{b*}$ is equivalent to $\bar{W}^{b*} = 0$.

A simple computation shows that

$$K(\lambda)[\bar{W}(\lambda) - \bar{W}(\mu)]K(\mu) = K(\lambda)[K(\lambda)^{-1} - K(\mu)^{-1}]K(\mu) = K(\mu) - K(\lambda) \tag{31}$$

Applying the above formula we can easily verify that (3.12) holds for the resolvent equation of $\psi(\lambda)$. We have pointed out that (8) is just the norm condition (3.13); therefore $\psi(\lambda)$ is a non-sticky Q process.

(c) Suppose that $\psi(\lambda)$ is of B-type, that is, the B condition is satisfied. We proceed to prove $J_H = \varnothing$. If Q is conservative, the conclusion is trivially true. If Q is non-conservative, by Theorem 2.12.1, $F^a(\lambda) = 0$ $(a\in H_e)$ holds in (2.10), and so $F^a(\lambda) = 0$ $(a\in J_H)$ in (3.10). But by (28), every $F^a(\mu) \neq 0$ $(a\in J)$. Hence surely $J_H = \varnothing$. Or else, if $J_H = \varnothing$, then for every $a\in J = J_B$, $(\lambda I - Q)Z^b(\lambda) = 0$ holds. Thus the B condition of $\psi(\lambda)$ is satisfied, that is, $\psi(\lambda)$ is of B-type.

The F condition of $\psi(\lambda)$ is equivalent to

$$0 = \sum_{a\in J}\sum_{b\in J} Z^a(\lambda)K^{ab}(\lambda)\bar{\eta}^b(\lambda)(\lambda I - Q) = \sum_{a\in J} Z^a(\lambda) \sum_{b\in J} K^{ab}(\lambda)\bar{\alpha}^b$$

Since $Z^a(\lambda)$ $(a\in J)$ are linearly independent, $K(\lambda)^{-1}$ exists. The above formula is equivalent to $\bar{\alpha}^b = 0$, $b\in J$. By (8) it is also equivalent to (12). The proof is completed.

QED

8.6 GENERAL CONSTRUCTION

Lemma 1. Suppose $a\in J$, and (3.11) and (3.12) hold. If for some $\lambda > 0$, we have

$$S^{aa}(\lambda) \equiv \lambda[F^a(\lambda), Z^a] = 1 \tag{1}$$

then (1) holds for all $\lambda > 0$ and there exists a non-zero and harmonic entrance family $\eta^a(\lambda)$, $\lambda > 0$, such that

$$F^a(\lambda) = \frac{\eta^a(\lambda)}{W^{aa}(\lambda)} \qquad \frac{F^a(\mu)A(\mu, \lambda)}{\| F^a(\mu)A(\mu, v_a)\|} = \eta^a(\lambda) \tag{2}$$

Here v_a is a positive number chosen such that $\|\eta^a(v_a)\| = 1$, and $W^{aa}(\lambda)$ is determined by (5,15).

Proof. By (1) and (3.11) we have $\lambda[F^a(\lambda), Z^b] = 0$ for $b \in J$, $b \neq a$, so that $\lambda[F^a(\lambda), Z^b(\mu)] = 0$. Consequently (3.12) becomes

$$F^a(\lambda)A(\lambda, \mu) = F^a(\mu) + (\mu - \lambda)[F^a(\lambda), Z^a(\mu)]F^a(\mu)$$

Since $F^a(\mu) \neq 0$, it follows that $1 + (\mu - \lambda)[F^a(\lambda), Z^b(\mu)] > 0$. By imitation of the proof of Theorem 3.2.1, we know there exist a non-zero harmonic entrance family $\eta^a(\lambda), \lambda > 0$, and a constant c^a such that

$$F^a(\lambda) = \frac{\eta^a(\lambda)}{c^a + W^{aa}(\lambda)} \qquad \lambda > 0$$

By (1) we know $c^a = 0$ so (1) and the first expression in (2) hold for all $\lambda > 0$. Normalizing $\eta^a(\lambda)$ appropriately so that $\|\eta^a(v_a)\| = 1$, the second expression in (2) holds. The proof is terminated. QED

Theorem 2. (i) Select the set J and $Z^*(\lambda)$, $Z^a(\lambda)$, $a \in J$ and $Z^*, Z^a, a \in J$ according to (i) and (ii) in Theorem 5.1.

(ii) Take a subset L (L may be empty) of J, and set

$$L^a = \begin{cases} 1 & \text{if } a \in L \\ 0 & \text{if } a \in J - L \end{cases} \tag{3}$$

(iii) Take a non-zero and harmonic entrance family $(\bar{\eta}^a(\lambda), \lambda > 0)$, $a \in J$, such that

$$\lim_{\lambda \to \infty} \lambda[\bar{\eta}^a(\lambda), \mathbf{1} - Z^a] < \infty \qquad a \in J \tag{4}$$

or equivalently, according to the notation in (5.3)–(5.6),

$$\bar{W}^{a*} < \infty \qquad \bar{W}^{ab} < \infty \qquad a, b \in J,\ a \neq b \tag{5}$$

(iv) Take a non-negative $J \times J$ matrix $\bar{S} = (\bar{S}^{ab})$ such that

$$\bar{S}^{aa} = 0 \qquad \bar{S}^{a*} \geqslant \bar{W}^{a*} \qquad \bar{S}^{ab} \geqslant \bar{W}^{ab} \qquad a, b \in J,\ a \neq b \tag{6}$$

$$\bar{\sigma}^a + \bar{W}^{a*} + \sum_{b \in J} \bar{S}^{ab} \leqslant L^a \qquad a \in J \tag{7}$$

(v) Furthermore, the following two equations should hold:

$$(L^b + \bar{W}^{bb})^{-1}\bar{\alpha}^b = 0 \qquad \text{if } b \in J_H \tag{8}$$

$$N^{ab}(L^b + \bar{W}^{bb})^{-1}\bar{\alpha}^b = 0 \qquad \text{if } a \in J_H, b \in J_B \tag{9}$$

Here $J \in J$ matrix $N(\lambda) \downarrow N = (N^{ab})$, $\lambda \uparrow \infty$,

$$N(\lambda) = [I - R(\lambda)]^{-1} = \sum_{n=0}^{\infty} R^n(\lambda) \tag{10}$$

$$R^{aa}(\lambda)=0 \qquad R^{ab}(\lambda)=\frac{\bar{S}^{ab}-\bar{W}^{ab}(\lambda)}{L^a+\bar{W}^{aa}(\lambda)} \qquad (a\neq b) \tag{11}$$

$$N=\sum_{n=0}^{\infty} R^n \qquad R=(R^{ab}) \tag{12}$$

$$R^{aa}=0 \qquad R^{ab}=\frac{\bar{S}^{ab}-\bar{W}^{ab}}{L^a+\bar{W}^{aa}} \qquad (a\neq b) \tag{13}$$

(vi) Let

$$\psi_{ij}(\lambda)=\phi_{ij}(\lambda)+\sum_{a,b\in J} Z_i^a(\lambda)\bar{K}^{ab}(\lambda)\bar{\eta}_j^b(\lambda) \tag{14}$$

where the $J\times J$ matrix $\bar{K}(\lambda)$ is given by

$$\bar{K}(\lambda)=(\bar{K}^{ab}(\lambda))=[\bar{I}-\bar{S}+\bar{W}(\lambda)]^{-1}=N(\lambda)D(\lambda) \tag{15}$$

and $\bar{I}$ is a diagonal matrix with L^a, $a\in J$, as its diagonal elements. $D(\lambda)$ is a diagonal matrix whose diagonal elements are

$$D^{aa}(\lambda)=\frac{1}{L^a+\bar{W}^{aa}(\lambda)} \qquad a\in J \tag{16}$$

Then $\psi(\lambda)$ is a Q process. Every Q process $\psi(\lambda)$ can be obtained in the way stated above. The process is honest if and only if

$$\bar{S}^{a*}=0 \qquad \bar{\sigma}^a+\sum_{b\in J}\bar{S}^{ab}=L^a \qquad a\in J \tag{17}$$

Remark 1

For the process $\psi(\lambda)$ in Theorem 2, if $a\in J-L$, then

$$\lambda[\bar{\eta}^a(\lambda),1]=\bar{W}^{aa}(\lambda)>0 \tag{18}$$

$$F^a(\lambda)=\frac{\bar{\eta}^a(\lambda)}{\bar{W}^{aa}(\lambda)} \tag{19}$$

Actually, from (6) and (7) follows (18). Furthermore, from

$$F^a(\lambda)=\sum_{b\in J} K^{ab}(\lambda)\bar{\eta}^b(\lambda)$$

we obtain

$$\bar{\eta}^a(\lambda)=\sum_{b\in J}[\bar{I}-\bar{S}+\bar{W}(\lambda)]^{ab}F^b(\lambda) \tag{20}$$

By (6) and (7), the above expression becomes $\bar{\eta}^a(\lambda)=\bar{W}^{aa}(\lambda)F^a(\lambda)$; therefore we get (19).

Proof of Theorem 2. (a) Assume $\psi(\lambda)$ to be a Q process. Then (5.13) and (5.14) still hold. By Lemma 1 we can write

$$L = \{a: a \in J, S^{aa}(\lambda) < 1\} \tag{21}$$

Let

$$\delta^a(\lambda) = \begin{cases} 1 - S^{aa}(\lambda) & \text{if } a \in L \\ \| F^a(\lambda) A(\lambda, v_a) \| & \text{if } a \in J - L \end{cases} \tag{22}$$

Then $\delta^a(\lambda) > 0$. Dividing (5.13) and (5.14) by $\delta^a(\lambda)$ we have

$$\sum_{b \in J} \bar{M}^{ab}(\lambda)\sigma^b + \bar{S}^{a*}(\lambda) + \sum_{b \in J} \bar{S}^{ab}(\lambda) \leqslant L^a \tag{23}$$

$$\sum_{b \in J} \bar{M}^{ab}(\lambda)\eta^b(\mu) = \sum_{b \in J} [\bar{I} - \bar{S}(\lambda) + \bar{M}(\lambda)W(\mu)]^{ab} F^b(\mu) \tag{24}$$

where

$$\bar{M}^{ab}(\lambda) = \frac{M^{ab}(\lambda)}{\delta^a(\lambda)} \qquad \bar{S}^{aa}(\lambda) = 0 \qquad \bar{S}^{a*}(\lambda) = \frac{S^{a*}(\lambda)}{\delta^a(\lambda)}$$
$$\bar{S}^{ab}(\lambda) = \frac{S^{ab}(\lambda)}{\delta^a(\lambda)} \qquad (b \neq a) \tag{25}$$

satisfies

$$\bar{S}^{ab}(\lambda) = \sum_{c \in J} \bar{M}^{ac}(\lambda) W^{cb}(\lambda) \qquad b \neq a$$
$$\bar{S}^{a*}(\lambda) = \sum_{c \in J} \bar{M}^{ac}(\lambda) W^{c*}(\lambda) \tag{26}$$

Select a subsequence $\lambda \to \infty$ from the sequence μ_n of Lemma 4.2, such that for all $a, b \in J$ the following hold:

$$\bar{M}^{ab}(\lambda) \to M^{ab} \qquad \bar{S}^{ab}(\lambda) \to \bar{S}^{ab} \qquad \bar{S}^{a*}(\lambda) \to \bar{S}^{a*} \tag{27}$$

By (23), (26) and Fatou's lemma we obtain

$$\sum_{b \in J} \bar{M}^{ab}\sigma^b + \bar{S}^{a*} + \sum_{b \in J} \bar{S}^{ab} \leqslant L^a \tag{28}$$

$$\bar{S}^{aa} = 0 \qquad \bar{S}^{ab} \geqslant \sum_{c \in J} \bar{M}^{ac} W^{cb} \qquad (b \neq a)$$
$$\bar{S}^{a*} \geqslant \sum_{c \in J} \bar{M}^{ac} W^{c*} \tag{29}$$

(setting $\infty \cdot 0 = 0 \cdot \infty = 0$).

Obviously $\bar{S}^{a*}$ $(a \in J)$ and the $J \times J$ matrix $\bar{S} = (\bar{S}^{ab})$ are finite. We proceed now to prove that the $J \times J$ matrix $\bar{M} = (\bar{M}^{ab})$ is finite.

When $a \in J - L$, by (4.6), $\bar{M}^{ab} = H^{ab} < \infty$ $(b \in J)$. Suppose $a \in L$, below. Let

$$\Lambda = \left\{ c: c \in J, \eta^c(\lambda) \neq 0, \sigma^c + W^{c*} + \sum_{b \neq c} W^{cb} = 0 \right\}$$

If $\eta^c(\mu)=0$, we have $\bar{M}^{ac}=0\,(a\in J)$ from the fact that $M^{ac}(\lambda)=0\,(a\in J)$. If $\eta^c(\mu)\neq 0$, and moreover $c\in J-\Lambda$, then by (5.25) and (5.26) we have

$$\sum_{c\in J}\bar{M}^{ac}\left(\sigma^c+W^{c*}+\sum_{b\neq c}W^{cb}\right)\leqslant\sum_{c\in J}\bar{M}^{ac}\sigma^c+\bar{S}^{a*}+\sum_{b\in J}\bar{S}^{ab}\leqslant L^a$$

so that $\bar{M}^{ac}<\infty$ for $c\in J-\Lambda$, and $a\in J$.

Note that when $c\in\Lambda$,

$$\mu[\eta^c(\mu),1-Z^a]=\sigma^c+W^{c*}(\mu)+\sum_{b\neq a}W^{cb}(\mu)=0$$

so that

$$\mu[\eta^c(\mu),1]=W^{ca}(\mu)>0$$

Hence

$$\mu[\eta^c(\mu)-W^{ca}(\mu)F^a(\mu),Z^a]=W^{ca}(\mu)\delta^a(\mu)>0 \tag{30}$$

Since

$$\sum_{b\in J}\sum_{c\in J}\bar{M}^{ac}(\lambda)W^{cb}(\mu)F^b(\mu)=\sum_{c\in J-\Lambda}\bar{M}^{ac}(\lambda)\sum_{b\in J}W^{cb}(\mu)F^b(\mu)+\sum_{c\in\Lambda}\bar{M}^{ac}(\lambda)W^{ca}(\mu)F^a(\mu)$$

(24) can be rewriten as

$$\sum_{c\in\Lambda}\bar{M}^{ac}(\lambda)[\eta^c(\mu)-W^{ca}(\mu)F^a(\mu)]$$

$$=F^a(\mu)-\sum_{b\neq a}\bar{S}^{ab}(\lambda)F^b(\mu)+\sum_{c\in J-\Lambda}\bar{M}^{ac}(\lambda)\left(\sum_{b\in J}W^{cb}(\mu)F^b(\mu)-\eta^c(\mu)\right)$$

Multiplying the above expression on both sides by μZ^a, we have

$$\sum_{c\in\Lambda}\bar{M}^{ac}(\lambda)W^{ca}(\mu)\delta^a(\mu)=S^{aa}(\mu)-\sum_{b\neq a}\bar{S}^{ab}(\lambda)S^{ba}(\mu)+\sum_{c\in J-\Lambda}\bar{M}^{ac}(\lambda)\sum_{b\in J}[W^{cb}(\mu)S^{ba}(\mu)-W^{ca}(\mu)]$$

When $\lambda\to\infty$, the limit on the right-hand side is finite, and so is the limit on the left. Noting (30) we have $\bar{M}^{ac}<\infty$ $(c\in\Lambda)$. Thus we have completed the proof of $(\bar{M}^{ab})$ being finite.

Letting $\lambda\to\infty$ in (24) we have

$$\sum_{b\in J}\bar{M}^{ab}\eta^b(\mu)=\sum_{b\in J}[\bar{I}-\bar{S}+\bar{M}W(\mu)]^{ab}F^b(\mu) \tag{31}$$

If we write

$$\bar{\eta}^a(\mu)=\sum_{b\in J}\bar{M}^{ab}\eta^b(\mu) \tag{32}$$

then (28), (29) and (31) become (6), (7) and

$$\bar{\eta}^a(\mu)=\sum_{b\in J}[\bar{I}-\bar{S}+\bar{W}(\mu)]^{ab}F^b(\mu) \tag{33}$$

We proceed now to prove that the inverse matrix

$$\bar{K}(\mu)=[\bar{I}-\bar{S}+\bar{W}(\mu)]^{-1} \tag{34}$$

exists and is non-negative.

Actually, the diagonal elements of $\bar{I}-\bar{S}+\bar{W}(\mu)$ are non-negative, while the non-diagonal elements are non-positive. In order to prove that the inverse matrix exists, we only need to prove that the row sums are bigger than zero. When $a\in J-L$, by (6) and (7) we know the row sum is equal to $\bar{W}^{aa}(\mu)=\mu[\bar{\eta}^a(\mu),1]>0$. When $a\in L$ surely $\bar{\eta}^a(\mu)\neq 0$. Because if $\bar{\eta}^a(\mu)=0$, then $F^a(\mu)=\sum_{b\in J}\bar{S}^{ab}F^b(\mu)$, that is, (3.7) and (3.8) hold for $h^{ab}=\bar{S}^{ab}$. But by Lemma 3.2, this is impossible. Therefore when $a\in L$, we have $\bar{\eta}^a(\mu)\neq 0$. Thus by (6) and (7), the row sum

$$1-\sum_{b\in J}\bar{S}^{ab}+\sum_{b\in J}\bar{W}^{ab}$$

$$\geqslant 1-\left(\bar{\eta}^a+\bar{S}^{a*}+\sum_{b\in J}\bar{S}^{ab}\right)+\bar{\sigma}^a+W^*+\sum_{b\in J}\bar{W}^{ab}(\mu)\geqslant\mu[\bar{\eta}^a(\mu),\mathbf{1}]>0$$

Hence (15) holds, and from (33) we find that

$$F^a(\lambda)=\sum_{b\in J}\bar{K}^{ab}(\lambda)\eta^b(\lambda) \tag{35}$$

So (14) is proved.

(b) By (15) and (10)–(13), the Q condition is equivalent to

$$\sum_{b\in J}N^{ab}(L^b+\bar{W}^{bb})^{-1}\bar{\alpha}^b=0 \qquad a\in J_H \tag{36}$$

Since $N^{aa}\geqslant 1$ $(a\in J_H)$, it follows that the above expression is equivalent to (8) and (9).

(c) Suppose $\psi(\lambda)$ is determined by (i)–(vi) in the theorem.

Noting $\bar{W}(\lambda)=K(\lambda)^{-1}-\bar{I}-\bar{S}$, and imitating the proof of the norm condition in Theorem 5.1, we know $\lambda(F^a(\lambda),\mathbf{1}]\leqslant 1$, and furthermore equality holds if and only if (17) holds.

But (5.31) still holds for $\bar{K}(\lambda)$ in (34), and hence we come to the conclusion that the resolvent equation of $\psi(\lambda)$ holds. We have already pointed out that (8) and (9) are equivalent to the Q condition. Therefore $\psi(\lambda)$ is a Q process. The theorem is proved. QED

Theorem 3. Suppose that $\psi(\lambda)$ is the Q process in Theorem 2. Then $\psi(\lambda)$ is of B-type if and only if $J_H=\varnothing$ and of F-type if and only if

$$\bar{\alpha}^a=0 \qquad (a\in J)$$

Proof. The same as in Theorem 5.1. QED

Theorem 4. If $\psi(\lambda)$ is the Q process of Theorem 2, then $\psi(\lambda)$ is non-sticky if and only if

$$\lim_{\lambda\to\infty} \lambda \| \bar{\eta}^a(\lambda) \| < \infty \qquad a\in J \tag{37}$$

Proof. Suppose (37) holds. According to Definition 5.1, $\psi(\lambda)$ is non-sticky. Conversely, suppose $\psi(\lambda)$ is non-sticky, that is $\psi(\lambda)$ has the representations (3.9) and (4.3). Moreover

$$\lim_{\lambda\to\infty} \lambda \| \eta^a(\lambda) \| < \infty \qquad a\in J \tag{38}$$

$\psi(\lambda)$ has the representations (5.9) and (6.14). As $Z^a(\lambda)$, $a\in J$, are linearly independent, so

$$\sum_{b\in J} K^{ab}(\lambda)\eta^b(\lambda) = \sum_{b\in J} \bar{K}^{ab}(\lambda)\bar{\eta}^b(\lambda) \qquad a\in J$$

Hence

$$\bar{\eta}^a(\lambda) = \sum_{b\in J} [\bar{K}(\lambda)^{-1}K(\lambda)]^{ab}\eta^b(\lambda)$$

Multiplying the above formula from the right by $A(\lambda, \mu)$, we find that

$$\begin{aligned} \bar{\eta}^a(\mu) &= \sum_{b\in J} [\bar{K}(\lambda)^{-1}K(\lambda)]^{ab}\eta^b(\mu) \\ \mu \| \bar{\eta}^a(\mu) \| &= \sum_{b\in J} [\bar{K}(\lambda)^{-1}K(\lambda)]^{ab}\mu \| \eta^b(\mu) \| \end{aligned} \tag{39}$$

By the above expression and (38) we obtain (37). The proof is completed.

QED

Remark 2

In the representation (14) of the Q process $\psi(\lambda)$ if $\bar{\eta}^a(\lambda)$, $a\in J$, and $\tilde{\eta}^a(\lambda)$, $a\in J$, correspond to the same $\psi(\lambda)$, from the proof (39) of Theorem 4 we see that there exists a $J\times J$ matrix $R=(r_{ab})$ such that the inverse matrix of R exists and, moreover,

$$\bar{\eta}^a(\lambda) = \sum_{b\in J} r_{ab}\tilde{\eta}^b(\lambda) \qquad \lambda > 0$$

8.7 EQUIVALENT CONSTRUCTION

Theorem 1. (i) The same as (i) in Theorem 6.2.

(ii) Take a non-zero entrance family $(\eta^a(\lambda), \lambda > 0)$, $a\in J$, such that

$$\lim_{\lambda\to\infty} \lambda[\eta^a(\lambda), \mathbf{1} - Z^a] < \infty \qquad a\in J \tag{1}$$

or equivalently

$$W^{a*} < \infty \qquad W^{ab} < \infty \ (b \neq a) \qquad a \in J \tag{2}$$

(iii) Take a $J \times J$ matrix $T = (T^{ab})$ such that

$$\sigma^a \leqslant -W^{a*} + \sum_{b \in J} T^{ab} \tag{3}$$

$$W^{ab} \leqslant -T^{ab} \qquad (b \neq a) \tag{4}$$

$$(T^{bb} + W^{bb})^{-1}\bar{\alpha}^b = 0 \qquad b \in J_H \tag{5}$$

$$F^{ab}(T^{bb} + W^{bb})^{-1}\bar{\sigma}^b = 0 \qquad a \in J_H, \quad b \in J_B \tag{6}$$

Here the $J \times J$ matrix

$$F(\lambda) = (F^{ab}(\lambda)) \downarrow F = (F^{ab}) \qquad \lambda \uparrow \infty$$

and the $J \times J$ matrix

$$F(\lambda) = [I - V(\lambda)]^{-1} = \sum_{n=0}^{\infty} V(\lambda)^n \downarrow F = \sum_{n=0}^{\infty} V^n \qquad \lambda \uparrow \infty$$

while

$$V(\lambda) = (V^{ab}(\lambda)) \downarrow V = (V^{ab}) \qquad \lambda \uparrow \infty$$

$$V^{aa}(\lambda) = 0 \qquad V^{ab}(\lambda) = \frac{(-T^{ab}) - W^{ab}(\lambda)}{T^{aa} + W^{aa}(\lambda)} \qquad (b \neq a)$$

$$V^{aa} = 0 \qquad V^{ab} = \frac{(-T^{ab}) - W^{ab}}{T^{aa} + W^{aa}} \qquad (b \neq a) \tag{7}$$

(iv) Set

$$\psi_{ij}(\lambda) = \phi_{ij}(\lambda) + \sum_{a \in J} \sum_{b \in J} Z_i^a(\lambda) K^{ab}(\lambda) \eta_j^b(\lambda) \tag{8}$$

where

$$K(\lambda) = [T + W(\lambda)]^{-1} = [I - V(\lambda)]^{-1} E(\lambda) \tag{9}$$

and

$$E(\lambda) = \operatorname{diag}\{[T^{aa} + W^{aa}(\lambda)]^{-1}\} \tag{10}$$

Then $\psi(\lambda)$ is a non-minimal Q process. Any non-minimal Q process can be obtained according to the above method.

Remark

When Q is conservative the $\psi(\lambda)$ in Theorem 1 is precisely the construction in Williams (1964, 1966).

Theorem 2. Assume $\psi(\lambda)$ to be the Q process having the form of Theorem 6.2.

Let

$$T = \bar{I} - \bar{S} \qquad \eta^a(\lambda) = \bar{\eta}^a(\lambda) \qquad a \in J \tag{11}$$

Then $\psi(\lambda)$ is a Q process having the form of Theorem 1. Conversely suppose that $\psi(\lambda)$ is a Q process having the form of Theorem 1. Let

$$L = \{a: a \in J,\ T^{aa} > 0\} \tag{12}$$

$$\begin{aligned} &\bar{S}^{a*} = 0 \qquad \bar{S}^{ab} = 0\ (b \in J) \qquad \text{if } a \in J - L \\ &\bar{S}^{a*} = \frac{W^{a*}}{T^{aa}} \qquad \bar{S}^{aa} = 0 \qquad \bar{S}^{ab} = \frac{-T^{ab}}{T^{aa}}(b \neq a) \qquad \text{if } a \in L \end{aligned} \tag{13}$$

$$\bar{\eta}^a(\lambda) = \begin{cases} \eta^a(\lambda) & \text{if } a \in J - L \\ \eta^a(\lambda)/T^{aa} & \text{if } a \in L \end{cases} \tag{14}$$

Then $\psi(\lambda)$ is a Q process having the form of Theorem 6.2.

Proof of Theorems 1 and 2. Suppose that $\psi(\lambda)$ has the form of Theorem 1. Determining L, $\bar{S}^{a*}$, $\bar{S} = (\bar{S}^{ab})$, $\bar{\eta}^a(\lambda)$ $(a \in J)$ according to (12)–(14), we find that $\psi(\lambda)$ has the form of Theorem 6.2. Thus $\psi(\lambda)$ is a Q process.

Provided $\psi(\lambda)$ is a Q process, according to Theorem 6.2, it surely has the form of Theorem 6.2. And by determining $T = (T^{ab})$ and $\eta^a(\lambda), a \in J$, according to (11), we see that $\psi(\lambda)$ has the form of Theorem 1. The proof is completed.

QED

8.8 REMARKS ON THE CONSTRUCTION OF THE NON-BIFINITE Q PROCESS

Remark 1

Suppose that $A_e = H_e \cup B_e$ is an infinite set. Take a decomposition $A_e = a_1 \cup a_2 \cup \cdots$. The set $A = \{a_1, a_2, \ldots\}$ may be an infinite set. Take a non-empty subset J of A. So long as the subset J is finite, or J is finite, but if only we define $K(\lambda)$ in Theorem 5.1 according to (5.27) and define $\bar{K}(\lambda)$ in Theorem 6.2 as follows:

$$\bar{K}(\lambda) = \left\{ \sum_{n=0}^{\infty} R^n(\lambda) \right\} D(\lambda)$$

$R(\lambda) = (R^{ab}(\lambda))$ as in (6.11), the diagonal form matrix $D(\lambda)$ as in (6.16), $K(\lambda)$ in Theorem 7.1 is defined by the formula below:

$$K(\lambda) = \left\{ \sum_{n=0}^{\infty} V(\lambda)^n \right\} E(\lambda)$$

$V(\lambda) = (V^{ab}(\lambda))$ as in (7.7), diagonal form $E(\lambda)$ as in (7.10), then the $\psi(\lambda)$ obtained according to Theorems 5.1, 6.2 and 7.1 are all non-minimal Q processes.

Remark 2

Suppose that H_e is infinite and B_e is finite, and that after taking a decomposition of A_e, the set $A = \{a_1, a_2, \ldots\}$ may still be infinite. If only we select a non-empty subset J of A such that $J_H = \varnothing$ (at this moment, J must be a finite set), the $\psi(\lambda)$ constructed according to Theorems 5.1, 6.2 and 7.1 are non-minimal Q processes satisfying the system of backward equations, and moreover at this moment, the Q processes constructed by Theorem 6.2 or Theorem 7.1 have already exhausted all the non-minimal Q processes satisfying the sysem of backward equations.

PART IV PATH STRUCTURE OF DENUMERABLE MARKOV PROCESSES

CHAPTER 9

The W Transformation and Strong Limit

9.1 INTRODUCTION

Professor Zi-kun Wang introduced the transforms g_n and f_n, and the concept of strong limit for the processes, during his study of construction theory of birth–death processes, and successfully solved the construction problem of birth–death processes by using probability methods. This will be seen in Zi-kun Wang (1958, 1962) and Zi-kun Wang and Xiang-qun Yang (1978, 1979). Their basic idea is to transform general processes with more complex paths into processes with simpler paths; then they proved that a general process is a strong limit of simpler processes. Professor Zhen-ting Hou applied transformations to general denumerable Markov processes. The W transformation is the further extension of the transformations g_n and f_n. Therefore, the results about the transformations g_n and f_n in Zi-kun Wang (1962), Zi-kun Wang and Xiang-qun Yang (1978, 1979) and Zhen-ting Hou (1975) may all be obtained as special cases of our present fundamental results. The results in this chapter are taken from Xiang-qun Yang (1978, 1980a).

9.2 DEFINITION OF W TRANSFORMATION

We first discuss the W transformation for a function. Let $X = \{x(t), t < \sigma\}$ be a function that has a domain of values $\bar{E} = E \cup \{\infty\}$ and is defined in $[0, \sigma)$ $(\sigma \leqslant \infty)$. Then $\{\tau, \beta\}$ is called a pair for X, if

$$0 = \tau_0 \leqslant \beta_0 \leqslant \tau_1 \leqslant \beta_1 \leqslant \cdots \leqslant \sigma \tag{1}$$

For a pair $\{\tau, \beta\}$, we set

$$\sigma_n = \begin{cases} 0 & \text{if } n = 0 \\ \sum_{k=1}^{n} (\tau_k - \beta_{k-1}) & \text{if } n > 0 \end{cases} \tag{2}$$

$$\delta_n = \sum_{k=0}^{n} (\beta_k - \tau_k) \qquad n \geqslant 0 \tag{3}$$

$$\bar{\sigma} = \lim_{n\to\infty} \sigma_n \qquad \bar{\delta} = \lim_{n\to\infty} \delta_n \tag{4}$$

whereas for $t \in [0, \bar{\sigma})$, we set

$$a_t = \beta_n + (t - \sigma_n) \qquad \text{if } t \in [\sigma_n, \sigma_{n+1}) \tag{5}$$

$$\bar{x}(t) = x(a_t) \qquad t \in [0, \bar{\sigma}) \tag{6}$$

Definition 1. The transformation that transforms the function $X = \{x(t), t < \sigma\}$ into the function $\bar{X} = \{\bar{x}(t), t < \bar{\sigma}\}$ is called a W transformation for X, to be denoted by $W_{\tau,\beta}$. Hence $\bar{X} = W_{\tau,\beta}(X)$.

Intuitively speaking, we throw out those sections of the function X corresponding to the intervals $[\tau_i, \beta_i)$, reserve those sections corresponding to $[\beta_i, \tau_{i+1})$ and move them towards the left so that the section on $[\beta_0, \tau_1)$ gets to $[0, \tau_1 - \beta_0)$, and link up all other sections on $[\beta_i, \tau_{i+1})$ $(i = 1, 2, \ldots)$ according to their original order without intersecting them, and the function thus obtained is called X.

We may define the W transformation for a process X, in the light of the W transformation for a function.

Definition 2. Let $X = \{x(t), t < \sigma\} \in \mathscr{H}$. Assume that for each $\omega \in \Omega$, there exists a pair $\{\tau(\omega), \beta(\omega)\}$, and we may determine $W_{\tau(\omega),\beta(\omega)}(X(\omega)) = \bar{X}(\omega) = \{\bar{x}(t,\omega), t < \bar{\sigma}(\omega)\}$. We call $\bar{X}$ the W transformation for the process X and write $\bar{X} = W_{\tau,\beta}(X)$.

9.3 STRONG LIMIT THEOREM

9.3.1 In the case that X is a function

Definition 1. A set A is assumed to be a Lebesgue-measurable set in $[0, \infty)$. Then $[\lambda, \eta)$ is called a composition interval of A if the following conditions are satisfied:

(i) $(\lambda, \eta) \subset A$

(ii) maximality: if $[\bar{\lambda}, \bar{\eta}) \supset [\lambda, \eta)$, and $(\bar{\lambda}, \bar{\eta}) \subset A$, then

$$[\bar{\lambda}, \bar{\eta}) = [\lambda, \eta)$$

We denote the collection of all the composition intervals of A by $\mathscr{U}(A)$. Put

$$C_1(A) = \bigcup_{[\lambda,\eta)\in\mathscr{U}(A)} (\lambda, \eta) \tag{1}$$

$$C_2(A) = \bigcup_{[\lambda,\nu)\in\mathscr{U}(A)} [\lambda, \eta) \tag{2}$$

$$\bar{C}_2^+(A) = \{t \mid \text{there exist strictly decreasing } t_n \in C_2(A) \text{ such that } t_n \downarrow t\}$$
$$= \{t \mid \text{there exist non-increasing } t_n \in C_2(A) \text{ such that } t_n \downarrow t\} \qquad (3)$$

$\bar{C}_2^+(A)$ is called the right closure of set A.

Definition 2. A transformation $\gamma: u \to \gamma(u)$ is said to be the one determined by the pair $\{\tau, \beta\}$ or by the sequence of pairs $\{\tau^n, \beta^n\}$ $(n \geqslant 1)$, if

$$\gamma(u) = L\{A \cap [0, u)\} \qquad (4)$$

where L is the Lebesgue measure, and

$$A = \begin{cases} \bigcup_{k=0}^{\infty} [\beta_k, \tau_{k+1}) & \text{for the pair } \{\tau, \beta\} \\ \bigcup_{n=1}^{\infty} \bigcup_{k=0}^{\infty} [\beta_k^n, \tau_{k+1}^n) & \text{for the sequence of pairs } \{\tau^n, \beta^n\} \end{cases} \qquad (5)$$

Definition 3. A function $X = \{x(t), t < \sigma\}$ is said to be right-continuous, if

$$\lim_{s \downarrow t} x(s) = x(t) \in \bar{E} \qquad \text{for all } t < \sigma \qquad (6)$$

Theorem 1. Let $X = \{x(t), t < \sigma\}$ be a right-continuous function, and that there exists a sequence of pairs $\{\tau^n, \beta^n\}$ $(n \geqslant 1)$ such that

$$A^n = \bigcup_{k=0}^{\infty} [\beta_k^n, \tau_{k+1}^n) \uparrow A \qquad (7)$$

Let v^n and v denote transformations determined by the pair $\{\tau^n, \beta^n\}$ for the given n and the sequence of pairs $\{\tau^n, \beta^n\}$ $(n \geqslant 1)$ respectively. Put

$$\tau_k^{n,m} = \gamma^n(\tau_k^m) \qquad \beta_k^{n,m} = \gamma^n(\beta_k^m) \qquad m < n \qquad (8)$$

$$W_{\beta^n, \tau^n}(X) = X^n = \{x^n(t), t < \sigma^n\} \qquad (9)$$

Then:

(i) $\{\tau^{n,m}, \beta^{n,m}\}$ $(m < n)$ is a pair for X^n, and

$$X^m = W_{\tau^{n,m}, \beta^{n,m}}(X^n). \qquad m < n \qquad (10)$$

(ii) As $n \to \infty$ there exist limits

$$\sigma^n \uparrow \bar{\sigma} = L(A) \qquad (11)$$

$$\lim_{n \to \infty} x^n(t) = x(\alpha_t) \qquad \text{for all } t \in [0, \bar{\sigma}) \qquad (12)$$

where α_t is the unique solution of the equation

$$L\{C_2(A)\cap[0,u)\}=t \qquad u\in\bar{C}_2^+(A) \tag{13}$$

(iii) $\bar{X}=\{x(\alpha_t), t<\bar{\sigma}\}$ is right-continuous.

The intuitive meaning of this theorem is as follows: Suppose X becomes X^m when transformation W_{τ^m,β^m} is performed and, moreover, for $m<n$, the sections reserved when W_{τ^m,β^m} transformation is performed are still reserved as W_{τ^n,β^n} transformation is carried out. Then (10) indicates that transformation of X into X^m can be completed in two steps. First, we throw out the sections of X corresponding to $\bigcup_{i=0}^{\infty}[\tau_i^n,\beta_i^n)$, reserve the sections corresponding to $\bigcup_{i=0}^{\infty}[\beta_i^n,\tau_{i+1}^n)$, and, moreover, translate them towards the left according to their original order, then connect them up without intersecting them so that we have X^n. In this process those sections corresponding to $[\beta_k^m,\tau_k^m)$ are preserved and after they move to the left β_k^m becomes $\beta_k^{n,m}$ while τ_k^m changes into $\tau_k^{n,m}$. Secondly, we abandon the sections of X^n corresponding to $\bigcup_{i=0}^{\infty}[\tau_i^{n,m},\beta_i^{n,m})$, and reserve the sections corresponding to $\bigcup_{i=0}^{\infty}[\beta_i^{n,m},\tau_{i+1}^{n,m})$, and, furthermore, translate them towards the left according to their original order and link them up without rendering them intersected; thus we obtain X^m.

Conclusion (iii) shows: reserve all the sections of X corresponding to $A=\bigcup_{n=1}^{\infty}\bigcup_{k=0}^{\infty}[\beta_k^n,\tau_{k+1}^n)$, abandon the remaining sections and, moreover, translate the reserved sections towards the left in their original order and connect them up without intersecting them, and the function obtained thus is precisely $\bar{X}$. In this process of translation α_t turns into t. $\bar{X}$ is also a right-continuous function.

Conclusion (ii) indicates that the limit function of X^n is $\bar{X}$.

The proof of the theorem is left to section 9.4.

9.3.2 In the case that X is a process

Definition 4. Let $X=\{x(t),t<\sigma\}$, $X^n=\{x^n(t),t<\sigma^n\}$ $(n\geqslant 1)$ belong to $\mathscr{H}$. X is called the strong limit of a sequence of X^n and we write $X=\lim_{n\to\infty}X^n$, if for almost all ω

$$\sigma^n(\omega)\uparrow\sigma(\omega)$$

$$\lim_{n\to\infty}x^n(t,\omega)=x(t,\omega) \qquad \text{for all } t<\sigma(\omega) \tag{14}$$

By definition, obviously, we have the following.

Theorem 2. If $X, X^n\in\mathscr{H}$, $\lim_{n\to\infty}X^n=X$, then for all $i,j\in E$, $t\geqslant 0$,

$$\lim_{n\to\infty} P^n_{ij}(t) = P_{ij}(t) \tag{15}$$

where $\{P^n_{ij}(t)\}$, $\{P_{ij}(t)\}$ are the transition probabilities of X^n and X, respectively.

Definition 5. Suppose that $X\in\mathscr{H}$, that $\beta_0 \leqslant \tau_1 \leqslant \beta_1$ are Markov times relative to X, satisfying $P(\beta_0 < \sigma) > 0$, and that the following hold:

$$\beta_0 + \theta_{\beta_0}\tau_1 = \tau_1, \qquad \beta_0 + \theta_{\beta_0}\beta_1 = \beta_1 \tag{16}$$

on $(\beta_0 < \tau_1)$, where θ is a transition operator. Assume that τ_n, β_n have been determined already. If $\beta_n = \sigma$, we define $\tau_{n+1} = \beta_{n+1} = \sigma$, otherwise we define

$$\tau_{n+1} = \beta_n + \theta_{\beta_n}\tau_1 \qquad \beta_{n+1} = \beta_n + \theta_{\beta_n}\beta_1 \tag{17}$$

We call such a pair $\{\tau, \beta\}$ a canonical one.

Recall the class $\mathscr{H}_s$ of processes in which the Q matrices are finite and the class $\mathscr{H}_D$ of D-type processes, and we have the following.

Theorem 3. Let $X\in\mathscr{H}_s$, and $\{\tau, \beta\}$ be a canonical pair satisfying the following conditions:

(a) $P\{t + \theta_t\tau_1 = \tau_1 \text{ for all } t\in[\beta_0, \tau_1)\} = 1$

$$P\{t + \theta_t\beta_1 = \beta_1 \text{ for all } t\in[\beta_0, \tau_1)\} = 1 \tag{18}$$

(b) For α_t determined by (2.5), we have

$$P\{x(\alpha_t) = \infty\} = 0 \qquad t \geqslant 0 \tag{19}$$

Then the following conclusion holds true:

(i) A sufficient condition for (b) is that the following condition (c) holds, that is:

(c) $P\{x(t)\in E \text{ for all } t\in[\beta_0, \tau_1)\} = 1$

$$P\{x(\beta_1) = \infty\} = 0 \tag{20}$$

(ii) If conditions (a) and (b) are satisfied, then $\bar{X} = W_{\tau,\beta}(X)\in\mathscr{H}_s$.

(iii) If conditions (a) and (c) are satisfied, then $\bar{X} = W_{\tau,\beta}(X)\in\mathscr{H}_D$.

The proof of the theorem is left to section 9.6.

Theorem 4. Suppose that $X\in\mathscr{H}_s$, that for every $n \geqslant 1$ there exists a canonical pair $\{\tau^n, \beta^n\}$ satisfying the conditions (a) and (c), that

$$P\{x(\inf \beta^n_0) = \infty\} = 0 \tag{21}$$

and that for almost all $\omega\in\Omega$ the following holds:

$$A^n(\omega) = \bigcup_{k=0}^{\infty} [\beta^n_k(\omega), \tau^n_{k+1}(\omega)) \uparrow A(\omega) \tag{22}$$

Then

(i) $X^n = W_{\tau^n, \beta^n}(X) \in \mathscr{H}_D$ and (10) is satisfied, where $\tau_k^{n,m}$, $\beta_k^{n,m}$ $(m < n, k \geq 0)$ are

$$\tau_k^{n,m} = L\{A^n \cap [0, \tau_k^m)\} \qquad \beta_k^{n,m} = L\{A^n \cap [0, \beta_k^m)\} \tag{23}$$

(ii) There exists a strong limit of X^n $(n \geq 1)$

$$\lim_{n \to \infty} X^n = \bar{X} \tag{24}$$

where $\bar{X} = \{x(\alpha_t), t < \bar{\sigma}\} \in \mathscr{H}_s$, $\bar{\sigma}(\omega) = L\{A(\omega)\}$, $\alpha_t(\omega)$ is the unique solution of the equation

$$L\{C_2(A(\omega)) \cap [0, u)\} = t$$
$$u \in \bar{C}_2^+(A(\omega)) \tag{25}$$

The proof of the theorem is left to section 9.6.

9.4 THE PROOF OF THEOREM 3.1

Assume that $\{\tau, \beta\}$ is a pair of the function $X = \{x(t), t < \sigma\}$. Set

$$A = \bigcup_{k=0}^{\infty} [\beta_k, \tau_{k+1}) \qquad B = \bigcup_{k=0}^{\infty} [\tau_k, \beta_k) \tag{1}$$

$$\gamma(u) = L\{A \cap [0, u)\} \qquad \rho(u) = L\{B \cap [0, u)\} \qquad u < \sigma \tag{2}$$

Obviously, we have

Lemma 1. (i) $\gamma(u) + \rho(u) = u$

(ii) $\gamma(u)$ and $\rho(u)$ are non-decreasing continuous functions.

(iii) If $u \in A, u < t < \sigma$, then $\gamma(u) < \gamma(t)$. If $u \in B, u < t < \sigma$, then $\rho(u) < \rho(t)$.

(iv) If $u \in [\beta_k, \tau_{k+1})$, then $\rho(u) = \rho(\beta_k) = \delta_k$. If $u \in [\tau_k, \beta_k)$, then $\gamma(u) = \gamma(\tau_k) = \sigma_k$.

(v) For $t \in [0, \bar{\sigma})$, $\bar{\sigma} = L(A) = \gamma(\sigma)$, the quantity α_t in (2.5) is the unique solution of the equation

$$\gamma(u) = t \qquad u \in A \tag{3}$$

Lemma 2. When $u < \sigma$, the following conditions are equivalent to each other:

(i) $\alpha_t < u$

(ii) $\gamma(u) > t$

(iii) $\rho(u) < u - t$

Proof. By (i) in Lemma 1, we know that (ii) and (iii) in this lemma are equivalent.

Let $\alpha_t < u$; (v) in Lemma 1 leads to $\alpha_t \in A$, $\gamma(\alpha_t) = t$, whereas (iii) in Lemma 1 leads to $\gamma(u) > \gamma(\alpha_t) = t$. Conversely, let $u \leqslant \alpha_t$. From (ii) in Lemma 1 we have $\gamma(u) \leqslant \gamma(\alpha_t) = t$. Therefore, (i) and (ii) in this lemma are equivalent, and the proof is completed. QED

Lemma 3. Let $\{\tau, \beta\}$ and $\{\tau', \beta'\}$ be two pairs for the function X. What follows contains notations $'$ or $\tilde{}$ whose meanings are quite clear. Assume that

$$A \subset A' \qquad (\text{or } B \supset B') \tag{4}$$

Put $\bar{X} = W_{\tau,\beta}(X)$, $\bar{X}' = W_{\tau',\beta'}(X)$. Then

(i) $\bar{\sigma} \leqslant \bar{\sigma}'$ and for $t \in [0, \bar{\sigma})$, we have $\alpha_t' \leqslant \alpha_t$.

(ii) Set

$$\tilde{\tau}_k = L\{A' \cap [0, \tau_k)\} \qquad \tilde{\beta}_k = L\{A' \cap [0, \beta_k)\} \tag{5}$$

Then $\{\tilde{\tau}, \tilde{\beta}\}$ is a pair for $\bar{X}'$ and, moreover,

$$\bar{X} = W_{\tilde{\tau},\tilde{\beta}}(\bar{X}') \tag{6}$$

Remark

The intuitive meaning of this lemma goes as follows. If what is abandoned in the process of carrying out transformation $W_{\tau,\beta}$ for X is more than what is thrown out in the process of performing transformation $W_{\tau'\beta'}$, then $\bar{X} = W_{\tau,\beta}(X)$ can be derived by abandoning certain sections in two steps. To begin with, conduct transformation $W_{\tau',\beta'}$ for X, that is, abandon the sections of X corresponding to B', reserve those sections corresponding to A' and, furthermore, move them to the left, and link them up so that $\bar{X}'$ is obtained. In this process of shift, τ^k, β_k become $\tilde{\tau}_k, \tilde{\beta}_k$. Then apply transformation $W_{\tilde{\tau},\tilde{\beta}}$ to $\bar{X}'$, that is, abandon the sections of $\bar{X}'$ corresponding to $\tilde{B} = \bigcup_{k=0}^{\infty} [\tilde{\tau}_k, \tilde{\beta}_k)$, reserve those sections corresponding to $\tilde{A} = \bigcup_{k=0}^{\infty} [\tilde{\beta}_k, \tilde{\tau}_{k+1})$ and, moreover, translate them to the left, and connect them up so that we have $\bar{X}$.

Proof of Lemma 3. From (4) we have $\bar{\sigma} = L(A) \leqslant L(A') = \bar{\sigma}'$. If $\alpha_t < \alpha_t'$ by using (iii) and (v) in Lemma 1 we have $\alpha_t \in A \subset A', \alpha_t \in A'$, whence $t = L\{A \cap [0, \alpha_t)\} \leqslant L\{A' \cap [0, \alpha_t)\} < L\{A' \cap [0, \alpha_t')\} = t$.

There exists an inconsistency between them. Therefore $\alpha_t' \leqslant \alpha_t$.

Obviously, $\{\tilde{\tau}, \tilde{\beta}\}$ is a pair for $\bar{X}'$. Suppose $W_{\tilde{\tau},\tilde{\beta}}(\bar{X}') = \tilde{X} = \{\tilde{x}(t), t < \tilde{\sigma}\}$. Then

$$\tilde{\sigma}_n = \sum_{k=1}^{n} (\tilde{\tau}_k - \tilde{\beta}_{k-1}) = \sum_{k=1}^{n} L\{A' \cap [\beta_{k-1}, \tau_k)\} = L\left\{A' \cap \left(\bigcup_{k=1}^{n} [\beta_{k-1}, \tau_k)\right)\right\}$$

By (4), $\tilde{\sigma}_n = L\{\bigcup_{k=1}^{n} [\beta_{k-1}, \tau_k)\} = \sigma_n$, whence $\tilde{\sigma} = \bar{\sigma}$. In order to prove $\tilde{X} = \bar{X}$, it only remains to prove that $\tilde{x}(t) = \bar{x}(t)$, while $t \in [\tilde{\sigma}_n, \tilde{\sigma}_{n+1}) = [\sigma_n, \sigma_{n+1})$, i.e.

$$\bar{x}'(\tilde{\beta}_n + t - \tilde{\sigma}_n) = x(\beta_n + t - \sigma_n) \tag{7}$$

Put $u=\tilde{\beta}_n+t-\tilde{\sigma}_n$ $(<\bar{\sigma}')$. In order to prove (7), we need only to prove $\beta_n+t-\sigma_n=\alpha'_u$. By (v) in Lemma 1, we need only prove

$$L\{A'\cap[0,\beta_n+t-\sigma_n)\}=\tilde{\beta}_n+t-\tilde{\sigma}_n \quad \beta_n+t-\sigma_n\in A' \tag{8}$$

Since $0\leqslant t-\sigma_n<\sigma_{n+1}-\sigma_n=\tau_{n+1}-\beta_n$, then $\beta_n+t-\sigma_n\in[\beta_n,\tau_{n+1})\subset A\subset A'$. Moreover

$$\begin{aligned}L\{A'\cap[0,\beta_n+t-\sigma_n)\}&=L\{A'\cap[0,\beta_n)\}+L\{A'\cap[\beta_n,\beta_n+t-\sigma_n)\}\\&=\tilde{\beta}_n+L\{[\beta_n,\beta_n+t-\sigma_n)\}=\tilde{\beta}_n+t-\sigma_n=\tilde{\beta}_n+t-\tilde{\sigma}_n\end{aligned}$$

which proves (8), and the proof is completed. QED

Lemma 4. Suppose that the function $X=\{x(t),t<\sigma\}$ satisfies the condition in Theorem 3.1. Moreover,

$$A=\bigcup_{n=1}^{\infty}\bigcup_{k=0}^{\infty}[\beta_k^n,\tau_{k+1}^n) \tag{9}$$

$\gamma(u)=L\{A\cap[0,u)\}$ and the expressions $C_1(A)$, $C_2(A)$, $\bar{C}_2^+(A)$ are determined by (3.1)–(3.3). Then

(i) The conclusion (i) in Theorem 3.1 holds.

(ii)
$$C_1(A)\subset A\subset C_2(A)\subset\bar{C}_2^+(A)\subset[0,\sigma) \tag{10}$$

$$L\{C_1(A)\}=L\{A\}=L\{C_2(A)\}=L\{\bar{C}_2^+(A)\}=\bar{\sigma} \tag{11}$$

(iii) (a) $\gamma(u)$ is non-decreasing on $[0,\sigma)$.

(b) If $[a,b)\cap C_2(A)=\varnothing$, then for $u\in[a,b)$, we have $\gamma(u)=\gamma(a)$.

(c) If $u\in\bar{C}_2^+(A)$ and $u<s<\sigma$, then $\gamma(u)<\gamma(s)$. Especially $\gamma(u)$ is strictly increasing on $\bar{C}_2^+(A)$.

(d) For transformation γ we have $\gamma[0,\sigma)=[0,\bar{\sigma})$, $\bar{\sigma}=L(A)$.

(e) γ as a transformation from $\bar{C}_2^+(A)$ onto $[0,\bar{\sigma})=\gamma[\bar{C}_2^+(A)]$ is one-to-one. Therefore, it has the inverse transformation γ^{-1}. Furthermore, γ and γ^{-1} are measure-preserving transformations. Particularly, if $[\lambda,\eta)\in\mathscr{U}(A)$, $t\in[\gamma(\lambda),\gamma(\eta))$, then

$$\gamma^{-1}(t)=\lambda+t-\gamma(\lambda) \tag{12}$$

(iv) As $n\to\infty$ there exist limits

$$\sigma^n=L(A^n)\uparrow\bar{\sigma}=L(A) \tag{13}$$

$$\alpha_t^n\downarrow t\equiv\gamma^{-1}(t) \qquad t\in[0,\bar{\sigma}) \tag{14}$$

Proof. By conclusion (ii) in Lemma 3, follows (i) in this lemma. Inclusions (10) are obvious. It is easily shown that $\mathscr{U}(A)$ is a denumrable set, therefore, sets $C_1(A), A, C_2(A)$ are different from each other by one countable set at most. Thus, the first two equalities in (11) are proved.

(iii) (a), (b), (d) in this lemma are obvious. In order to prove (iii) (c), first notice that if $t \in C_2(A), t < s < \sigma$, then $\gamma(t) < \gamma(s)$. Therefore, for $u \in \bar{C}_2^+(A), u < s < \sigma$, there exists $t_n \in C_2(A), t_n \downarrow u$. Hence we have $u \leqslant t_n < s < \sigma$, for sufficiently large n. From (iii) (a) and what was mentioned just now, we have $\gamma(u) \leqslant \gamma(t_n) < \gamma(s)$. Thus (iii) (c) in this lemma is proved.

We now proceed to prove (iii) (d). By (iii) (c), γ is a one-to-one transformation from $\bar{C}_2^+(A)$ onto $\gamma(\bar{C}_2^+(A))$ so that γ^{-1} exists. In order to prove measure-preservation, we assume $\mathscr{L}$ to be the collection of Lebesgue sets F satisfying $F \subset \bar{C}_2^+(A)$ and $L\{\gamma(F)\} = L\{F\}$. Put $\mathscr{C} = \{[a, b) | [a, b) \subset C_2^+(A)\}$. It is easily found that $\mathscr{L}$ is a λ-system, $\mathscr{C}$ is a π-system, and $\mathscr{C} \subset \mathscr{L}$, whence $\mathscr{L}$ contains all Borel sets in $\bar{C}_2^+(A)$; see Dynkin (1959, Lemma 1.1). Consequently, $\mathscr{L}$ contains all Lebesgue-measurable sets in $\bar{C}_2^+(A)$.

From conclusion (i) in Lemma 3, it follows that the limits in (13) and (14) exist. For $t \in [0, \bar{\sigma})$, when n is sufficiently large, we have $t \in [0, \sigma^n)$. By conclusion (v) in Lemma 1, we have

$$L\{A^n \cap [0, \alpha_t^n)\} = t \qquad \alpha_t^n \in A^n \tag{15}$$

Since $A^n \subset A \subset C_2(A)$, letting $n \to \infty$, we find that the limit $\alpha_t = \lim_{n \to \infty} \alpha_t^n$ satisfies

$$\gamma(\alpha_t) = L\{A \cap [0, \alpha_t)\} = t \qquad \alpha_t \in \bar{C}_2^+(A) \tag{16}$$

Noticing (10) and (11), we know that α_t is a solution of (3.13). By (iii) (c) in this lemma, α_t is a unique solution, and $\gamma^{-1}(t)$ is a solution, too. Whence $\alpha_t = \gamma^{-1}(t)$, $\gamma(\bar{C}_2^+(A)) = [0, \bar{\sigma})$ and $L\{\bar{C}_2^+(A)\} = L\{\gamma(\bar{C}_2^+(A))\} = \bar{\sigma}$. The proof is terminated.

QED

Proof of Theorem 3.1. From the right-continuity of X, and (13) and (14), (3.11) and (3.12) can be proved, and hence by Lemma 4, conclusions (i) and (ii) in Theorem 3.1 can be verified. We are now going to prove conclusion (iii) in Theorem 3.1. Let $t \in [0, \bar{\sigma})$ and $s \downarrow t$. From (iii) (c) and (e) in Lemma 4, we have $\gamma^{-1}(s) \downarrow$. Let the limit be u. By (3.13), we know

$$L\{C_2(A) \cap [0, \gamma^{-1}(s))\} = s \qquad \gamma^{-1}(s) \in \bar{C}_2^+(A)$$

Putting $s \downarrow t$, we find that u satisfies (3.13); therefore $u = \gamma^{-1}(t)$, i.e. $\gamma^{-1}(s) \downarrow \gamma^{-1}(t)$. Consequently,

$$\lim_{s \downarrow t} \bar{x}(s) = \lim_{s \downarrow t} x(\alpha_s) = \lim_{s \downarrow t} x(\gamma^{-1}(s)) = x(\gamma^{-1}(t)) = x(\alpha_t) = \bar{x}(t)$$

and the proof is concluded. QED

9.5 SEVERAL LEMMAS

Let $X = \{x(t), t < \sigma\} \in \mathscr{H}_s$, β be a Markov time relative to X, $\mathscr{F}_\beta$ denote the pre-β field and $\mathscr{F}'_\beta$ denote the post-β field. Set

$$x'_\beta(t) = x(\beta + t) \qquad t < \sigma'_\beta = \sigma - \beta \tag{1}$$

We call $X'_\beta = \{x'_\beta(t), t < \sigma'_\beta\}$, or more simply X' the post-β process. Let symbols $\mathscr{L}^0_t(\beta)$, or simply $\mathscr{L}^0_t$, denote the complete Borel field which is on $\Omega_\beta = (\beta < \sigma)$ and is generated by $\{x'_\beta(u), u \leqslant t\}$, $\mathscr{L}^0_\infty(\beta) = \mathscr{F}'_\beta$.

Obviously, if $\tau \leqslant \beta$ are two Markov times relative to X, then

$$\mathscr{F}_\tau \cap \Omega_\beta \subset \mathscr{F}_\beta \qquad \mathscr{F}'_\tau \supset \mathscr{F}'_\beta. \tag{2}$$

Lemma 1. Let β be a Markov time relative to X. Then

$$\theta_\beta \mathscr{F}^0_t = \mathscr{L}^0_t(\beta) \subset \mathscr{F}'_\beta \cap \mathscr{F}_{\beta+t} \tag{3}$$

Proof. Let $\Lambda \in \mathscr{F}^0_t$. By Dynkin (1959, Lemma 1.5), there exist $t_k \in [0, t]$, and a Borel set Γ in $E^\infty = E \times E \times E \times \cdots$ such that

$$\Lambda = \{[x(t_1), x(t_2), \ldots] \in \Gamma\} \tag{4}$$

Therefore by Dynkin (1963, p. 122).

$$\theta_\beta \Lambda = \{[x'_\beta(t_1), x'_\beta(t_2), \ldots] \in \Gamma\} \in \mathscr{L}^0_t(\beta) \tag{5}$$

Hence we obtain $\theta_\beta \mathscr{F}^0_t \subset \mathscr{L}^0_t(\beta)$. Similarly it can be proved that the inverse inclusion holds. Consequently $\theta_\beta \mathscr{F}^0_t = \mathscr{L}^0_t(\beta)$.

Obviously, $\mathscr{L}^0_t(\beta) \subset \mathscr{F}'_\beta$. Now we need to prove that $\{x'(s) = i\} \in \mathscr{F}_{\beta+t}$ $(s \leqslant t, i \in E)$, that is, for arbitrary $u \geqslant 0$, $s \leqslant t$, we have

$$\{x(\beta + s) = i \qquad \beta + t < u < \sigma\} \in \mathscr{F}^0_u$$

When $u \leqslant t$, the left is empty. We can reasonably assume $u > t$. Set

$$\beta_n = \frac{k+1}{2^n}(u - t) \qquad \text{if} \qquad \frac{k}{2^n}(u - t) \leqslant \beta < \frac{k+1}{2^n}(u - t)$$

We have $P\{x(\beta_n + s) = \infty\} = 0$ by (7.6.1). From the right-continuity of X and $(\beta + t < u) \subset (\beta_n + t \leqslant u)$, it follows that

$$\{x(\beta + s) = i, \beta + t < u < \sigma\} = \left\{\lim_{n \to \infty} x(\beta_n + s) = i, \beta < u - t,\ u < \sigma\right\} \in \mathscr{F}^0_u$$

and the proof is completed. QED

Lemma 2. Let τ and β be two Markov times relative to X. Then $\theta_\beta \tau$ is a Markov time relative to X'_β.

Proof.

$$\{\theta_\beta\tau < u\} = \bigcup_{t\geqslant 0}\{\theta_t\tau < u\}\cap(\beta = t) = \bigcup_{t\geqslant 0}\theta_t(\tau < u)\cap(\beta = t) = \theta_\beta(\tau < u)$$

Since $(\tau < u)\in\mathscr{F}^0_u$, we have $(\theta_\beta\tau < u)\in\mathscr{L}^0_u$, by Lemma 1, and hence $(\theta_\beta\tau < u < \sigma'_\beta)\in\mathscr{L}^0_u$. QED

Lemma 3. Let $\{\tau,\beta\}$ be a canonical pair for X. Then τ_n and β_n are Markov times relative to X. For a fixed n, set

$$\tau'_0 = \beta'_0 = 0 \qquad \tau'_k = \tau_{n+k} - \beta_n \qquad \beta'_k = \beta_{n+k} - \beta_n \qquad k\geqslant 1 \tag{6}$$

Then $\{\tau',\beta'\}$ is a canonical pair for X'_{β_n}.

Proof. It is known that β_0, τ_1 and β_1 are Markov times relative to X. Assume that τ_n, β_n are Markov times relative to X. By Lemmas 1 and 2 we have

$$\begin{aligned}(\theta_{\beta_n}\tau_1 < t < \sigma'_{\beta_n})\in\mathscr{L}^0_t(\beta_n)\subset\mathscr{F}_{\beta_n+t}\\(\theta_{\beta_n}\beta_1 < t < \sigma'_{\beta_n})\in\mathscr{L}^0_t(\beta_n)\subset\mathscr{F}_{\beta_n+t}\end{aligned} \tag{7}$$

By Theorem 7.6.3, $\tau_{n+1} = \beta_n + \theta_{\beta_n}\tau_1$ and $\beta_{n+1} = \beta_n + \theta_{\beta_n}\beta_1$ are Markov times relative to X.

According to Lemma 2, $\tau'_1 = \tau_{n+1} - \beta_n = \theta_{\beta_n}\tau_1$, $\beta'_1 = \beta_{n+1} - \beta_n = \theta_{\beta_n}\beta_1$ are Markov time relative to X'_{β_n}. Moreover,

$$\beta'_1 + \theta_{\beta'_1}\tau'_1 = \beta_{n+1} - \beta_n + \theta_{\beta_{n+1}-\beta_n}\theta_{\beta_n}\tau_1 = \beta_{n+1} + \theta_{\beta_{n+1}}\tau_1 - \beta_n = \beta_{n+2} - \beta_n = \tau'_2$$

In a similar manner, it can be shown that (3.17) holds for $\{\tau',\beta'\}$, that is, $\{\tau',\beta'\}$ is a canonical pair for X'_{β_n}. QED

Lemma 4. Let $\{\tau,\beta\}$ be a canonical pair for X. Then random variables σ_n and δ_n determined by (2.2) and (2.3) are $\mathscr{F}_{\tau_n}$ and $\mathscr{F}_{\beta_n}$ measurable, respectively.

Proof. It is required to prove that for any $u\geqslant 0$, $t\geqslant 0$, we have

$$\begin{aligned}(\sigma_n < u, \tau_n < t < \sigma)\in\mathscr{F}^0_t\\(\delta_n < u, \beta_n < t < \sigma)\in\mathscr{F}^0_t\end{aligned}$$

In fact, let R denote the set of all rational numbers on $[0,\infty)$; then

$$\begin{aligned}(\sigma_n < u, \tau_n < t < \sigma) &= \left\{\sum_{k=1}^{n}(\tau_k - \beta_{k-1}) < u, \tau_n < t < \sigma\right\}\\&= \bigcup_{r_k\in R}(\tau_k - \beta_k < r_k, \tau_k < t)\ (\tau_n < t < \sigma)\end{aligned}$$

$$\sum_{k=1}^{n} r_k < u$$

and

$$(\delta_n < u, \beta_n < t < \sigma) = \left\{ \sum_{k=0}^{n} (\beta_k - \tau_k) < u, \beta_n < t < \sigma \right\}$$
$$= \bigcup_{\substack{r_k \in R \\ \sum_{k=1}^{n} r_k < u}} (\beta_k - \tau_k < r_k, \beta_k < t) \qquad (\beta_n < t < \sigma)$$

From Zhen-ting Hou and Qin-fen Guo (1978, Lemma 1.4.2) the above sets belong to $\mathscr{F}_t^0$ and the proof is concluded. QED

Lemma 5. Let $\{\tau, \beta\}$ be a canonical pair for X. α_t is determined by (2.5). For each fixed $t \geqslant 0$, α_t is a Markov time relative to X.

Proof. Since $t \leqslant \alpha_t < \sigma$ we may as well assume $u > t$. By Lemma 2

$$(\alpha_t < u < \sigma) = \bigcup_{k=1}^{\infty} (\beta_{k-1} \leqslant u < \tau_k, \alpha_t < u < \sigma) + \bigcup_{k=1}^{\infty} (\tau_k \leqslant u < \beta_k, \alpha_t < u < \sigma)$$

$$= \bigcup_{k=1}^{\infty} (\beta_{k-1} \leqslant u < \tau_k, \rho(u) < u - t, u < \sigma) + \bigcup_{k=1}^{\infty} (\tau_k \leqslant u < \beta_k, \gamma(u) > t, u < \sigma)$$

$$= \bigcup_{k=1}^{\infty} (\delta_{k-1} < u - t, \beta_{k-1} \leqslant u < \sigma)(u < \tau_k) + \bigcup_{k=1}^{\infty} (\sigma_k > t, \tau_k \leqslant u < \sigma)(u < \beta_k)$$

From Lemma 4 we know the above expression belongs to $\mathscr{F}_u^0$, and the proof is finished. QED

Lemma 6. Let $\{\tau, \beta\}$ be a canonical pair for X. α_t is determined according to (2.5). If the condition (c) in Theorem 3.3 is satisfied, then the condition (b) in Theorem 3.3 holds. More precisely

$$P\{x(t) \in E \text{ for all } t \in A\} = 1 \tag{8}$$

where A is just the same as in (4.1). In particular,

$$P\{x(\alpha_t) \in E \text{ for all } t \in [0, \bar{\sigma})\} = 1 \tag{9}$$

$$P\{x(\beta_k) = \infty \text{ for some } k\} = 0 \tag{10}$$

If the conditions (a) and (c) in Theorem 3.3 are satisfied, then for almost all $\omega \in \Omega_{\beta_k} = (\beta_k < \sigma)$ and all $u \in [\beta_k(\omega), \tau_{k+1}(\omega)]$, we have

$$\begin{aligned} u + (\theta_u \tau_1)(\omega) &= \tau_{k+1}(\omega) \\ u + (\theta_u \beta_1)(\omega) &= \beta_{k+1}(\omega) \end{aligned} \tag{11}$$

Proof. (a) First suppose $P\{\beta_0=0\}=1$. Put

$$\Lambda_k=\{x(u)\in E \text{ for all } u\in[\beta_k,\tau_{k+1})\}\cap\Omega_{\beta_k}$$

By (3.17) and $\beta_0=0$, it is evident that $\Lambda_k=\theta_{\beta_k}\Lambda_0$. From the condition (c) in Theorem 3.3 and the strong Markov property, we have

$$P\{\Lambda_1\}=P\{\theta_{\beta_1}\Lambda_0\}=\sum_{k\in E}P\{x(\beta_1)=k\}P\{\Lambda_0|x(0)=k\}=\sum_{k\in E}P\{x(\beta_1)=k\}=P\{\Omega_{\beta_1}\} \tag{12}$$

Similarly, it can be proved that $P\{\Lambda_k\}=P\{\Omega_{\beta_k}\}$. This verifies (8).

Set

$$\Delta_k=\{u+\theta_u\tau_1=\tau_{k+1} \text{ for all } u\in[\beta_k,\tau_{k+1})\}\cap\Omega_{\beta_k}$$

Similarly, we have $\Delta_k=\theta_{\beta_k}\Delta_0$. From conditions (a) and (c) in Theorem 3.3 and the strong Markov property, and by following the discussion in (12), we can prove that $P(\Delta_k)=P(\Omega_{\beta_k})$, which verifies the first formula in (11). The second formula can be proved in similar way.

(b) Suppose $P\{\beta_0=0\}\leqslant 1$. In this case consider the post-β_0 process $X'=X'_{\beta_0}$ By Lemma 2, $\{\tau',\beta'\}$ is a canonical pair for X', where

$$\tau'_0=\beta'_0=0 \qquad \tau'_l=\tau_l-\beta_0 \qquad \beta'_l=\beta_l-\beta_0 \qquad l\geqslant 1 \tag{13}$$

Since X and $\{\tau,\beta\}$ satisfy the conditions (a) and (c) in Theorem 3.3, so do X' and $\{\tau',\beta'\}$. By (a) it follows that the conclusions in this lemma hold for X' and $\{\tau',\beta'\}$. On the other hand, from (13), we have

$$\{x'(t)\in E \text{ for all } t\in A'\}=\{x(t)\in E \text{ for all } t\in A\}$$

So $\Omega_{\beta'_k}=\Omega_{\beta_R}u+\theta_u\tau'_1=\tau'_{k+1}$ for all $u\in[\beta_k,\tau'_{k+1})$ is equivalent to $u+\theta_u\tau_1=\tau_{k+1}$ for all $u\in[\beta_k,\tau_{k+1})$. Therefore, the conclusions in this lemma hold for X and $\{\tau,\beta\}$ and the proof is terminated. QED

Lemma 7. Let $\{\tau,\beta\}$ be a canonical pair for X, $P\{\beta_0=0\}=1$. The conditions (a) and (c) in Theorem 3.3 are satisfied. Then for any fixed $s,t\geqslant 0$, we have

$$\alpha_{s+t}=\alpha_s+\theta_{\alpha_s}(\alpha_t) \tag{14}$$

Proof. Put $X'=X'_{\alpha_s}$. By Lemma 2

$$\beta'_0=0 \qquad \tau'_1=\theta_{\alpha_s}\tau_1 \qquad \beta'_1=\theta_{\alpha_s}\beta_1 \tag{15}$$

are Markov times relative to X', so that we can determine the canonical pair $\{\tau',\beta'\}$ for X' and the corresponding α'_t, Since $P\{\beta_0=0\}=1$, it is evident that $\alpha_t=\theta_{\alpha_s}(\alpha_t)$.

From (v) in Lemma 1, $\alpha_s\in A=\bigcup_{k=0}^{\infty}[\beta_k,\tau_{k+1})$. To be specific, we assume $\alpha_s\in[\beta_k,\tau_{k+1})$. By Lemma 6, $\alpha_s+\theta_{\alpha_s}\tau_1=\tau_{k+1}$, $\alpha_s+\theta_{\alpha_s}\beta_1=\beta_{k+1}$. Hence from (15) it follows that $\tau'_1=\tau_{k+1}-\alpha_s$, $\beta'_1=\beta_{k+1}-\alpha_s$. Following the proof of Lemma 3, we have

$$\tau_l' = \tau_{k+1} - \alpha_s \qquad \beta_l' = \beta_{k+l} - \alpha_s \tag{16}$$

By Lemma 1 (v) $\alpha_t' = \theta_{\alpha_s}(\alpha_t) \in A' = \bigcup_{k=0}^{\infty} [\beta_k', \tau_{k+1}')$.

To be specific, now we assume $\alpha_t' \in [\beta_l', \tau_{l+1}')$. By (16), that is, $\beta_{k+l} - \alpha_s \leqslant \theta_{\alpha_s}(\alpha_t) < \tau_{k+l} - a_s$. Hence, $\beta_{k+l} \leqslant \alpha_s + \theta_{\alpha_s}(\alpha_t) < \tau_{k+l}$, that is, $\alpha_s + \theta_{\alpha_s}(\alpha_t) \in A$. Furthermore, by (v) in Lemma 1

$$t = L\{A' \cap [0, \alpha_t')\} = L\left\{\bigcup_{l=0}^{\infty} [\beta_l', \tau_{l+1}') \cap [0, \alpha_t')\right\}$$

$$= L\left\{\bigcup_{l=0}^{\infty} [\beta_{k+1} - \alpha_s, \tau_{k+1} - \alpha_s) \cap [0, \theta_{\alpha_s}(\alpha_t'))\right\}$$

$$= L\left\{\bigcup_{l=0}^{\infty} [\beta_{k+l}, \tau_{k+l}) \cap [\alpha_\delta, \alpha_s + \theta_{\alpha_s}(\alpha_t))\right\}$$

$$= L\{A \cap [\alpha_s, \alpha_s + \theta_{\alpha_s}(\alpha_t))\}$$

Therefore, $L\{A \cap [0, \alpha_s + \theta_{\alpha_s}(\alpha_t))\} = L\{A \cap [0, \alpha_s)\} + t = s + t$. Consequently, $\alpha_s + \theta_{\alpha_s}(\alpha_t)$ is the solution of the equation

$$\gamma(u) = s + t \qquad u \in A$$

But α_{s+t} is the unique solution of the above equation. Hence (14) holds, and the proof is terminated. QED

9.6 PROOF OF THE STRONG LIMIT THEOREM

Proof of Theorem 3.3. (a) Suppose $P\{\beta_0 = 0\} = 1$. Assume $t_1 < t_2 < \cdots < t_{n+1}$, $i_1, i_2, \ldots, i_{n+1} \in E$. By Lemma 5.5 and 5.1, and by (5.2), we have

$$\Lambda_1 = \bigcap_{k=1}^{n-1} \{x(\alpha_{t_k}) = i_k\} \cap (\alpha_{t_n} < \sigma) \in \mathscr{F}_{\alpha_{t_n}}$$

$$\Lambda_2 = \{x(\alpha_{t_{n+1}}) = i_{n+1}\} \in \mathscr{F}'_{\alpha_{t_n}}$$

From Lemma 7 and the strong Markov property, we obtain

$$\begin{aligned} P\{x(\alpha_{t_{n+1}}) = i_{n+1} | x(\alpha_{t_k}) = i_k, 1 \leqslant k \leqslant n\} &= P\{x(\alpha_{t_{n+1}}) = i_{n+1} | \Lambda_1, \alpha_{t_n} < \sigma, x(\alpha_{t_n}) = i_n\} \\ &= P\{x[\alpha_{t_n} + \theta_{\alpha_{t_n}}(\alpha_{t_{n+1}} - t_n)] = i_{n+1} | x(\alpha_{t_n}) = i_n\} \\ &= P\{x(\alpha_{t_{n+1} - t_n}) = i_{n+1} | x(0) = i_n\} \end{aligned}$$

Noticing (3.19), we could prove that $\bar{X} = W_{\tau,\beta}(X)$ is a homogeneous Markov process.

(b) Suppose $P\{\beta_0 = 0\} \leqslant 1$. Now we consider $X' = X^{\wedge}_{\beta_0}$, its canonical pair $\{\tau', \beta'\}$ determined by (5.6) with $n = 0$ and the corresponding α_t'. Then (5.13) holds and, moreover,

$$\alpha'_t = \alpha_t - \beta_0 \qquad x'(\alpha'_t) = x(\alpha_t) \tag{1}$$

Since the conditions (a) and (b) in Theorem 3.3 are satisfied for X and $\{\tau, \beta\}$, they are satisfied for X' and $\{\tau', \beta'\}$, too. Consequently according to (a), $W_{\tau',\beta'}(X')$ possesses the homogeneous Markov property. But $W_{\tau,\beta}(X) = W_{\tau',\beta'}(X')$. Therefore $\bar{X} = W_{\tau,\beta}(X)$ possesses the homogeneous Markov property.

(c) Since $X \in \mathscr{H}_s$, we have $\bar{X} \in \mathscr{H}_s$. By Lemma 6, we know that the condition (c) in Theorem 3.3 implies the condition (b), and hence we prove the conclusion (iii) in Theorem 3.3. The proof is concluded. QED

Proof of Theorem 3.4. Except that it remains to prove that $\bar{X}$ in Theorem 3.4 is a process, the rest of the conclusions of Theorem 3.4 may all be deduced from Theorems 3.1 and 3.3.

Let us now prove $P\{\bar{x}(t) = \infty\} = 0$ $(t \geqslant 0)$.

By Lemma 6, for almost all ω, $x(t, \omega) \in E$ for all $t \in A^n(\omega)$, so that

$$x(t, \omega) \in E \qquad \text{for all } t \in A(\omega)$$

Therefore, by (iii) (e) in Lemma 4.4, when $t \in \gamma[C_1(A(\omega))]$, $\gamma^{-1}(t) \in C_1[A(\omega)] \subset A(\omega)$. From (2), we get $\bar{x}(t, \omega) = x(\gamma^{-1}(t), \omega) \in E$. Consequently if we set $\bar{S}_i(\omega) = \{t \mid \bar{x}(t, \omega) = i\}$ $(i \in \bar{E})$, then

$$\gamma[C_1(A(\omega))] \subset \{t \mid \bar{x}(t, \omega) \in E\} = \bigcup_{i \in E} \bar{S}_i(\omega) \subset \bigcup_{i \in E} \bar{S}_i(\omega) \cup \bar{S}_\infty(\omega) = [0, \bar{\sigma}(\omega))$$

By (ii) in Lemma 4.4 $\bar{\sigma}(\omega) = L\{\gamma[C_1(A(\omega))]\}$, hence $L\{\bar{S}_\infty(\omega)\} = 0$. By Fubini's theorem it follows that there exists a null Lebesgue-measurable set T such that $P\{\bar{x}(t) = \infty\} = 0$ when $t \notin T$.

From (3.22), $\beta_0^n \downarrow \inf_n \beta_0^n = \alpha_0$, so that (3.21) becomes $P\{\bar{x}(0) = \infty\} = 0$. Therefore $0 \notin T$.

Now let us assume $t_0 \in T$. Clearly $t_0 > 0$. Since $L(T) = 0$, we may choose $0 < t < t_0$ such that both $t_0 - t$ and $t \notin T$. Notice that for each $\omega \in \Omega$, we have $\lim_{n\to\infty} x^n(t) = \lim_{n\to\infty} x(\alpha_t^n) = x(\alpha)$ in $\bar{E}$. Accordingly we set $\bar{D} = \bar{E} - D$ for a finite set $D \subset E$. It follows that

$$P\{\bar{x}(t_0) \in \bar{D}\} = P\left\{\lim_{n\to\infty} x(\alpha_{t_0}^n) \in \bar{D}\right\} = \lim_{n\to\infty} P\{x(\alpha_{t_0}^n) \in \bar{D}\}$$

Write $X' = X'_{\beta_0^n}$, $\alpha'_t = \alpha_t^n - \beta_0^n$. From (1), Lemma 7 and the strong Markov property for X', we obtain

$$\begin{aligned} P\{x(\alpha_{t_0}^n) \in \bar{D}\} &= P\{x'(\alpha'_{t_0}) \in \bar{D}\} = P\{x'[\alpha'_t + \theta_{\alpha'_t}(\alpha'_{t_0 - t})] \in \bar{D}\} \\ &= E\{P_{x'(\alpha'_t)}[x'(\alpha'_{t_0 - t}) \in \bar{D}]\} \\ &= E\{P_{x(\alpha_t^n)}[x(\alpha_{t_0 - t}^n) \in \bar{D}]\} \end{aligned}$$

Therefore, $P\{\bar{x}(t_0)\in\bar{D}\}=\lim_{n\to\infty}E\{P_{x(\alpha_t^n)}[x(\alpha_{t_0-t}^n)\in\bar{D}]\}$. Since $t\in T$, we have $x(\alpha_t,\omega)\in E$ for almost all ω. As a result, when n is sufficiently large, we have $x(\alpha_t^n,\omega)=x(\alpha_t,\omega)$. Hence

$$\begin{aligned}P\{\bar{x}(t_0)\in\bar{D}\}&=E\left\{\lim_{n\to\infty}P_{x(\alpha_t)}[x(\alpha_{t_0-t}^n)\in\bar{D}]\right\}\\&=E\{P_{x(\alpha_t)}[x(\alpha_{t_0-t})\in\bar{D}]\}\\&=\sum_{k\in E}P\{x(\alpha_t)=k\}P_k\{x(\alpha_{t_0-t})\in\bar{D}\}\end{aligned}$$

Letting $D\uparrow E$, we obtain

$$\begin{aligned}P\{\bar{x}(t_0)=\infty\}&=\sum_{k\in E}P\{x(\alpha_t)=k\}P_k\{x(\alpha_{t_0-t})=\infty\}\\&=\sum_{k\in E}P\{x(\alpha_t)=k\}\cdot 0=0\end{aligned}$$

Since $P\{\bar{x}(t)=\infty\}=P\{x^n(t)=\infty\}=0$ $(t\geqslant 0)$, by Lemma 5.3, we know that as $n\to\infty$, $\{x^n(t_k)=i_k,\ 1\leqslant k\leqslant l\}$ converges to $\{\bar{x}(t_k)=i_k,\ 1\leqslant k\leqslant l\}$ in distribution. Hence from the homogeneous Markov property for X, we obtain the same property for $\bar{X}$. The proof is completed. QED

9.7 SEVERAL KINDS OF SPECIAL STRONG LIMIT THEOREMS

Theorem 1. Let $X\in\mathscr{H}_{\mathrm{D}}$. We fix a state $i\in E$. Let β_0 be a Markov time relative to X, satisfying $P\{\beta_0<\sigma\}>0$. Set

$$\begin{aligned}\tau_1&=\inf\{t\mid\beta_0\leqslant t<\sigma,\ x(t)=i\}\\\beta_n&=\inf\{t\mid\tau_n\leqslant t<\sigma,\ x(t)\neq i\}\\\tau_{n+1}&=\inf\{t\mid\beta_n\leqslant t<\sigma,\ x(t)=i\}\end{aligned}$$

We set $\inf\varnothing=\sigma$, $\varnothing$ being an empty set. Then $\bar{X}=W_{\tau,\beta}(X)\in\mathscr{H}_{\mathrm{D}}$, and $\bar{X}$ does not take the value i.

Proof. Evidently $\{\tau,\beta\}$ is a canonical pair, and the condition (a) in Theorem 3.3 is satisfied. Since $X\in\mathscr{H}_{\mathrm{D}}$, the condition (b) in the same theorem is, of course, satisfied. Hence by quoting Theorem 3.3 we obtain Theorem 1. QED

Theorem 2. Let $X\in\mathscr{H}_{\mathrm{D}}$, and the subsets M and N of E be disjoint. Put

$$\begin{aligned}\beta_0&=\inf\{t\mid 0\leqslant t<\sigma,\ x(t)\in N\}\\\tau_1&=\inf\{t\mid\beta_0\leqslant t<\sigma,\ x(t)\in M\}\\\beta_n&=\inf\{t\mid\tau_n\leqslant t<\sigma,\ x(t)\in N\}\\\tau_{n+1}&=\inf\{t\mid\beta_n\leqslant t<\sigma,\ x(t)\in M\}\end{aligned}$$

Then $\bar{X}=W_{\tau,\beta}(X)\in\mathscr{H}_{\mathrm{D}}$, and the value of $\bar{X}$ belongs to $E-M$.

Proof. It is the same as Theorem 1. QED

Theorem 3. Let $X \in \mathscr{H}_s$, and finite sets $D_n \uparrow E$. Put

$$\begin{aligned} \beta_0^n &= \inf\{t \mid 0 \leqslant t < \sigma, x(t) \in D_n\} \\ \tau_1^n &= \inf\{t \mid \beta_0^n \leqslant t < \sigma, x(t) \notin D_n\} \\ \beta_k^n &= \inf\{t \mid \tau_k^n \leqslant t < \sigma, x(t) \in D_n\} \\ \tau_{k+1}^n &= \inf\{t \mid \beta_k^n \leqslant t < \sigma, x(t) \notin D_n\} \end{aligned}$$

Then $X^n = W_{\tau^n, \beta^n}(X)$ is a minimal process, whose state space is D_n, and

$$\lim_{n \to \infty} X^n = X. \tag{1}$$

Proof. Referring to Theorem 3.4, we easily see that $\{\tau^n, \beta^n\}$ is a canonical pair for X, and the conditions (a) and (c) in Theorem 3.3 hold for X and $\{\tau^n, \beta^n\}$. Since $P\{x(0) = \infty\} = 0$, it follows that $\inf_n \beta_0^n = 0$, so that (3.21) holds. As $D_n \uparrow$, (3.22) holds, too.

Notice

$$\begin{aligned} x(t) &\in D_n & \text{if } t \in A^n &= \bigcup_{k=0}^{\infty} [\beta_k^n, \tau_{k+1}^n) \\ x(t) &\notin D_n & \text{if } t \in B_n &= \bigcup_{k=0}^{\infty} [\tau_k^n, \beta_k^n) \end{aligned} \tag{2}$$

By (1) in Theorem 3.4, we obtain $X^n = W_{\tau^n, \beta^n}(X) \in \mathscr{H}_D$. By the definition of X^n, we know that X^n does not have any leaping point, hence X^n is a minimal process.

Let us now proceed to prove

$$\{t \mid x(t) \in E\} = A = \bigcup_{n=1}^{\infty} A^n \tag{3}$$

Assume $x(t) = i \in E$; then there exists a number n such that $i \in D_n$, and therefore $x(t) \in D_n$. By (2), we get $t \in A^n \subset A$. Hence $\{t \mid x(t) \in E\} \subset A$. On account of (2), the inverse inclusion relation holds too.

By (3), $[0, \sigma)$ and A are differ by a null Lebesgue-measurable set $S_\infty = \{t \mid x(t) = \infty\}$. So we have

$$\gamma(u) = L\{A \cap [0, u)\} = L\{[0, \sigma) \cap [0, u)\} = u$$

Therefore, $\alpha_t = \gamma^{-1}(t) = t, \bar{\sigma} = \sigma$. Consequently, (3.14) becomes (1). The proof is completed. QED

Theorem 4. Let $X \in \mathscr{H}_s$ be a non-minimal process and finite sets $D_n \uparrow E$. Put $\beta_0^n = 0$,

$$\begin{aligned} &\tau_1^n \text{ being the first leaping point for } X \\ &\beta_k^n = \inf\{t \mid \tau_k^n \leqslant t < \sigma, x(t) \in D_n\} \\ &\tau_{k+1}^n \text{ being the first leaping point for } X \text{ after } \beta_k^n \end{aligned} \tag{4}$$

Then $X^n = W_{\tau^n, \beta^n}(X) \in \mathscr{H}_1$ (collection of all first-order processes), (3.10) holds and, moreover, (1) is still valid.

Proof. Following the proof of Theorem 3 and quoting Theorem 3.4 we need only explain that, in the present case, (3) still holds. For this, let us notice that

$$\begin{aligned} &x(t) \in E && \text{for all } t \in A^n \\ &x(t) \notin D_n && \text{for all } t \in B^n \end{aligned} \tag{5}$$

And following the proof of Theorem 3, we know that (3) holds. QED

Remark

The transformation X to X^n in Theorem 4 is just the transformation g_n. Theorem 4 is precisely the fundamental result in Zhen-ting Hou (1975), i.e. the first construction theorem in Zhen-ting Hou and Qin-fen Guo (1978).

CHAPTER 10

Leaping Interval and Entrance Decomposition

10.1 INTRODUCTION

In this chapter we study the entrance of a process. We have introduced the concept of leaping interval, namely, U interval, and studied the relation between leaping intervals and set of leaping points, and we have found the necessary and sufficient condition which is expressed by leaping intervals and under which the Kolmogorov equations hold. We have found that it is effective to apply leaping intervals to the study of the entrance of a process. Especially, we have derived the entrance decomposition theorem of a process. Making use of this decomposition, we can investigate distinct entrances independently in the case that various kinds of entrances present themselves. By using the results of leaping intervals, we have obtained two kinds of strong limit theorems of a process, that is, transformations ${}_{M}g_n$ and ${}_{M}f_n$, and the strong limit theorems corresponding to them. As a simple deduction of these theorems, they are precisely the basic results in Zheng-ting Hou (1975), namely, the first construction theorem in Zheng-ting Hou (1974) and Theorem 5.3 in Zi-kun Wang (1962).

10.2 DEFINITION OF LEAPING INTERVAL

Suppose $X = \{x(t),\ t < \sigma\} \in \mathscr{H}_s$. Owing to (7.6.1), without loss of generality, we shall suppose henceforth for every $\omega \in \Omega$,

$$x(r, \omega) \in E \qquad \text{for all } r \in R \cap [0, \sigma(\omega))$$

Hereafter, R represents the set of rational numbers in $[0, \infty)$.

Definition 1. A point $t \in (0, \sigma(\omega))$ is called a continuity point of $X(\omega)$ if $\lim_{s \uparrow t} x(s, \omega) = x(t, \omega) \in E$. The set of continuity points is written as $G(\omega)$. $D(\omega) = [0, \sigma(\omega)] - G(\omega)$ is called the set of discontinuity points of $X(\omega)$.

Recall Definition 7.8.1, denote the set of jumping points of $X(\omega)$ by $T(\omega)$, and denote the set of leaping points by $\Gamma(\omega)$. We appoint $0 \in T(\omega) \cap \Gamma(\omega)$, but call $t=0$ the zeroth jumping point and zeroth leaping point. $X(\omega)$ has the first discontinuity point $\tau_1(\omega)>0$, and the first leaping point $\tau(\omega)\geqslant\tau_1(\omega)>0$.

Obviously, if $\sigma(\omega)<\infty$ and $\sigma(\omega)\in T(\omega)$, then $x(\sigma(\omega)-0,\ \omega)\in H$ (non-conservative state set). It can be easily seen that $D(\omega)$ and $\Gamma(\omega)$ are closed sets.

For every $s\in(0,\sigma(\omega))$, we can define

$$\begin{aligned}\mu_s(\omega)&=\max\{D(\omega)\cap[0,s]\},\\ v_s(\omega)&=\min\{D(\omega)\cap[s,\sigma(\omega)]\}\\ \lambda_s(\omega)&=\max\{\Gamma(\omega)\cap[0,s]\}\\ \eta_s(\omega)&=\min\{\Gamma(\omega)\cap[s,\sigma(\omega)]\}\end{aligned}\tag{1}$$

For $a=0$, define

$$v_0(\omega)=\tau_1(\omega)\qquad \eta_0(\omega)=\tau(\omega)\qquad \mu_0(\omega)=0\qquad \lambda_0(\omega)=0\tag{2}$$

For $s\geqslant 0$, call $\mu_s(\omega)$, $\lambda_s(\omega)$, $v_s(\omega)$, $\eta_s(\omega)$ respectively the last discontinuity point before s, the last leaping point before s, the first discontinuity point after s, and the first leaping point after s. Suppose $s>0$. If $s\in D(\omega)$ or $s\in\Gamma(\omega)$, then obviously $\mu_s(\omega)=v_s(\omega)=s$ or $\lambda_s(\omega)=\eta_s(\omega)=s$. If $s\notin D(\omega)$ or $s\notin\Gamma(\omega)$, then we have $\mu_s(\omega)<s<v_s(\omega)$ or $\lambda_s(\omega)<s<\eta_s(\omega)$.

Definition 2. Call $[\lambda,\eta)$ a leaping interval, namely, U interval, of $X(\omega)$ if $\lambda,\eta\in\Gamma(\omega)$, and $(\lambda,\eta)\cap\Gamma(\omega)=\varnothing$ (empty set). The collection of U intervals of $X(\omega)$ is written as $\mathscr{U}(\omega)$.

It is quite plain that U intervals in $\mathscr{U}(\omega)$ have the order of 'before' and 'after' according to its place in $[0,\sigma]$

From Zi-kun Wang (1965a, Theorem 3.2.4, corollary) we know that if $[\lambda,\eta]$ is a U interval, then the discontinuity points in (λ,η) are jumping points and, moreover,

$$x(t,\omega)\in E\qquad \text{for all } t\in(\lambda,\eta)\tag{3}$$

Definition 3. suppose $[\lambda,\eta)\in\mathscr{U}(\omega)$, $M\subset\bar{E}$, $N\subset\bar{E}$. If $x(\lambda,\omega)\in M$, call $[\lambda,\eta)$ a ${}_MU$ interval. If $x(\eta-0,\ \omega)\in N$, call $[\lambda,\eta)$ a U_N interval. Similarly, we can define ${}_MU_N$ interval, ${}_iU$ interval, etc. Write the collection of ${}_MU$ intervals of $X(\omega)$ as ${}_M\mathscr{U}(\omega)$. Similar notations are $\mathscr{U}_N(\omega)$, ${}_i\mathscr{U}(\omega)$, etc., whose meanings are clear.

By Theorem 7.7.2, $X(\omega)$ has no U_{E-H} interval; in other words, if $X(\omega)$ has a U_E interval, it must be a U_H interval.

By Zi-kun Wang (1965a, Theorem 3.2.3), for any fixed $t>0$ for almost all $\omega\in(t<\sigma)$ we have

$$\lambda_t(\omega)\leqslant\mu_t(\omega)<t<v_t(\omega)\leqslant\eta_t(\omega)\tag{4}$$

thus we know easily that for almost all $\omega\in\Omega$

$$\lambda_r(\omega)<r<\eta_r(\omega)\qquad \text{for all } r\in R\cap[0,\sigma(\omega))\tag{5}$$

$$\{[\lambda_r(\omega), \eta_r(\omega)) | r \in R \cap [0, \sigma(\omega))\} = \mathscr{U}(\omega) \tag{6}$$

Hence $\mathscr{U}(\omega)$ is a denumerable set.

Write

$$\mathscr{U}^\tau(\omega) = \{[\lambda, \eta) | [\lambda, \eta) \in \mathscr{U}(\omega), \lambda \geqslant \tau(\omega)\}. \tag{7}$$

Similar notations are ${}_M\mathscr{U}^\tau(\omega), \mathscr{U}^\tau_N(\omega)$, etc., whose meanings are clear. Write

$$C_2(\omega) = \bigcup_{[\lambda,\eta) \in \mathscr{U}(\omega)} [\lambda, \eta) \tag{8}$$

$$C_1(\omega) = \bigcup_{[\lambda,\eta) \in \mathscr{U}(\omega)} (\lambda, \eta) \tag{9}$$

Similar notations are ${}_MC_2(\omega)$, $C^\tau_{1N}(\omega)$, etc., whose meanings are clear.

10.3 LEAPING POINT AND LEAPING INTERVAL

Definition 1. Suppose $t \in \Gamma(\omega)$. Call t a right isolated leaping point of $X(\omega)$ if there exists $\varepsilon > 0$ such that $(t, t + \varepsilon) \cap \Gamma(\omega) = \varnothing$. The set of all the right isolated leaping points is written as $\Gamma^r(\omega)$. Similarly we define the set $\Gamma^l(\omega)$ of left isolated leaping points.

For $M \subset \bar{E}$, let

$$\Gamma^r_M(\omega) = \{t | t \in \Gamma^r(\omega), x(t, \omega) \in M\} \tag{1}$$

$$\bar{\Gamma}^{r+}_M(\omega) = \{t | \text{there exist non-increasing } t_n \in \Gamma^r_M(\omega), \text{ such that } t_n \downarrow t\} \tag{2}$$

Furthermore, call $\bar{\Gamma}^{r+}_M$ the right closure of $\Gamma^r_M(\omega)$. Similarly, let

$$\Gamma^1_M(\omega) = \{t | t \in \Gamma^l(\omega), x(t - 0, \omega) \in M\} \tag{3}$$

and the left closure

$$\bar{\Gamma}^{l-}_M(\omega) = \{t | \text{there exist non-decreasing } t_n \in \Gamma^l_M(\omega), \text{ such that } t_n \uparrow t\} \tag{4}$$

Obviously, $0 \in \Gamma^l(\omega) \cap \Gamma^r(\omega)$. If $\sigma(\omega) < \infty$ then $\sigma(\omega) \in \Gamma^r(\omega)$.

Theorem 1. For almost all $\omega \in \Omega$,

$$\Gamma^l(\omega) = \{\eta | [\lambda, \eta) \in \mathscr{U}(\omega)\} \cup \{0\} \tag{5}$$

$$\Gamma^r(\omega) = \{\lambda | [\lambda, \eta) \in \mathscr{U}(\omega)\} \cup \{\sigma(\omega) | \sigma(\omega) < \infty\} \tag{6}$$

Proof. It suffices to prove (6). It is obvious that the right-hand set is included in $\Gamma^r(\omega)$. Suppose $t \in \Gamma^r(\omega)$ and $t \neq \sigma(\omega)$. By the definition, there exists $\varepsilon > 0$ such that $(t, t + \varepsilon) \cap \Gamma(\omega) = \varnothing$, and therefore taking arbitrarily $r \in (t, t + \varepsilon) \cap R$, $t = \lambda_r(\omega) < r < \eta_r(\omega)$. By (2.6) we know t belongs to the right-hand set in (6), and the proof is completed. QED

Theorem 2. For almost all ω,

$$\Gamma(\omega) = \bar{\Gamma}^{r+}(\omega) = \bar{\Gamma}^{l-}(\omega) \tag{7}$$

Proof. $\bar{\Gamma}^{r+}(\omega) \subset \Gamma(\omega)$ is obvious. Suppose $t \in \Gamma(\omega)$. If $t = 0$ or $t = \sigma(\omega)$, then obviously $t \in \Gamma^r(\omega) \subset \bar{\Gamma}^{r+}(\omega)$. If $t \in (0, \sigma(\omega))$ and $t \notin \Gamma^r(\omega)$, take arbitrarily strictly decreasing $r_n \downarrow t$, $r_n \in R \cap [0, \sigma(\omega))$. By the definition of $\lambda_{r_n}(\omega)$ we have $t \leqslant \lambda_{r_n}(\omega) < r_n$. Since $t \notin \Gamma^r(\omega)$ it follows that $t < \lambda_{r_n}(\omega) < r_n$. Consequently there exists a subsequence r'_n of r_n such that $\lambda_{r'}(\omega)$ strictly decreases. When $r'_n \downarrow t$, $\lambda_{r'_n}(\omega) \downarrow t$. By Theorem 1, $\lambda_{r'}(\omega) \in \Gamma^r(\omega)$. Hence $t \in \bar{\Gamma}^{r+}(\omega)$. Therefore $\Gamma(\omega) = \bar{\Gamma}^{r+}(\omega)$. Similarly we can prove $\Gamma(\omega) = \bar{\Gamma}^{l-}(\omega)$, and the proof is concluded. QED

Theorem 3. For almost all $\omega \in \Omega$,

$$S_E(\omega) = C_1(\omega) \cup \{\lambda \mid \lambda \in \Gamma^r(\omega), x(\lambda, \omega) \in E\} \subset C_2(\omega) \tag{8}$$

$$[0, \sigma(\omega)) - C_2(\omega) \subset S_\infty(\omega) \subset \Gamma(\omega) \tag{9}$$

$$L\{C_1(\omega)\} = L\{C_2(\omega)\} = \sigma(\omega) \tag{10}$$

where

$$S_M(\omega) = \{t \mid t \in [0, \sigma(\omega)), x(t, \omega) \in M\} \qquad M \subset \bar{E} \tag{11}$$

Proof. The inclusion relation in (8) is clear. $S_\infty(\omega) \subset \Gamma(\omega)$ is also clear. Suppose $t \in S_E(\omega)$, then there must exist $i \in E$, such that t belongs to some i-interval $[a, b)$ of $X(\omega)$. We take arbitrarily $r \in [a, b) \cap R$. Then $t \in [a, b) \subset [\lambda_r(\omega), \eta_r(\omega))$. Now, either $t \in (\lambda_r(\omega), \eta_r(\omega)) \subset C_1(\omega)$ or $t = \lambda_r(\omega) \in \Gamma^r(\omega)$, and $x(t, \omega) = i \in E$. Accordingly $S_E(\omega) \subset C_1(\omega) \cup \{\lambda \mid \lambda \in \Gamma^r(\omega), x(\lambda, \omega) \in E\}$. The inverse inclusion relation is clear. Thus (8) is proved and hence so is (9). Because $[0, \sigma(\omega)) = S_E(\omega) \cup S_\infty(\omega)$ and $L\{S_\infty(\omega)\} = 0$, from (8) and (9), follows (10). The proof is terminated QED

Corollary

$$L\{\Gamma(\omega)\} = 0.$$

Proof. Because $\Gamma(\omega) \subset [0, \sigma(\omega)) - C_1(\omega)$, by (10) we obtain $L\{\Gamma(\omega)\} = 0$. QED

Let

$$A(\omega) = S_E(\omega) - \Gamma(\omega) \simeq [0, \sigma(\omega)) - \Gamma(\omega) \tag{12}$$

$\mathscr{U}[A(\omega)]$ represents the whole of the composition interval in $A(\omega)$.

Theorem 4. For almost all $\omega \in \Omega$,

$$\mathscr{U}[A(\omega)] = \mathscr{U}(\omega) \tag{13}$$

Proof. Suppose $[\lambda, \eta) \in \mathscr{U}[A(\omega)]$. By definition, $(\lambda, \eta) \subset A(\omega)$. Then by (12),

$(\lambda,\eta)\cap\Gamma(\omega)=\varnothing$. Moreover taking arbitrarily $t\in(\lambda,\eta)$, we have $x(t,\omega)\in E$, $(\lambda,\eta)\subset(\lambda_t(\omega),\eta_t(\omega))\subset A(\omega)$. By the maximality of $[\lambda,\eta)$, $[\lambda,\eta)=[\lambda_t(\omega),\eta_t(\omega))\in\mathscr{U}(\omega)$.

Suppose $[\lambda,\eta)\in\mathscr{U}(\omega)$. By definition, $(\lambda,\eta)\cap\Gamma(\omega)=\varnothing$. Noting (2.3) we have $(\lambda,\eta)\subset A(\omega)$. If it holds that $[\lambda',\eta')\supset[\lambda,\eta)$ and $(\lambda',\eta')\subset A(\omega)$, then $\lambda'\leqslant\lambda<\eta\leqslant\eta'$, and for any $s\in(\lambda',\eta')$ we have $s\notin\Gamma(\omega)$. Thus for any $t\in(\lambda,\eta)$ we have $\lambda_t(\omega)\leqslant\lambda'$, $\eta'\leqslant\eta_t(\omega)$. However, $\lambda,\eta\in\Gamma(\omega)$, so that we also have $\lambda=\lambda_t(\omega),\eta=\eta_t(\omega)$. Thus $\lambda'=\lambda=\lambda_t(\omega)$, $\eta'=\eta=\eta_t(\omega)$. The maximality of $[\lambda,\eta)$ is proved. Thereby $[\lambda,\eta)\in\mathscr{U}[A(\omega)]$. The proof is completed. QED

Theorem 5. Suppose $M\subset\bar{E}$, $N\subset\bar{E}$, and H is a non-conservative state set. Then all the following sets are $\mathscr{F}^0_\infty$-measurable sets:

$F_1=\{\omega|\mathscr{U}(\omega)$ has a last U interval and it is a U_N interval$\}$
$F_2=\{\omega|\mathscr{U}(\omega)=\mathscr{U}_N(\omega)\}$
$F_3=\{\omega|$there is a last U interval in $\mathscr{U}(\omega)$, and except for that last one, the remaining intervals in $\mathscr{U}(\omega)$ are all U_N intervals$\}$
$F_4=\{\omega|\mathscr{U}^\tau(\omega)={}_M\mathscr{U}^\tau(\omega)\}$
$F_5=\{\omega|$there is at most one U_H interval in $\mathscr{U}(\omega)$, and if there is one such, it is the last U interval$\}$
$F_6=\{\omega|$there is at least one U_H interval in $\mathscr{U}(\omega)$, and such a U_H interval is not the last U interval$\}$

Proof. Obviously first, for $r\geqslant 0$, $k\in\bar{E}$,

$$\{x(\eta_r-0)=k\}\in\mathscr{F}^0_\infty \tag{14}$$

Secondly we proceed to prove

$$\{x(\lambda_r)=k\}\in\mathscr{F}^0_r \tag{15}$$

Actually, when $k\neq\infty$, by (2.1),

$$\{x(\lambda_r)=k\}=\{x(\lambda_r)=k,\lambda_r<r\}=\bigcup_{i=1}^{\infty}\bigcap_{n=i}^{\infty}\bigcup_{m=0}^{2^i-2}\left\{x\left(\frac{m+1}{2^n}r\right)=k,\left[\frac{m+1}{2^n}r,r\right)\cap T(\omega)\right.$$

$$\left.\text{is a finite set},\left[\frac{m-1}{2^n}r,\frac{m+1}{2^n}r\right)\cap T(\omega)\text{ is an infinite set}\right\}\in\mathscr{F}^0_r$$

$$\{x(\lambda_r)=\infty\}=(r<\sigma)-\bigcup_{k\in E}\{(\lambda_r)=k\}\in\mathscr{F}^0_r$$

It follows that

$$F_1=\bigcap_{r\in R}\{r<\sigma,\eta_r=\sigma,x(\eta_r-0)\in N\}\in\mathscr{F}^0_\infty$$

$$F_2=\bigcap_{r\in R}\{(\sigma\leqslant r)\cup(r<\sigma,x(\eta_r-0)\in N)\}\in\mathscr{F}^0_\infty$$

$$F_3=\bigcap_{r\in R}\{(\sigma\leqslant r)\cup(r<\sigma,\eta_r=\sigma)\cup(r<\sigma,\eta_r<\sigma,\ x(\eta_r-0)\in N)\}\in\mathscr{F}^0_\infty$$

$$F_4=\bigcap_{r\in R}\{(\sigma\leqslant r)\cup(r<\tau)\cup(\tau\leqslant r<\sigma,x(\lambda_r)\in M)\}\in\mathscr{F}^0_\infty$$

$$F_5=\Omega-F_6\in\mathscr{F}^0_\infty$$

$$F_6=\bigcup_{r\in R}\{\eta_r<\sigma,x(\eta_r-0)\in H\}\in\mathscr{F}^0_\infty$$

The proof is over. QED

10.4 LEAPING INTERVAL AND KOLMOGOROV EQUATIONS

Suppose $M\subset\bar{E}$, $N\subset\bar{E}$, $S\subset E$. Let

$$\xi_{MS}=\inf\{t\,|\,\tau\leqslant t<\sigma,x(\lambda_t)\in M,x(t)\in S\}\tag{1}$$

$$\rho_{SN}=\sup\{t\,|\,0\leqslant t<\sigma,x(t)\in S,t(\eta_t-0)\in N\}\tag{2}$$

$$\delta_S=\inf\{t\,|\,\tau\leqslant t<\sigma,t\in\Gamma,x(t)\in S\}\tag{3}$$

Here τ is the first leaping point of X and Γ the set of leaping points. For empty set $\varnothing$, set $\inf\varnothing=\sigma$ and $\sup\varnothing=0$.

Lemma 1. Suppose S is a finite set, then the 'inf' in (1) and (3) can be replaced by 'min'.

Proof. Suppose $t_n\downarrow\xi_{MS}$, $\tau\leqslant t_n<\sigma$, $x(\lambda_{t_n})\in M,x(t_n)\in S$. Since S is finite, by the right-continuity property of X,

$$x(\xi_{MS})=\lim_{n\to 1}x(t_n)\in S$$

Therefore ξ_{MS} belongs to some i ($\in S$) interval $[a,b)$. When n is large enough, $t_n\in[a,b)$. Thus $x(\xi_{MS})=x(t_n)$, $\lambda_{\xi_{MS}}=\lambda_{t_n}$. Hence $x(\lambda_{\xi_{MS}})=x(\lambda_{t_n})\in M$. So we have proved that 'inf' in (1) can be replaced by 'min'. The remaining proof is similar, and the proof is terminated. QED

Lemma 2.

$$\xi_{MS}=\inf\{\xi_{Mj}|\,j\in S\}=\inf\{\xi_{kS}|k\in M\}=\inf\{\xi_{kj}|k\in M,\ j\in S\}\tag{4}$$

$$\rho_{SN}=\sup\{\rho_{Sj}|\,j\in N\}=\sup\{\rho_{kN}|k\in S\}=\sup\{\rho_{kj}|k\in S,\ j\in N\}\tag{5}$$

$$\delta_S=\inf\{\delta_k|k\in S\}\tag{6}$$

Proof. Let

$$A_{MS}=\{t|\tau\leqslant t<\sigma,x(\lambda_t)\in M,x(t)\in S\}\tag{7}$$

Obviously, $A_{Mj} \subset A_{MS}$ $(j \in S)$ so that $\xi_{Mj} \geqslant \xi_{MS}$ $(j \in S)$. On the other hand, for any $\varepsilon > 0$, there exists $t \in A_{MS} = \bigcup_{j \in S} A_{Mj}$ such that $t < \xi_{MS} + \varepsilon$. Thus there exists $j \in S$ such that $t \in A_{Mj}$. Hence $\xi_{Mj} \leqslant t < \xi_{MS} + \varepsilon$. So we have proved that $\xi_{MS} = \inf\{\xi_{Mj} | j \in S\}$. The remainder can be proved similarly. The proof is concluded. QED

Lemma 3. ξ_{MS}, δ_S are Markov times of X, and ρ_{SN} is a random variable.

Proof. By Lemma 2, it suffices to prove, for $k \in \bar{E}$, $j \in E$, that ξ_{kj}, δ_j are Markov times and ρ_{kj} is a random variable.

Actually, noting (3.15), for any $u \geqslant 0$

$$(\xi_{kj} < u < \sigma) = \bigcup_{\substack{r < u \\ r \in R}} \{\tau \leqslant r < \sigma, x(\lambda_r) = k, x(r) = j, u < \sigma\} \in \mathscr{F}_u^0$$

Secondly,

$$(\delta_j < u < \delta) = \bigcup_{\substack{r < u \\ r \in R}} \left\{ (\tau \leqslant r < \sigma, x(r) = j) \cap \left[\bigcap_{n=1}^{\infty} \bigcup_{m=n}^{\infty} \bigcup_{v=1}^{2^n - 1} \right. \right.$$

$$\left(X \text{ has infinitely many jumping points in } \left[\frac{v-1}{2^m} r, \frac{v}{2^m} r \right) \right.$$

$$\left. \left. \left. \text{and takes a constant value } j \text{ in } \left[\frac{v}{2^m} r, r \right) \right) \right) \right] \right\} \in \mathscr{F}_u^0$$

and

$$(\rho_{kj} > u) = \bigcup_{\substack{r \in R \\ r > u}} \{0 \leqslant \sigma < 0, x(r) = k, x(\eta_r - 0) = j\} \in \mathscr{F}_\infty^0$$

The proof is completed. QED

Definition 1. Assume $X \in \mathscr{H}_s$. We call X pure entrance from M, if $P_i\{\mathscr{U}^\tau = {}_M\mathscr{U}^\tau\} = 1$ $(i \in E)$; we call X pure exit to N, if $P_i\{\mathscr{U} = \mathscr{U}_N\} = 1$ $(i \in E)$; we call X quasi-exit to N, if $P_i(\Omega_0) = 1$ $(i \in E)$. Here

$$\Omega_0 = (F_2 - F_1) \cup F_3$$
$$= \{\omega | \text{the last } U \text{ interval does not exist in } \mathscr{U}(\omega) \text{ and, moreover, } \mathscr{U}(\omega) = \mathscr{U}_N(\omega)\}$$
$\cup \{\omega |$ there exists the last U interval in $\mathscr{U}(\omega)$; except for the last U interval, the others are all U_N intervals$\}$

F_1, F_2 and F_3 are determined by Theorem 3.5.

Theorem 4. Suppose $X \in \mathscr{H}_s$, then the following conclusions are equivalent to each other:

(i) The process X is pure entrance from ∞, that is, $P_i\{\mathscr{U}^\tau = {}_\infty\mathscr{U}^\tau\} = 1$ $(i \in E)$.

(ii) $P_i\{\xi_{EE} < 0\} = 0$ $(i \in E)$, ξ_{EE} is determined by (4.1).

(iii) The process X satisfies the system of forward equations.

(iv) For any $t > 0$, $P_i\{\tau < t < \sigma, x(\lambda_t) \in E\} = 0$ $(i \in E)$. If one of the above conclusions holds, then

(v) $P_i\{\xi_{\infty E} = \tau\} = 1$ $(i \in E)$. More precisely, for any $i \in E$, P_i almost all $\omega \in (\tau < \sigma)$, we have $\tau(\omega) \in \bar{\Gamma}_{\infty}^{r+}(\omega)$.

Proof. Clearly, (i)⇒(iv). Conversely by (iv) we deduce $P_i\{x(\lambda_r) \in E$ for all $r \in R \cap [\tau, \sigma)\} = 0$ $(i \in E)$. From this and noting (2.6) we obtain $P_i\{{}_E\mathscr{U}^\tau = \mathscr{U}^\tau\} = 0$ $(i \in E)$. This is just the condition (i).

(i)⇒(iii). Suppose $t > 0$. Since $P_i\{x(t) = \infty\} = 0$, thus for P_i almost all $\omega \in (t < \sigma)$, we have $t \in S_E(\omega)$. By Theorem 3.3, $t \in C_2(\omega)$. Hence there exists $[\lambda, \eta) \in \mathscr{U}(\omega)$ such that $t \in [\lambda, \eta)$. If $[\lambda, \eta) = [0, \tau(\omega))$, then evidently there is the last discontinuity point $\mu_t(\omega) \in T(\omega)$, before t. Or else, by conclusion (i), $[\lambda, \eta)$ is a ${}_\infty U$ interval, $x(\lambda) = \infty$. Thus $t \in (\lambda, \eta)$ so that $\lambda < \mu_t(\omega) < \eta$, $\mu_t(\omega) \in T(\omega)$. By Theorem 7.8.3, X satisfies the system of forward equations.

(iii)⇒(ii). By Theorem 7.8.3, if conclusion (iii) holds, then $p_k(\Omega_1) = 1$ $(k \in E)$, where

$$\Omega_1 = \{\omega \mid \text{for all } r \in R \cap [0, \sigma(\omega)),\ x(r, \omega) \in E, \mu_r(\omega) \in T(\omega)\}$$

If $\omega \in (\xi_{ij} < \sigma)$ $(i, j \in E)$ then $x(\lambda_{\xi_{ij}}, \omega) = i$. So that there exists $r \in R \cap [0, \sigma(\omega))$ such that $\lambda_{\xi_{ij}}(\omega)$ and r are in the same i-interval; thus $\mu_r(\omega) = \lambda_{\xi_{ij}(\omega)}(\omega) \notin T(\omega)$. Therefore $\omega \notin \Omega_1$, namely, $(\xi_{ij} < \sigma) \subset \Omega - \Omega_1$. Thus by Lemma 2,

$$(\xi_{EE} < \sigma) = \bigcup_{i, j \in E} (\xi_{ij} < \sigma) \subset \Omega - \Omega_1$$

Conclusion (ii) is proved.

(ii)⇒(i). If $\omega(\mathscr{U}^\tau \neq {}_\infty\mathscr{U}^\tau)$, then there exists ${}_E U$ interval $[\lambda, \eta) \in \mathscr{U}^\tau(\omega)$. Hence there exists some $i \in E$ such that $\xi_{Ei}(\omega) < \sigma(\omega)$. Consequently,

$$(\mathscr{U}^\tau \neq {}_\infty\mathscr{U}^\tau) \subset \bigcup_{i \in E} (\xi_{Ei} < \sigma) = (\xi_{EE} < \sigma)$$

Conclusion (i) is proved.

(i)⇒(v). By conclusion (i), for P_i almost all ω, $\Gamma_E^r(\omega) - \{0\} = \phi$. Moreover, by Theorem 3.2, if $\tau(\omega) < \sigma(\omega)$, then either $\tau(\omega)$ is the left end-point of some U interval, and then $x(\tau(\omega), \omega) = \infty$, $\tau(\omega) \in \Gamma_\infty^r(\omega)$; or there exists strictly decreasing $\lambda_n \downarrow \tau(\omega)$, λ_n being the left end-point of some U interval. By conclusion (i), $x(\lambda_n, \omega) = \infty$ so that $\tau(\omega) \in \bar{\Gamma}_\infty^{r+}(\omega)$, and the proof is concluded. QED

Theorem 5. Suppose $X \in \mathscr{H}_s$, then the following conclusions are equivalent to each other:

(i) The process X satisfies the system of backward equations.

(ii) The process is quasi-exit to ∞.

(iii) $P_i\{\omega|\mathscr{U}(\omega)$ there is at most one U_H interval in $\mathscr{U}(\omega)$; if there is one, it is the last U interval$\} = 1$ $(i \in E)$.

(iv) $P_i\{\tau_1 \in \Gamma(\omega), \tau_1 < \sigma\} = 0$ $(i \in H)$, where τ_1 is the first jumping point, and H is the non-conservative state set.

(v) For any $t \geqslant 0$, $P_i\{t < \sigma, x(\eta_t - 0) \in H, \eta_t < \sigma\} = 0$ $(i \in E)$

Proof. (ii) and (iii) have the same meaning.

(i)$\Rightarrow$(iii). By Theorem 7.8.3 $P_i(\Omega_2) = 1$, where $\Omega_2 = \{\omega|$ for all $r \in R \cap [0, \sigma(\omega))$, $v_r(\omega) \in T(\omega)\}$. By Theorem 7.7.2, $P_i(\Omega_3) = 1$, where $\Omega_3 = \{\omega|$ for all $r \in R \cap [0, \sigma(\omega))$, if $x(\eta_r - 0, \omega) \in H$ and $\eta_r(\omega) < \sigma(\omega)$, then $x(\eta_r, \omega) = \infty\}$. We are going to prove $\Omega_2 \cap \Omega_3 \subset F_5$. Actually, suppose $\omega \in \Omega_2 \cap \Omega_3$. If $X(\omega)$ has a U_H interval $[\lambda, \eta)$, namely, $x(\eta - 0, \omega) \in H$, then there exists $r \in R \cap [0, \sigma(\omega))$ such that $\eta(\omega) = v_r(\omega) = \eta_r(\omega)$. If $\eta(\omega) < \sigma(\omega)$, since $\omega \in \Omega_2$ it follows that $\eta(\omega) = v_r(\omega) \in T(\omega)$. Because $\omega \in \Omega_3$ we have $\eta = \eta_r(\omega) \in \Gamma(\omega)$. This is a contradiction. Therefore surely $\eta(\omega) = \sigma(\omega)$, that is $[\lambda, \eta)$ is the last U interal. So we have proved $\Omega_2 \cap \Omega_3 \subset F_5$ and, hence, we have derived conclusion (iii).

(iii)$\Rightarrow$(iv) is obvious. By conclusion (iv) and the strong Markov property we obtain conclusion (v) easily.

(v)$\Rightarrow$(i). By conclusion (v) we have $P_i\{\Omega_4\} = 1$ $(i \in E)$, where $\Omega_4 = \{\omega|$ for any $r \in R \cap [0, \sigma(\omega))$, if $x(\eta_r - 0, \omega) \in H$, then surely $\eta_r(\omega) = \sigma(\omega)\}$.

Fixing $t \geqslant 0$, because $P_i\{x(t) = \infty\} = 0$, by Theorem 3.3, for P_i almost all $\omega \in (t < \sigma) \cap \Omega_4$, we have $t < S_E(\omega) \subset C_2(\omega)$. Thus there exists $[\lambda, \eta) \in \mathscr{U}(\omega)$ such that $t \in [\lambda, \eta)$. If $x(\eta - 0, \omega) = \infty$, then $\lambda \leqslant t < v_t(\omega) < \eta$, so that $v_t(\omega) \in T(\omega)$. Otherwise $x(\eta - 0, \omega) \in H$. If $v_t(\omega) < \eta$, obviously $v_t(\omega) \in T(\omega)$. If $v_i(\omega) = \eta$, then there exists $r \in R \cap [0, \sigma(\omega))$ such that $\eta = v_t(\omega) = v_r(\omega) = \eta_r(\omega)$. Since $\omega \in \Omega_4$, it follows that $\eta = \sigma(\omega)$. Thus $v_t(\omega) = \sigma(\omega) \in T(\omega)$. Therefore there exists constantly $v_t(\omega) \in T(\omega)$. By Theorem 7.8.3, X satisfies the system of backward equations, and the proof is terminated. QED

By Theorems 4 and 5 we obtain the following two theorems immediately.

Theorem 6. Suppose $X \in \mathscr{H}_s$, then the following conclusions are equivalent to each other:

(i) X satisfies the systems of backward and forward equations simultaneously.

(ii) X makes its pure entrance from ∞ and quasi-exist to ∞.

(iii) $P_i\{\xi_{EE} < \sigma\} = 0$ $(i \in E)$, $P_i\{\tau_1 \in \Gamma, \tau_1 < \sigma\} = 0$ $(i \in H)$.

(iv) $P_i\{\mathscr{U}^\tau = {}_\infty\mathscr{U}^\tau\} = 1$, $P_i\{$there is at most one U_H interval in $\mathscr{U}$; if there is one, it is the last U interval$\} = 1$ $(i \in E)$.

(v) $P_i\{\tau < t < \sigma, x(\lambda_t) \in E\} = 0$ $(t > 0, i \in E)$
$P_i\{t < \sigma, x(\eta_t - 0) \in H, \eta_t < \sigma\} = 0$ $(t \geqslant 0, i \in E)$

Theorem 7. Suppose $X \in \mathscr{H}_s$, then the following conclusions are equivalent to each other:

(i) X satisfies neither backward nor forward equations.

(ii) There exist $i \in E$, $k \in H$ such that $P_i\{\xi_{EE} < \sigma\} > 0$, $P_k\{\tau_1 \in \Gamma, \tau_1 < \sigma\} > 0$.

(iii) There exist $i, k \in E$ such that $P_i\{\mathscr{U}^\tau = {}_E\mathscr{U}^\tau\} > 0$ and P_k {there is a U_H interval in $\mathscr{U}$, but it is not the last U interval} > 0.

(iv) There exist $t_1 > 0$, $t_2 \geqslant 0$, $i, k \in E$ such that

$$P_i\{\tau < t_1 < \sigma, x(\lambda_{t_1}) \in E\} > 0 \qquad P_k\{t_2 < \sigma, X(\eta_{t_2} - 0) \in H, \eta_{t_2} < \sigma\} > 0.$$

Theorem 8. Suppose $X \in \mathscr{H}_s$, $M \subset \bar{E}$. Then the following conclusions are equivalent to each other:

(i) The process X makes its pure entrance from M, that is

$$P_i\{\mathscr{U}^\tau = {}_M\mathscr{U}^\tau\} = 1\,(i \in E)$$

(ii) $P_i\{\xi_{\bar{M}E} < \sigma\} = 0\,(i \in E, \bar{M} = \bar{E} - M)$

(iii) For any $t > 0$, $P_i\{\tau < t < \sigma, x(\lambda_t) \in \bar{M}\} = 0$ $(i \in E)$. Or equivalently, for any $i \in E$, $t > 0$, for P_i almost all $\omega \in (\tau < t < \sigma)$, we have $x[\lambda_t(\omega), \omega] \in M$. If one of the above conclusions holds, then

(iv) $P_i\{\xi_{ME} = \tau\} = 1$ $(i \in E)$. More precisely, for any $i \in E$, for P_i almost all $\omega \in (\tau < \sigma)$, $\tau(\omega) \in \bar{\Gamma}^{r+}_M(\omega)$.

Proof. (i)$\Rightarrow$(iii). This will be done by following the proof in Theorem 5 that (i) and (iv) are equivalent.

(i)$\Rightarrow$(ii). Suppose $\omega \in (\xi_{\bar{M}E} < \sigma) = \bigcup_{k \in E}(\xi_{\bar{M}k} < \sigma)$. Then there exists some $k \in E$ such that $\xi_{\bar{M}E}(\omega) < \sigma(\omega)$. So $[\lambda_{\xi_{\bar{M}k}}(\omega), \eta_{\xi_{\bar{M}k}}\omega)) \in \mathscr{U}^\tau(\omega)$, and by $x[\lambda_{\xi_{\bar{M}i}}(\omega), \omega] \in \bar{M}$ we know it is a ${}_{\bar{M}}U$ interval so that $\omega \in (\mathscr{U}^\tau \neq {}_M\mathscr{U}^\tau)$.

(ii)$\Rightarrow$(i). By the homogeneity of X and the conclusion (ii) we have for all $r \in R$, $P_i\{\theta_r(\xi_{\bar{M}E} < \sigma)\} = 0$, θ being the translation operator. Let $\Omega_0 = \bigcap_{r \in R}\{\Omega - \theta_r(\xi_{\bar{M}E} < \sigma)\}$, then $P_i(\Omega_0) = 1$. When $\omega \in \Omega_0$, for all $r \in [0, \sigma(\omega))$, there is no ${}_{\bar{M}}U$ interval in $[\max(r, \tau(\omega)), \sigma(\omega))$, so that all the U intervals in $[\tau(\omega), \sigma(\omega))$ are ${}_MU$ intervals, namely, $\Omega_0 \subset (\mathscr{U}^\tau = {}_M\mathscr{U}^\tau)$.

Suppose one of the conclusions (i)–(iii) holds, then for P_i almost all ω, $\Gamma^r_{\bar{M}}(\omega) - \{0\}$ is an empty set, so that $\tau(\omega) \notin \bar{\Gamma}^{r+}_{\bar{M}}(\omega)$. By Theorem 3.2, $\tau(\omega) \in \Gamma(\omega) = \bar{\Gamma}^+(\omega) = \bar{\Gamma}^{r+}_M(\omega)$, and the proof is completed. QED

10.5 THE ${}_Mg_n$ TRANSFORMATION AND ITS STRONG LIMIT THEOREM

Suppose $X = \{x(t), t < \sigma\} \in \mathscr{H}_s$, $M \subset \bar{E}$, and finite sets $D_n \uparrow E$. Let

$$\begin{aligned} &{}_M\beta_0^n = 0 \\ &{}_M\tau_1^n \text{ be the first leaping point of } X \\ &{}_M\beta_k^n = \min\{t \mid {}_M\tau_k^n \leqslant t < \sigma, X(\lambda_t)\in M, X(t)\in D_n\} \\ &{}_M\tau_{k+1}^n \text{ be the first leaping point after } {}_M\beta_k^n \end{aligned} \tag{1}$$

λ_t is the last leaping point before t. By Lemma 4.1, the 'min' in (1) exists, so long as the set after it is not empty, or else let it take value σ.

By Lemma 4.3, ${}_M\tau_k^n$, ${}_M\beta_k^n$ are Markov times of X, and we can easily see that for every n, $\{{}_M\tau^n, {}_M\beta^n\}$ is the canonical pair of X, and satisfies the conditions (a), (c) in Theorem 9.3.3. Thus by theorem 9.3.3, ${}_MX^n = W_{{}_M\tau^n, {}_M\beta^n}(X)\in\mathscr{H}_D$. Actually, ${}_MX^n\in\mathscr{H}_1$, that is, ${}_MX^n$ is a first-order process.

Definition 1. The transformation $W_{{}_M\tau^n, {}_M\beta^n}$ obtained by transforming $X\in\mathscr{H}_s$ into ${}_MX^n\in\mathscr{H}_1$ in the same way as stated above is called the ${}_Mg_n$ transformation. When $M = \bar{E}$, we simply write ${}_{\bar{E}}g_n$ as g_n.

Thus,

$$ {}_MX^n = {}_Mg_n(X) \tag{2}$$

Obviously, when $n\uparrow\infty$,

$$ {}_MA^n = \bigcup_{k=0}^{\infty} [{}_M\beta_k^n, {}_M\tau_{k+1}^n)\uparrow {}_MA $$

$$ {}_MA_1^n = \bigcup_{k=0}^{\infty} ({}_M\beta_k^n, {}_M\tau_{k+1}^n)\uparrow {}_MA_1 \tag{3}$$

Theorem 1. For almost all $\omega\in\Omega$ we have

$$ \mathscr{U}[{}_MA_1^\tau(\omega)] = {}_M\mathscr{U}^\tau(\omega) \tag{4}$$

$$ C_2[{}_MA^\tau(\omega)] = C_2[{}_MA_1^\tau(\omega)] = {}_MC_2^\tau(\omega) \tag{5}$$

where ${}_MA_1^\tau = {}_MA_1 - (0,\tau)$, ${}_MA^\tau = {}_MA - [0,\tau)$, τ is the first leaping point. Here $\mathscr{U}(A)$, $C_2(A)$ are understood according to Definition 9.3.1 and ${}_M\mathscr{U}(\omega)$, ${}_MC_2^\tau(\omega)$ are understood according to Definition 2.3 and (2.8).

Proof. Note first that according to the definition, for any $t\in {}_MB^n = \bigcup_{k=1}^{\infty}[{}_M\tau_k^n, {}_M\beta_k^n)$, there must exist

$$ x(t)\bar{\in} D_n \qquad \textit{or} \qquad x(t)\in D_n, x(\lambda_t)\bar{\in} M \tag{6}$$

while for $\tau\leqslant t\in {}_MA^n$, there surely exists

$$ x(\lambda_t)\in M \tag{7}$$

We proceed to prove (4) so that (5) follows from (4).

(a) Suppose $[\lambda,\eta)\in\mathscr{U}[{}_MA_1^\tau(\omega)]$. Since $(\lambda,\eta)\subset{}_MA_1^\tau(\omega)$, and ${}_MA_1^\tau(\omega)\cap\Gamma(\omega)=\varnothing$, it follows that $(\lambda,\eta)\cap\Gamma(\omega)=\varnothing$. Therefore take arbitrarily $t\in(\lambda,\eta)$, and we have $(\lambda_t,\eta_t)\supset(\lambda,\eta)$. If $\lambda_t<\lambda$, then we can take arbitrarily $s\in(\lambda_t,\lambda)$. Since $x(s,\omega)\in E$, there exists n such that $x(s,\omega)\in D_n$. But $\lambda_s=\lambda_t$, since $t\in(\lambda,\eta)\subset{}_MA_1^\tau(\omega)\subset{}_MA^\tau(\omega)$. So by (6) and (7), $x(\lambda_s,\omega)=x(\lambda_t,\omega)\in M$, and moreover $\eta_s=\eta_t\geqslant\eta$. On account of (6), $s\bar{\in}{}_MB^n(\omega)$, $s\in{}_MA^n(\omega)$. Namely, there exists $l\geqslant 1$ such that $s\in[{}_M\beta_l^n(\omega),{}_M\tau_{l+1}^n(\omega))$. Because $\eta_s={}_M\tau_{l+1}^n(\omega)\geqslant\eta$, $(s,\eta)\subset({}_M\beta_l^n(\omega),{}_M\tau_{l+1}^n(\omega))\subset{}_MA_1^\tau(\omega)$. This contradicts the fact that $[\lambda,\eta)$ is a composite interval of ${}_MA_1^\tau(\omega)$. Thus $\lambda_t=\lambda$. That $\eta_t=\eta$ can be proved similarly. Hence $\lambda,\eta\in\Gamma(\omega)$.

Summing up what precedes we get $[\lambda,\eta)\in{}_M\mathscr{U}^\tau(\omega)$.

(b) Suppose $[\lambda,\eta)\in{}_M\mathscr{U}^\tau(\omega)$, that is, $\lambda\geqslant\tau(\omega)$, $\lambda,\eta\in\Gamma(\omega)$, $(\lambda,\eta)\cap\Gamma(\omega)=\varnothing$, $x(\lambda,\omega)\in M$.

Take arbitrarily $s\in(\lambda,\eta)$, then $\lambda=\lambda_s$, $\eta=\eta_s$, and $x(s,\omega)\in E$. Thus there exists n such that $x(s,\omega)\in D_n$, and $x(\lambda_s,\omega)=x(\lambda,\omega)\in M$. By (6) and (7) we obtain $s\in{}_MA_n$, so that there exists $k\geqslant 1$ such that $s\in[{}_M\beta_k^n(\omega),{}_M\tau_{k+1}^n(\omega))$, thus $\eta=\eta_s={}_M\tau_{k+1}^n$. Hence $(s,\eta)\subset({}_M\beta_k^n(\omega),{}_M\tau_{k+1}^n(\omega))\subset{}_MA_1^\tau(\omega)$. By the arbitrariness of s, $(\lambda,\eta)\subset{}_MA_1^\tau(\omega)$.

Next, if $[\lambda',\eta')\supset[\lambda,\eta)$, and $(\lambda',\eta')\subset{}_MA_1^\tau(\omega)$, then we can take arbitrarily $t\in(\lambda,\eta)$. Since λ, $\eta\in\Gamma(\omega)$ it follows that $\lambda=\lambda_t$, $\eta=\eta_t$. On the other hand, since $(\lambda',\eta')\subset{}_MA_1^\tau(\omega)$ and ${}_MA_1^\tau(\omega)\cap\Gamma(\omega)=\varnothing$, so $\lambda_t\leqslant\lambda'$, $\eta'\leqslant\eta_t$. Thus $\lambda'=\lambda$, $\eta'=\eta$. Namely, $[\lambda',\eta')=[\lambda,\eta)$. Thus we have proved the maximal property of $[\lambda,\eta)$.

Summarizing what precedes, we obtain $[\lambda,\eta)\in\mathscr{U}[{}_MA_1^\tau(\omega)]$, and the proof is completed. QED

Remark

$\mathscr{U}[{}_MA^\tau(\omega)]={}_M\mathscr{U}^\tau(\omega)$ is not necessarily true. For instance, if $X\in\mathscr{H}_1$, then ${}_EA(\omega)=[0,\sigma(\omega))$, so $\mathscr{U}[{}_EA^\tau(\omega)]=[\tau(\omega),\sigma(\omega))$. But ${}_E\mathscr{U}^\tau(\omega)=\{[\lambda_n,\lambda_{n+1})|n=1,2,\ldots\}$, λ_n is the nth leaping point of $X(\omega)$.

On account of (3), according to Theorems 9.3.1 and 9.3.4,

$$ {}_MX^n={}_Mg_n(X)\in\mathscr{H}_1 \tag{8} $$

$$ {}_Mg_m({}_MX^n)={}_MX^m \qquad m<n \tag{9} $$

Furthermore the strong limit of ${}_MX^n$ exists and it is just ${}_MX\in\mathscr{H}_s$, namely,

$$ \lim_{n\to\infty}{}_MX^n={}_MX\in\mathscr{H}_s \tag{10} $$

where

$$ {}_MX=\{x({}_M\gamma^{-1}(t)),t<{}_M\sigma\} \qquad {}_M\sigma=L\{{}_MA\} \tag{11} $$

Here ${}_M\gamma$ is the transformation determined by the sequence of pairs $\{{}_M\tau^n, {}_M\beta^n\}$ $(n \geqslant 1)$, that is,

$$ {}_M\gamma(u) = L\{{}_MA \cap [0, u)\} \qquad u \in [0, \sigma) \tag{12} $$

By Theorem 1 and the conclusion (ii) of Lemma 9.4.4,

$$ {}_M\gamma(u) = L\{C_2[{}_MA(\omega)] \cap [0, u)\} = L\{{}_MC_2(\omega) \cap [0, u)\} \tag{13} $$

$$ {}_M\sigma = L\{C_2[{}_MA(\omega)]\} = L\{{}_MC_2(\omega)\} \tag{14} $$

By Lemma 9.4.4, the inverse transformation ${}_M\gamma^{-1}$ maps $[0, {}_M\sigma)$ onto $\bar{C}_2^+({}_MA) = {}_M\bar{C}_2^+(\omega)$. Thus we obtain the preceding part of the following theorem.

Theorem 2. Suppose $X \in \mathscr{H}_s$. Then (8)–(10) hold, and ${}_MX$ is the process making its pure entrance from M. If X satisfies the system of backward equations, then X^n, ${}_MX$ all satisfy the system of backward equations, and have the same Q matrix as X.

Remark

The intuitive meaning of ${}_MX$ is as follows. Reserve all the parts of Xwhich correspond to $[0, \tau)$ and ${}_MU$ interval, abandon all the remaining parts,and translate the reserved parts towards the left in the original order, linking them up in a disjoint manner and, thus we obtain the process ${}_MX \in \mathscr{H}_s$.

Proof. Because among the U intervals of ${}_MX$, except for the first U interval $[0, \tau(\omega))$, the remainder are all ${}_MU$ intervals, ${}_MX$ makes its pure entrance from M. Quoting Theorem 4.5(iii) we derive the conclusion about equations in this theorem. As X, X^n, ${}_MX$ have the same first leaping point $\tau(\omega)$, their Q matrices are the same, and the proof is concluded. QED

Corollary 1

Assume $x \in \mathscr{H}_s$, and let $P\{{}_M\mathscr{U}^\tau = \mathscr{U}^\tau\} = 1$. Then

$$ \lim_{n \to \infty} {}_MX^n = X \tag{15} $$

Proof. This is quite obvious, for under the assumption of Corollary 1, ${}_MX = X$. QED

Especially, when $M = \bar{E}$, the assumption in Corollary 1 is always satisfied. Thus we have

Corollary 2

Suppose $X \in \mathscr{H}_s$. Then

$$\lim_{n \to \infty} X^n = X \tag{16}$$

where

$$X^n = g_n(X) \in \mathscr{H}_1 \tag{17}$$

$$X^m = g_m(X^n) \qquad m < n \tag{18}$$

If X satisfies the system of backward equations, so does X^n; furthermore X and X^n have the same Q matrix.

Corollary 2 is just Theorem 9.7.4, that is the basic result, in Zheng-ting Hou (1974), namely the first construction theorem.

10.6 ENTRANCE DECOMPOSITION OF A PROCESS

Theorem 1. Suppose $X \in \mathscr{H}_s$, finite set $M \subset E$. If

$$P\{x(\tau) \in M \mid \tau < \sigma\} = 1 \tag{1}$$

where τ is the first leaping point, then $X \in \mathscr{H}_1$, that is, X is a one-order process. Especially when X makes its pure entrance from M, (1) holds, thus X is a one-order process.

Proof. When X makes its pure entrance from M, all U intervals of X (except the first one) are ${}^{+}_{M}U$ intervals. Suppose $\tau < \sigma$. By Theorem 3.2, $\tau \in \Gamma \subset \bar{\Gamma}^{r+}_{M}$. If $\tau \in \Gamma^r_M$, then $x(\tau) \in M$, or else there exist strictly decreasing $\lambda_n \in \Gamma^r_M$ such that $\lambda_n \downarrow \tau$. But $x(\lambda_n) \in M$; by the right-continuity of X and the finiteness of M, surely $x(\tau) = \lim_{n \to \infty} x(\lambda_n) \in M$. Therefore it is always true that $x(\tau) \in M$. Thus (1) is proved.

By (1) and the strong Markov property of X we can obtain the first, second, third,... leaping points $\tau^1, \tau^2, \tau^3, \ldots$, and moreover if $\tau^n < \sigma$, then $x(\tau^n) \in M$. If for $t < \sigma(\omega)$, $X(\omega)$ has infinitely many leaping points in $[0, t]$, then in $[0, t]$, $X(\omega)$ takes values in M infinitely many times. Because M is finite, there exists $i \in M$ such that $X(\omega)$ takes the value i inifnitely many times in $[0, t]$. Thus $X(\omega)$ has infinitely many i-intervals in $[0, t]$. This contradicts Theorem 7.7.1. Consequently X has only finitely many leaping points in $[0, t]$ $(t < \sigma)$. Therefore X is a first-order process, and the proof is completed. QED

Theorem 2. Assume $X \in \mathscr{H}_s$. For any $i \in \bar{E}$, let

$${}_iX = \{x({}_i\gamma^{-1}(t)), t < {}_i\sigma\} \tag{2}$$

where ${}_iC_2$ is determined by (2.8), and

$$ {}_i\sigma = L\{{}_iC_2\}, {}_i\gamma(u) = L\{{}_iC_2 \cap [0, u)\} \tag{3}$$

${}_i\gamma^{-1}$ is the inverse transformation of ${}_i\gamma$ and it maps $[0, {}_i\sigma)$ onto ${}_i\bar{C}_2^+$. Then:

(i) For $i \in E$, ${}_iX \in \mathscr{H}_1$ is the process making its pure entrance from i. If X satisfies the system of backward equations, then so does ${}_iX$; and what is more, X and ${}_iX$ have the same Q matrix.

(ii) ${}_\infty X \in \mathscr{H}_s$ is the process making its pure entrance from ∞, satisfying the system of forward equations. ${}_\infty X$ and X have the same Q matrix.

Proof. Letting $M = \{i\}$ $(i \in \bar{E})$ in Theorem 5.2, noting Theorem 1 and applying Theorems 4.4 and 4.5, we can derive the theorem. QED

Remark

For the intuitive meaning of Theorem 2 see the remark in Theorem 5.2.

10.7 THE ${}_Mf_n$ TRANSFORMATION AND ITS STRONG LIMIT THEOREM

Suppose $X \in \mathscr{H}_s$, Γ is a set of leaping points, $M \subset E$ and finite set $D_n \uparrow E$. Let

$$\begin{aligned}
&{}_\mu\bar{\beta}_0^n = 0 \\
&{}_M\bar{\tau}_1^n \text{ be the first leaping point of } X \\
&{}_M\bar{\beta}_k^n = \min\{t \mid {}_M\bar{\tau}_k^n \leqslant t < \sigma, t \in \Gamma, x(t) \in D_n \cap M\} \\
&{}_M\bar{\tau}_{k+1}^n \text{ be the first leaping point after } {}_M\bar{\beta}_k^n
\end{aligned} \tag{1}$$

We can easily prove ${}_M\bar{\tau}_k^n$, ${}_M\bar{\beta}_k^n$ are Markov times of X, for every n, $\{{}_M\bar{\tau}^n, {}_M\bar{\beta}^n\}$ is the canonical pair of X, and the conditions (a) and (c) in Theorem 9.3.3 are satisfied. Thus according to Theorem 9.3.3, ${}_M\bar{X}^n = W_{{}_M\bar{\tau}^n, {}_M\bar{\beta}^n} \in \mathscr{H}_D$. Actually ${}_M\bar{X}^n \in \mathscr{H}_1$ and is a one-order process.

Definition 1. The transformation $W_{{}_M\bar{\tau}^n, {}_M\bar{\beta}^n}$ obtained by transforming $X \in \mathscr{H}_s$ to ${}_M\bar{X}^n$ according to the above method is called the ${}_Mf_n$ transformation. When $M = E$, we simply write ${}_Ef_n$ as f_n. Thus

$$ {}_M\bar{X}^n = {}_Mf(X) \tag{2}$$

Obviously, when

$$
\begin{aligned}
{}_M\bar{A}^n &= \bigcup_{k=0}^{\infty} [{}_M\bar{\beta}_k^n, {}_M\bar{\tau}_{k+1}^n) \uparrow {}_M\bar{A} \\
{}_M\bar{A}_1^n &= \bigcup_{k=0}^{\infty} ({}_M\bar{\beta}_k^n, {}_M\bar{\tau}_{k+1}^n) \uparrow {}_M\bar{A}_1
\end{aligned} \tag{3}
$$

Following Theorem 5.1, we can prove:

Theorem 1. For almost all $\omega \in \Omega$,

$$\mathscr{U}[{}_M\bar{A}_1^\tau(\omega)] = {}_M\mathscr{U}^\tau(\omega) \tag{4}$$

$$C_2[{}_M\bar{A}^\tau(\omega)] = C_2[{}_M\bar{A}_1^\tau(\omega)] = {}_MC_1^\tau(\omega) \tag{5}$$

where ${}_M\bar{A}_1^\tau = {}_M\bar{A}_1 - (0, \tau)$, ${}_M\bar{A}^\tau = {}_M\bar{A} - [0, \tau)$, all the $\mathscr{U}$-quality and the C-quality are still to be understood by Definitions 9.3.1 and 2.3 and (2.8).

As a result of (3), according to Theorems 9.3.1 and 9.3.4,

$${}_M\bar{X}^n = {}_Mf_n(X) \in \mathscr{H}_1 \tag{6}$$

$${}_Mf_m({}_M\bar{X}^n) = {}_M\bar{X}^n \qquad m < n \tag{7}$$

The strong limit process of ${}_M\bar{X}^n$ exists, and by (4) and (5), the strong limit process is just the process ${}_MX \in \mathscr{H}_s$ in Theorem 5.2. Applying Theorem 4.5(iii) we derive the following theorem.

Theorem 2. Suppose $X \in \mathscr{H}_s$, then (6) and (7) hold, and

$$\lim_{n \to \infty} {}_M\bar{X}^n = {}_MX \in \mathscr{H}_s \tag{8}$$

If X satisfies the system of backward equations, then so do ${}_M\bar{X}^n$ and ${}_MX$; moreover X, ${}_M\bar{X}^n$, ${}_MX$ have the same Q matrix.

Corollary 1

Suppose $X \in \mathscr{H}_s$. If $P\{{}_M\mathscr{U}^\tau = \mathscr{U}^\tau\} = 1$, then

$$\lim_{n \to \infty} {}_M\bar{X}^n = X \tag{9}$$

Corollary 2

Assume $X \in \mathscr{H}_s$. If $P\{{}_E\mathscr{U}^\tau = \mathscr{U}^\tau\} = 1$, that is it is impossible for the process X to make its entrance from infinity, then

$$\lim_{n \to \infty} \bar{X}^n = X \tag{10}$$

where $\bar{X}^n = f_n(X) \in \mathscr{H}_1$. If X satisfies the system of backward equations, then so does $\bar{X}$, and it has the same Q matrix as X.

Remark

When X is a birth–death process, Corollary 2 is precisely the case considered in Zi-kun Wang (1962, Theorem 5.3) when $S = \infty$.

CHAPTER 11

Extension of Processes

11.1 INTRODUCTION

The construction problem of processes is basically an extension one. If the minimal Q process is a stopping process, construction of all Q processes is equivalent to that of an extension process of the minimal process, such that the extension process has the same Q matrix as the minimal process. When Q is conservative, Doob (1945) introduces for the first time extension processes of the minimal Q process, i.e. the so-called Doob processes. Chung (1967) provides a strict proof for Doob's construction. For conservative Q, under some restrictions on processes, Xiang-qun Yang (1966a) and Kunita (1962) take into account extensions which are more general than Doob processes.

In this chapter we impose no restriction on Q processes or instantaneous-return processes of order k. For example, we do not require the conservativeness of Q or the existence of 'a centre' and so on. Furthermore, we even do not require the finiteness of the Q matrix in section 11.2.

We mainly consider non-sticky extension. Because the D-type extension may change Q matrices, we introduce the D*-type extension in order that Q matrices remain unchanged. Of course, we can consider extensions of other types. For instance, for some minimal processes we can introduce the V-type extension so that for such an extended process $X=\{x(t), t<\sigma\}$ has only a finite number of leaping intervals in $[0,t]$ for any $t<\sigma(\omega)$, and all leaping intervals except the first one are ${}_{\infty}U$ intervals. Also we can introduce extensions of mixing D-type and V-type, but in this chapter we shall only give a simple and elementary discussion of it.

11.2 D-TYPE EXTENSION

Suppose $\bar{P}(t)=\{\bar{p}_{ij}(t)\}$ $(i,j\in E, t\geqslant 0)$ satisfies conditions (2.2.A–C) and

$$\sum_j \bar{p}_{ij}(t)<1 \qquad \text{for some } i\in E \tag{1}$$

Suppose that we are given a distribution $\pi=\{\pi_i, i\in E\}$ satisfying

$$0<\sum_j \pi_j \leqslant 1 \tag{2}$$

Lemma 1. There exists a probability space $(\Omega, \mathscr{F}, P)$ on which may be defined a sequence of processes $X^n = \{x^n(t), t < \sigma^n\}$ $(n \geqslant 0)$ with the following properties:

(i) $X^n \in \mathscr{H}$ $(n \geqslant 0)$, and they have the same transition probability matrix $\bar{P}(t)$.

(ii) $\{\sigma^n = 0\} \cup \{\sigma^n = \infty\} \subset \{\sigma^{n+1} = 0\}$[1], $n \geqslant 0$.

(iii) $P\{x^{n+1}(0) = j | 0 < \sigma^n < \infty\} = \pi_j, P\{\sigma^{n+1} = 0 | 0 < \sigma^n < \infty\} = 1 - \sum_j \pi_j$.

(iv) Given that $(0 < \sigma^n < \infty)$ or $\{x^{n+1}(0) = i\}, X^m (m \leqslant n)$ and $X^m (m > n)$ are conditionally independent, i.e. for $0 \leqslant t_{m1} < t_{m2} < \cdots < t_{ml_m}, j_{m1}, j_{m2}, \ldots, j_{ml_n} \in E$, if we put

$$\Lambda_m = \{x^m(t_{mk}) = j_{mk}, 1 \leqslant k \leqslant l_m\} \tag{3}$$

then for any $l \geqslant 1$ and $n \geqslant 0$ we have

$$P\left\{\bigcap_{a=0}^{n+i} \Lambda_a | \Delta\right\} = P\left\{\bigcap_{a=0}^{n} \Lambda_a | \Delta\right\} P\left\{\bigcap_{a=n+1}^{n+l} \Lambda_a | \Delta\right\} \tag{4}$$

where $\Delta = \{0 < \sigma^n < \infty\}$ or $\Delta = \{x^{n+1}(0) = i\}$.

Proof. Utilizing the technique of making an independent product space, with no difficulty we can prove that there exists a probability space $(\Omega, \mathscr{F}, P)$ on which may be defined a sequence of processes $X^0 = \{x^0(t), t < \sigma^0\}, X_i^n = \{x_i^n(t), t < \sigma_i^n\}$ $(n \geqslant 1, i \in E)$ and a family of random variables $f^n (n \geqslant 0)$, taking values in $\bar{E}$ with the following properties:

(1°) X^0 and $X_i^n \in \mathscr{H}$ $(n \geqslant 1, i \in E)$ and they have the same transition probability $\bar{P}(t)$.

(2°) $P\{x^0(0) \in E\} = P\{\sigma^0 > 0\} = 1$

$$P\{x_i^n(0) = i\} = 1 \qquad (n \geqslant 1, i \in E)$$

$$P\{f^n = i\} = \pi_i \qquad (i \in E) \qquad P\{f^n = \infty\} = 1 - \sum_i \pi_i$$

(3°) All $X^0, X_i^n (n \geqslant 1, i \in E)$ and $f^n, n \geqslant 0$, are independent.

Let $C(t)$ be the indicator function of the set $\{t | 0 < t < \infty\}$. If $\omega \in \{f^0 = i, C(\sigma^0) = 1\}$ $(i \in E)$, let $\sigma^1(\omega) = \sigma_i^1(\omega)$; and $x^1(t, \omega) = x_i^1(t, \omega)$ for $t < \sigma'(\omega)$; otherwise let $\sigma'(\omega) = 0$. If $\omega \in \{f^1 = i, C(\sigma^1) = 1\}$ $(i \in E)$, let $\sigma^2(\omega) = \sigma_i^2(\omega)$ and $x^2(t, \omega) = x_i^2(t, \omega)$ for $t < \sigma(\omega)$; otherwise let $\sigma^2(\omega) = 0$. Continuing the procedure we can obtain a sequence of processes $X^n = \{x^n(t), t < \sigma^n\}$ $(n \geqslant 0)$. It is easy to see that $X^n \in \mathscr{H}$ $(n \geqslant 0)$ and has the transition probability $\bar{P}(t)$ for each $n \geqslant 0$, and so property (i) holds. It follows from the structure of X^n that they also have property (ii).

[1] Generally speaking, the initial distribution of a process has total mass 1, but in this lemma for the processes $X^n (n \geqslant 1)$, we drop the restriction, i.e. it is permitted that $P\{0 < \sigma^n\} < 1$ $(n \geqslant 1)$.

Because σ^n only depends on $X^0, X_i^m(m \leqslant n, i \in E)$ and $f^m(m < n)$, by (1°) and (3°) for $j \in E$

$$P\{x^{n+1}(0) = j, 0 < \sigma^n < \infty\} = P\{f^n = j, 0 < \sigma^n < \infty\}$$
$$= P\{f^n = j\}P\{0 < \sigma^n < \infty\} = \pi_j P\{0 < \sigma^n < \infty\}$$

This is just property (iii).

For the proof of property (iv), fix n and $j \in E$ temporarily. Let $\bar{X}^0 = X_j^{n+1}$, $\bar{X}_i^m = X_i^{n+m+1}, \bar{f}^m = f^{n+m+1}$ As we defined $X^m(m \geqslant 0)$ by $X^0, X_i^m(m \geqslant 1, i \in E)$ and $f^m(m \geqslant 0)$ just now according to $\bar{X}^0, \bar{X}_i^m(m \geqslant 1, i \in E)$ and $\bar{f}^m(m \geqslant 0)$, we can determine $\bar{X}^m = \{\bar{x}^m(t), t < \bar{\sigma}^m\}(m \geqslant 0)$. Obviously, $\bar{X}^m(m \geqslant 0)$ only depend on $X_i^m(m > n, i \in E)$ and $f^m(m > n)$ and if we put

$$\bar{\Lambda}_{n+m+1} = \{\bar{x}^m(t_{n+m+1,k}) = j_{n+m+1,k}, 1 \leqslant k \leqslant l_{n+m+1}\}$$

$$\bar{N} = \bigcap_{a=n+1}^{n+l} \bar{\Lambda}_a \qquad M = \bigcap_{a=0}^{n} \Lambda_a \qquad N = \bigcap_{a=n+1}^{n+l} \Lambda_a$$

we can easily see that

$$\{x^{n+1}(0) = j\} \cap \bar{N} = \{x^{n+1}(0) = j\} \cap N$$

and

$$P\{\bar{N}\} = P\{N | x^{n+1}(0) = j\}$$

So from (1°), (3°), property (iii) and the fact that $\Delta \cap \{x^{n+1}(0) \in E\} = \Delta$ for $\Delta = \{0 < \sigma^n < \infty\}$ we obtain that

$$\begin{aligned}
P\{MN\Delta\} &= \sum_j P\{M\Delta, x^{n+1}(0) = j, N\} \\
&= \sum_j P\{M\Delta, f^n = j, \bar{N}\} \\
&= \sum_j P\{M\Delta\}P\{f^n = j\}P\{\bar{N}\} \\
&= \sum_j P\{M\Delta\}\pi_j P\{N | x^{n+1}(0) = j\} \\
&= P(M\Delta)\sum_j P\{x^{n+1}(0) = j | \Delta\}P\{N | x^{n+1}(0) = j\} \\
&= P\{M\Delta\}P\{N | \Delta\}
\end{aligned}$$

From this we are sure that property (iv) is true for $\Delta = (0 < \sigma^n < \infty)$. For $\Delta = \{x^{n+1}(0) = i\}$ it can be proved similarly. The proof is completed. QED

Theorem 2. Assume that the sequence of processes $X^n = \{x^n(t), t < \sigma^n\}(n \geqslant 0)$ defined on the same probability space has the properties (i)–(iv) in Lemma 1.

Let

$$\tau^0 = 0 \qquad \tau^{n+1} = \sum_{a=0}^{n} \sigma^a \qquad \sigma = \sum_{a=0}^{\infty} \sigma^a \tag{5}$$

For $0 \leqslant t \leqslant \sigma$ let

$$x(t) = x^n(t - \tau^n) \qquad \tau^n \leqslant t < \tau^{n+1} \tag{6}$$

Then:

(i) $$X = \{x(t), t \leqslant \sigma\} \in \mathscr{H}.$$

(ii) The transition probability $\{p_{ij}(t)\}$ of X is given by

$$p_{ij}(t) = \bar{p}_{ij}(t) + \int_0^t \pi_j(t-s)\,\mathrm{d}K_i(s) \tag{7}$$

where

$$\pi_j(t) = \sum_i \pi_i \bar{p}_{ij}(t) \tag{8}$$

and $K_i(t)$ is defined as follows

$$L_i(t) = 1 - \sum_j \bar{p}_{ij}(t) \qquad L = \sum_i \pi_i L_i$$

$$L^0(t) = \begin{cases} 0 & \text{if } t \leqslant 0 \\ 1 & \text{if } t > 0 \end{cases}$$

$$L^{n+1} = L^n * L$$

$$K_i = \sum_{n=0}^{\infty} L_i * L^n \tag{9}$$

where $*$ represents convolution.

(iii) X is an honest process if and only if

$$\sum_j \pi_j = 1 \tag{10}$$

Proof. Because X^n are canonical processes, if only we have shown that X is a homogeneous Markov process, we can confirm that X is also a canonical one, and so (i) is proved.

For $0 \leqslant t_1 < t_2 < \cdots < t_l < t_{l+1}, j_1, j_2, \ldots, j_{l+1} \in E$, let

$$\Lambda_k = \bigcap_{a=1}^{k} \{x(t_a) = j_a\} \tag{11}$$

$$\Delta(m_1, m_2, \ldots, m_k) = \bigcap_{a=1}^{k} \{x^{m_a}(t_a - \tau^{m_a}) = j_a, \tau^{m_a} \leqslant t_a < \tau^{m_a+1}\} \tag{12}$$

It is clear that

$$P\{\Lambda_{l+1}\} = \sum_{0 \leqslant m_1 \leqslant \cdots \leqslant m_{l+1}} P\{\Delta(m_1, m_2, \ldots, m_{l+1})\} \tag{13}$$

Abbreviate $\Delta(m_1, \ldots, m_k) = \Delta_k$.

Suppose $m_l = m_{l+1} = m$. Then there exists $k < 1$ such that $m_1 \leqslant \cdots \leqslant m_k < m_{k+1} = \cdots = m_l = m_{l+1} = m$. Hence

$$\begin{aligned} &P\{\Delta(m_1, \ldots, m_{l+1})\} \\ &\quad = \sum_i P\{\Delta_k, \tau^m \leqslant t_{k+1}, x^m(0) = i, x^m(t_a - \tau^m) = j_a, k+1 \leqslant a \leqslant l+1\} \\ &\quad = \sum_i \int_0^{t_{k+1}} P\{x^m(t_a - \tau^m) = j_a, k+1 \leqslant a \leqslant l+1 \mid \Delta_k, \tau^m = s, x^m(0) = i\}\, \mathrm{d}s \\ &\qquad \times P\{\Delta_k, \tau^m \leqslant s, x^m(0) = i\} \end{aligned} \tag{14}$$

Since Δ_k and τ^m only depend on $X^n\,(n \leqslant m-1)$ it follows that by Lemma 1(iv) the integrand in (14) is equal to

$$\begin{aligned} &P\{x^m(t_a - s) = j_a, k+1 \leqslant a \leqslant l+1 \mid \Delta_k, \tau^m = s, x^m(0) = i\} \\ &\quad = P\{x^m(t_a - s) = j_a, k+1 \leqslant a \leqslant l+1 \mid x^m(0) = i\} \\ &\quad = P\{x^m(t_a - s) = j_a, k+1 \leqslant a \leqslant l \mid x^m(0) = i\} \bar{p}_{j_l j_{l+1}}(t_{l+1} - t_l) \\ &\quad = P\{x^m(t_a - s) = j_a, k+1 \leqslant a \leqslant l \mid \Delta_k, \tau^m = s, x^m(0) = i\} \bar{p}_{j_l j_{l+1}}(t_{l+1} - t_l) \end{aligned}$$

Substituting this into (14) and reversing the above operation, for $m_1 = m_{l+1} = m$ we arrive at

$$P\{\Delta(m_1, \ldots, m_{l+1})\} = P\{\Delta_l\} \bar{p}_{j_l j_{l+1}}(t_{l+1} - t_l) \tag{15}$$

Now suppose $m_l = m < m_{l+1} = \gamma$. By Lemma 1(iv),

$$\begin{aligned} &P\{\Delta(m_1, \ldots, m_{l+1})\} \\ &\quad = \sum_i P\{\Delta_l, x^r(0) = i, x^r(t_{l+1} - \tau^r) = j_{l+1}, t_l < \tau^r \leqslant t_{l+1}\} \\ &\quad = \sum_i \int_{t_l}^{t_{l+1}} P\{x^r(t_{l+1} - \tau^r) = j_{l+1} \mid \Delta_l, \tau^r = s, x^r(0) = i\}\, \mathrm{d}s \\ &\qquad \times P\{\Delta_l, \tau^r \leqslant s, x^r(0) = i\} \end{aligned} \tag{16}$$

Also by property (iv) in Lemma 1 the integrand is equal to

$$P\{x^r(t_{l+1} - s) = j_{l+1} \mid x^r(0) = i\} = \bar{p}_{ij_{l+1}}(t_{l+1} - s) \tag{17}$$

Again by Lemma 1(ii)–(iv),

$$P\{\Delta_l, \tau^r \leqslant s, x^r(0)=i\} = P\{\Delta_l, \tau^r \leqslant s, 0<\sigma^{r-1}<\infty, x^r(0)=i\}$$
$$= P\{\Delta_l, \tau^r \leqslant s, 0<\sigma^{r-1}<\infty\}\pi_i \quad (18)$$

If it can be shown that for

$$P\{\Delta_l, \tau^r \leqslant s, 0<\sigma^{r-1}<\infty\} = P\{\Delta_l\}(L_{jl} * \mathrm{L}^{r-m-1})(s-t_l) \quad (19)$$

then substituting (15)–(19) into (13) we obtain that

$$P\{\Lambda_{l+1}\} = \sum_{0 \leq m_1 \leq \cdots \leq m_l} P\{\Delta_l\}\bigg\{\bar{p}_{j_l j_{l+1}}(t_{l+1}-t_l)$$
$$+ \sum_{m_l+1=m_{l+1}}^{\infty} \int_{t_l}^{t_{l+1}} \pi_{j_{l+1}}(t_{l+1}-s)\,\mathrm{d}s\,(L_{j_l} * L^{m_{l+1}-m_l-1})(s-t_l)\bigg\}$$
$$= \sum_{0 \leq m_1 \leq \cdots \leq m_l} p\{\Delta_l\}\bigg\{\bar{p}_{j_l j_{l+1}}(t_{l+1}-t_l)$$
$$+ \int_{t_l}^{t_{l+1}} \pi_{j_{l+1}}(t_{l+1}-s)\,\mathrm{d}s\,K_{j_l}(s-t_l)\bigg\}$$
$$= P\{\Lambda_l\}p_{j_l j_{l+1}}(t_{l+1}-t_l)$$

This demonstrates that X is a homogeneous Markov process with the transition probability $\{p_{ij}(t)\}$.

Proof of (19): We have to prove that

$$P\{\Delta_1, \tau^r \leqslant s+t_l, 0<\sigma^{r-1}<\infty\} = P\{\Delta_l\}(L_{j_l} * L^{r-m-1})(s) \qquad s \geqslant 0 \quad (20)$$

Assume $\gamma = m+1$. The left-hand side of the above is equal to

$$\sum_i P\{\Delta_{l-1}, \tau^m + \sigma^m \leqslant s+t_l, x^m(0)=i, x^m(t_l-\tau^m)=j_l, \tau^m \leqslant t_l\}$$
$$= \sum_i \int_0^{t_l} P\{x^m(t_1-\tau^m)=j_l, \tau^m+\sigma^m \leqslant s+t_l \mid \Delta_{l-1}, \tau^m=u, x^m(0)=i\}\,\mathrm{d}u$$
$$\times P\{\Delta_{l-1}, x^m(0)=i, \tau^m \leqslant u\}$$

By Lemma 1(iv) the integrand is equal to

$$P\{x^m(t_l-u)=j_l, \sigma^m \leqslant s+t_l-u \mid x^m(0)=i\}$$
$$= P\{x^m(t_l-u)=j_l \mid x^m(0)=i\}L_{jl}(s)$$

Substituting this for the original and reversing the above calculation, we obtain that the left-hand side of (20) equals $P\{\Delta_l\}L_{jl}(s)$, i.e. (20) holds for $r=m+1$.

We will prove (20) by induction. Suppose $r>m+1$. Then the left-hand side of (20) is equal to

$$\sum_i P\{\Delta_l, \tau^{r-1}+\sigma^{r-1}\leqslant s+t_l, x^{r-1}(0)=i\}$$

$$=\sum_i \int_{t_l}^{s+t_l} P\{\tau^{r-1}+\sigma^{r-1}\leqslant s+t_l|\Delta_l, \tau^{r-1}=u, x^{r-1}(0)=i\}\,\mathrm{d}u$$

$$\times P\{\Delta_l, \tau^{r-1}\leqslant u, x^{r-1}(0)=i\}$$

$$=\sum_i \int_{t_l}^{s+t_l} P\{\sigma^{r-1}\leqslant s+t_l-u|x^{r-1}(0)=i\}\,\mathrm{d}u\, P\{\Delta_l, \tau^{r-1}\leqslant u, x^{r-1}(0)=i\}$$

But $P\{\sigma^{r-1}\leqslant s+t_l-u|x^{r-1}(0)=i\}=L_i(s+t_l-u)$, and by Lemma 1(ii)–(iv) and the hypothesis of induction, for $u\geqslant t_l$ we have

$$\begin{aligned}P\{\Delta_l, \tau^{r-1}\leqslant u, x^{r-1}(0)=i\} &= P\{\Delta_l, \tau^{r-1}\leqslant u, 0<\sigma^{r-2}<\infty, x^{r-1}(0)=i\}\\ &= P\{\Delta_l, \tau^{r-1}\leqslant u, 0<\sigma^{r-2}<\infty\}\pi_i\\ &= P\{\Delta_l\}(L_{jl}*L^{r-1-m-1})(u-t_l)\pi_i\end{aligned}$$

Thus the left-hand side of (20) is equal to

$$\sum_i \int_{t_l}^{s+t_l} \pi_i L_i(s+t_l-u)P\{\Delta_l\}\,\mathrm{d}u(L_l*L^{r-1-m-1})(u-t_l)$$

$$=P\{\Delta_l\}\int_0^s L(s-v)\mathrm{d}v\,(L_l*L^{r-1-m-1})(v)$$

$$=P\{\Delta_l\}(L_l*L^{r-m-1})(s)$$

Thus (20) is verified.

Summing over $j\in E$ on (7) we obtain that

$$\sum_j p_{ij}(t)=1-L_i(t)+\int_0^t \sum_j \pi_j\{1-L_j(t-s)\}\mathrm{d}K_i(s)$$

$$=1-L_i(t)+\int_0^t \left(\sum_j \pi_j - L(t-s)\right)\mathrm{d}K_i(s)$$

$$=1-L_i(t)+\left(\sum_j \pi_j\right)K_i(t)-\sum_{n=0}^{\infty}(L_i*L^{n+1})(t)$$

$$=1-K_i(t)\left(1-\sum_j \pi_j\right)$$

On account of this, there exists at least one i such that $K_i\neq 0$. Therefore X is honest if and only if (10) holds. The proof is completed. QED

Remark 1

The first part $X^0 = \{x^0(t), t < \sigma^0\}$ of the process $X = \{x(t), t < \sigma\}$ in Theorem 2 is a stopping process and satisfies

$$P\{x(\tau^1) = j \mid \tau^1 < \infty\} = \pi_j \tag{21}$$

Moreover their transition probabilities have the relation (7). We call X a $\pi = \{\pi_j, j \in E\}$ D-type extension process of the stopping process X^0.

Remark 2

Taking the Laplace transform of (7) we obtain that

$$\psi_{ij}(\lambda) = \bar{\psi}_i(\lambda) + \bar{\xi}_i(\lambda) \frac{\sum_k \pi_k \bar{\psi}_{kj}(\lambda)}{1 - \sum_k \pi_k \bar{\xi}_k(\lambda)} \tag{22}$$

where $\psi(t)$ and $\bar{\psi}(t)$ are the Laplace transforms of $P(t)$ and $\bar{P}(t)$ respectively and

$$\bar{\xi}_i(\lambda) = 1 - \lambda \sum_j \bar{\psi}_{ij}(\lambda) \tag{23}$$

11.3 D*-TYPE EXTENSION

In the previous section we did not require the finiteness of the Q matrices. In this section we assume that all Q matrices are finite.

In the D-type extension, if the Q matrix of X^0 is conservative, from the sample paths of the process we know that X^0 and its D-type extension process X have the same Q matrix. But if the Q matrix of X^0 is non-conservative, then the Q matrices of X and X^0 may be different. In order to preserve the Q matrix we introduce the D*-type extension.

Definition 1. Suppose $X = \{x(t), t < \sigma\} \in \mathscr{H}_s$. We call $\sigma(\omega)$ the T-tail of $X(\omega)$ if $0 < \sigma(\omega) < \infty$ and $\sigma(\omega)$ is a jump point (see Definition 7.8.1); otherwise we call $\sigma(\omega)$ the P-tail.

Obviously, if the transition probability of the process $X^0 = \{x^0(t), t < \sigma^0\} \in \mathscr{H}_s$ is $\bar{P}(t)$ then the following quantities

$$\begin{aligned} M_{ij}(t) &= P_i\{x^0(t) = j, \sigma^0 \text{ is a } P\text{-tail}\} \\ N_{ij}(t) &= P_i\{x^0(t) = j, \sigma^0 \text{ is a } T\text{-tail}\} \\ M_i(t) &= \sum_j M_{ij}(t) \qquad N_i(t) = \sum_j N_{ij}(t) \\ R_i(t) &= M_i(0) - M_i(t) = P_i\{\sigma^0 \leqslant t, \sigma^0 \text{ is a } P\text{-tail}\} \end{aligned} \tag{1}$$

and uniquely determined by $\bar{P}(t)$.

Imitating Lemma 2.1, we have the following:

Lemma 1. Assume that $\bar{P}(t)$ satisfies conditions (2.2.A–C), $R_i(t) > 0$ for some $i \in E$ and $t > 0$, and the distribution π satisfies (2.2). Then there exist a probability space $(\Omega, \mathscr{F}, P)$ and defined on it a sequence of processes $X^n = \{x^n(t), 0 \leqslant t < \sigma^n\}$ $(n \geqslant 0)$ with the following properties:

(i*) $X^n \in \mathscr{H}_s (n \geqslant 0)$ and they have the same transition probability $\bar{P}(t)$.

(ii*) $\{\sigma^n = 0\} \cup \{\sigma^n = \infty\} \cup \{\sigma^n$ is the T-tail of $X^n\} \subset \{\sigma^{n+1} = 0\} \, (n \geqslant 0\}$.

(iii*) $P\{x^{n+1}(0) = j | \Delta\} = \pi_j, P\{\sigma^{n+1} = 0 | \Delta\} = 1 - \sum_j \pi_j$, where $\Delta = (0 < \sigma^n < \infty$ and σ^n is the P-tail of X^n).

(iv*) Given that $\{0 < \sigma^n < \infty$ and σ^n is the P-tail of $X^n\}$ or $\{X^{n+1}(0) = i\}$, then $X^m \, (m \leqslant n)$ and $X^m \, (m > n)$ are conditionally independent.

Theorem 2. Assume that processes $X^n = \{x^n(t), t < \sigma^n\} \, (n \geqslant 0)$, defined on the same probability space, have the properties (i*)–(iv*) in Lemma 1. Define $X = \{x(t), t < \sigma\}$ according to (2.5) and (2.6). Then:

(i) $X \in \mathscr{H}_s$ and $X^n \, (n \geqslant 0)$ have the same Q matrix.

(ii) The transition probability $P^*(t) = \{p_{ij}^*(t)\}$ is given by

$$p_{ij}^*(t) = \bar{p}_{ij}(t) + \int_0^t \pi_j(t-s) \mathrm{d}s \, K_i^*(t) \tag{2}$$

where $\pi_j(t)$ is determined by (2.8) and K_i^* by (2.9) but $L_i(t)$ in it should be replaced by $R_i(t)$ in (1).

(iii) X is an honest process if and only if Q is conservative and (2.10) holds.

Proof. Imitate the proof of Theorem 2.1. But the last conclusion in the theorem requires a new proof. By (2) it follows that

$$\sum_j p_{ij}^*(t) = 1 - \{N_i(0) - N_i(t)\} - \left\{1 - \sum_k \pi_k + \sum_k \pi_k N_k(0)\right\} K_i^*(t)$$

$$+ \int_0^t \sum_k \pi_k N_k(t-s) \mathrm{d}K_i^*(s) \tag{3}$$

If Q is conservative, clearly $N_i(t) = 0$, and if (2.10) holds, then by (3) $\sum_j p_{ij}^*(t) = 1$, i.e. X is an honest process. Conversely, if X is an honest process, necessarily $N_i(t) = 0$. Otherwise, for some i we have $N_i(t) = P_i\{t < \sigma^0, \sigma^0$ is the T-tail of $X^0\} > 0$. Then according to Lemma 1(ii*) on the set $\{t < \sigma^0, \sigma^0$ is the T-tail of $X^0\}$ of positive probability, $\sigma = \sigma^0 < \infty$, which contradicts the hypothesis that X is honest. So by $N_i(t) = 0$ and (3), (2.10) holds. We claim that Q is conservative. In fact, if for some $i, d_i = q_i - \sum_{j \neq i} q_{ij} > 0$ then we have

$$P_i\{0 < \sigma^0 < \infty, \sigma^0 \text{ is the } T\text{-tail of } X^0\} \geqslant d_i / q_i > 0$$

and so $N_i(t) > 0$, which is impossible. QED

Remark 1

Taking the Laplace transformation in (2), we get

$$\psi_{ij}^*(\lambda) = \bar{\psi}_{ij}(\lambda) + \xi_i(\lambda)\frac{\sum_k \pi_k \bar{\psi}_{kj}(\lambda)}{1 - \sum_k \pi_k \xi_k(\lambda)} \tag{4}$$

where $\psi^*(\lambda)$ and $\bar{\psi}(\lambda)$ are respectively the Laplace transforms of $P^*(t)$ and $\bar{P}(t)$, and $\xi_i(\lambda)$ is the Laplace transform of $R_i(t)$ in (1).

Remark 2

We call the process $X \in \mathscr{H}_s$ in Theorem 2 the πD*-type extension process of $X^0 \in \mathscr{H}_s$. The process X^0 and its D*-type extension have the same Q matrix. If X^0 satisfies the system of backward equations, so does X.

Remark 3

When Q is conservative, because there is no T-tail the D*-type extensions and D-type extension are the same.

11.4 DOOB PROCESSES

In the D*-type extension, if X^0 or $\{\bar{p}_{ij}(t)\}$ is the minimal Q process, using the representation of Martin exit boundary $B_e, R_i(t)$ in (3.1) becomes

$$L_i(t) = P_i\{\sigma^0 \leqslant t, x(\sigma^0 - 0) \in B_e\} \tag{1}$$

and Theorem 3.2 becomes:

Theorem 1. Suppose $\{f_{ij}(t)\}$ is the minimal Q process. For some i and $t > 0$, $L_i(t)$, defined according to (1), is positive. Let π satisfy (2.2). Suppose $X^n \in \mathscr{H}_s(Q)(n \geqslant 0)$ are minimal Q processes defined on the same probability space with properties (ii*)–(iv*) in Lemma 3.1. Let X be defined according to (2.5)–(2.6). Then $X \in \mathscr{H}_1(Q)$ and its transition function is given by

$$p_{ij}(t) = f_{ij}(t) + \int_0^t \pi_j(t-s)\mathrm{d}K_i(s) \tag{2}$$

where $\pi_j(t) = \sum_i \pi_i f_{ij}(t)$, and $K_i(t)$ is determined by (2.9), but $L_i(t)$ in it should be understood as (1). Moreover, X is an honest process if and only if Q is conservative and (2.10) holds.

Definition 1. A πD*-type extension process of the minimal Q process is called a (Q, π) Doob process.

For the (Q, π) Doob process $X = \{x(t), t < \sigma\}$,

$$P\{x(\tau) = i | x(\tau - 0) \in B_e\} = \pi_i \qquad i \in E \tag{3}$$

$$P\{\sigma = \tau | x(\tau - 0) \in B_e\} = 1 - \sum_i \pi_i \tag{4}$$

where τ is the first leaping point. The Laplace transform of (2) is

$$\psi_{ij}(\lambda) = \phi_{ij}(\lambda) + \xi_i(\lambda) \frac{\sum_k \pi_k \phi_{kj}(\lambda)}{1 - \sum_k \pi_k \xi_k(\lambda)}$$

$$= \phi_{ij}(\lambda) + \xi_i(\lambda) \frac{\sum_k \pi_k \phi_{kj}(\lambda)}{(1 - \sum_k \pi_k) + \sum_k \pi_k [1 - \xi_k(\lambda)]} \tag{5}$$

where $\xi_i(\lambda)$ is the Laplace transform of $L_i(t)$ in (1). Particularly, when Q is conservative,

$$\xi_i(\lambda) = 1 - \lambda \sum_i \phi_{ij}(\lambda) \tag{6}$$

11.5 GENERALIZED D-TYPE EXTENSION

The D-type and D*-type extensions do not involve the boundaries of processes. We will consider in this section an extension that depends on boundaries of processes.

For a finite Q matrix Q, according to Chapter 7 we can determine the essential Martin boundary B, the exit boundory B_e and the passive boundary B_p. Recall that H denotes the non-conservative state set.

Assume that the minimal Q process $X = \{x(t), t < \tau\}$ is stopping, i.e. for some $i, P_i\{\tau < \infty\} < 0$ or, equivalently, $\mu\{H \cup B_e\} > 0$, where measure μ is defined according to (7.12.2).

Clearly, for a Borel set $\Gamma \subset H \cup B_e$, the quantities

$$L_i(\Gamma, t) = P_i\{x(\tau - 0) \in \Gamma, \tau \leqslant t\} \qquad t \geqslant 0 \tag{1}$$

$$h_i(\Gamma, \lambda) = E_i\{e^{-\lambda\tau}, x(\tau - 0) \in \Gamma\} = \int_0^\infty e^{-\lambda t} \mathrm{d}t\, L_i(\Gamma, t) \qquad \lambda > 0 \tag{2}$$

are uniquely determined by Q.

We are given a family of distributions $\Pi(a, \cdot)(a \in H \cup B_e)$, satisfying

$$\Pi(a, E) \leqslant 1 \qquad \int_{H \cup B_e} \Pi(a, E) \mu(\mathrm{d}a) > 0 \tag{3}$$

Lemma 1. There exists a probability space $(\Omega, \mathscr{F}, P)$ on which a sequence of minimal Q processes $X^n = \{x^n(t), t < \sigma^n\} \in \mathscr{H}_s(Q) (n \geqslant 0)$ can be defined, with the following properties:

(i) $\{\sigma^n = 0\} \cup \{\sigma^n = \infty\} \subset \{\sigma^{n+1} = 0\}$.

(ii) On $\{x^n(\sigma^n - 0) \in H \cup B_e\}$, almost surely

$$P\{x^{n+1}(0) = j \mid x^n(\sigma^n - 0)\} = \Pi(x^n(\sigma^n - 0), j) \qquad j \in E$$
$$P\{\sigma^{n+1} = 0 \mid x^n(\sigma^n - 0)\} = 1 - \Pi(x^n(\sigma^n - 0), E)$$

(iii) For Λ_m determined by (2.3)

$$P\left\{\bigcap_{a=0}^{n+l} \Lambda_a \middle| x^{n+1}(0) = i\right\}$$
$$= P\left\{\bigcap_{a=0}^{n} \Lambda_a \middle| x^{n+1}(0) = i\right\} P\left\{\bigcap_{a=n+1}^{n+l} \Lambda_a \middle| x^{n+1}(0) = i\right\} \qquad (i \in E, l \geqslant 1)$$

(iv) On the set $\{x^n(\sigma^n - 0) \in H \cup B_e\}$, almost surely

$$P\left\{\bigcap_{a=0}^{n+l} \Lambda_a \middle| x^n(\sigma^n - 0)\right\} = P\left\{\bigcap_{a=0}^{n} \Lambda_a \middle| x^n(\sigma^n - 0)\right\} P\left\{\bigcap_{a=n+1}^{n+l} \Lambda_a \middle| x^n(\sigma^n - 0)\right\}$$

where Λ_a are also given by (2.3).

Proof. By the technique of making an independent product space, it is not difficult to show that there exists a probability space $(\Omega, \mathscr{F}, P)$ on which can be defined a sequence of minimal processes $X^0 = \{x^0(t), t < \sigma^0\} \in \mathscr{H}_s$, $X_i^n = \{x_i^n(t), t < \sigma_i^n\} \in \mathscr{H}_s$ $(n \geqslant 1, i \in E)$ and an $\bar{E}$-valued random variable $f^n(a)$ $(n \geqslant 0, a \in H \cup B_e)$, with the following properties:

(1°) $P\{x^0(0) \in E\} = P\{x_i^n(0) = i\} = 1$.

(2°) $P\{f^n(a) = i\} = \Pi(a, i)$ $(i \in E)$; $P\{f^n(a) = \infty\} = 1 - \Pi(a, E)$.

(3°) All $X^0, X_i^n (n \geqslant 1, i \in E), f^n(a)$ $(n \geqslant 0, a \in H \cup B_e)$ are independent.

If $\omega \in \{x^0(\sigma^0 - 0) \in H \cup B_e, f^0[x^0(\sigma^0 - 0)] = i\}$ $(i \in E)$, let $\sigma^1(\omega) = \sigma_i^1(\omega)$ and $x^1(t, \omega) = x_i^1(t, \omega)$, $t < \sigma^1(\omega)$; otherwise, let $\sigma^1(\omega) = 0$. It is easy to see that $X^1 = \{x^1(t), t < \sigma^1\} \in \mathscr{H}_s$ is a minimal Q process. Therefore $\sigma^1 > 0$, and one can define $x^1(\sigma^1 - 0)$. If $\omega \in (x^1(\sigma^1 - 0) \in H \cup B_e$, $f^1[x^1\sigma^1 - 0)] = i\}$ $(i \in E)$, let $\sigma^2(\omega) = \sigma_i^2(\omega)$ and $x^2(t, \omega) = x_i^2(t, \omega), t < \sigma^2(\omega)$. Otherwise, let $\sigma^2(\omega) = 0$. Continuing with this step, we can obtain a sequence of processes X^n $(n \geqslant 0)$.

Just as done in Theorem 2.1, one can prove that $X^n \in \mathscr{H}_s$ $(n \geqslant 0)$ are all minimal Q processes and have properties (i)–(iv) in this lemma. QED

Theorem 2. Suppose that the minimal Q processes $X^n = \{x^n(t), t < \sigma^n\} \in \mathscr{H}_s$ $(n \geqslant 0)$ defined on the same probability space possess properties (i)–(iv) in Lemma 1. Define $X = \{x(t), t < \sigma\}$ according to (2.5)–(2.6). Then:

(i) If $\int_{B_e} \Pi(a, E)\mu(da) = 0$, then $X \in \mathscr{H}_s$ is a minimal process and its Q matrix is different from that of X^0. If $\int_{B_e}\Pi(a,E)\mu(da)>0$, then $X\in\mathscr{H}_1$ is a process of order 1.

(ii) The transition probability $\{p_{ij}(t)\}$ of X is given by

$$p_{ij}(t)=f_{ij}(t)+\sum_k\int_0^t f_{kj}(t-s)\,dK_{ik}(s) \tag{4}$$

where $K_{ik}(t)$ are determined by

$$\begin{aligned} T_{ij}^1(t)&=\int_{H\supset B_e}\Pi(a,j)L_i(da,t)\\ T_{ij}^{n+1}(t)&=\int_{H\cup B_e}\Pi(a,j)\sum_k[T_{ik}^n(\cdot)*L_k(da,\cdot)](t)\\ K_{ij}(t)&=\sum_{n=1}^{\infty}T_{ij}^n(t)\end{aligned} \tag{5}$$

Here $*$ represents convolution, i.e.

$$[T_{ik}^n(\cdot)*L_k(\Gamma,\cdot)](t)=\int_0^t T_{ik}^n(t-s)L_k(\Gamma,ds) \tag{6}$$

Making some changes in (4) we obtain that

$$p_{ij}(t)=f_{ij}(t)+\int_{H\cup B_e}\int_0^t\pi_j(a,t-s)K_i(da,ds) \tag{7}$$

where

$$\pi_j(a,t)=\sum_k\Pi(a,k)f_{kj}(t) \tag{8}$$

$$L_i^1(\Gamma,t)=L_i(\Gamma,t) \qquad \text{(see (1))} \tag{9}$$

$$L_i^{n+1}(\Gamma,t)=\int_0^t\int_{H\cup B_e}L_i^n(da,ds)\sum_k\Pi(a,k)L_k(\Gamma,t-s) \tag{10}$$

$$K_i(\Gamma,s)=\sum_{n=1}^{\infty}L_i^n(\Gamma,s) \tag{11}$$

We can imitate the proof of Theorem 2.2 to prove Theorem 2. The only difference is that the latter is more complicated in writing. Although the proofs of the theorems in sections 11.6 and 11.7, which we have omitted, are similar to this theorem, we still give a brief clue to the proof of Theorem 2 here.

Proof. We need only verify that X is a homogeneous Markov process with transition probability (4). The rest is clear.

Determine Λ_l and $\Delta(m_1,\dots,m_k)=\Delta_k$ according to (2.11)–(2.12) and the formula (2.13) still holds. Following the proof of (2.15), in the present case (2.15) still holds for $m_l = m_{l+1}$.

Assume $m_l < m_{l+1}$. Abbreviate $m_l = m$, $m_{l+1} = \gamma$. Then by Lemma 1(iii),

$$P\{\Delta_{l+1}\} = \sum_k P\{\Delta_l, t_l < \tau' \leqslant t_{l+1} < \tau^r + \sigma^r, x^r(0) = k, x^r(t_{l+1} - \tau^r) = j_{l+1}\}$$

$$= \sum_k \int_{t_l}^{t_{l+1}} P\{x^r(t_{l+1} - \tau^r) = j_{l+1} | \Delta_l, \tau^r = s, x^r(0) = k\}\, \mathrm{d}s$$

$$\times P\{\Delta_l, \tau' \leqslant s, x^r(0) = k\}$$

$$= \sum_k \int_{t_l}^{t_{l+1}} f_{kj_{l+1}}(t_{l+1} - s)\, \mathrm{d}s\, P\{\Delta_l, \tau' \leqslant s, x^r(0) = k\} \tag{12}$$

If it can be proved that for $s \geqslant t_l$,

$$P\{\Delta_l, \tau^r \leqslant s, x^r(0) = k\} = T^{r-m}_{j_l k}(s - t_l) \tag{13}$$

combining (12) we obtain

$$P\{\Delta_{l+1}\} = \sum_k P\{\Delta_l\} \int_{t_l}^{t_{l+1}} f_{kj_{l+1}}(t_{l+1} - s)\, \mathrm{d}s\, T^{m_{l+1} - m_l}_{j_l k}(s - t_l)$$

$$= P\{\Delta_l\} \sum_k \int_0^{t_{l+1} - t_l} f_{kj_{l+1}}(t_{l+1} - t_l - s)\, \mathrm{d}s\, T^{m_{l+1} - m_l}_{j_l k}(s) \tag{14}$$

and substituting (2.15) and (14) into (2.13) we find that $P\{\Lambda_{l+1}\} = P\{\Lambda_l\} p_{j_l j_{l+1}} \times (t_{l+1} - t_l)$, where $p_{ij}(t)$ is determined by (4). Therefore X is a homogeneous Markov process.

To prove (13) is to verify

$$P\{\Delta_l, \tau^r \leqslant s + t_l, x^r(0) = k\} = P\{\Delta_l\} T^{r-m}_{j_l k}(s) \qquad s \geqslant 0 \tag{15}$$

By Lemma 1(iv), the left-hand side of the above is equal to

$$\int_{H \cup B_e} P\{\Delta_l, x^{r-1}(\sigma^{r-1} - 0) \in \mathrm{d}a, x^r(0) = k, \tau^r \leqslant s + t_l\}$$

$$= \int_{H \cup B_e} \Pi(a, k) P\{\Delta_l, x^{r-1}(\sigma^{r-1} - 0) \in \mathrm{d}a, \tau^r \leqslant s + t_l\} \tag{16}$$

First we deal with $r = m + 1$. Then

$$P\{\Delta_l, x^{r-1}(\sigma^{r-1} - 0) \in \mathrm{d}a, \tau^r \leqslant s + t_l\}$$

$$= \sum_i P\{\Delta_{l-1}, x^m(0) = i, x^m(t_l - \tau^m)$$

$$= j_l, \tau^m \leqslant t_l < \tau^m + \sigma^m, x^m(\sigma^m - 0) \in \mathrm{d}a, \tau^m + \sigma^m \leqslant s + t_l\} \tag{17}$$

By Lemma 1(iii), the summand in the above equals

$$\int_0^{t_l} P\{x^m(t_l-u)$$

$$= j_l, x^m(\sigma^m-0)\in \mathrm{d}a, \sigma^m \leqslant s+t_l-u \mid x^m(0)=i\}\,\mathrm{d}uP\{\Delta_{l-1}, x^m(0)=i, \tau^m \leqslant u\}$$

$$= \int_0^{t_l} f_{ij_l}(t_l-u)L_{j_l}(\mathrm{d}a, s)\,\mathrm{d}uP\{\Delta_{l-1}, x^m(0)=i, \tau^m \leqslant u\}$$

Substituting into (17) and reversing the above calculation we see that the left-hand side of (17) equals $P\{\Delta_l\}L_{j_l}(\mathrm{d}a, s)$. Thereby substituting into (16) we know that (13) holds for $r = m+1$. By a similar consideration and by induction we can confirm that (13) holds for all $r > m$. The proof is terminated. QED

Denote the Laplace transform of $T^n_{ij}(t)$ by $G^n_{ij}(\lambda)$. Define $h_i(\Gamma, \lambda)$ by (2). Then from (5),

$$G^n_{ij}(\lambda) = \sum_k G^n_{ik}(\lambda)G^1_{ik}(\lambda)$$

Thus it follows from (4) that

$$\psi_{ij}(\lambda) = \phi_{ij}(\lambda) + \sum_k \left(\sum_{n=1}^{\infty} G^n_{ik}(\lambda)\right)\phi_{kj}(\lambda) \tag{18}$$

When Q is conservative, the above expression is precisely formula (10.2.25) in Kunita (1962). We shall use matrix symbols. Write $G(\lambda) = G^1(\lambda)$. Then (18) becomes

$$\psi(\lambda) = \phi(\lambda) + \sum_{n=1}^{\infty} G^n(\lambda)\phi(\lambda) \tag{19}$$

Setting

$$h^t_i(\Gamma, \lambda) = \int_0^{\infty} \mathrm{e}^{-\lambda t}\,\mathrm{d}t L^n_i(\Gamma, t) \tag{20}$$

it follows from (9) and (10) that

$$h^1_i(\Gamma, \lambda) = h_i(\Gamma, \lambda)$$

$$h^{n+1}_i(\Gamma, \lambda) = \int_{H\cup B_e} h^n_i(\mathrm{d}a, \lambda)\sum_k \Pi(a, k)h_k(\Gamma, \lambda) \tag{21}$$

Hence (7) becomes

$$\psi_{ij}(\lambda) = \phi_{ij}(\lambda) + \int_{H\cup B_e}\left(\sum_{n=1}^{\infty} h^n_i(\mathrm{d}a, \lambda)\right)\sum_k \Pi(a, k)\phi_{kj}(\lambda) \tag{22}$$

Letting

$$ {}_0V(a,\Gamma,\lambda)=\begin{cases}0 & a\notin\Gamma\\ 1 & a\in\Gamma\end{cases} $$

$$ V^n(a,\Gamma,\lambda)=\sum_k \Pi(a,k)h_k^n(\Gamma,\lambda) \tag{23} $$

we have

$$ V^{n+1}(a,\Gamma,\lambda)=\int_{H\cup B_e} V^n(a,\mathrm{d}b,\lambda)V^1(b,\Gamma,\lambda) \tag{24} $$

$$ h_i^{n+1}(\Gamma,\lambda)=\int_{H\cup B_e} h_i(\mathrm{d}a,\lambda)V^n(a,\Gamma,\lambda) \tag{25} $$

Consequently (22) becomes

$$ \psi_{ij}(\lambda)=\phi_{ij}(\lambda)+\int_{H\cup B_e} h_i(a,\lambda)\int_{H\cup B_e}\left(\sum_{n=0}^{\infty} V^n(a,\mathrm{d}b,\lambda)\right)\left(\sum_k \Pi(b,k)\phi_{kj}(\lambda)\right) \tag{26} $$

When Q is conservative, this is just formula (10) in Kunita (1962).

Definition 1. Call X in Theorem 2 the $\{\Pi(a,\cdot), a\in H\cup B_e\}$ generalized D-type extension process of X^0.

Clearly, for X^0 and X in Definition 1, on the set $\{x(\sigma^0-0)\in H\cup B_e\}$ almost surely holds

$$ P\{x(\sigma^0)=j\,|\,x(\sigma^0-0)\}=\Pi(x(\sigma^0-0),j) \qquad j\in E $$

11.6 GENERALIZED D*-TYPE EXTENSION

When Q is conservative, the generalized D-type extension preserves the Q matrix. But when Q is not conservative, things may be different. Hence we introduce the generalized D*-type extension.

Suppose the exit boundary B_e of the minimal Q process is non-empty, i.e. $\mu\{B_e\}>0$. Given a family of distributions $\Pi(a,\cdot)$ $(a\in B_e)$ satisfying $\Pi(a,E)\leqslant 1$ and $\int_{B_e}\Pi(a,E)\mu(\mathrm{d}a)>0$.

Let

$$ \Pi(a,E)=0 \qquad a\in H \tag{1} $$

Then the basic conditions in section 11.5 are satisfied and so Lemma 5.1 and Theorem 5.2 are still valid.

Definition 1. Suppose X^0 is the minimal Q process. A $\{\Pi(a,\cdot),\ a\in H\cup B_e\}$ generalized D-type extension of X^0 that satisfies (1) is called a generalized D*-type extension.

Clearly, when Q is conservative, the generalized D-type and generalized D*-type extensions are identical. The following theorem is obvious.

Theorem 1. The minimal Q process X^0 and its $\{\Pi(a,\cdot),\ a\in B_e\}$ generalized D*-type extension process X have the same Q matrix. More precisely, X satisfies the system of backward equations and, moreover, on $\{x(\tau-0)\in B_e\}$ almost surely holds

$$P\{x(\tau)=j\,|\,x(\tau-0)\}=\Pi\{x(\tau-0),j\} \tag{2}$$

11.7 EXTENSION OF INSTANTANEOUS-RETURN PROCESSES

In this section we shall give a brief discussion of extensions of instantaneous-return processes. Because the proofs are similar to those in sections 11.5 and 11.6, they are omitted.

Suppose $X=\{x(t),t<\sigma\}\in\mathscr{H}_s$ is a kth-order instantaneous-return process with transition probability ${}_kp_{ij}(t)$ and ith-order exit boundary ${}_iB_e$. Instead of $H\cup B_e$ in section 11.5 we consider ${}_kH\cup{}_kB_e$, where

$${}_kH=H+{}_0B_e+{}_1B_e+\cdots+{}_{k-1}B_e \tag{1}$$

The measure on ${}_kH\cup{}_kB_e$ is ${}_k\mu$ for order k. Assume that X is a stopping process, i.e. ${}_k\mu\{{}_kH\cup{}_kB_e\}>0$. We are arbitrarily given a family of distributions ${}_k\Pi(a,\cdot)(a\in{}_kH\cup{}_kB_e)$ satisfying

$${}_k\Pi(a,E)\leqslant 1 \qquad \int_{{}_kH\cup{}_kB_e}{}_k\Pi(a,E){}_k\mu(\mathrm{d}a)>0 \tag{2}$$

Lemma 1. There exists a probability space $(\Omega,\mathscr{F},P)$ on which a sequence of processes $X^n=\{x^n(t),\ t<\sigma^n\}\in\mathscr{H}_k$ $(n\geqslant 0)$ can be defined, with the following properties

(i) X^n $(n\geqslant 0)$ have the same transition probability ${}_kp_{ij}(t)$.

(ii) $(\sigma^n=0)\cup(\sigma^n=\infty)\subset(\sigma^{n+1}=0)$.

(iii) On the set $\{x^n(\sigma^n-0)\in{}_kH\cup{}_kB_e\}$ almost surely hold

$$P\{x^{n+1}(0)=j\,|\,x^n(\sigma^n-0)]={}_k\Pi(x^n(\sigma^n-0),j)$$
$$P\{\sigma^{n+1}=0\,|\,x(\sigma^n-0)\}=1-{}_k\Pi(x^k(\sigma^n-0),E)$$

(iv) The same as Lemma 5.1(iii)

(v) Obtained by changing $H\cup B_e$ in Lemma 5.1(iv) to ${}_kH\cup{}_kB_e$.

Theorem 2. Suppose that processes $X^n\in\mathscr{H}_k$ $(n\geqslant 0)$, which are defined on the same probability space, have properties (i)–(v) in Lemma 1. Define $X=\{x(t),$ $t<\sigma\}$ according to (2.5)–(2.6). Then:

(i) if $\int_{{}_kB_e} {}_k\Pi(a, E)_k\mu(\mathrm{d}a) = 0$ then $X \in \mathscr{H}_k$ but the kth-order exit boundaries of X and X^0 are different, and if the above integral is positive,

(ii) the transition function ${}_{k+1}p_{ij}(t)$ of X is given by

$$ {}_{k+1}p_{ij}(t) = {}_kp_{ij}(t) + \sum_l \int_0^t {}_kp_{lj}(t-s)\,\mathrm{d}s_k K_{il}(s) \tag{3}$$

or

$$ {}_{k+1}p_{ij}(t) = {}_kp_{ij}(t) + \int_{{}_kH \cup {}_kB_e} \int_0^t {}_k\pi_j(a, t-s)_k K_i(\mathrm{d}a\,\mathrm{d}s) \tag{4}$$

where

$$ {}_k\pi_j(a, t) = \sum_i {}_k\Pi(a, i)_k p_{ij}(t) \tag{5}$$

and ${}_kK_{ij}(t)$, ${}_kK_i(\Gamma, t)$ are still defined according to (5.5)–(5.6) and (5.9)–(5.11), but the following changes should be made. To each of $L_i(\Gamma, t)$, $\Pi(a, \cdot)$, H, B_e, $T_{ij}^n(t)$, $K_{ij}(t)$, $L_i^n(\Gamma, t)$ and $K_i(\Gamma, t)$ should be added a left subscript k, and for $\Gamma \subset {}_kH \cup {}_kB_e$, moreover,

$$ {}_kL_i(\Gamma, t) = P_i\{x^0(\sigma^0 - 0) \in \Gamma, \sigma^0 \leqslant t\} \tag{6}$$

Correspondingly, after the obvious supplements to those lower left corners in (5.18)–(5.26) we have

$$ {}_{k+1}\psi(\lambda) = {}_k\psi(\lambda) + \sum_{n=1}^{\infty} ({}_kG(\lambda))^n {}_k\psi(\lambda) \tag{7}$$

$$ {}_{k+1}\psi(\lambda) = {}_k\psi(\lambda) + \int_{{}_kH \cup {}_kB_e} \left\{ \sum_{n=1}^{\infty} {}_kh_i^n(\mathrm{d}a, \lambda) \right\} \sum_l {}_k\Pi(a, l)_k\psi_{lj}(\lambda) \tag{8}$$

$$ {}_{k+1}\psi_{ij}(\lambda) = {}_k\psi_{ij}\lambda) + \int_{{}_kH \cup {}_kB_e} {}_kh_i(\mathrm{d}a, \lambda)$$

$$ \times \int_{{}_kH \cup {}_kB_e} \left(\sum_{n=0}^{\infty} {}_kV^n(a, \mathrm{d}b, \lambda) \right) \left(\sum_l {}_k\Pi(b, l)_k\psi_{lj}(\lambda) \right) \tag{9}$$

Definition 1. Call the process X in Theorem 1 the $\{{}_k\Pi(a, \cdot),\ a \in {}_kH \cup {}_kB_e\}$ generalized D-type extension process of X^0.

For X and X^0 given in Definition 1, on $\{x(\sigma^0 - 0) \in {}_kH \cup {}_kB_e\}$ almost surely holds

$$ P\{x(\sigma^0) = j \,|\, x(\sigma^0 - 0)\} = {}_k\Pi(x(\sigma^0 - 0), j) \tag{10}$$

Note that X and X^0 do not necessarily have the same kth-order exit boundary. Of course, their Q matrices may be different. Therefore it is necessary to introduce the generalized D*-type extension.

Assume that the kth-order instantaneous-return process $X \in \mathscr{H}_k$ satisfies ${}_k\mu\{{}_kB_e\} > 0$. Suppose we are given a family of distributions ${}_k\Pi(a,\cdot)$ $(a \in {}_kB_e)$ satisfying ${}_k\Pi(a, E) \leqslant 1$ and $\int_{{}_kB_e} \Pi(a, E)_k\mu(\mathrm{d}a) > 0$.

$$ {}_k\Pi(a, E) = 0 \qquad a \in {}_kH \tag{11} $$

Then the basic conditions at the beginning of this section are satisfied and so Lemma 1 and Theorem 2 are valid.

Definition 2. Suppose $X^0 \in \mathscr{H}_k$. The $\{{}_k\Pi(a,\cdot),\ a \in {}_kH \cup {}_kB_e\}$ generalized D-type extension of X^0 satisfying (11) is called a generalized D*-type extension process.

Theorem 3. A kth-order instantaneous-return process $X^0 \in \mathscr{H}_k$ and its $\{{}_k\Pi(a,\cdot),\ a \in {}_kB_e\}$ generalized D*-type extension process have the same lth-order exit boundary ${}_lB_e$ $(l \leqslant k)$. Particuarly, they have the same Q matrix and X satisfies the system of backward equations.

11.8 ON NON-STICKY EXTENSIONS

From now on, we turn to the V-type extension. The D-type extension is of instantaneous-return type: after a process moves to infinity it returns to finite states instantly. Of course, a process may return to finite states slowly from infinity. This is just the V-type extension. We may also have the mixed DV-type extension.

When Chung (1963, 1966) studied the boundary theory of Markov chains, he analysed meticulously their sample paths, introduced the concept of sticky and non-sticky boundary points and at last under the hypothesis of finiteness of exit boundaries obtained the analytic expressions for transition probabilities of processes. But he did not directly construct their sample paths. The D-type, V-type and mixed DV-type extension processes are all non-sticky return processes, i.e. processes whose leaping points may be arranged into a sequence in order of magnitude. A left-open and right-closed interval with its adjacent end-points being leaping points is a leaping interval, i.e. U interval. The construction of sample paths on leaping intervals is the key to the construction of non-sticky return processes.

The motion of a D-type extension process in each leaping interval is the motion of a minimal Q process starting from finite states. So an imbedded chain describing its jump behaviour is an ordinary discrete-parameter Markov chain $(x_n,\ n \geqslant 0)$ with its starting time and an initial distribution concentrated in the state space E. But for a V-type extension process its motion on each leaping interval is the motion of a minimal Q process starting from infinity. An imbedded chain describing its jump behaviour should be $(x_n,\ -\infty < n < +\infty)$. However, it is not a Markov chain and nor is it a stationary sequence. It belongs to the class of approximating Markov chains introduced by Hunt (1960) for the first time. Hence approximating Markov chains are precisely the foundation stones for our construction of non-sticky returning processes. But an approximating Markov chain is defined on a measure space which may have infinite total measure.

From now on until the end of this chapter, we will first introduce approximating Markove chains and their characteristic measures and then study the relation between characteristic measure and the initial time and life-time of an approximating Markov chain. After that we shall use approximating chains are imbedding chains to construct sample paths on leaping intervals. Thus the so-called approximating minimal Q processes arise and therefore the minimal Q processes starting from infinity can be described. We establish the correspondence between entrance families and approximating minimal Q processes. Finally, on the basis of approximating minimal Q processes, we shall construct the sample paths of a class of non-sticky returning processes, i.e. we shall obtain the DV-type extension processes of minimal Q processes. In terms of entrance families we derive the analytic expressions of transition probabilities for this class of Q processes. Utilizing the boundary theory we can also construct other non-sticky processes that are more complicated, i.e. the so-called generalized DV-type and (DV)*-type extension processes.

11.9 STOCHASTIC CHAINS AND CHARACTERISTIC MEASURES

The concept of stochastic chains was introduced by Hunt (1960). We will give a brief description of their definition and results. Let E be a denumerable set and $\Pi = (\Pi_{ij}, i, j \in E)$ a sub-stochastic matrix, i.e. a matrix that has non-negative entries and row sums that are at most 1. When the row sum is equal to 1, it is called a stochastic matrix. Let $(\Omega, \mathscr{F}, P)$ be a measure space, i.e. Ω is a non-empty set of abstract points, $\mathscr{F}$ is a Borel field composed of some subsets of Ω and P is a measure on $\mathscr{F}$. We emphatically point out that we do not require that $(\Omega, \mathscr{F}, P)$ be a totally finite measure space. Henceforth $\mathbf{N}$ will denote the set of all integers.

Definition 1. Suppose on $(\Omega, \mathscr{F}, P)$ a triple (x, α, β) is given:

(i) α is an $\mathscr{F}$-measurable function with values in $\{-\infty\} \cup \mathbf{N}$, β is an $\mathscr{F}$-measurable function with values in $\mathbf{N} \cup \{+\infty\}$, and for all $\omega \in \Omega$, $\alpha(\omega) \leqslant \beta(\omega)$;

(ii) for each $\omega \in \Omega$ and $n \in \mathbf{N}$ satisfying $\alpha(\omega) \leqslant n \leqslant \beta(\omega)$, there exists a unique $x(n, \omega) \in E$ and for $n \in \mathbf{N}$, $i \in E$,

$$\Lambda(n, i) \equiv (x(n) = i) = (\omega : x(n, \omega) = i, \alpha(\omega) \leqslant n \leqslant \beta(\omega)) \in \mathscr{F}$$

We call (x, α, β) a quasi-stochastic chain.

For a quasi-stochastic chain (x, α, β), write $\mathscr{F}(x, \alpha, \beta)$ for the minimal Borel field which contains all sets $\Lambda(n, i)$ $(n \in \mathbf{N}, i \in E)$ and call it the Borel field generated by (x, α, β). Write $\mathscr{F}_k(x, \alpha, \beta)$ for the minimal Borel field which contains all sets $\Lambda(n, i)$ $(n \leqslant k, i \in E)$.

Definition 2. Suppose that (x, α, β) is a quasi-stochastic chain. If for all $n \in \mathbf{N}$ and $i \in E$, $P[\Lambda(n, i)] < +\infty$, we call (x, α, β) a stochastic chain.

Because $\Omega = \bigcup_{n \in \mathbf{N}} \bigcup_{i \in E} \Lambda(n, i)$, it follows that for a measure space $(\Omega, \mathscr{F}, P)$ on which a stochastic chain is defined, Ω must be σ-finite.

We can define $x(n, \omega)$ for any $n \in \mathbf{N}$. For example, take arbitrarily two indices $\Delta \notin E$, $\theta \notin E$ and $\Delta \neq \theta$. For $n < \alpha(\omega)$ define $x(n, \omega) = \Delta$, and for $n > \beta(\omega)$ define $x(n, \omega) = \theta$. Obviously, sets $(x(n) = \Delta) = (n < \alpha)$ and $(x(n) = \theta) = (\beta < n)$ are $\mathscr{F}$-measurable. But even for a stochastic chain (x, α, β), their measures are not necessarily finite.

Definition 3. Suppose that (x, α, β) is a stochastic chain. If given 'the present', its 'past' and 'future' are conditionally independent, i.e. for any k, m, $n \in \mathbf{N}$, $k < m < n$ and $i_k, i_{k+1}, \ldots, i_n \in E$. Putting

$$\Lambda = \Lambda(m, i_m) \qquad \Lambda' = \prod_{h \leqslant j \leqslant m} (j, i_j) \qquad \Lambda'' = \prod_{m \leqslant j \leqslant n} \Lambda(j, i_j)$$

$$\frac{P(\Lambda'\Lambda'')}{P(\Lambda)} = \frac{P(\Lambda')}{P(\Lambda)} \frac{P(\Lambda'')}{P(\Lambda)}$$

as long as $P(\Lambda) > 0$. We call (x, α, β) a Markov chain.

Definition 4. Suppose (x, α, β) is a Markov chain. If for any $n \in \mathbf{N}$, $i \in E$, we have

$$P(x(n) = i, x(n+1) = j) = P(x(n) = i)\Pi_{ij}$$

we call (x, α, β) a Π-chain. The matrix Π is called the (one-step) transition matrix of the chain.

Definition 5. Suppose (x, α, β) is a stochastic chain and σ is an $\mathscr{F}$-measurable function which is defined on Ω and takes values in $\{-\infty\} \cup \mathbf{N} \cup \{+\infty\}$. Let

$$\Omega' = (\sigma \in \mathbf{N}, \alpha \leqslant \sigma \leqslant \beta) \qquad \mathscr{F}' = \Omega' \cap \mathscr{F} \qquad P'(\Lambda') = P(\Lambda') \qquad \Lambda' \in \mathscr{F}'$$

For $\omega \in \Omega'$ let

$$\gamma(\omega) = \beta(\omega) - \sigma(\omega) \qquad y(n, \omega) = x(\sigma(\omega) + n, \omega)$$

If $(y, 0, \gamma)$ is a Π-chain on $(\Omega', \mathscr{F}', P')$, we say that the random time σ leads the stochastic chain (x, α, β) to the Π-chain $(y, 0, \gamma)$. Write

$$f_\sigma(x, \alpha, \beta) = (y, 0, \gamma)$$

Definition 6. Suppose (x, α, β) is a stochastic chain. If there exists a sequence of random times α_n, $n \geqslant 1$, taking values in $\mathbf{N} \cup \{+\infty\}$ such that on Ω almost surely $\alpha_n \downarrow \alpha$ and each α_n leads the stochastic chain (x, α, β) to some Π-chain $f_{\alpha_n}(x, \alpha, \beta)$, we call (x, α, β) an approximating Π-chain.

Definition 7. Suppose (x, α, β) is a stochastic chain and σ is an $\mathscr{F}$-measurable function with values in $\mathbf{N} \cup \{+\infty\}$. If for any $n \in \mathbf{N}$, $(\sigma \leqslant n) \in \mathscr{F}_n(x, \alpha, \beta)$, we call σ a random time independent of the future or a wide-stopping time.

Definition 8. Suppose that (x, α, β) is an approximating Π-chain. If (x, α, β) is a Markov chain and if the sequence of random times α_n which leads (x, α, β) to a Π-chain can be chosen as a sequence of wide-stopping times of the chain, we call (x, α, β) a strong approximating Π-chain.

Write $\Pi^{(n)} = \Pi^n$, and $G = \sum_{n=0}^{+\infty} \Pi^n$. Then $G_{ij} \leqslant G_{jj}$. According to Π we can divide the states in E into recurrent and non-recurrent states. A state i is Π-recurrent if and only if $G_{ii} = +\infty$.

Definition 9. Assume that $\eta = (\eta_j, j \in E)$ is a measure on E, i.e. $0 \leqslant \eta_j \leqslant +\infty$ $(j \in E)$. We call η finite if each η_j is finite, and totally finite if $\sum_j \eta_j < +\infty$. We say that η is a Π-excessive measure if $\eta\Pi \leqslant \eta$, i.e. $\sum_i \eta_i \Pi_{ij} \leqslant \eta_j$ $(j \in E)$, and that η is a Π-harmonic measure if $\eta\Pi = \eta$.

Theorem 1. Suppose (x, α, β) is an approximating Π-chain. Let $C(i)$ be the measure concentrated on i with unit mass, i.e. $C_i(i) = 1$ and $C_j(i) = 0$ for $j \neq i$. Let

$$\eta_j = \int_\Omega \sum_{\alpha \leqslant n \leqslant \beta} C_j[x(n)]\, dP = \sum_{n \in N} P(x(n) = j) \tag{1}$$

Then $\eta = (\eta_j, j \in E)$ is a Π-excessive measure, which is called the characteristic measure of the Π-chain

Proof. Suppose $(x, 0, \beta)$ is a Π-chain with the initial distribution ν. Then the characteristic measure determined by $(x, 0, \beta)$ is $\eta = \nu G$. Of course, it is a Π-excessive measure. Suppose (x, α, β) is an approximating chain. Denote by η^n the characteristic measure of the Π-chain $f_{\alpha_n}(x, \alpha, \beta)$. Then η^n is Π-excessive, i.e. $\eta^n\Pi \leqslant \eta^n$. But $\eta^n \uparrow \eta$, so that $\eta\Pi \leqslant \eta$, i.e. η is Π-excesssive. The proof is completed.

QED

Theorem 2. Suppose that (x, α, β) is an approximating Π-chain with a finite characteristic measure η. Let

$$\alpha' = -\beta \qquad \beta' = -\alpha \qquad x'(n) = x(-n)$$

Then the inverse chain (x', α', β') is an approximating Q-chain, where for $\eta_j > 0$

$$Q_{ji} = \eta_i \Pi_{ij} / \eta_j \tag{2}$$

and for $\eta_j = 0$, Q_{ji} may be chosen arbitrarily as long as Q_{ji} is nonnegative and the row sums of Q are not more than 1.

Proof. Suppose D is finite subset of E. Let τ be the last exit time of the chain (x, α, β) from D, i.e.

$$\tau = \begin{cases} \sup\{n : \alpha \leqslant n \leqslant \beta, x(n) \in D\} \\ -\infty & \text{if the above set is empty} \end{cases} \tag{3}$$

Clearly, $-\tau$ is the hitting time of (x', α', β') at D. Because $\eta(D) < \infty$ by (1), we know that there are only a finite number of $n \in \mathbf{N}$ such that $x(n) \in D$. Hence $\tau \leqslant \beta$ and $\tau < \infty$.

First of all suppose that (x, α, β) is a Π-chain, $\alpha = 0$ and its initial distribution is v. Determine $L_D(i)$ according to (7.3.12). Then (7.3.16) is true, i.e.

$$\begin{aligned} &P(x(\tau) = i_0, x(\tau - 1) = i_1, \ldots, x(\tau - k) = i_k) \\ &\quad = \eta(i_k)\Pi(i_k, i_{k-1})\Pi(i_{k-1}, i_{k-2}) \cdots \Pi(i_1, i_0) L_D(i_0) \end{aligned} \tag{4}$$

where $\eta = vG$.

The equality (4) also holds for the approximating Π-chain (x, α, β). In fact, let

$$f_{\alpha_n}(x, \alpha, \beta) = (x_n, 0, \beta_n) \tag{5}$$

and let τ_n be the last exit time of the Π-chain $(x_n, 0, \beta_n)$ from D and η^n be the characteristic measure. Then $\eta^n \uparrow \eta$ and for almost every $\omega \in \{\tau < \infty\}$ and sufficiently large n,

$$\alpha_n(\omega) + \tau_n(\omega) = \tau(\omega) \qquad x_n(\tau_n(\omega) - j) = x(\tau(\omega) - j)$$

Since (4) holds for $(x_n, 0, \beta_n)$, passing to the limit we are sure that (4) is also true for (x, α, β).

For the inverse chain, (4) can be rewritten as

$$\begin{aligned} &P(x'(\varepsilon') = i_0, x'(\varepsilon' + 1) = i_1, \ldots, x'(\varepsilon' + k) = i_k) \\ &\quad = \eta(i_0) L_D(i_0) Q(i_0, i_1) Q(i_1, i_2) \cdots Q(i_{k-1}, i_k) \end{aligned}$$

where ε' is the hitting time for the inverse chain (x', α', β') at D. The above equality indicates that ε' leads (x', α', β') to a Q-chain. Take $D = D_n \uparrow E$. Then $\tau(D_n) \uparrow \beta$ and $\varepsilon'(D_n) = -\tau(D_n) \downarrow \alpha'$. Therefore, (x', α', β') is an approximating Q-chain, and the proof is completed. QED

Theorem 3. Suppose that (x, α, β) is an approximating Π-chain with a finite characteristic measure η. Let $\alpha(D)$ be the hitting time of the chain (x, α, β) at a finite subset D of E, i.e.

$$\alpha(D) = \begin{cases} \inf\{n : \alpha \leqslant n \leqslant \beta, x(n) \in D\} \\ +\infty & \text{if the above set is empty} \end{cases} \tag{6}$$

Then $\alpha \leqslant \alpha(D)$, $-\infty < \alpha(D)$, $\alpha(D)$ is a wide-stopping time, $\alpha(D)\downarrow\alpha$ as $D\uparrow E$ and $\alpha(D)$ leads (x, α, β) to the Π-chain

$$f_{\alpha(D)}(x, \alpha, \beta) = (x_D, 0, \beta_D) \tag{7}$$

Proof. Consider the last exit time $\tau'(D)$ of the inverse chain (x', α', β') from D. According to Theorem 2, $\tau'(D) \leqslant \beta'$, $\tau'(D) < +\infty$ and $\tau'(D)\uparrow\beta'$ as $D\uparrow E$. Obviously, $\alpha(D) = -\tau'(D)$. Hence, $\alpha \leqslant \alpha(D)$, $-\infty < \alpha(D)$ and $\alpha(D)\downarrow\alpha$ as $D\uparrow E$. It follows from Theorem 2 that $\alpha(D)$ leads (x, α, β) to a Π-chain. Finally, by the definition of $\alpha(D)$ it is easy to verify that $\alpha(D)$ is a wide-stopping time of (x, α, β) and we conclude the proof. QED

Remark

Theorem 3 demonstrates that for any approximating Π-chain (x, α, β) with a finite characteristic measure we can always choose a sequence of wide-stopping times $\alpha_n \downarrow \alpha$ such that $f_{\alpha_n}(x\, \alpha, \beta)$ are Π-chains. However, (x, α, β) is not necessarily a strong approximating chain, for (x, α, β) itself may not be a Markov chain.

Corollary

Assume that (x, α, β) is an approximating Π-chain with a finite characteristic measure η. Write ν_D for the distribution that (x, α, β) hits the finite set D, i.e.

$$\nu_D(j) = P(x(\alpha(D)) = j) \tag{8}$$

where $\alpha(D)$ is defined according to (6). Then the support of ν_D is contained in D and, moreover,

$$\nu_D(j) = \eta(j) L_D^\eta(j)$$

where

$$L_D^\eta(i) = P_i(y_0 \in D,\ y(n) \notin D \text{ for } 0 < n \leqslant \delta)$$

Here $(y, 0, \delta)$ is a Q-chain and Q is determined by (2).

Proof. Letting $k = 0$ in (4) we have

$$P(x(\tau) = j) = \eta(j) L_D(j)$$

Applying this to the inverse chain (x', α', β') we obtain

$$P(x'(\tau') = j) = \eta(j) L_D^\eta(j)$$

where τ' is the last exit time of $(x', \alpha', \beta)'$ from D, i.e. the hitting time of (x, α, β) at D, and $x'(\tau') = x(\alpha(D))$. Then the above equality is just (9). The proof is completed. QED

Theorem 4. Assume that η is a finite Π-excessive measure. Fix a sequence D_n of finite subsets of E, $D_n \uparrow E$. Then there exists a sequence of measures ν^n with the following properties:

(i) ν^n is totally finite with its support contained in D_n;

(ii) $\nu^n G \leqslant \eta$ and equality holds on D_n;

(iii) $\nu^n G \uparrow \eta$ as $n \uparrow \infty$;

(iv) suppose $n < m$; then ν^n is the hitting distribution at D of the Π-chain with the initial distribution ν^m, i.e. if $(x, 0, \beta)$ is a Π-chain with the initial distribution ν^m and the hitting times $\alpha(D_n)$ lead (x, α, β) to Π-chains $(x_n, 0, \beta_n)$, then the initial distribution of $(x_n, 0, \beta_n)$ is ν^n for each $n < m$.

(v) there exists a measure space $(\Omega, \mathscr{F}, P)$ on which an approximating Π-chain (x, α, β) is defined with its characteristic measure coincident with η and with ν_{D_n} determined by (8) coincident with ν^n.

The proof of the theorem is a bit complicated. The reader is referred to Kemeny *et al.* (1966, Chapter 10, section 12). We point out that by (9) ν^n in Theorem 4 is determined by η and Π according to

$$\nu^n(j) = \eta(j) L^{\eta}_{D_n}(j) \tag{10}$$

11.10 INITIAL TIME AND LIFETIME OF APPROXIMATING Π-CHAINS

Let (x, α, β) be a quasi-stochastic chain. We call α the initial time and β the lifetime. Note that from the statement under Definition 9.2, for all $n \in \mathbf{N}$, $x(n)$ are defined but they take values in $E \bigcup \{\Delta, \theta\}$. We agree that $C_j(\Delta) = C_j(\theta) = 0$ for $j \in E$. Then, the Π-excessive measure determined by the approximating Π-chain (x, α, β) is

$$\eta_j = \int_{\Omega} \sum_{n \in N} C_j[x(n)]\, \mathrm{d}P = \int_{\Omega} \sum_{\alpha \leqslant n < +\infty} C_j[x(n)]\, \mathrm{d}p \qquad j \in E \tag{1}$$

Theorem 1. Let (x, α, β) be an approximating Π-chain with the characteristic measure η. Then the following are true:

(i) If Π is a stochastic matrix, almost surely $\beta = +\infty$.

(ii) If $\alpha = -\infty$, almost surely η is Π-harmonic.

(iii) Suppose η is finite and its Riesz decomposition is

$$\eta = \eta^1 + \eta^2 \qquad \eta^1 = \nu G \qquad \nu = \eta - \eta\Pi \qquad \text{and} \qquad \lim_{l \to +\infty} \eta \Pi^l = \eta^2 \tag{2}$$

Then the excessive quantity v can be expressed as

$$v_j = P(-\infty < \alpha, x(\alpha) = j) \tag{3}$$

the potential measure η^1 can be represented by

$$\eta_j^1 = \int_{-\infty<\alpha} \sum_{n\in\mathbf{N}} C_j[x(n)]\,\mathrm{d}P \tag{4}$$

and the harmonic measure has the representation

$$\eta^2 = \int_{-\infty=\alpha} \sum_{n\in\mathbf{N}} C_j[x(n)]\,\mathrm{d}P \tag{5}$$

(iv) Suppose that η is finite. Then η is a potential if and only if $-\infty < \alpha$ almost surely, and η is harmonic if and only if $-\infty = \alpha$ almost surely.

Proof. Suppose that random times α_n, $n \geqslant 1$, lead (x, α, β) to Π-chains $(x_n, 0, \beta_n)$ on $\Omega_n = (\alpha_n \in \mathbf{N}, \alpha \leqslant \alpha_n \leqslant \beta)$ respectively.

(i) Assume that Π is a stochastic matrix. It is well known that for any Π-chain, its lifetime is almost surely equal to $+\infty$. Hence it almost goes without saying that $\beta_n = \beta - \alpha_n = +\infty$, i.e. $\beta = +\infty$ on Ω_n. So $\beta = +\infty$ almost surely on Ω since $\Omega_n \uparrow \Omega$.

(ii) Since

$$\begin{aligned}
\eta_i &= \int_\Omega \sum_{\alpha\leqslant k<+\infty} C_i[x(k)]\,\mathrm{d}P \\
&= \lim_{n\to+\infty} \int_{\Omega_n} \sum_{\alpha_n\leqslant k<+\infty} C_i[x(k)]\,\mathrm{d}P \\
&= \lim_{n\to+\infty} \int_{\Omega_n} \sum_{0\leqslant n<+\infty} C_i[x_n(m)]\,\mathrm{d}P \\
&= \lim_{n\to+\infty} \sum_{m=0}^{+\infty} P(x_n(m) = i)
\end{aligned}$$

it follows that

$$\begin{aligned}
\eta_i \Pi_{ij} &= \lim_{n\to+\infty} \sum_{m=0}^{+\infty} P(x_n(m) = i)\Pi_{ij} \\
&= \lim_{n\to+\infty} \sum_{0\leqslant m<+\infty} P(x_n(m) = i, x_n(m+1) = j) \\
&= \lim_{n\to+\infty} \int_{\Omega_n} \sum_{0\leqslant m<+\infty} C_i[x_n(m)] C_j[x_n(m+1)]\,\mathrm{d}P
\end{aligned}$$

$$= \lim_{n \to +\infty} \int_{\Omega_n} \sum_{\alpha_n \leqslant k < +\infty} C_i[x(k)]C_j[x(k+1)]\,\mathrm{d}P$$

$$= \int_{\Omega} \sum_{\alpha \leqslant k < +\infty} C_i[x(k)]C_j[x(k+1)]\,\mathrm{d}P$$

Taking summation over $i \in E$, we have

$$\sum_i \eta_i \Pi_{ij} = \int_{\Omega} \sum_{a \leqslant k < +\infty} C_j[x(k+1)]\,\mathrm{d}P = \int_{\Omega} \sum_{\alpha+1 \leqslant m < \infty} C_j[x(m)]\,\mathrm{d}P \tag{6}$$

If $\alpha = -\infty$ almost surely, then the right-hand side of (6) is equal to η_j. Therefore η is harmonic.

(iii) If η is finite, by (6) we have

$$(\eta\Pi)_j = \eta_j - P(-\infty < \alpha, x(\alpha) = j) \tag{7}$$

Then (3) follows.

Similarly to (6) we have

$$(\eta\Pi^l)_j = \int_{\Omega} \sum_{\alpha+l \leqslant m < +\infty} C_j[x(m)]\,\mathrm{d}P$$

$$= \int_{-\infty < \alpha} \sum_{\alpha+l \leqslant m < +\infty} C_j[x(m)]\,\mathrm{d}P + \int_{-\infty = \alpha} \sum_{n \in N} C_j[x(m)]\,\mathrm{d}P \tag{8}$$

Noting

$$\eta_j = \int_{-\infty < \alpha} \sum_{\alpha \leqslant m < +\infty} C_j[x(m)]\,\mathrm{d}P + \int_{-\infty = \alpha} \sum_{m \in N} C_j[x(m)]\,\mathrm{d}P < +\infty \tag{9}$$

(8) becomes

$$(\eta\Pi^l)_j = \eta_j - \int_{-\infty < \alpha} \sum_{\alpha \leqslant m < \alpha + l} C_j[x(m)]\,\mathrm{d}P \uparrow \eta_j$$

$$= \int_{-\infty < \alpha} \sum_{\alpha \leqslant m < +\infty} C_j[x(m)]\,\mathrm{d}P \qquad l \uparrow +\infty$$

Combining with (9) we obtain (5) and so (4) follows.

(iv) When $-\infty < \alpha$ almost surely, by (5) we have $\eta^2 = 0$. Hence $\eta = \eta^1$ is a potential. Conversely, if η is a finite potential, then $\eta\Pi^l \downarrow 0$. So that the second term of the right-hand side of (8) equals zero, i.e.

$$\sum_{m \in N} P\{-\infty = \alpha, x(m) = i\} = 0$$

Sum over $i \in E$, and we obtain $-\infty < \alpha$ almost surely.

If $-\infty = \alpha$ almost surely, from (4) we have $\eta^1 = 0$. Hence $\eta = \eta^2$ is harmonic. Conversely, if η is finite and harmonic, then by (7) $P\{-\infty < \alpha, x(\alpha) = j\} = 0$. Summing up over $j \in E$, we obtain $P\{-\infty < \alpha\} = 0$. We conclude the proof.

QED

Theorem 1 shows that the approximating Π-chain (x, α, β) can be decomposed into two stochastic chains which are defined respectively on $\Omega_1 = (-\infty < \alpha)$ and $\Omega_2 = (-\infty = \alpha)$; nevertheless, they may not be approximating Π-chains. Fortunately, we can make them become approximating Π-chains very easily.

Theorem 2. Assume that η is a finite Π-excessive measure with the Riesz decomposition (2). Then there exist a measure space $(\Omega, \mathscr{F}, P)$ and an approximating Π-chain (x, α, β) defined on it with the characteristic measure η. Furthermore, let $\Omega_1 = (-\infty < \alpha)$, $\Omega_2 = (-\infty = \alpha)$ and $(x_a, \alpha_a, \beta_a) = (x, \alpha, \beta)$ on Ω_a $(a = 1, 2)$. Then (x_a, α_a, β_a) $(a = 1, 2)$ are approximating Π-chains on Ω_a and they determine Π-excessive measures η^a, respectively.

Proof. According to Theorem 9.4, for each $a = 1$ and 2 there exists a measure space $(\Omega_a, \mathscr{F}_a, P_a)$ on which an approximating Π-chain (x_a, α_a, β_a) is defined with the characteristic measure η^a. By Theorem 1 we can take $\Omega_1 = (-\infty < \alpha_1)$ and $\Omega_2 = (-\infty = \alpha_2)$.

We can consider that Ω_1 and Ω_2 have no common points. Write $\Omega = \Omega_1 \cup \Omega_2$ and $\mathscr{F} = \{A : A \subset \Omega,\ A \cap \Omega_a \in \mathscr{F}_a,\ a = 1, 2\}$. Clearly $\mathscr{F}$ is a borel field on Ω and $\mathscr{F}_a \subset \mathscr{F}$. For $A \in \mathscr{F}$, let $P(A) = P_1(A \cap \Omega_1) + P_2(A \cap \Omega_2)$. Let $\omega \in \Omega$, if $\omega \in \Omega_a$, and define $\alpha(\omega) = \alpha_a(\omega)$, $\beta(\omega) = \beta_a(\omega)$, $x(n, \omega) = x_a(n, \omega)$. Then, it is easily seen that (x, α, β) is an approximating Π-chain, which is defined on $(\Omega, \mathscr{F}, P)$ and has all the properties presented in this theorem. The proof is completed. QED

11.11 IMBEDDED CHAINS

A matrix $A = (a_{ij}, i, j \in E)$ is called a Q matrix if its entries in the diagonal are non-positive and finite, if the entries outside the diagonal are non-negative and finite, and if the row sums are non-positive and finite. When the row sums are equal to zero, the Q matrix A is called conservative. Suppose we are given a Q matrix $Q = (q_{ij}, i, j \in E)$. Then

$$0 \leqslant q_{ij} < +\infty \ (i \neq j) \qquad \sum_{j \neq i} q_{ij} \leqslant -q_{ii} < +\infty \tag{1}$$

Put $q_i = -q_{ii}$. We call $d_i = q_i - \sum_{j \neq i} q_{ij}$ the non-conservative quantity of i. Recall that $C(i)$ is the measure concentrated on i with a unit mass. Let[1]

[1] For convenience, the imbedded matrix Π here is slightly different from that in (2.9.7).

$$\Pi_{ij} = \begin{cases} [1 - C_j(i)]q_{ij}/q_i & \text{if } q_i > 0 \\ 0 & \text{if } q_i = 0 \end{cases}$$

We call the sub-stochastic matrix $\Pi = (\Pi_{ij}, i, j \in E)$ the imbedded matrix of Q and a Π-chain the imbedded chain of Q. Recall that $\theta \notin E$. Write $E_\theta = E \cup \{\theta\}$. Let $q_{i\theta} = d_i$ $(i \in E)$ and $q_\theta = -q_{\theta\theta} = 0$. Then $Q_\theta = (q_{ij}, i, j \in E_\theta)$ is a conservative Q matrix and the imbedded matrix Π of Q is a submatrix of the imbedded matrix $\Pi_\theta = (\Pi_{ij}, i, j \in E_\theta)$. Clearly

$$\Pi_{i\theta} = \begin{cases} d_i/q_i & \text{if } q_i > 0 \\ 0 & \text{if } q_i = 0 \end{cases} \qquad \Pi_{\theta j} = 0 \qquad j \in E_\theta \tag{2}$$

For $\lambda > 0$ let

$$q_{ij}(\lambda) = q_{ij} (i \neq j) \qquad q_i(\lambda) = -q_{ii}(\lambda) = \lambda + q_i$$

Then $\Pi(\lambda) = (q_{ij}(\lambda), i, j \in E)$ is a non-conservative Q matrix. Its imbedded matrix is denoted by $\Pi(\lambda) = (\Pi_{ij}(\lambda), i, j \in E)$ and is called a λ-imbedded matrix of Q. Denote the λ-imbedded matrix of Q_θ by $\Pi_\theta(\lambda) = \Pi_{ij}(\lambda), i, j \in E_\theta)$. Then $\Pi(\lambda)$ is a submatrix $\Pi_\theta(\lambda)$ and, moreover,

$$\Pi_{ij}(\lambda) = \frac{[1 - C_i(j)]q_{ij}}{\lambda + q_i} \qquad i, j \in E$$

$$\Pi_{i\theta}(\lambda) = \frac{d_i}{\lambda + q_i} (j \in E) \qquad \Pi_{\theta j}(\lambda) = 0 (j \in E_\theta) \tag{3}$$

Note that all elements in the θth row of Π_θ are equal to zero and the other row sums are equal to 1, and that all elements in the θth row of $\Pi_\theta(\lambda)$ are equal to zero and the other row sums are less than 1. Hence, for any approximating Π-chain or approximating $\Pi_\theta(\lambda)$-chain (x, α, β), if at time δ the chain visits θ, i.e. $x(\delta) = \theta$, then δ must be the lifetime, i.e. $\beta = \delta$.

Write

$$G(\lambda) = \sum_{n=0}^{\infty} [\Pi(\lambda)]^n \tag{4}$$

$$\phi_{ij}(\lambda) = G_{ij}(\lambda)(\lambda + q_j)^{-1} \tag{5}$$

Then $\phi(\lambda) = \{\phi_{ij}(\lambda), i, j \in E\}$ is the Laplace transform of the Feller minimal solution $f(t) = \{f_{ij}(t), i, j \in E\}$ $(t \geqslant 0)$. Because $\lambda \sum_j \phi_{ij}(\lambda) \leqslant 1$, it follows that $G(\lambda) < \infty$ and, therefore, all states in E are $\Pi\omega$-non-recurrent.

Lemma 1. Assume that $(y, 0, \delta)$ is a $\Pi_\theta(\lambda)$-chain with its initial distribution concentrated in E. Let $\delta(E)$ be the first exit time of the chain from E:

$$\delta(E) = \sup\{n: y(n) \notin E, 0 \leqslant n \leqslant \delta\}$$

Then

$$\delta(E)=\begin{cases}\delta & \text{if } \delta=+\infty, \text{or } \delta<+\infty,\ y(\delta)\notin E\\ \delta-1 & \text{if } \delta<+\infty,\ y(\delta)=\theta\end{cases} \tag{6}$$

and $(y, 0, \delta(E))$ is a $\Pi(\lambda)$-chain.

Proof. Since $y(0)\notin E$, it follows that $0\leqslant\delta(E)$. And since all elements in the θth row of $\Pi_\theta(\lambda)$ are equal to zero, the only possible time for the chain to reach θ is the lifetime $\delta<\infty$. Hence (6) is justified. Because $(y, 0, \delta)$ is a $\Pi_\theta(\lambda)$-chain, it is clear that $(y, 0, \delta(E))$ is a $\Pi(\lambda)$-chain. The proof is completed. QED

Lemma 2. Assume that (y, α, δ) is an approximating $\Pi_\theta(\lambda)$-chain and that $\delta(E)$ is the first exit time from E,

$$\delta(E)=\begin{cases}\sup\{n: y(n)\in E, \alpha\leqslant n\leqslant\delta\}\\ -\infty & \text{if the above set is empty}\end{cases} \tag{7}$$

Then

$$\delta(E)=\begin{cases}-\infty & \text{if } \alpha=\delta,\ y(\delta)=\theta\\ \delta-1 & \text{if } \alpha<\delta<+\infty,\ y(\delta)=\theta\\ \delta & \text{if } \delta=-\infty, \text{or } \alpha\leqslant\delta<+\infty,\ y(\delta)\in E\end{cases}$$

and $(y, \alpha, \delta(E))$ is on approximating $\Pi(\lambda)$-chain on $\Omega(E)=(-\infty<\delta(E))=(-\infty=\alpha)\cup(-\infty<\alpha,\ y(a)\in E)$.

Proof. Suppose the random times $\alpha_n\downarrow\alpha$ and α_n leads (y, α, δ) to a $\Pi_\theta(\lambda)$-chain $(y_n, 0, \delta_n)$ on $\Omega_n=\{\alpha_n\in N, \alpha\leqslant\alpha_n\leqslant\delta\}$ respectively. Restricted to $\Omega(E)\cap\Omega_n$, the chain $(y_n, 0, \delta_n)$ is a $\Pi_\theta(\lambda)$-chain with its initial distribution concentrated in E. On $\Omega(E)$, let $\alpha_n(E)=\alpha_n$. Then on $\Omega(E)$, $\alpha_n(E)\downarrow\alpha$ and for each n, $\alpha_n(E)$ leads the stochastic chain $(y, \alpha, \delta(E))$ on $\Omega(E)$ to the chain $(y_n, 0, \delta(E))$ on $\Omega_n(E)\equiv\{\alpha_n(E)\in\mathbf{N}, \alpha\leqslant\alpha_n(E)\leqslant\delta\}$. Notice that $\Omega(E)\cap\Omega_n=\Omega_n(E)$ and $\delta_n(E)$ is just the first exit time from E of the $\Pi_\theta(\lambda)$-chain $(y_n, 0, \delta_n)$ on $\Omega_n(E)$. According to Lemma 1 the chains $(y_n, 0, \delta_n(E))$ on $\Omega_n(E)$ are all $\Pi(\lambda)$-chains. Therefore $(y, \alpha, \delta(E))$ is an approximating $\Pi(\lambda)$-chain on $\Omega(E)$. The lemma is proved. QED

In the following it will be pointed out that we can, if necessary, enlarge the measure space and always extend a $\Pi(\lambda)$-chain or an approximating $\Pi(\lambda)$-chain to a Π-chain or an approximating Π-chain. For this purpose, we first introduce the concept of conditional distributions and conditional independence about stochastic chains. We remark that 'distribution' in the following definition means a probability measure which is deduced from a random variable or a stochastic process on its path space. For example, the exponential distribution with the parameter q is the probability measure on the real line deduced from the distribution function

$$F(t)=\begin{cases}0 & \text{if } t<0\\ 1-e^{-qt} & \text{if } t\geq 0\end{cases}$$

Another example: the distribution of the Π-chain starting from i is the probability measure on the infinite-dimensional space $E\times E\times E\times\cdots$ generated by the consistent family of finite-dimensional distributions $\{R(i,i_1,i_2,\ldots,i_n)\}$, where

$$R(i,i_1,i_2,\ldots,i_n)=\Pi_{ii_1}\Pi_{i_1i_2}\cdots\Pi_{i_{n-1}i_n} \tag{8}$$

Definition 1. Let $(\Omega,\mathscr{F},P)$ be a measure space and $\mathscr{A}$ a Borel subfield of $\mathscr{F}$. Suppose ρ_n $(n\geq 1)$ are $\mathscr{F}$-measurable functions and F_n $(n\geq 1)$ are distributions on the real line. If for any $\Lambda\in\mathscr{A}$ and Borel sets B_n $(n\geq 1)$ on the real line,

$$P(\Lambda,\rho_1\in B_1,\ldots,\rho_n\in B_n)=P(\Lambda)F_1(B_1)\cdots F_n(B_n)$$

we say that given $\mathscr{A}$, ρ_n $(n\geq 1)$ are conditionally independent and have the conditional distributions F_n, respectively.

Similarly, if a quasi-stochastic chain (z,ε,δ) or a sequence of quasi-stochastic chains $(z_n,\varepsilon_n,\delta_n)$ $(n\geq 1)$ are defined on $(\Omega,\mathscr{F},P)$, we can state the definition that $\mathscr{A}$, all ρ_n, $n\geq 1$, (z,ε,δ), $(z_n,\varepsilon_n,\delta_n)$, $n\geq 1$, are conditionally independent and the definition that given $\mathscr{A}$, the conditional distribution of (z,ε,δ) is some distribution, and so on.

Lemma 3. Assume that ζ is a finite $\Pi(\lambda)$-excessive measure. Then there exist a measure space $(\Omega,\mathscr{F},P)$ and an approximating $\Pi_\theta(\lambda)$-chain (y,α,β) defined on it such that if $-\infty<\alpha$ then $y(\alpha)\in E$ and, moreover,

$$\zeta_j=\int_\Omega\sum_{\alpha\leq n\leq\delta}C_j[y(n)]\,\mathrm{d}P \qquad j\in E \tag{9}$$

Proof. According to Theorem 9.4, there exist a measure space $(\Omega_1,\mathscr{F}_1,P_1)$ and an approximating $\Pi(\lambda)$-chain (y_1,α_1,δ_1) defined on it with the characteristic measure ζ. It is well known that there exist a probability space $(\Omega_2,\mathscr{F}_2,P_2)$ and a family of independent random variables $\xi_2(i)$ $(i\in E)$, defined on it such that for each i, $\xi_2(i)$ take values 0 and 1 with probabilities $\lambda/(\lambda+d_i)$ and $d_i/(\lambda+d_i)$ respectively, where d_i is the non-conservative quantity of the state i. Take the product space $(\Omega,\mathscr{F},P)=(\Omega_1\times\Omega_2,\mathscr{F}_1\times\mathscr{F}_2,P_1\times P_2)$ and for $w=(w_1,w_2)\in\Omega$ let

$$\xi(i,w)=\xi_2(i,w_2) \qquad \alpha(w)=\alpha_1(w_1)$$

$$\delta(E,w)=\delta_1(w_1)$$

$$y(n,w)=y_1(n,w_1) \qquad \text{if } \alpha(w)\leq n\leq\delta(E,w)$$

$$\delta=\begin{cases}+\infty & \text{if } \delta(E)=+\infty\\ \delta(E)+\xi\{y[\delta(E)]\} & \text{if } \delta(E)<+\infty\end{cases}$$

When $\delta(E) < +\infty$ and $\xi\{y[\delta(E)]\} = 1$ put $y(\delta) = \theta$. Then it is easy to prove that (y, α, δ) is what we require, and the proof is completed. QED

Lemma 4. Assume that ξ is a finite $\Pi(\lambda)$-excessive measure. Then there exists a measure space $(\Omega, \mathscr{F}, P)$ on which are defined an approximating $\Pi_\theta(\lambda)$-chain (y, α, δ), a sequence of quasi-stochastic chains $(z_i, 0, q_i)$ $(i \in E)$ and a family of $\mathscr{F}$-measurable functions s_u, $u \in U$, with U being a given subset of $(-\infty, +\infty)$, such that

(i) on $(-\infty, +\infty)$, $y_\alpha \in E$ and (9) is satisfied;

(ii) given $\mathscr{A} = \mathscr{F}(y, \alpha, \delta)$ for each $u \in U$ the conditional distribution of S_u is a given distribution F_u;

(iii) given $\mathscr{A}$, for each i the conditional distribution of $(z_i, 0, \varepsilon_i)$ is the distribution of the Π-chain starting from i;

(iv) given $\mathscr{A}$, all S_u, $u \in U$; $(z_i, 0, \varepsilon_i)$, $i \in E$, are conditionally independent.

Proof. This can be verified by an application of Lemma 3 and the technique of making independent product spaces. QED

Definition 2. Suppose stochastic chains (y, ε, v) and (x, α, β) are defined on the same probability space. If $\varepsilon = \alpha$, $v \leqslant \beta$ and for $n \in \mathbf{N}$ and $\varepsilon \leqslant n \leqslant v$, $y(n) = x(n)$ holds, we call (x, α, β) an extension of (y, ε, v) and (y, ε, v) the front part of (x, α, β).

Theorem 5. Suppose that $\lambda > 0$ is fixed and assume that $\zeta(\lambda)$ is a finite $\Pi(\lambda)$-excessive measure. Then, there exist a measure space $(\Omega, \mathscr{F}, P)$ and defined on it an approximating $\Pi(\lambda)$-chain (y, α, v) and an approximating Π-chain (x, α, β) such that (x, α, β) is an extension of (y, α, v). The characteristic measure of (y, α, v) is $\zeta(\lambda)$ and the characteristic measure of (x, α, β) is

$$\zeta_j = \zeta_j(\lambda) + \lambda \sum_i \zeta_i(\lambda)(\lambda + q_i)^{-1} G_{ij} \tag{10}$$

Proof. It follows from Lemma 4 that there exist a measure space $(\Omega, \mathscr{F}, P)$ and an approximating $\Pi_\theta(\lambda)$-chain (y, α, δ) defined on it such that

$$\zeta_j(\lambda) = \int_\Omega \sum_{\alpha \leqslant n \leqslant \delta} C_j[y(n)]\, dP \qquad j \in E$$

Removing $(-\infty < \alpha,\ y(\alpha) = \theta)$ if necessary, we can always assume $\{-\infty < \alpha,\ y(\alpha) = \theta\}$ is empty. On $(\Omega, \mathscr{F}, P)$ we can also define a sequence of quasi-stochastic chains $\{z_i, 0, \varepsilon_i\}$ $(i \in E)$ such that given $\mathscr{F}(y, \alpha, \delta)$ all $(z_i, 0, \varepsilon_i)$, $i \in E$, are conditionally independent and for each i, given $\mathscr{F}(y, \alpha, \delta)$, the conditional distribution of $(z_i, 0, \varepsilon_i)$ is the distribution of the Π-chain starting from i.

Define the first exist time $\delta(E)$ of the approximating $\Pi_\theta(\lambda)$-chain from E according to (7). Abbreviate $\delta(E) = \gamma^1$. Then $\Omega = (-\infty < \gamma) = (\alpha \leqslant \gamma)$ and (y, α, γ) is an approximating $\Pi(\lambda)$-chain with the characteristic measure $\zeta(\lambda)$.

Let

$$\beta = \begin{cases} \gamma & \text{if } \gamma = +\infty \text{ or } \gamma = \delta - 1 < +\infty \\ \gamma + \varepsilon_{y(\delta)} & \text{if } \gamma = \delta < +\infty \end{cases} \tag{11}$$

For $n \in \mathbf{N}$, $\alpha \leqslant n \leqslant \beta$, define

$$x(n) = \begin{cases} y(n) & n \leqslant \gamma \\ z_{y(\delta)}(n-\delta) & \gamma < +\infty, \gamma < n \leqslant \beta \end{cases} \tag{12}$$

Obviously, (x, α, β) takes its values in E and is the extension of (y, α, γ). Next we will prove that (x, α, β) is an approximating Π-chain.

First, suppose the approximating $\Pi_\theta(\lambda)$-chain (y, α, δ) is a $\Pi_\theta(\lambda)$-chain and $\alpha = 0$. We will prove that $(x, 0, \beta)$ defined as above is a Π-chain. That is, we prove $A_n = A_{n-1}\Pi_{i_{n-1}i_n}$, where $i_0, i_1, \ldots, i_n \in E$, $n \geqslant 1$,

$$A_n \equiv P(x(0) = i_0, \ldots, x(n) = i_n) \tag{13}$$

Define $R(i_0, i_1, \ldots, i_n)$ according to (8) and similarly define

$$R(\lambda; i_0, i_1, \ldots, i_n) = \Pi_{i_0 i_1}(\lambda)\Pi_{i_1 i_2}(\lambda)\cdots\Pi_{i_{n-1}i_n}(\lambda) \tag{14}$$

Note that

$$(y(0) = i_0) \subset (0 \leqslant \gamma) \qquad (y(n) = i_n) \subset (n \leqslant \delta) \subset (n \leqslant \delta)$$

hence

$$\begin{aligned} A_n &= P(x(0) = i_0, \ldots, x(n) = i_n, n \leqslant \gamma) + P(x(0) = i_0, \ldots, x(n) = i_n, \gamma < n) \\ &= P(y(0) = i_0, \ldots, y(n) = i_n, n \leqslant \delta) \\ &\quad + P(y(0) = i_0, \ldots, y(\delta) = i_\delta, z_{y(\delta)}(1) = i_{\delta+1}, \ldots, z_{y(\delta)}(n-\delta) = i_n, \delta < n) \\ &= P(y(0) = i_0)R(\lambda, i_0, i_1, \ldots, i_n) \\ &\quad + \sum_{k=0}^{n-1} P(y(0) = i_0, y(1) = i_1, \ldots, y(k) = i_k, \delta = k, z_{i_k}(1) \\ &= i_{k+1}, \ldots, z_{i_k}(n-k) = i_n) \end{aligned}$$

By conditional independence, the summand in $\sum_{k=0}^{m-1}$ is equal to

$$\begin{aligned} &P(y(0) = i_0, y(1) = i_1, \ldots, y(k) = i_k, \delta = k)R(i_k, i_{k+1}, \ldots, i_n) \\ &= P(y(0) = i_0)R(\lambda; i_0, i_1, \ldots, i_k)\left(1 - \sum_{j \in E_\theta} \Pi i_{kj}(\lambda)\right)R(i_k, i_{k+1}, \ldots, i_n) \end{aligned}$$

Consequently,

$$\begin{aligned} A_n = P(y(0) = i_0)\Big\{ &R(\lambda; i_0, i_1, \ldots, i_n) \\ &+ \sum_{k=0}^{n-1} R(\lambda; i_0, i_1, \ldots, i_k)\frac{\lambda}{\lambda + q_{i_k}}R(i_k, i_{k+1}, \ldots, i_n)\Big\} \end{aligned}$$

$$= P(y(0) = i_0)\Big\{R(\lambda; i_0, i_1, \ldots, i_{n-1})\Pi_{i_{n-1}i_n}(\lambda)$$

$$+ R(\lambda; i_0, i_1, \ldots, i_{n-1})\lambda(\lambda + q_{i_{n-1}})^{-1}\Pi_{i_{n-1}i_n}$$

$$+ \sum_{k=0}^{n-1} R(i_0, \ldots, i_k)\lambda(\lambda + q_{i_k})^{-1}R(i_k, \ldots, i_{n-1})\Pi_{i_{n-1}i_n}\Big\}.$$

Noting

$$\Pi_{i_{n-1}i_n}(\lambda) + \lambda(\lambda + q_{i_{n-1}})^{-1}\Pi_{i_{n-1}i_n} = \Pi_{i_{n-1}i_n}$$

we find that $A_n = A_{n-1}\Pi_{i_{n-1}i_n}$.

On the other hand,

$$\zeta_j = \int_\Omega \sum_{0 \leqslant n \leqslant \beta} C_j[x(n)]\,\mathrm{d}P = \int_\Omega \sum_{0 \leqslant n \leqslant \gamma} C_j[y(n)]\,\mathrm{d}P + \int_{\gamma < +\infty} \sum_{\gamma < n \leqslant \beta} C_j[x(n)]\,\mathrm{d}P$$

$$= \zeta_j(\lambda) + \sum_i \int_{\substack{\delta < +\infty \\ y(\delta) = i}} \sum_{0 \leqslant n \leqslant \varepsilon_i} C_j[z_i(n)]\,\mathrm{d}P$$

By conditional independence the summand in Σ is equal to $P(\delta < +\infty,\ y(\delta) = i)G_{ij}$, and

$$P(\delta < +\infty,\ y(\delta) = i) = \sum_{n=0}^{+\infty} P(y(n) = i, \delta = n) = \sum_{n=0}^{+\infty} P(y(n) = i)\frac{\lambda}{\lambda + q_1}$$

$$= \lambda\zeta_i(\lambda)(\lambda + q_i)^{-1} \tag{15}$$

Therefore (10) is verified.

Now suppose that (y, α, δ) is an approximating $\Pi_\theta(\lambda)$-chain and that $\alpha_n \downarrow \alpha$ and α_n leads (y, α, δ) to $\Pi_\theta(\lambda)$-chains $(y_n, 0, \delta_n)$ respectively. Suppose α_n leads (x, α, β) to $(x_n, 0, \beta_n)$ for each n. Let γ be the first exit time of (y, α, δ) from E. Just as we have defined the extension process (x, α, β) of (y, α, γ) from (y, α, δ) and $(z_i, 0, \varepsilon_i)$ $(i \in E)$ in the fashion of (11)–(12), for each n we define the extension process of $(y_n, 0, \gamma_n)$ from $(y_n, 0, \delta_n)$ and $(z_i, 0, \varepsilon_i)$ $(i \in E)$ in the fashion of (11)–(12). Here γ_n is the first exit time of $(y_n, 0, \delta_n)$ from E. Then this extension process is just $(x_n, 0, \beta_n)$. Since, given $\mathscr{F}(y, \alpha, \delta)$, $(z_i, 0, \varepsilon_i)$ $(i \in E)$ are conditionally independent and the conditional distribution is that of the Π-chain starting from i, it follows that given $\mathscr{F}(y_n, 0, \delta_n)$ the same assertion remains true. From what we have proved, $(x_n, 0, \beta_n)$ are Π-chains. Therefore (x, α, β) is an approximating Π-chain.

Let ζ^n and $\zeta^n(\lambda)$ be the characteristic measure determined by $(x_n, 0, \beta_n)$ and $(y_n, 0, \gamma_n)$, respectively. Then we have proved that

$$\zeta_j^n = \zeta_j^n(\lambda) + \lambda \sum_i \zeta_i^n(\lambda)(\lambda + q_i)^{-1}G_{ij}$$

Since $\zeta^n \uparrow \zeta$ and $\zeta^n(\lambda) \uparrow \zeta(\lambda)$, (10) follows. And the proof is completed. QED

11.12 Q PROCESS ON A MEASURE SPACE

We will extend the concept of denumerable Markov processes defined on a probability space to that on a measure space $(\Omega, \mathscr{F}, P)$. Write T for $[0, +\infty)$ or $(0, +\infty)$. Suppose that E has the discrete topology. Let $\bar{E} = E \cup \{\infty\}$ $(\infty \notin E)$ be the one-point compactification of E.

Definition 1. Let $(\Omega, \mathscr{F}, P)$ be a measure space and σ be an $\mathscr{F}$-measurable function valued in $[0, +\infty)$. Assume that for every fixed $t \in T$ and almost every $\omega \in \{t < \sigma\}, x(t, \omega)$ is defined an takes a value in E. If for any $t \in T$ and $i \in E$, $P(x(t) = \infty) = 0$ and, moreover,

$$\Delta(t, i) \equiv (x(t) = i) = (\omega : x(t, \omega) = i, t < \sigma(\omega)) \in \mathscr{F} \tag{1}$$

we call $X = \{x(t),\ t \in T \cap [0, \sigma)\}$ a quasi-random process. If $P(\Delta(t, i)) < +\infty$ for every $t \in T$ and $i \in E$ we say that $X = \{x(t),\ t \in T \cap [0, \sigma)\}$ is a random process and that the minimal Borel field containing all $\Delta(t, i)$ $(t \in T,\ i \in E)$, denoted by $\mathscr{F}(X)$ or $\mathscr{F}\{X(t),\ t \in T \cap [0, \sigma)\}$, is the Borel field generated by X. When $T = (0, +\infty)$, a random process $X = \{x(t),\ 0 < t < \delta\}$ is said to be open.

Definition 2. Let $P(t) = (p_{ij}(t),\ i, j \in E)$, $t \geqslant 0$, be a standard generalized transition matrix and assume $X = \{x(t),\ t \in T \cap [0, \sigma)\}$ to be a random process. If for any $n \geqslant 1$, $t_0 \in T$, $t_0 < t_1 < \cdots < t_{n+1}$ and $i_0, i_1, \ldots, i_{n+1} \in E$,

$$P(x(t_0) = i_0, \ldots, x(t_{n+1}) = i_{n+1}) = P(x(t_0) = i_0, \ldots, x(t_n) = i_n) p_{i_n i n+1}(t_{n+1} - t_n) \tag{2}$$

we call X a Markov process with the transition probability matrix $P(t)$. X is called a Q process if the Q matrix of $P(t)$ is the matrix Q. X is said to be a minimal process if $P(t)$ is the minimal solution.

Theorem 1. Let $X = \{x(t), t \in T \cap [0, \sigma)\}$ be a Markov process with the transition probability matrix $P(t)$. Suppose S is an $\mathscr{F}$-measurable function and given $\mathscr{F}(X)$, its conditions distribution is the exponential distribution with parameter 1. For $\lambda > 0$ let $p_{ij}^{\lambda}(t) = e^{-\lambda t} p_{ij}(t)$ and $\sigma(\lambda) = \min(\sigma, S/\lambda)$. Then $X^{\lambda} = \{x(t), t \in T \cap [0, \sigma(\lambda))\}$ is a Markov process with the transition probability matrix $P^{\lambda}(t) = (p_{ij}^{\lambda}(t), i, j \in E)$. In particular, if X is a minimal Q process, then X^{λ} is a minimal $Q(\lambda)$ process.

Proof. It is easy by making use of the conditional independence. QED

Lemma 2. Given a totally finite measure $v = (v_i, i \in E)$, there exist a totally finite measure space $(\Omega, \mathscr{F}, P)$, a Π-chain $(x, 0, \beta)$ and $\mathscr{F}$-measurable functions ρ_i^n $(n \geqslant 0, i \in E)$ defined on it, such that the initial distribution of $(x, 0, \beta)$ is v; given $\mathscr{F}(x, 0, \beta)$, all ρ_i^n $(n \geqslant 0,\ i \in E)$ are conditionally independent, and for each i the conditional distribution of ρ_i^n is the exponential distribution with the parameter q_i.

Noting the above, the characteristic measure of $(x,0,\beta)$ is $\zeta = vG$ while vG is not necessarily finite. So Lemma 2 is not a direct corollary of Lemma 11.4. Since v is totally finite, it follows that with the aid of the ordinary existence theorem for Markov processes, there exists a Π-chain that is defined on a totally finite measure space and has the initial distribution v. From this, imitating the proof of Lemma 11.4, we can prove Lemma 2.

The following theorem follows immediately from the construction of minimal processes.

Theorem 3. Assume that v, $(x,0,\beta)$ and ρ_i^n $(n \geqslant 0,\ i \in E)$ are the same as in Lemma 2. For $0 \leqslant n \leqslant \beta$,, let $\rho^n = \rho^n_{x(\tau_n)}$, $\tau_0 = 0$, $\tau_n = \sum_{i=0}^{n-1} \rho^i$ and $\sigma = \tau_{\beta+1}$. For $0 \leqslant t < \sigma$, let

$$X(t) = x(n) \qquad \text{if } \tau_n \leqslant t < \tau_{n+1} \tag{3}$$

Then $X = \{x(t),\ 0 \leqslant t \leqslant \sigma\}$ is a minimal Q process.

Similarly the following Lemma 4 and Theorem 5 are also clear.

Lemma 4. Assume that $v = (v_i, i \in E)$ is a totally finite measure. Then there exist a totally finite measure space $(\Omega, \mathscr{F}, P)$ and defined on it a $\Pi_\theta(\lambda)$-chain $(y,0,\delta)$ and $\mathscr{F}$-measurable functions ρ^θ, ρ_i^n $(n \geqslant 0, i \in E)$ such that the initial distribution of $(y,0,\delta)$ is γ. Given $\mathscr{F}(y,0,\delta)$ all ρ^θ, ρ_i^n $(n \geqslant 0, i \in E)$ are conditionally independent and, furthermore, the conditional distributions are exponential distributions with parameters λ and $\lambda + q_i$ respectively.

Theorem 5. Assume that v, $(y,0,\delta)$ and ρ^θ, $\rho_i^n(n \geqslant 0,\ i \in E)$ are the same as in Lemma 4. Let $\delta(E)$ be the first exit time of the $\Pi_\theta(\lambda)$-chain $(y,0,\delta)$ from E. For $0 \leqslant n \leqslant \delta(E)$, let

$$\rho^n = \rho^n_{y(n)} \qquad \tau_0 = 0 \qquad \tau_{n+1} = \sum_{i=0}^{n} \rho^i$$

$$\sigma = \begin{cases} \sum_{i=0}^{+\infty} \rho^i & \text{if } \delta(E) = +\infty \\ \tau_{\delta(E)+1} & \text{if } \delta(E) < +\infty \end{cases}$$

$$\sigma_\theta = \begin{cases} \sum_{i=0}^{+\infty} \rho^i & \text{if } \delta(E) = +\infty \\ \tau_{\delta(E)+1} & \text{if } \delta(E) = \delta < +\infty \\ \tau_{\delta(E)+1} + \rho^\theta & \text{if } \delta(E) = \delta - 1 < +\infty \end{cases}$$

For $0 \leqslant t < \sigma_\theta$, let

$$X(t) = y(n) \qquad \text{if } \tau_n \leqslant t < \tau_{n+1}$$

Then $X = \{x(t), 0 \leq t < \sigma\}$ is a minimal $Q(\lambda)$ process and $X_\theta = \{x(t), 0 \leq t < \sigma_\theta\}$ is a minimal $Q_\theta(\lambda)$ process. Their initial distributions are all γ.

11.13 APPROXIMATING MINIMAL Q PROCESSES

Recall that E is endowed with the discrete topology and is compactified by the one point '∞'. Let finite sets $D_n \subset E$ and $D_n \uparrow E$ as $n \uparrow \infty$.

Definition 1. Assume that $0 < \sigma \leq +\infty$. A function $X = \{x(t), 0 \leq t < \sigma\}$ defined on $[0, \sigma)$ is called a jump function of type U if the following conditions are satisfied:

(i) $x(0) \in E \cup \{\infty\}$, $x(t) \in E$ $(0 < t < \sigma)$ and X is right-continuous;

(ii) for any $[c, d] \subset (0, \sigma)$, X has only a finite number of jump points in $[c, d]$;

(iii) for any $i \in E$ and any $d \in (0, \sigma)$, X has only a finite number of i intervals in $(0, d)$.

Definition 2. A random process $X = \{x(t), 0 \leq t < \sigma\}$ is called an approximating minimal Q process defined on a measure space $(\Omega, \mathscr{F}, P)$ if for any $t \geq 0$ and $i \in E$,

$$(x(t) = i) \in \mathscr{F} \qquad P(x(t) = i) < +\infty$$

and the following conditions are satisfied:

(i) Equation (12.2) holds for $0 \leq t_0 < t_1 < \cdots < t_{n+1}, i_0, i_1, \ldots, i_{n+1} \in E$, where $P(t)$ is a minimal solution $\bar{P}(t)$.

(ii) All paths of X are jump functions of type U.

(iii) There exists a sequence of $[0, +\infty]$-valued random times $\tau_n \downarrow 0$ such that $X_n = \{x(\tau_n + t), \ 0 \leq t < \sigma - \tau_n\}$ is a minimal Q process defined on $\Omega_n = (\tau_n < \sigma)$ and that the support of the initial distribution of X is contained in D_n.

Note that a minimal Q process is an approximating minimal Q process. In this case, τ_n may be taken as 0 or as the hitting time at D_n. When we consider the restriction on $\Omega' = (\sigma = 0) \cup (\sigma > 0, x(0) \neq \infty)$ of the approximating minimal Q process, we see it is a minimal Q process. In particular, if $P\{x(0) = \infty\} = 0$ the approximating Q process X is just a Q process. If we still work on $(\Omega, \mathscr{F}, P)$ but restrict the parameter set to $(0, \infty)$, then $X' = \{x(t), 0 < t < \sigma\}$ is an open minimal Q process. However, an approximating minimal Q process may not necessarily be a minimal Q process because for an approximating minimal Q process $X = \{x(t), 0 \leq t < \sigma\}$, the measure of the set $\{x(0) = \infty\}$ may be infinite although it is $\mathscr{F}$-measurable. But if $P(x(0) = \infty) = 0$, then an approximating minimal Q process is precisely a minimal Q process.

Definition 3. Assume that $X = \{x(t), 0 \leq t < \sigma\}$ is an approximating minimal Q

process. Let

$$\eta_j(\lambda) = \int_\Omega \int_0^\sigma e^{-\lambda t} C_j[x(t)] \mathrm{d}P \qquad \lambda > 0, \quad j \in E \tag{1}$$

$$\eta_j = \int_\Omega \int_0^\sigma C_j[x(t)] \mathrm{d}P \qquad j \in E \tag{2}$$

We call η the characteristic measure of X and $\eta(\lambda)$ its λ-characteristic measure. Theorem 14.3 will prove that $(\eta(\lambda), \lambda > 0)$ is an entrance family (see Definition 2.11.1). So we call it the characteristic entrance family of X.

Theorem 1. Suppose τ_n are a sequence of random times which are given in the definition of the approximating minimal Q process $X = \{x(t), 0 \leqslant t < \sigma\}$. Suppose for each n, τ_n leads X to a minimal Q process $X_n = \{x_n(t), 0 \leqslant t < \sigma_n\}$ on $\Omega_n = (\tau_n < \sigma)$. Then on Ω for any $t \in [0, \sigma)$,

$$\lim_{n \to +\infty} X_n(t) = X(t) \qquad t \in [0, \sigma) \tag{3}$$

Proof. Since $\tau_n \downarrow 0$, $\sigma_n = \sigma - \tau_n \uparrow \sigma$. Then by the right-continuity of X, for $t \in [0, \sigma)$ we have

$$\lim_{n \to +\infty} X_n(t) = \lim_{n \to +\infty} X(\tau_n + t) = X(t) \qquad \text{QED}$$

11.14 ENTRANCE FAMILIES AND APPROXIMATING MINIMAL PROCESSES

Fix $\lambda > 0$. Assume that a totally finite measure $\eta(\lambda)$ satisfies the inequality

$$\lambda u - uQ \geqslant 0 \tag{1}$$

Then $\zeta_j(\lambda) = \eta_j(\lambda)(\lambda + q_j)$ is a finite $\Pi(\lambda)$-excessive measure. According to Theorem 9.4, there exist totally finite measures $v^n(\lambda)$ such that:

(i) the support of $v^n(\lambda)$ is contained in D_n;

(ii) $v^n(\lambda)G(\lambda) \uparrow \zeta(\lambda)$ as $n \uparrow \infty$ and $(v^n(\lambda)G(\lambda))_j = \zeta_j$, $j \in D_n$;

(iii) for $m < n$, the hitting distribution that the $\Pi(\lambda)$-chain with the initial distribution $v^n(\lambda)$ hits D_m is $v^m(\lambda)$.

Note that (ii) is equivalent to

$$\begin{aligned} &v^n(\lambda)\phi(\lambda) \uparrow \eta(\lambda) \qquad n \uparrow +\infty \\ &(v^n(\lambda)\phi(\lambda))_j = \eta_j(\lambda) \qquad j \in D_n \end{aligned} \tag{2}$$

Theorem 1. Fix $\lambda > 0$. Assume that the totally finite measure $\eta(\lambda)$ satisfies (1).

Then there exist a measure space $(\Omega, \mathscr{F}, P)$ and an approximating minimal $Q_\theta(\lambda)$ process $X = \{x(t), 0 \leqslant t < \sigma(\theta)\}$ defined on it, such that:

(i) if we let τ_n be the hitting time of X at D_n, then τ_n leads X to the minimal $Q_\theta(\lambda)$ process with the initial distribution $v^n(\lambda)$ for each n;

(ii)

$$\eta_j(\lambda) = \int_\Omega \int_0^{\sigma(\theta)} C_j[X(t)]\,dt\,dP \qquad j \in E \tag{3}$$

$$P(X(0) = \theta) = 0 \qquad P(x(0) = i) = v_i(\lambda) \tag{4}$$

where

$$v(\lambda) = \lambda\eta(\lambda)\eta(\lambda)Q \tag{5}$$

Particularly, if $\eta(\lambda)$ is the solution of the equation

$$\lambda u - uQ = 0 \tag{6}$$

then $x(0) = \infty$ almost surely on Ω.

Proof. Because $\zeta_j(\lambda) = \eta_j(\lambda)(\lambda + q_j)$ $(j \in E)$ is a $\Pi(\lambda)$-excessive measure, on the basis of Lemma 11.4 there exists a measure space $(\Omega, \mathscr{F}, P)$ on which an approximating $\Pi_\theta(\lambda)$-chain (y, α, δ) can be defined such that

$$P(-\infty < \alpha, y(\alpha) = \theta) = 0 \qquad P(-\infty < \alpha, y(\alpha) = i) = v_i(\lambda)$$

and on $(-\infty = \alpha)$ almost surely

$$\lim_{n \to -\infty} y(n) = \infty$$

and

$$\zeta_j(\lambda) = \int_\Omega \sum_{\alpha \leqslant n \leqslant \delta} C_j[y(n)]\,dP \qquad j \in E \tag{7}$$

and the hitting distribution that the chain (y, α, δ) hits D_n is $v^n(\lambda)$. On $(\Omega, \mathscr{F}, P)$ a family of $\mathscr{F}$-measurable functions ρ_θ and ρ_i^n $(n \geqslant 0, i \in E)$ can also be defined such that given $\mathscr{F}(y, \alpha, \delta)$, all ρ_θ, ρ_i^n $(n \geqslant 0, i \in E)$ are conditionally independent and the conditional distributions of ρ_θ and ρ_i^n are exponential distributions with parameters λ and $\lambda + q_i$ respectively. Let $\delta(E)$ be the first exit time of the chain (y, α, δ) from E. Then $(-\infty < \delta(E)) = (\alpha \leqslant \delta(E)) = \Omega$. If $\alpha \leqslant n \leqslant \delta(E)$, we let $\rho^n = \rho^n_{y(n)}$; and if $\delta(E) = \delta - 1 < +\infty$, we let $\rho^\delta = \rho_\theta$. Then we proceed to prove

$$\int_\Omega \sum_{\alpha \leqslant n \leqslant \delta} \rho^n \mathrm{d}P < +\infty \tag{8}$$

In fact

$$\int_\Omega \sum_{\alpha \leqslant n \leqslant \delta} \rho^n \mathrm{d}P = \sum_{n \in N} \int_{\alpha \leqslant n \leqslant \delta(E)} \rho^n_{y(n)} \mathrm{d}P + \int_{\delta(E) = \delta - 1 < +\infty} \rho^\delta \mathrm{d}P$$

By conditional independence the first part of the above formula is equal to

$$\sum_{n\in N}\sum_{i\in E}\int_{y(n)=i}\rho_i^n\mathrm{d}P=\sum_{n\in N}\sum_{i\in E}P(y(n)=i)(\lambda+q_i)^{-1}=\sum_{i\in E}\sum_{n\in N}P(y(n)=i)(\lambda+q_i)^{-1}$$
$$=\sum_{i\in E}\zeta_i(\lambda)(\lambda+q_i)^{-1}=\sum_{i\in E}\eta_i(\lambda)<+\infty$$

By a deduction similar to (27) the second part is equal to

$$\int_{\substack{\delta<+\infty\\ y(\delta)=\theta}}\rho_\theta\mathrm{d}P=P(\delta<+\infty,y(\delta)=\theta)^{\lambda-1}$$
$$=\lambda^{-1}\sum_{i\in E}P(\delta(E)<+\infty,y(\delta(E))=i,y(\delta(E)+1)=\theta)$$
$$=\lambda^{-1}\sum_{i\in E}P(\delta(E)<+\infty,y(\delta(E))=i)\Pi_{i\theta}(\lambda)$$
$$=\lambda^{-1}\sum_{i\in E}\lambda\zeta_i(\lambda)(\lambda+q_i)^{-1}\Pi_{i\theta}(\lambda)$$
$$=\sum_{i\in E}\eta_i(\lambda)\Pi_{i\theta}(\lambda)\leqslant\sum_{i\in E}\eta_i(\lambda)<+\infty$$

Now we define $X=\{x(t),0\leqslant t<\sigma(\theta)\}$. Let

$$\tau_\alpha=0\qquad\tau_n=\sum_{\alpha\leqslant k<n}\rho^k\qquad(\alpha<n<\delta)\qquad\sigma(\theta)=\tau_{\delta+1}\tag{9}$$

For $0\leqslant t<\sigma(\theta)$ let

$$\begin{aligned}x(t)&=y(n)&&\text{if }\tau_n\leqslant t<\tau_{n+1}\\ x(0)&=\infty&&\text{if }-\infty=\alpha\end{aligned}\tag{10}$$

We shall prove that $X=\{x(t),0\leqslant t<\sigma(\theta)\}$ is what we want.

First, (4) is clear. For each path of X, (i)–(ii) in Definition 1 hold (substitute E_θ for E). From (7) and $\zeta_j(\lambda)<+\infty$ we have almost surely $\sum_{\alpha\leqslant n<\delta}C_j[y(n)]<+\infty$. Hence condition (iii) in Definition 1 also holds. Therefore the paths of X are all jump functions of type U.

Next, for $j\in E$, (3) follows from

$$\int_\Omega\int_0^{\delta(\theta)}C_j[x(t)]\mathrm{d}t\,\mathrm{d}P=\int_\Omega\sum_{\alpha\leqslant n\leqslant\delta}C_j[y(n)]\rho^n\mathrm{d}P$$
$$=\int_\Omega\sum_{\alpha\leqslant n\leqslant\delta}C_j[y(n)]\rho_j^n\mathrm{d}P$$
$$=\sum_{n\in E}\int_{y(n)=j}\rho_j^n\mathrm{d}P=\sum_{n\in N}P(y(n)=j)(\lambda+q_j)^{-1}$$
$$=\zeta_j(\lambda)(\lambda+q_j)^{-1}=\eta_j(\lambda)$$

Moreover, for each n let α_n be the hitting time of the approximating $\Pi_\theta(\lambda)$- chain (y, α, δ) at D_n. Then the hitting time of X at D_n is

$$b_n = \begin{cases} \tau_{\alpha_n} & \text{if } \alpha_n < +\infty \\ +\infty & \text{if } \alpha_n = +\infty \end{cases}$$

Suppose for each n, α_n leads (y, α, δ) to the $\Pi_\theta(\lambda)$-chain $(y_n, 0, \delta_n)$ on $\Omega_n = (\alpha_n \leqslant \delta)$. On Ω_n let $\sigma_n = \sigma(\theta) - b_n$ and $x_n(t) = x(b_n + t)$ for $0 \leqslant t < \sigma_n$. Then $X_n = \{x_n(t), 0 \leqslant t < \sigma_n\}$ is a Markov chain defined on Ω_n with the initial distribution $v^n(\lambda)$. As α_n is a random time of the chain (y, α, δ) which is independent of the future, given $\mathscr{F}(y_n, 0, \delta_n)$, ρ^θ, $\bar{\rho}_i^m \equiv \rho_i^{\alpha_n + m}$ $(m \geqslant 0, i \in E)$ are conditionally independent and given $\mathscr{F}(y_n, 0, \delta_n)$ the conditional distributions of ρ^θ and $\bar{\rho}_i^m$ are exponential distributions with parameters λ and $\lambda + q_i$ respectively. In the fashion of Theorem 12.5, from $(y_n, 0, \delta_n)$ and $\rho^\theta, \bar{\rho}_i^m$ $(m \geqslant 0, i \in E)$ we can obtain a minimal $Q_\theta(\lambda)$ process that has the same initial distribution as $(y_n, 0, \delta_n)$. This minimal $Q_\theta(\lambda)$ process is none other than X_n. But $(\alpha_n < +\infty) = (b_n < \sigma(\theta))$, hence for each n, X_n is a $Q_\theta(\lambda)$ process defined on $\Omega_n = (b_n < \sigma(\theta))$ and the support of the initial distribution of X_n is contained in D_n.

Finally we are going to prove that (12.2) holds for $0 \leqslant t_\theta < t_1 < \cdots < t_{n+1}, i_0, i_1, \ldots, i_{n+1} \in E$ and the minimal solution $\bar{P}^\lambda(t) = (\bar{p}_{ij}^\lambda(t), i, j \in E_\theta)$ to the Q matrix $Q_\theta(\lambda)$.

In fact, $\alpha_n \downarrow \alpha$ since $y(\alpha) \in E$ on $(-\infty < \alpha)$. Hence $\Omega_n = (\alpha_n < +\infty) \uparrow \Omega$. Since on $\Omega, \sigma(\theta) < +\infty$, it follows that $b_n = \tau_n \downarrow 0$ and so $\sigma(\theta) - b_n \uparrow \sigma(\theta)$. Therefore for all $t \in [0, \sigma(\theta))$, $x_n(t) = x(b_n + t) \to x(t)$. Hence

$$(x(t_0) = i_0, \ldots, x(t_{m+1}) = j_{m+1}) = \lim_{n \to +\infty} \Omega_n \cap (X_n(t_0) = i_0, \ldots, X_n(t_{m+1}) = i_{m+1})$$

so that

$$\begin{aligned} &P(X(t_0) = i_0, \ldots, X(t_{m+1}) = i_{m+1}) \\ &\quad = \lim_{n \to +\infty} P(\Omega_n, X_n(t_0) = i_0, \ldots, X_n(t_{m+1}) = i_{m+1}) \\ &\quad = \lim_{n \to +\infty} P(\Omega_n, X_n(t_0) = i_0) \bar{p}_{i_0 i_1}^\lambda(t_1 - t_0) \cdots \bar{P}_{i_m i_{m+1}}^\lambda(t_{m+1} - t_m) \\ &\quad = P(X(t_0) = i) \bar{P}_{i_0 i_1}^\lambda(t_1 - t_0) \cdots \bar{P}_{i_m i_{m+1}}^\lambda(t_{m+1} - t_m) \end{aligned}$$

Thus, $X = \{x(t), 0 \leqslant t < \sigma(\theta)\}$ is an approximating minimal $Q_\theta(\lambda)$ process. The proof is terminated. QED

The following theorem is obvious.

Theorem 2. Fix $\lambda > 0$. Assume that $X = \{x(t), 0 \leqslant t < \sigma(\theta)\}$ is an approximating minimal $Q_\theta(\lambda)$ process. Define $\eta_j(\lambda)$ $(j \in E)$ according to (3). Let $\sigma(E)$ be the last exit time of X from E, that is,

$$\sigma(E)=\begin{cases}\sup\{t:0\leqslant t<\sigma(\theta),x(t)\in E\} & \\ -\infty & \text{if the above set is empty}\end{cases} \tag{11}$$

Then $X_E=\{X(t),0\leqslant t<\sigma(E)\}$ is an approximating minimal $Q(\lambda)$ process which is defined on $\Omega(E)=(-\infty<\sigma(E))$ and has the λ-characteristic measure $\eta(\lambda)$.

Theorem 3. Assume that $(\eta(\lambda),\lambda>0)$ is an entrance family and the number $\mu>0$ is fixed. Then there exists a measure space $(\Omega,\mathscr{F},P)$ on which are defined an approximating minimal $Q(\mu)$ process $Y=\{y(t),0\leqslant t<\sigma(E)\}$ and an approximating minimal Q process $X=\{x(t),0\leqslant t<\sigma\}$, such that:

(i) X is an extension of Y, i.e. $\sigma(E)\leqslant\sigma$ and $y(t)=x(t)$ for $t\in[0,\sigma(E))$;

(ii) $P\{y(0)=i\}=P\{x(0)=i\}\equiv v_i$, $i\in E$; in particular, if $v=0$, then $y(0)=x(0)=\infty$ almost surely;

(iii) for each n, let τ_n be the hitting time of Y at D_n, then $\tau_n\downarrow 0$ $(n\uparrow+\infty)$ and τ_n leads Y and X respectively to a minimal $Q(\mu)$ process Y_n and to a minimal Q process X_n on $\Omega_n=(\tau_n<+\infty)$, which have the same initial distribution $v^n(\lambda)$;

(iv) the characteristic measure of Y is $\eta(\mu)$;

(v) the characteristic entrance family of X is $(\eta(\lambda),\lambda>0)$;

(vi) the characteristic measure of X is the standard image of $(\eta(\lambda),\ \lambda>0)$ (see (2.11.9)).

Proof. From Theorem 1, there exists a measure space $(\Omega,\mathscr{F},P)$ on which can be defined an approximating $Q_\theta(\mu)$ process $Y_\theta=\{y(t),0\leqslant t<\sigma(\theta)\}$ such that $P\{y(0)=\theta\}=0$, $P\{y(0)=i\}=v_i$ and, moreover,

$$\eta_j(\mu)=\int_\Omega\int_0^{\sigma(\theta)}C_j[y(t)]\,\mathrm{d}t\,\mathrm{d}P \qquad j\in E \tag{12}$$

By the technique of independent product spaces, we can also suppose that on $(\Omega,\mathscr{F},P)$ a family of quasi-random processes $Z_i=\{z_i(t),0\leqslant t<\varepsilon_i\}$ $(i\in E)$ are defined such that given $\mathscr{F}(Y_\theta)$, all Z_i, $i\in E$, are conditionally independent and for each i the conditional distribution of Z_i is the distribution of the minimal Q process starting from i.

Let $\sigma(E)$ be the last exit time of Y_θ from E. Then, clearly $Y=\{y(t),0\leqslant t<\sigma(E)\}$ is an approximating minimal $Q(\lambda)$ process and the characteristic measure of Y is $\eta(\mu)$, i.e. (iv) holds.

Let

$$\sigma=\begin{cases}\sigma(E) & \text{if } \sigma(E)<\sigma(\theta) \text{ or } \sigma(E)=+\infty\\ \sigma(E)+\varepsilon_{y(\sigma(E)-0)} & \text{if } \sigma(E)=\sigma(\theta)<+\infty\end{cases} \tag{13}$$

For $0 \leqslant t < \sigma$, let

$$X(t) = \begin{cases} y(t) & \text{if } 0 \leqslant t < \sigma(E) \\ Z_{y(\sigma(E)-0)}(t - \sigma(E)) & \text{if } \sigma(E) = \sigma(\theta) < +\infty \end{cases} \tag{14}$$

$$\sigma(E) \leqslant t < \sigma.$$

Then X is an extension of Y, i.e. (i) holds and so does (ii). Suppose the characteristic measure of X is η. Then

$$\eta_j = \int_\Omega \int_0^\sigma C_j[x(t)]\,\mathrm{d}t\,\mathrm{d}p$$

$$= \int_\Omega \int_0^{\sigma(E)} C_j[y(t)]\,\mathrm{d}t\,\mathrm{d}p + \int_{\sigma(E)=\sigma(\vartheta)<+\infty} \int_{\sigma(E)}^{\sigma} C_j[Z_{y(\sigma(E)-0)}(t - \sigma(E))\,\mathrm{d}t\,\mathrm{d}P$$

$$= \eta_j(\mu) + \sum_{i \in E} \int_{\substack{\sigma(E)=\sigma(\vartheta)<+\infty \\ y(\sigma(E))=i}} \int_0^{\varepsilon_i} C_j[Z_i(u)]\,\mathrm{d}u\,\mathrm{d}p$$

By conditional independence and (11.5) the above is equal to

$$\eta_j(\mu) + \sum_{i \in E} P(y(\sigma(E)) = i, \sigma(E) = \sigma(\theta) < +\infty)\Gamma_{ij}$$

$$= \eta_j(\mu) + \sum_{i \in E} P(y(\sigma(\theta)) = i)\Gamma_{ij}$$

$$= \eta_j(\mu) + \mu \sum_{i \in E} \eta_i(\mu)\Gamma_{ij}$$

This is just the standard image (2.11.9) of $(\eta(\lambda), \lambda > 0)$ and so (vi) is verified. We now prove claim (iii). For each n, let τ_n be the hitting time of Y at D_n. Then $\tau_n \downarrow 0$. Suppose τ_n leads Y_θ, Y and X respectively to $Y_{n\theta}$, Y_n and X_n on $\Omega_n = (\tau_n < +\infty)$. By Theorem 1, Y_n is a minimal $Q(\mu)$ process with the initial distribution $v^n(\mu)$ for each n. Clearly, X_n is an extension of Y_n and, furthermore, they have the same initial distribution. Hence if we can prove that X_n $(n \geqslant 1)$ are minmal Q processes, then X is an approximating minimal Q process.

Just as we obtain X from Y_θ and Z_i $(i \in E)$ in the fashion (13)–(14), on Ω_n we can derive X_n from $Y_{n\theta}$ and Z_i $(i \in E)$ in the same fashion. Hence we need only prove that if Y_θ is a minimal $Q_\theta(\lambda)$ process, then the process X obtained in the fashion of (13)–(14) is also a minimal Q process. Notice that under the above hypothesis y is a minimal $Q(\lambda)$ process.

For $i_0, i_1, \ldots, i_{n+1} \in E$ and $0 \leqslant t_0 < t_1 < \cdots < t_{n+1}$, assume that

$$\Delta_k = \bigcap_{j=0}^{k} (x(t_j) = i_j) \qquad \Lambda_k = \bigcap_{j=0}^{k} (y(t_j) = i_j)$$

We are to prove

$$R_k(i,u)=\sum_{j=k}^{n+1}(Z_i(t_j-u)=i_j, 0\leqslant t_j-u\leqslant \varepsilon_i)$$

$$P(\Delta_{n+1})=P(\Delta_n)\bar{P}_{i_n i_{n+1}}(t_{n+1}-t_n) \tag{15}$$

Clearly

$$P(\Delta_{n+1})=A_{n+1}+B_{n+1}+C_{n+1} \tag{16}$$

where

$$A_{n+1}=P(\Delta_{n+1}, t_{n+1}<\sigma(E))$$

$$B_{n+1}=P(\Delta_{n+1}, \sigma(E)\leqslant t_0)$$

$$C_{n+1}=\sum_{k=0}^{n}P(\Delta_{n+1}, t_k<\sigma(E)\leqslant t_{k+1})$$

It follows that

$$(\Delta_{n+1}, t_{n+1}<\sigma(E))=(\Lambda_{n+1}, t_{n+1}<\sigma(E))=\Lambda_{n+1}$$

$$A_{n+1}=P(\Lambda_{n+1})=P(\Lambda_n)\bar{P}^{\lambda}_{i_n i_{n+1}}(t_{n+1}-t_n)$$

$$=A_n\bar{P}_{i_n i_{n+1}}(t_{n+1}-t_n)\exp[-\lambda(t_{n+1}-t_n)] \tag{17}$$

By conditional independence

$$B_{n+1}=\sum_{i\in E}P(y(\sigma(E)-0)=i, R_0(i,\sigma(E)))$$

$$=\sum_i\int_0^{t_0}\bar{P}_{i_0 i_1}(t_1-t_0)\cdots\bar{P}_{i_n i_{n+1}}(t_{n+1}-t_n)\,\mathrm{d}u\,P(y(\sigma(E)-0)$$

$$=i, \sigma(E)\leqslant u, Z_i(t_0-\sigma(E))=i_0)$$

$$=\sum_i\int_0^{t_0}\bar{P}_{i_0 i_1}(t_1-t_0)\cdots\bar{P}_{i_{n-1} i_n}(t_n-t_{n-1})\,\mathrm{d}u\,P(y(\sigma(E)-0)$$

$$=i, \sigma(E)\leqslant u, Z_i(t_0-\sigma(E))=i_0)\bar{P}_{i_n i_{n+1}}(t_{n+1}-t_n)$$

$$=B_n\bar{P}_{i_n i_{n+1}}(t_{n+1}-t_n)$$

By a similar calculation we have

$$P(\Delta_{n+1}, t_k<\sigma(E)\leqslant t_{k+1})=P(\Delta_n, t_k<\sigma(E)\leqslant t_{k+1})\bar{P}_{i_n i_{n+1}}(t_{n+1}-t_n)$$

so that

$$C_{n+1}=C_n\bar{P}_{i_n i_{n+1}}(t_{n+1}-t_n)+P(\Delta_{n+1}, t_n<\sigma(E)\leqslant t_{n+1})$$

Therefore, to prove (15) it suffices to verify

$$P(\Delta_{n+1}, t_n<\sigma(E)\leqslant t_{n+1})=A_n\bar{P}_{i_n i_{n+1}}(t_{n+1}-t_n)\{1-\exp[-\lambda(t_{n+1}-t_n)]\} \tag{18}$$

Note that

$$L_{ki} \equiv P(y(\sigma(E)-0)=i\,|\,y(0)=k)$$
$$= \sum_{m=0}^{+\infty} \Pi_{ki}^{(m)}(\lambda)\left(1-\sum_{j\in E_\theta}\Pi_{ij}(\lambda)\right)$$
$$= G_{ki}(\lambda)(\lambda+q_i)^{-1}\lambda = \lambda\phi_{ki}(\lambda) = \lambda\int_0^{+\infty} \mathrm{e}^{-\lambda t}\bar{P}_{ki}(t)\,\mathrm{d}t$$

and

$$L_{ki}(u) \equiv P(y(\sigma(E)-0)=i,\ \sigma(E)\leqslant u\,|\,y(0)=k)$$
$$= L_{ki} - \sum_{l\in E} P(y(u)=l,\ u<\sigma(E),\ y(\sigma(E)-0)=i\,|\,y(0)=k)$$
$$= L_{ki} - \sum_{l\in E} \bar{P}_{kl}^{\lambda}(u)L_{li}$$

But

$$\sum_{l\in E} \bar{P}_{kl}^{\lambda}(u)L_{li} = \sum_{l\in E} \mathrm{e}^{-\lambda t}\bar{P}_{kl}(u)\lambda\int_0^{+\infty} \mathrm{e}^{-\lambda t}\bar{P}_{li}(t)\,\mathrm{d}t$$
$$= \lambda\int_0^{+\infty} \mathrm{e}^{-\lambda(u+t)}\bar{P}_{ki}(u+t)\,\mathrm{d}t$$
$$= \lambda\int_u^{+\infty} \mathrm{e}^{-\lambda t}\bar{P}_{ki}(t)\,\mathrm{d}t$$

so that

$$L_{ki}(u) = \lambda\int_0^u \mathrm{e}^{-\lambda t}\bar{P}_{ki}(t)\,\mathrm{d}t$$
$$\mathrm{d}L_{ki}(u) = \lambda\mathrm{e}^{-\lambda u}\bar{P}_{ki}(u)\,\mathrm{d}u$$

Abbreviate $\delta = t_{n+1} - t_n$. By conditional independence,

$$P(\Delta_{n+1}, t_n < \sigma(E) \leqslant t_{n+1})$$
$$= \sum_i P(\Delta_{n+1}, y(\sigma(E)-0)=i,\ t_n<\sigma(E)\leqslant t_{n+1})$$
$$= \sum_i P(\Lambda_n, y(\sigma(E)-0)=i,\ Z_i(t_{n+1}-\sigma(E))=i_{n+1},\ t_n<\sigma(E)\leqslant t_{n+1})$$
$$= \sum_i \int_{t_n}^{t_{n+1}} \bar{P}_{ii_{n+1}}(t_{n+1}-u)\,\mathrm{d}u\,P(\Lambda_1, y(\sigma(E)-0)=i,\ \sigma(E)\leqslant u)$$
$$= \sum_i \int_{t_n}^{t_{n+1}} \bar{P}_{ii_{n+1}}(t_{n+1}-u)\,\mathrm{d}u\,P(\Lambda_n)L_{i_n i}(u-t_n)$$

$$= A_n \sum_i \int_0^\delta \bar{P}_{ii_{n+1}}(\delta - u)\mathrm{d}u L_{i_1 i}(u)$$

$$= A_n \sum_i \int_0^\delta \bar{P}_{ii_{n+1}}(\delta - u)\lambda \mathrm{e}^{-\lambda u}\bar{P}_{i_n i}(u)\mathrm{d}u$$

$$= A_n \sum_i \int_0^\delta \lambda \mathrm{e}^{-\lambda u}\bar{P}_{i_n i_{n+1}}(\delta)\mathrm{d}u$$

$$= A_n \bar{P}_{i_n i_{n+1}}(\delta)(1 - \mathrm{e}^{-\lambda\delta})$$

Hence, (18) is proved

We now prove claim (v). To begin with, suppose X is a minimal Q process with the initial distribution v. Then

$$\int_\Omega \int_0^\sigma e^{-\lambda t} C_j[x(t)]\mathrm{d}t\,\mathrm{d}p = \int_0^{+\infty} \mathrm{e}^{-\lambda t} \int_{t<\sigma} C_j[x(t)]\mathrm{d}P\,\mathrm{d}t$$

$$= \int_0^{+\infty} \mathrm{e}^{-\lambda t} \sum_i v_i \bar{P}_{ij}(t)\mathrm{d}t$$

$$= \sum_i v_i \phi_{ij}(\lambda) = [v\phi(\lambda)]_j$$

Now suppose X is an approximating minimal Q process. Then

$$\int_\Omega \int_0^\sigma \mathrm{e}^{-\lambda t} C_j[x(t)]\mathrm{d}t\,\mathrm{d}p = \lim_{n\to+\infty} \int_{\Omega_n} \int_0^{\sigma_n} \mathrm{e}^{-\lambda t} C_j[x_n(t)]\mathrm{d}t\,\mathrm{d}p$$

$$= \lim_{n\to+\infty} [v^n(\mu)\phi(\lambda)]$$

But by the resolvent equation for $\phi(\lambda)$,

$$\phi(\lambda) = \phi(\mu) + (\mu - \lambda)\phi(\mu)\phi(\lambda)$$

and by (2), we have

$$v^n(\mu)\phi(\lambda) = v^n(\mu)\phi(\mu) + (\mu - \lambda)v^n(\mu)\phi(\mu)\phi(\lambda) \to \eta(\mu) + (\mu - \lambda)\eta(\mu)\phi(\lambda) = \eta(\lambda)$$

Therefore (v) is proved. Thus we conclude the proof. QED

11.15 APPROXIMATING MINIMAL Q PROCESSES ON A TOTALLY FINITE MEASURE SPACE

Recall Definition 8.5.1. Let $(\eta(\lambda), \lambda > 0)$ be an entrance family. If

$$\lim_{\lambda\to+\infty} \lambda\eta(\lambda)\mathbf{1} = M < +\infty \tag{1}$$

where $\eta(\lambda)1 = \sum_j \eta_j(\lambda)$, we call $(\eta(\lambda), \lambda > 0)$ a non-sticky entrance family. When $M \leqslant 1$ the entrance family is said to be probabilistic.

Theorem 1. Assume that $X = \{x(t), 0 \leqslant t < \sigma\}$ is an approximating minimal Q process defined on the measure space $(\Omega, \mathscr{F}, P)$ and for each $\lambda > 0$ the λ-characteristic measure of X is $\eta(\lambda)$. Then $(\eta(\lambda), \lambda > 0)$ is an entrance family.

Proof. Imitate the proof of Theorem 2.4.3(v). QED

Theorem 2. Assume that $X = \{x(t), 0 \leqslant t < \sigma\}$ is an approximating minimal Q-process which is defined on the totally finite measure space $(\Omega, \mathscr{F}, P)$ and has the characteristic entrance family $(\eta(\lambda), \lambda > 0)$. Then $(\eta(\lambda), \lambda > 0)$ is a non-sticky entrance family. If $(\Omega, \mathscr{F}, P)$ is a probability space, then $(\eta(\lambda), \lambda > 0)$ is also a probabilistic entrance family.

Proof.

$$\lambda\eta(\lambda)1 = \int_\Omega \int_0^\sigma \lambda e^{-\lambda t} \, dt \, dP$$

$$= \int_{\sigma > 0} (1 - e^{-\lambda\sigma}) dp \uparrow p(\sigma > 0) < +\infty \qquad \lambda \uparrow \infty \tag{2}$$

Theorem 3. Assume that $(\eta(\lambda), \lambda > 0)$ is a non-sticky entrance family. Then there exists an approximating minimal Q process $X = \{x(t), 0 \leqslant t < \sigma\}$ defined on a finite measure space $(\Omega, \mathscr{F}, P)$ and its characteristic entrance family is just $(\eta(\lambda), \lambda > 0)$.

Proof. From Theorem 14.3, there exist a measure space $(\Omega, \mathscr{F}, P)$ and defined on it an approximating minimal Q process X, the characteristic measure of which is precisely $(\eta(\lambda), \lambda > 0)$. By the non-stickiness and (2) we have $P\{\sigma > 0\} < \infty$. removing $(\sigma = 0)$, we can assume that $\Omega = (\sigma > 0)$. Therefore $(\Omega, \mathscr{F}, P)$ is a totally finite measure space. QED

Theorem 4. Assume that $(\eta(\lambda), \lambda > 0)$ is an entrance family (correspondingly, probabilistic entrance family). Then there exist a measure space (correspondingly, probability space) $(\Omega, \mathscr{F}, P)$ and defined on it an approximating minimal Q process $X = \{x(t), 0 \leqslant t < \sigma\}$, such that

$$P(\sigma > 0) = \lim_{\lambda \to \infty} \lambda\eta(\lambda)1 \tag{3}$$

$$\eta_j(\lambda) = \int_\sigma \int_0^\Omega e^{-\lambda t} C_j[x(t)] \, dt \, dP$$

$$= \int_0^{+\infty} e^{-\lambda t} H_j(t) \, dt \qquad \lambda > 0 \tag{4}$$

where

$$H_j(t) = P(x(t) = j) \qquad t \geqslant 0 \tag{5}$$

is the inverse Laplace transform of $\eta_j(\lambda), \lambda \geqslant 0$. Suppose the Riesz decomposition of $(\eta(\lambda), \lambda > 0)$ is

$$\eta(\lambda) = \alpha\phi(\lambda) + \bar{\eta}(\lambda) \tag{6}$$

Then

$$\alpha_i = P(x(0) = i) \qquad P(x(0) = \infty) = \lim_{\lambda \to +\infty} \lambda\bar{\eta}(\lambda)\mathbf{1} \tag{7}$$

Proof. We need only to prove the second equality of (7). In fact,

$$\begin{aligned} P(x(0) = \infty) &= 1 - \sum_i P(x(0) = i) \\ &= \lim_{\lambda \to +\infty} \lambda\eta(\lambda)\mathbf{1} - \lim_{\lambda \to +\infty} \lambda v\phi(\lambda)\mathbf{1} \\ &= \lim_{\lambda \to +\infty} \lambda\bar{\eta}(\lambda)\mathbf{1} \end{aligned} \tag{8}$$

11.16 PATH STRUCTURE OF NON-STICKY RETURN PROCESSES: DV-TYPE AND (DV)*-TYPE EXTENSIONS

In this section we assume that the minimal solution $(\bar{p}_{ij}(t))$ determined by a Q matrix Q is stopping. Let $\bar{X} = \{\bar{x}(t), t < \bar{\sigma}\}$ be a minimal Q process defined on a probability space $(\Omega, \mathscr{F}, P)$. Then

$$L_i(t) = P_i(\bar{\sigma} \leqslant t) = 1 - \sum_i \bar{P}_{ij}(t) \tag{1}$$

$$R_i(t) = P_i(\bar{\sigma} \leqslant t,\ \bar{x}(\bar{\sigma} - 0) = \infty) \tag{2}$$

are determined uniquely by Q, where $P_i(\cdot) = P\{\cdot \mid \bar{x}(0) = i\}$.

In the following we always assume that $(\eta(\lambda), \lambda > 0)$ is a given probabilistic entrance family and $H_i(t)$ is the inverse Laplace transform of $(\eta_i(\lambda), \lambda > 0)$. Let

$$H(t) = \sum_{i \in E} H_i(t) \qquad H^0(t) = \begin{cases} 0 & \text{if } t < 0 \\ 1 & \text{if } t \geqslant 0 \end{cases}$$

$$H^{n+1} = H^n * H$$

$$M_i = \sum_{n=0}^{+\infty} (L_i * H^n) \tag{3}$$

$$P_{ij}(t) = \bar{P}_{ij}(t) + \int_0^t H_j(t-s)\,\mathrm{d}M_i(s) \tag{4}$$

Here $*$ represents convolution.

Lemma 1. There exists a probability space $(\Omega, \mathscr{F}, P)$ on which can be defined a minimal Q process $X^0 = \{x^0(t), 0 \leqslant t < \sigma^0\}$ and a sequence of approximating minimal Q processes $X^n = \{x^n(t), 0 \leqslant t < \sigma^n\}$, every X^n $(n \geqslant 1)$ having the same characteristic entrance family $(\eta(\lambda), \lambda > 0)$. They have the following properties:

(i) $(\sigma_n = 0) \cup (\sigma_n = +\infty) \subset (\sigma_{n+1} = 0)$.

(ii) Write $\Delta_n = (0 < \sigma_n < +\infty)$, $\Omega_n = (\sigma_n > 0)$, $P(\Omega_{n+1} | \Delta_n) = M$, where M is determined by (15.1).

(iii) Given Δ_n or Ω_{n+1}, all X^m $(m \leqslant n)$ and X^m $(m \geqslant n+1)$ are conditionally independent.

Proof. Imitate the proof of Lemma 2.1.

Theorem 2. For X^0 and X^n $(n \geqslant 1)$ in Lemma 1, let

$$\tau^0 = 0 \qquad \tau^{n+1} = \sum_{m=0}^{n} \sigma^m \qquad \sigma = \sum_{m=0}^{+\infty} \sigma^m \tag{5}$$

For $0 \leqslant t < \sigma$ let

$$X(t) = X^n(t - \tau^n) \qquad \text{if } \tau^n \leqslant t < \tau^{n+1} \tag{6}$$

Then $X = \{x(t), 0 \leqslant t < \sigma\}$ is a Markov process and its transition probability is given by (3).

Proof. Imitate the proof of Theorem 11.2.2.

Remark 1

Taking Laplace transforms on both sides of (4) we obtain

$$\psi_{ij}(\lambda) = \phi_{ij}(\lambda) + Z_i(\lambda) \frac{\eta_j(\lambda)}{(1 - M) + \lambda \eta(\lambda) \mathbf{1}} \tag{7}$$

where

$$Z(\lambda) = 1 - \lambda \phi(\lambda) \mathbf{1}$$

Remark 2

When Q is conservative or $(\eta(\lambda), \lambda > 0)$ is a harmonic entrance family, the process X in Theorem 2 is a Q process. But in general it may not necessarily be so.

Remark 3

The Markov process X in Theorem 2 is called a DV-type extension of the minimal Q process X^0. Suppose the Riesz decomposition of the entrance family

$(\eta(\lambda), \lambda > 0)$ is (15.6). When $\bar{\eta}(\lambda) = 0$, the DV-type extension of X^0 becomes the D-type extension in section 11.2. When $\alpha = 0$, we call the DV-type extension the V-type extension. In this case, the leaping intervals of X, except the first one, are all ${}_{\infty}U$ intervals.

Lemma 3. Assume that for some $i \in E$ and $t > 0$, $R_i(t)$ is given by (2), and $R_i(t) > 0$. Then there exist a probability space $(\Omega, \mathscr{F}, P)$ and defined on it a minimal Q process $X^0 = \{x^0(t), 0 \leqslant t < \sigma^0\}$ and a sequence of approximating minimal Q processes $X^n = \{x^n(t), 0 \leqslant t < \sigma^n\}, n \geqslant 1$, every X^n $(n \geqslant 1)$ having the same characteristic entrance family $(\eta(\lambda), \lambda > 0)$. They satisfy the following conditions:

(i) $(\sigma^n = 0) \cup (\sigma^n = +\infty) \cup (X^n(\sigma^n - 0) \in E) \subset (\sigma^{n+1} = 0), n \geqslant 0$[1].

(ii) Put $\Delta_n = (0 < \sigma^n < +\infty, X^n(\sigma^n - 0) = \infty), \Omega_n = (\sigma_n > 0)$. Then

$$P(\Omega_{n+1} | \Delta_n) = M$$

where M is determined by (15.1).

(iii) Given Δ_n or Ω_{n+1}, all X^m $(m \leqslant n)$ and X^m $(m \geqslant n+1)$ are conditionally independent.

Proof. Follow the proof of Lemma 2.1. QED

Theorem 4. For processes X^0 and X^n in Lemma 3, define $X = \{x(t), 0 \leqslant t < \sigma\}$ according to (5)–(6). Then X is a Q process with transition probabilities

$$p_{ij}(t) = \bar{p}_{ij}(t) + \int_0^t H_j(t-s)\, \mathrm{d}N_i(s) \tag{8}$$

where

$$N_i = \sum_{n=0}^{+\infty} (R_i * H^n) \tag{9}$$

Proof. Imitiate the proof of Theorem 11.2.2. QED

Remark 1

Take Laplace transforms on both sides of (8) to obtain

$$\psi_{ij}(\lambda) = \phi_{ij}(\lambda) + \zeta_i(t) \frac{\eta_j(\lambda)}{1 - M + \lambda \eta(\lambda) \mathbf{1}} \tag{10}$$

where $\zeta_i(\lambda)$ are the Laplace transforms of $R_i(t)$, i.e.

$$\zeta(\lambda) = \mathbf{1} - \lambda \phi(\lambda) \mathbf{1} - \phi(\lambda) \boldsymbol{d} \tag{11}$$

where $\boldsymbol{d} = -Q\mathbf{1}$.

[1] when $\sigma^n = 0$, $x^n(\sigma^n - 0)$ are not defined.

Remark 2

The first leaping point τ of X in Theorem 4 is just σ^0 and if α is determined by (15.6) then

$$P(X(\tau)=i\,|\,\tau<\infty, X(\tau-0)=\infty)=\begin{cases}\alpha_i & \text{if } i\in E,\\ \lim_{\lambda\to+\infty}\lambda\bar{\eta}(\lambda)\mathbf{1} & \text{if } i=\infty\end{cases} \qquad (12)$$

$$P(\sigma=\tau\,|\,\tau<\infty, X(\tau-0)=\infty)=1-M$$

$$P(\sigma=\tau\,|\,X(\tau-0)\in H_e)=1$$

where H_e is the set of non-conservative states, and as long as the above conditional probabilities are meaningful.

Remark 3

The Q process X in Theorem 4 is called a (DV)*-type extension of the minimal Q process X^0. The (DV)*-type extension preserves the Q matrix. As in Remark 3 after Theorem 2, when $\bar{\eta}(\lambda)=0$, the (DV)*-type extension becomes a D*-type extension. When $\alpha=0$, the (DV)*-type extension is called the V*-type extension. The D-type and D*-type extensions, generally speaking, are different extensions, but the extensions of type V and type V* are the same. Therefore, we only use the phrase 'the V*-type extension'.

11.17 GENERALIZED DV-TYPE AND GENERALIZED (DV)*-TYPE EXTENSIONS

We keep the hypotheses and symbols used in the beginning of section 11.5 to (5.2). Write $A_e=H\cup B_e$. Denote by $\mathscr{B}_e$ the class of all Borel sets in A_e. For $L_i(\Gamma,t)$ and $h_i(\Gamma,\lambda)$ given in (5.1)–(5.2), we have $\lim_{t\to\infty}L_i(\Gamma,t)=\lim_{\lambda\downarrow 0}h_i(\Gamma,\lambda)$. Denote this limit by $h_i(\Gamma)$.

Suppose we are given a mapping $G(\cdot,\cdot)$ from $A_e\times\mathscr{B}_e$ to $[0,1]$ satisfying:

(i) for each $a\in A_e$, $G(a,\cdot)$ is a measure on $\mathscr{B}_e$ and $G(a,A_e)\leqslant 1$;

(ii) for each $\Gamma\in\mathscr{B}_e$, $G(\cdot,\Gamma)$ is a $\mathscr{B}_e$-measurable function.

Suppose we are also given a mapping from $A_e\times(0,\infty)$ to the Banach space l satisfying:

(i) for each $a\in A_e$, $(\eta(a,\lambda),\lambda>0)$ is a probabilistic entrance family;

(ii) for each $\lambda>0$, $\eta(\cdot,\lambda)$ is a measurable mapping from A_e to l.

Let $H_j(a,t)$ be the inverse Laplace transform of $\eta_j(a,\lambda)$, $\lambda>0$. Set

$$W(a,\Gamma,\lambda)=\lambda\sum_j\eta_j(a,\lambda)h_j(\Gamma) \qquad (1)$$

By Lemma 2.11.4 there exist limits

$$W(a, \Gamma, \lambda) \uparrow W(a, \Gamma) \qquad \lambda \uparrow \infty$$

Let

$$C(a, \Gamma, t) = W(a, \Gamma) - \sum_j H_j(a, t) h_j(\Gamma) \tag{2}$$

The following lemma illuminates the probability meaning of $W(a, \Gamma)$ and $C(a, \Gamma, t)$.

Lemma 1. Fix some $a \in A_e$. Assume that the characteristic entrance family of the approximating minimal Q process $X = \{x(t), 0 \leqslant t < \sigma\}$ is $(\eta(a, \lambda), \lambda > 0)$. Then

$$P(X(\sigma - 0) \in \Gamma) = W(a, \Gamma) \tag{3}$$

$$P(X(\sigma - 0) \in \Gamma, \sigma \leqslant t) = C(a, \Gamma, t) \tag{4}$$

Proof. According to the note after Definition 13.2 we know that $\{x(t), 0 < t < \sigma\}$ is a minimal Q process. Hence

$$P(t < \sigma, x(\sigma - 0) \in \Gamma) = \sum_j P(X(t) = j, t < \sigma, X(\sigma - 0) \in \Gamma)$$

$$= \sum_j H_j(a, t) h_j(\Gamma) \tag{5}$$

Then letting $t \downarrow 0$ in the above we obtain

$$P(X(\sigma - 0) \in \Gamma) = \lim_{t \downarrow 0} \sum_j H_j(a, t) h_j(\Gamma)$$

$$= \lim_{\lambda \to \infty} \lambda \sum_j \eta_j(a, \lambda) h_j(\Gamma)$$

$$= \lim_{\lambda \to \infty} W(a, \Gamma, \lambda) = W(a, \Gamma)$$

Therefore (4) follows from (3) and (5) and we conclude the proof. QED

Remark

When $(\eta(a, \lambda), \lambda > 0)$ is a general (i.e. probabilistic or non-probabilistic) entrance family, Lemma 1 also holds.

Let

$$F(a, \Gamma, \lambda) = \int_0^\infty e^{-\lambda t} C(a, \Gamma, dt) = \lambda \int_0^\infty e^{-\lambda t} C(a, \Gamma, t) \, dt \tag{6}$$

By (2) we have

$$F(a, \Gamma, \lambda) = W(a, \Gamma) - W(a, \Gamma, \lambda) \tag{7}$$

Define

$$L_i^1(\Gamma, t) = L_i(\Gamma, t) \qquad \text{(see (5.1))} \tag{8}$$

$$L_i^{n+1}(\Gamma, t) = \int_0^t \int_{A_e} \int_{A_e} G(a, \mathrm{d}b) C(b, \Gamma, t-s) L_i^n(\mathrm{d}a, \mathrm{d}s) \tag{9}$$

$$K_i(\Gamma, t) = \sum_{n=0}^{\infty} L_i^{n+1}(\Gamma, t) \tag{10}$$

$$p_{ij}(t) = f_{ij}(t) + \int_0^t \int_{A_e} \int_{A_e} G(a, \mathrm{d}b) H_j(b, t-s) K_i(\mathrm{d}a, \mathrm{d}s) \tag{11}$$

where $(f_{ij}(t))$ is the minimal solution. The above $P(t) = (p_{ij}(t))$ is determined completely by Q, $G(\cdot, \cdot)$ and $\eta(\cdot, \cdot)$.

Considering Theorem 15.4 and using the technique of independent product spaces, we can obtain the following lemma easily.

Lemma 2. There exists a probability space $(\Omega, \mathscr{F}, P)$ on which can be defined a minimal Q process $X^0 = \{x^0(t), 0 \leqslant t < \sigma^0\}$, a family of approximating minimal Q processes $X_a^n = \{x^n(a, t), 0 \leqslant t < \sigma^n(a)\}$, $a \in A_e$, $n \geqslant 1$, and a family of A_e-valued random variables $f^n(a)$, $a \in A_e$, $n \geqslant 0$, such that

(i) $P(X^0(0) \in E) = 1$;

(ii) for each a, the characteristic entrance family of all X_a^n $(n \geqslant 1)$ is $(\eta(a, \lambda), \lambda > 0)$;

(iii) $P(f^n(a) \in \Gamma) = G(a, \Gamma)$;

(iv) all X^a, X_a^n $(a \in A_e, n \geqslant 1)$, $f^n(a)$ $(a \in A_e, n \geqslant 0)$ are independent.

When $\sigma^0 > 0$ and $X^0(\sigma^0 - 0) = a \in A_e$, we let $\sigma^1 = \sigma^1(f^0(a))$, and $X^1(t) = X^1(f^0(a), t)$ for $t < \sigma^1$; otherwise let $\sigma^1 = 0$. Obviously, $X^1 = \{X^1(t), 0 \leqslant t < \sigma^1\}$ is an approximating minimal Q process. When $\sigma^1 > 0$ and $X^1(\sigma^1 - 0) = a \in A_e$, we let $\sigma^2 = \sigma^2(f^1(a))$ and $X^2(t) = X^2(f^1(a), t)$ for $t < \sigma^2$; otherwise let $\sigma^2 = 0$. Continuing with the procedure, we can obtain a minimal Q process X^0 and a sequence of approximating minimal Q processes $X^n = \{X^n(t), 0 \leqslant t < \sigma^n\}$ $(n \geqslant 1)$. Define $X = \{X(t), 0 \leqslant t < \sigma\}$ according to (16.5)–(16.6).

Theorem 3. X is a homogeneous Markov process with transition probability matrix $P(t) = (p_{ij}(t))$ given by (11), the resolvent matrix $\psi(\lambda) = (\psi_{ij}(\lambda))$ of which is given by

$$\psi_{ij}(\lambda) = \phi_{ij}(\lambda) + \int_{A_e} \int_{A_e} h_i(\mathrm{d}a, \lambda) \left(\sum_{n=0}^{\infty} V^n(a, \mathrm{d}b, \lambda) \right) \eta_j(b, \lambda) \tag{12}$$

where

$$V^0(a,\Gamma,\lambda)=\begin{cases}0 & \text{if } a\notin\Gamma\\ 1 & \text{if } a\in\Gamma\end{cases}$$

$$V^1(a,\Gamma,\lambda)=\int_{A_e} G(a,db)F(b,\Gamma,\lambda) \tag{13}$$

$$V^{n+1}(a,\Gamma,\lambda)=\int_{A_e} V^n(a,db,\lambda)V^1(b,\Gamma,\lambda)$$

Proof. We can prove the first conclusion by following the proof of Theorem 5.2. Here we only point out how to calculate $p_{ij}(t)$ and how to apply Lemma 2.

Clearly

$$p_i(X(t)=j)=P_i(X^0(t)=j,t<\sigma^0)+\sum_{n=0}^{\infty}R_{ij}^{n+1}(t)$$

where the first term equals $f_{ij}(t)$ while

$$R_{ij}^{n+1}(t)=P_i(x^{n+1}(t-\tau^{n+1})=j,\tau^{n+1}\leqslant t<\tau^{n+2})$$

$$=\int_0^t\int_{A_e}\int_{A_e} P_i(\tau^{n+1}\in ds, X^n(\sigma^n-0)\in da, f^n(a)\in db, X^{n+1}(b,t-s)$$

$$=j,t-s<\sigma^{n+1})$$

Apply Lemma 2 and the above becomes

$$R_{ij}^{n+1}(t)=\int_0^t\int_{A_e}\int_{A_e} G(a,db)H_j(b,t-s)L_i^{n+1}(da,ds)$$

where

$$L_i^{n+1}(\Gamma,t)=P_i(\tau^{n+1}\leqslant t, X^n(\sigma^n-0)\in\Gamma) \tag{14}$$

Hence

$$L_i^1(\Gamma,t)=L_i(\Gamma,t)$$

$$L_i^{n+1}(\Gamma,t)=\int_0^t\int_{A_e}\int_{A_e} P_i(\tau^n\in ds, X^{n-1}(\sigma^{n-1}-0)\in da, f^{n-1}(a)\in db,$$

$$X^n(b,\sigma^n(b)-0)\in\Gamma, \sigma^n(b)\leqslant t-s)$$

Applying Lemma 2 again we obtain (9). Therefore $P_i\{X(t)=j\}$ equals $p_{ij}(t)$, which is determined by (11).

We now perform the calculation for the Laplace transform $\psi_{ij}(\lambda)$. Take Laplace transforms on both sides of (11) to obtain

$$\psi_{ij}(\lambda)=\phi_{ij}(\lambda)+\int_{A_e}\int_{A_e} G(a,db)\eta_j(b,\lambda)\sum_{n=0}^{\infty}h_i^{n+1}(da,\lambda) \tag{15}$$

where

$$h_i^{n+1}(\Gamma,\lambda)=\int_0^\infty e^{-\lambda t}\Gamma_i^{n+1}(\Gamma,dt) \tag{16}$$

Taking Laplace transforms on both sides of (9) we obtain

$$h_i^{n+1}(\Gamma,\lambda)=\int_{A_e}\int_{A_e}G(a,db)F(b,\Gamma,\lambda)h_i^n(da,\lambda) \tag{17}$$

It follows from this that

$$h_i^{n+1}(\Gamma,\lambda)=\int_{A_e}V^n(a,\Gamma,\lambda)h_i(da,\lambda) \tag{18}$$

In fact, the above holds for $n=0$ obviously. Suppose it holds for $n=m-1$. Then

$$\begin{aligned}\int_{A_e}V^m(a,\Gamma,\lambda)h_i(da,\lambda)&=\int_{A_e}\left(\int_{A_e}V^{m-1}(a,db,\lambda)V^1(b,\Gamma,\lambda)\right)h_i(da,\lambda)\\&=\int_{A_e}\left(\int_{A_e}V^{m-1}(a,db,\lambda)h_i(da,\lambda)\right)V^1(b,\Gamma,\lambda)\\&=\int_{A_e}h_i^m(db,\lambda)V^1(b,\Gamma,\lambda)\\&\qquad\text{(by the hypothesis of induction)}\\&=\int_{A_e}\int_{A_e}h_i^m(db,\lambda)G(b,dc)F(c,\Gamma,\lambda)\\&=h_i^{m+1}(\Gamma,\lambda)\qquad\text{(by (17))}\end{aligned}$$

i.e. (18) holds for $n=m$.

Substituting (18) into (15) we obtain (12). The proof is completed. QED

Remark 1

We call X in Theorem 3 a generalized (DV)-type extension process of X^0. The generalized (DV)-type extension may change the Q matrix because σ^0 is not necessarily the first leaping point of X according to the fact that the probability

$$P(X(\sigma^0)\in E\,|\,\sigma^0<\infty,X(\sigma^0-0)=a) \tag{19}$$

may be positive for some $a\in H$.

Remark 2

Suppose that the Riesz decomposition of $\eta(a,\lambda)$ is

$$\eta(a,\lambda)=\alpha(a)\phi(\lambda)+\bar{\eta}(a,\lambda) \tag{20}$$

Then, the generalized (DV)-type extension preserves the Q matrix Q if and only if the probability in Remark 1 is equal to zero, i.e.

$$\int_{A_e} G(a, db)\sum_i \alpha_i(b) = 0 \qquad a\in H_e \tag{21}$$

or equivalently,

$$\int_{A_e} G(a, db)\alpha(b) = 0 \qquad a\in H_e \tag{22}$$

The generalized (DV)-type extension satisfying the above conditions is called a generalized (DV)*-type extension. Obviously, when Q is conservative, generalized (DV)-type and generalized (DV)*-type extensions are identical. For the generalized (DV)*-type extension σ^0 is the first leaping point of X and, furthermore,

$$P(X(\sigma^0) = i|\sigma^0 < \infty, X(\sigma^0 - 0)) = \begin{cases} \int_{A_e} G(X(\sigma^0 - 0), db)\alpha_i(b) & \text{if } i\in E \\ \int_{A_e} G(X(\sigma^0 - 0), db)\bar{M}(b) & \text{if } i = \infty \end{cases} \tag{23}$$

where

$$\bar{M}(b) = \lim_{\lambda\to\infty} \lambda \|\bar{\eta}(b, \lambda)\|$$

Remark 3

Suppose $\alpha(b) = 0$ for $b\in A_e$. Then (22) holds. Thus the generalized (DV)-type extensions and generalized (DV)*-type extensions are identical; hence the Q matrix remains unchanged. Therefore, when $\alpha(a) = 0$, $a\in A_e$, we call the generalized (DV)*-type a generalized V*-type extension. As pointed out in Remark 3 after Theorem 16.2, we only use the phrase 'generalized V*-type extension'. For the generalized V*-type extension σ^0 is the first leaping point and (22) becomes

$$P(X(\sigma^0) = i|\sigma^0 < \infty, X(\sigma^0 - 0)) = \begin{cases} 0 & \text{if } i\in E \\ \int_{A_e} G(X(\sigma^0 - 0), db)M(b) & \text{if } i = \infty \end{cases} \tag{24}$$

where

$$M(a) = \lim_{\lambda\to\infty} \lambda \|\eta(a, \lambda)\|$$

Remark 4

Suppose $\psi(\lambda)$ is a Q process in Theorem 8.5.1 (remark: H in this section is denoted by H_e in Chapter 8). For $a \in A_e$, let

$$\eta(a, \lambda) = \begin{cases} \bar{\eta}^{a_i}(\lambda) & \text{if } a \in \text{some } a_i \in J \\ 0 & \text{if } a \in \text{some } a_i \in A - J \end{cases} \tag{25}$$

and

$$G(a, b) = \begin{cases} G^{a_i a_k} & \text{if } a \in \text{some } a_i \in J \text{ and } b \in \text{some } a_k \in J \\ 0 & \text{otherwise} \end{cases} \tag{26}$$

Then $G(\cdot, \cdot)$ and $\eta(\cdot, \cdot)$ satisfy the requirement presented at the beginning of this section. For these $G(\cdot, \cdot)$ and $\eta(\cdot, \cdot)$, $\psi(\lambda)$ determined by (12) and that by Theorem 8.5.1 are the same. Therefore, for both $G(\cdot, \cdot)$ and $\eta(\cdot, \cdot)$ given by (25)–(26), Lemma 2 and Theorem 3 give the paths of Q processes in Theorem 8.5.1.

PART V CONSTRUCTION THEORY OF BIRTH–DEATH PROCESSES: PROBABILITY METHOD

CHAPTER 12

Probability Structure of Birth–Death Processes

12.1 INTRODUCTION

When $E = \{0, 1, 2, \ldots\}$, and Q has the form (6.1.1), the process $X \in \mathscr{H}_s(Q)$ is called a birth–death process. In this chapter, Q as used below will always refer to a birth–death matrix. In Chapter 6 we have constructed all birth–death processes by using the analytic method.

In section 2.3, we have already pointed out that Professor Zi-kun Wang introduced the method for solving the construction problem for conservative birth–death processes in 1958, that is, the limit transition method. The advantage of this method is that the path structure of the processes constructed is quite clear and their probability meaning is obvious. In order to study the properties of the processes we may first make a clear study of simple processes and, then, it suffices to turn to the investigation of the limits. Therefore, this method has important potential value in terms, of theory and practice. For examples, see Xiang-qun Yang (1965b, 1966b) and Zhen-ting Hou (1975).

The logical foundation of this method is published in Zi-kun Wang (1965a). In the paper all honest birth–death processes are constructed. But in construction of paths, transformation g_n and transformation f_n are employed respectively to handle such cases as $S < \infty$ and $S = \infty$. Zi-kun Wang (1980) provides a unified treatment for the two cases with respect to the honest processes. Zi-kun Wang and Xiang-qun Yang (1978, 1979) consistently handle the two cases and the case that processes are allowed to be stopping processes. The content of this chapter is derived from these two papers.

12.2 PROBABILITY EXPLANATION OF THE CHARACTERISTIC NUMBERS

We shall adopt the characteristic numbers and notations in section 6.2. In this section, we may assume $a_0 \geqslant 0$. Let $X = \{x(t), t < \sigma\} \in \mathscr{H}_s(Q)$, τ be the first leaping point and τ_n be the nth leaping point. Obviously

$$x(\tau - 0) = 0 \text{ or } \infty \tag{1}$$

and we have $|x(\tau_n)-x(\tau_n-0)|=1$ for the imbedded chain $X_T=\{x(\tau_n), \tau_n<\tau\}$ $(\tau_n<\tau)$.

Theorem 1

$$X_i^1 \equiv P_i\{x(\tau-0)=0\}=\frac{a_0(z-z_i)}{a_0(z-z_0)+1} \tag{2}$$

$$X_i^2 = P_i\{x(\tau-0)=\infty\}=\frac{a_0(z_i-z_0)+1}{a_0(z-z_0)+1} \tag{3}$$

We introduce the convention $\infty/\infty=1$ and $0\cdot\infty=0$.

Proof. By Theorem 7.8.2, X^1 satisfies equation (6.2.19) where $f_0=-a_0$ and $f_i=0$ $(i>0)$. Therefore by Lemma 6.2.5, we have

$$X_i^1=[a_0(z_i-z_0)+1]X_0^1-a_0(z_i-z_0) \tag{4}$$

Similarly, X^2 satisfies equation (6.2.19) where $f_i\geqslant 0$ $(i\geqslant 0)$. Hence by Lemma 6.2.5, we have

$$X_i^2=[a_0(z_i-z_0)+1]X_0^2 \tag{5}$$

Noticing $X^1+X^2=1$, it follows from the above expression that if $X_0^2=0$ then $X^2=0$, thus $X^1=1$; if $X_0^2>0$, then $X^2>0$, and consequently by the Martingale convergence theorem, we have

$$X^2_{x(\tau_n)}=P\{x(\tau-0)=\infty\,|\,x(\tau_0),x(\tau_1),\ldots,x(\tau_n)\}\to 1$$

on the positive probability set $\{x(\tau-0)=\infty\}$. Therefore, $X_i^2\to 1$ $(i\to\infty)$, so that $X_i^1\to 0$ $(i\to\infty)$. Hence by (4) and (5), it follows that

$$0=[a_0(z-z_0)+1]X_0^1-a_0(z-z_0) \qquad 1=[a_0(z-z_0)+1]X_0^2$$

Hence when $a_0z<\infty$,

$$X_0^1=\frac{a_0(z-z_0)}{a_0(z-z_0)+1} \qquad X_0^2=\frac{1}{a_0(z-z_0)+1}$$

Substituting the above formula into (4) and (5), we obtain that (2) and (3) hold for $a_0z<\infty$. By convention $\infty/\infty=1$, (2) and (3) likewise hold for $a_0z=\infty$ and the proof is completed. QED

Set

$$\xi_i=\begin{cases}\inf\{t\,|\,x(t)=i, t<\tau\}\\ \infty \qquad \text{if the set above is empty}\end{cases} \tag{6}$$

Obviously,

$$P_i\{\xi_n\uparrow\tau(i\leqslant n\uparrow\infty)\,|\,x(\tau-0)=\infty\}=1 \tag{7}$$

Theorem 2. For $i \leqslant k \leqslant n$,

$$P_k\{\xi_i < \xi_n\} = \frac{z_n - z_k}{z_n - z_i} \qquad P_k\{\xi_n < \xi_i\} = \frac{z_k - z_i}{z_n - z_i} \tag{8}$$

Proof. By Theorem 7.8.2, $u_k = P_k\{\xi_i < \xi_n\}$ satisfies equation (5.4.11) where $f_i = 1, f_k = 0\ (i < k \leqslant n)$. By Theorem 6.2.3 and (5.4.12) we obtain the first expression in (8). The second expression follows from $P_k\{\xi_n < \xi_i\} + P_k\{\xi_i < \xi_n\} = 1$ QED

Theorem 3. For $i \leqslant k$,

$$P_k\{\xi_i < \tau\} = \frac{z - z_k}{z - z_i} \qquad P_i\{\xi_k < \tau\} = \frac{a_0(z_i - z_0) + 1}{a_0(z_k - z_0) + 1} \tag{9}$$

$$P_k\{\tau \leqslant \xi_i, x(\tau - 0) = \infty\} = \frac{z_k - z_i}{z - z_i} \tag{10}$$

Proof. Since on $\{x(0) = k\}$,

$$(\xi_i < \xi_n)\uparrow \bigcup_{n=k+1}^{\infty} (\xi_i < \xi_n) = (\xi_i < \tau)$$

$$(\xi_n < \xi_i)\downarrow \bigcap_{n=k+1}^{\infty} (\xi_n < \xi_i) = \{\tau \leqslant \xi_i, x(\tau - 0) = \infty\} \tag{11}$$

Taking the limit in (8), we obtain (10) and the first formula in (9). Secondly, since $u_i = P_i(\xi_k < \tau)\ (0 \leqslant i \leqslant k)$ satisfies equation (6.2.13) for $n = k, f_i = 0\ (i < k)$, and $f_k = 1$, the second formula in (9) follows from Lemma 6.2.4. QED

Theorem 4. If $i \leqslant k \leqslant n$, then

$$E_k\{\xi_i, \xi_i < \xi_n\} = \frac{z_n - z_k}{z_n - z_i} \sum_{j=i+1}^{k-1} \frac{z_n - z_j}{z_n - z_i}(z_j - z_i)\mu_j + \frac{z_k - z_i}{z_n - z_i} \sum_{j=k}^{n-1} \frac{z_n - z_j}{z_n - z_i}(z_n - z_j)\mu_j \tag{12}$$

$$E_k\{\xi_n, \xi_n < \xi_i\} = \frac{z_n - z_k}{z_n - z_i} \sum_{j=i+1}^{k-1} \frac{z_j - z_i}{z_n - z_i}(z_j - z_i)\mu_j + \frac{z_k - z_i}{z_n - z_i} \sum_{j=k}^{n-1} \frac{z_j - z_i}{z_n - z_i}(z_n - z_j)\mu_j \tag{13}$$

Proof. By Theorem 7.8.2 and Theorem 2, $u_k = E_k\{\xi_i, \xi_i < \xi_n\}$ satisfies the equation (5.4.11) for $f_i = f_n = 0$, $f_k = (z_n - z_k)/(z_n - z_i)\ (i < k < n)$. By Theorem 6.2.3 and (5.4.12) we obtain (12), and (13) can be derived similarly. QED

Theorem 5

$$E_k\{\tau, x(\tau-0)=\infty\} = \frac{z-z_k}{[a_0(z-z_0)+1]^2} + \frac{z-z_k}{a_0(z-z_0)+1}\sum_{j=1}^{k-1}\frac{a_0(z_j-z_0)+1}{a_0(z-z_0)+1}\mu_j + \frac{a_0(z_k-z_0)+1}{a_0(z-z_0)+1}\sum_{j=k}^{\infty}\frac{a_0(z_j-z_0)+1}{a_0(z-z_0)+1}(z-z_j)\mu_j \quad (14)$$

Proof. Let $u_k = E_k\{\xi_n, x(\tau-0)=\infty\}$ $(k \leqslant n)$. Then u_k satisfies the equation (6.2.13) for $f_k = P_k\{x(\tau-0)=\infty\}$ $(k<n)$, and $f_n = 0$. By Lemma 6.2.4,

$$E_k\{\xi_n, x(\tau-0)=\infty\} = \frac{z_n-z_k}{a_0(z_n-z_0)+1}\frac{1}{a_0(z-z_0)+1} + \frac{z_n-z_k}{a_0(z_n-z_0)+1}\sum_{j=1}^{k-1}\frac{a_0(z_j-z_0)+1}{a_0(z-z_0)+1}\mu_j + \frac{a_0(z_k-z_0)+1}{a_0(z_n-z_0)+1}\sum_{j=k}^{n-1}\frac{a_0(z_j-z_0)+1}{a_0(z-z_0)+1}(z_n-z_j)\mu_j \quad (15)$$

Noticing (7) and taking the limit in the above expression, we obtain (14). The proof is terminated. QED

Theorem 6. Let $a_0 = 0$.

(i) Let c_{kj} be the probability of going from k to j in finitely many ($\geqslant 0$) steps; then

$$c_{kj} = P_k\{\xi_j < \tau\} = \begin{cases} 1 & \text{if } k \leqslant j \\ (z-z_k)/(z-z_j), & \text{if } k > j \end{cases} \quad (16)$$

(ii) For the minimal Q process X to be recurrent it is necessary and sufficient that $z=\infty$. If $z=\infty$, then for it to be ergodic it is necessary and sufficient that $\sum_{i=0}^{\infty}\mu_i < \infty$.

(iii) m_i, N_i and R and determined by (6.2.4)–(6.2.7), so that

$$m_i = E_i\xi_{i+1} \qquad N_i = E_i\tau \qquad R = E_0\tau \quad (17)$$

(iv) The necessary and sufficient condition for $P_k\{\tau<\infty\} = 1$ $(k \in E)$ is $R<\infty$.

Proof. (i) follows from (9).

(ii) The probability starting from 0 and coming back in finitely many steps is

$$f_0^* = \frac{b_0}{a_0+b_0}P_1(\xi_0<\tau) = \frac{z-z_1}{z-z_0}$$

From this we know that $f_0^* = 1$, that is, for the minimal process to be recurrent it is necessary and sufficient that $z = \infty$.

(iii) When $a_0 = 0$, choosing $k = n-1$ in (15) we obtain $E_{n-1}\xi_n = m_{n-1}$. Equation (14) $a_0 = 0$ is just the second and third formulae in (17).

(iv) By (17) and $N_k \leqslant R$ we know that if $R < \infty$ then $P_k\{\tau < \infty\} = 1$ $(k \in E)$. Conversely, if $P_k\{\tau < \infty\} = 1$ for some (hence all) k then by Theorem 7.8.2, $u_k(\lambda) = E_k(\mathrm{e}^{-\lambda\tau}) \neq 0$ $(\lambda > 0)$ satisfies $D_\mu u^+(\lambda) = \lambda u(\lambda)$, and so by Lemma 6.2.5

$$u_k(\lambda) = u_0(\lambda) + \lambda \sum_{j=0}^{i-1} (z_i - z_j) u_j(\lambda) \mu_j$$

Moreover, obviously, $u_i(\lambda)\uparrow$, $(i\uparrow)$, and therefore $1 \geqslant u_i(\lambda) \geqslant \lambda u_0(\lambda) \sum_{j=0}^{i-1}(z_i - z_j)\mu_j$, $1 \geqslant \lambda u_0(\lambda) R$, and hence $R < \infty$. The proof is completed. QED

Theorem 7. Let $a_0 = 0$, $S = \infty$ (see (6.2.7)). Suppose that $X = \{x(t), t < \sigma\} \in \mathscr{H}_s(Q)$ and that ξ_0 is determined by (6). Then for $\lambda > 0$, $E_i\{\mathrm{e}^{-\lambda\xi_0}\} \downarrow 0$ $(i \uparrow \infty)$.

Proof. Since X must visit $i-1$ while going from i to 0, therefore, $u_i(\lambda) = E_i\{\mathrm{e}^{-\lambda\xi_0}\} \downarrow \alpha \geqslant 0$ $(i \uparrow \infty)$. By Theorem 7.8.2, we obtain $u_0(\lambda) = 1$, $D_u u_i^+(\lambda) = \lambda u_i(\lambda)$ $(i > 0)$, that is,

$$u_{i-1}(\lambda) - u_i(\lambda) = \frac{b_i}{a_i}[u_i(\lambda) - u_{i+1}(\lambda)] + \frac{\lambda}{a_i} u_i(\lambda) \qquad i > 0$$

Applying the above formula again, we have

$$\begin{aligned} u_{i-1}(\lambda) - u_i(\lambda) &= \frac{b_i b_{i+1} \cdots b_{i+j+1}}{a_i a_{i+1} \cdots a_{i+j+1}} [u_{i+j+1}(\lambda) - u_{i+j+2}(\lambda)] \\ &\quad + \lambda\left(\frac{u_i(\lambda)}{a_i} + \sum_{l=0}^{j} \frac{b_i b_{i+1} \cdots b_{i+l} u_{i+l+1}(\lambda)}{a_i a_{i+1} \cdots a_{i+l} a_{i+l+1}}\right) \\ &\geqslant \lambda\left(\frac{1}{a_i} + \sum_{l=0}^{j} \frac{b_i b_{i+1} \cdots b_{i+l}}{a_i a_{i+1} \cdots a_{i+l} a_{i+l+1}}\right)\alpha \end{aligned}$$

Letting $j \to \infty$ we obtain $u_{i-1}(\lambda) - u_i(\lambda) \geqslant \lambda e_i \alpha$ $(i > 0)$, so that

$$1 \geqslant u_0(\lambda) \geqslant u_0(\lambda) - u_j(\lambda) \geqslant \lambda\left(\sum_{k=1}^{j} e_k\right)\alpha$$

Letting $j \to \infty$ we obtain $1 \geqslant \lambda S \alpha$. Since $S = \infty$, it follows that $\alpha = 0$. The proof is over. QED

Theorem 8. If $a_0 = 0$, $S = \infty$, then $P\{\xi_{\infty k} < \sigma\} = 0$, where $\xi_{\infty k}$ is determined by (10.4.1). In other words, each Q process $X \in \mathscr{H}_s(Q)$ is pure entrance from E, or we may say X cannot be entrance from '∞'.

Proof. First we prove $P\{\xi_{\infty 0} < \sigma\} = 0$. We shall use reduction to absurdity. Put $\Omega_i = (\xi_{\infty i} < \sigma)$, and suppose $P(\Omega_0) > 0$. Obviously, $\xi_{\infty i} \downarrow$ $(i\uparrow)$. We write $\Omega_i \uparrow \Omega_\infty$ $(i\uparrow\infty)$. By Lemma 10.4.3 and the strong Markov property, we may consider the process $X_i = \{X(\xi_{\infty i} + t), t < \sigma - \xi_{\infty i}\}$ on the probability space $(\Omega_i, \Omega_i \mathscr{F}, P(\cdot|\Omega_i))$. X_i and X have the same transition probability matrix, and $P\{x_i(0) = i|\Omega_i\} = 1$. We denote by ξ_i^j the random variable defined by (6) for X_i. Since

$$u_i(\lambda) = E\{\exp(-\lambda\xi_0^i)|\Omega_i\} \tag{18}$$

is determined only by the initial distribution and transition probability, therefore, the quantity given by (18) and the quantity in theorem 7 are equal, so that $u_i(\lambda)\downarrow 0$ $(i\uparrow\infty)$.

On the other hand,

$$u_i(\lambda) = \frac{1}{P(\Omega_i)}\int_{\Omega_i} \exp(-\lambda\xi_0^i)\,\mathrm{d}P \geqslant \frac{1}{P(\Omega_\infty)}\int_{\Omega_0} \exp(-\lambda\xi_0^i)\,\mathrm{d}P$$

Since $\xi_0^i \leqslant \xi_{\infty i} < \infty$ on Ω_0, it follows that

$$u_i(\lambda) \geqslant \frac{1}{P(\Omega_\infty)}\int_{\Omega_0} \exp(-\lambda\xi_{\infty 0})\,\mathrm{d}P > 0$$

This contradicts $u_i(\lambda)\downarrow 0$. Therefore $P(\xi_{\infty 0} < \sigma) = 0$. For $i > 0$,

$$P\{\xi_{\infty 0} < \sigma\} \geqslant P\{\xi_{\infty i} < \sigma\}\prod_{k=1}^{i}\frac{a_k}{a_k + b_k}$$

Consequently, $P(\xi_{\infty i} < \sigma) = 0$. The proof is completed. QED

12.3 AN EXTENDED DYNKIN LEMMA

We shall prove the following extended Dynkin lemma, where the processes may not necessarily be birth–death processes.

Lemma 1. Let $X = \{x(t), t < \sigma\} \in \mathscr{H}$, ξ be a non-negative random variable and θ be a translation operator satisfying the following conditions:

(i) For arbitrary $s \geqslant 0$ and $t \geqslant 0$, the set $A_s = \{\xi > s\} \in \mathscr{F}_s^0$ and $A_{s+t} \subseteq A_s \cap \theta_s A_t$.

(ii) There exist positive numbers $T > 0$ and $\alpha > 0$ such that for all $i \in E$ we have $P_i(A_T) \leqslant 1 - \alpha$.

Then for any initial distribution, all moments $E\xi^l$ of ξ are finite, and the distribution function $P\{\xi \leqslant t\}$ is uniquely determined by its moments $E\xi^l$ $(l = 0, 1, 2, \ldots)$.

Proof. Since $A_s \in \mathscr{F}_s^0$, $\theta_s A_T \in \mathscr{F}_\infty^s$, by the Markov property, we obtain

$$\begin{aligned} P_i\{A_{s+T}\} \leqslant P_i\{A_s \cap \theta_s A_T\} &= \int_{A_s} P_{x(s)}(A_T)\,\mathrm{d}P_i \\ &\leqslant (1-\alpha)P_i(A_s) \end{aligned}$$

Therefore $P_i(A_{nT}) \leqslant (1-\alpha)^n$. From this, we get $P(A_{nT}) \leqslant (1-\alpha)^n$ and $P(\xi < \infty) = 1$. Thus,

$$E\{\xi^l\} = \sum_{n=0}^{\infty} \int_{nT<\xi\leqslant(n+1)T} \xi^l \,\mathrm{d}P \leqslant \sum_{n=0}^{\infty} \{(n+1)T\}^l P\{\xi > nT\}$$

$$\leqslant \sum_{n=0}^{\infty} \{(n+1)T\}^l (1-\alpha)^n < \infty$$

Next take a positive number r such that $\mathrm{e}^{Tr}(1-\alpha) < 1$, then

$$\sum_{l=0}^{\infty} \frac{E\{\xi\}^l}{l!} r^l \leqslant \sum_{l=0}^{\infty} \sum_{n=0}^{\infty} \frac{\{(n+1)T\}^l r^l}{l!} (1-\alpha)^n$$

$$= \sum_{n=0}^{\infty} \mathrm{e}^{(n+1)Tr}(1-\alpha)^n = \frac{e^{Tr}}{1-e^{Tr}(1-\alpha)} < \infty$$

By a theorem in Cramer (1946, section 15.4), the distribution function of ξ is uniquely determined by its moments. The proof is concluded. QED

12.4 RECURRENCE AND ERGODIC PROPERTY OF THE HONEST PROCESS

From now on, we shall always consider conservative birth–death processes, that is, $a_0 = 0$. By Theorem 2.6, the minimal process is honest if and only if $R < \infty$, hence we further suppose $R < \infty$. Then $P\{\tau < \infty\} = 1$.

Theorem 1. Let $X = \{x(t), t < \sigma\} \in \mathscr{H}_s(Q)$ be honest, and τ be the first leaping point. Put

$$\beta_1^n = \inf\{t \mid \tau \leqslant t < \sigma, x(t) \leqslant n\} \tag{1}$$

Then for any initial distribution, all moments $E[\beta_1^n]^l$ $(l \geqslant 0)$ of β_1^n are finite, and its distribution $P\{\beta_1^n \leqslant t\}$ is uniquely determined by its moments $E[\beta_1^n]^l$ $(l \geqslant 0)$.

Proof. Since $P_i(\tau < \infty) = 1$ and X is honest, it follows that there exists $s > 0$ such that

$$0 < P_0(\tau < s) = P_0\{\tau < s, x(s) \in E\} = \sum_j P_0\{\tau < s, x(s) = j\}$$

So there exists $j \in E$ such that $P_0(\tau < s, x(s) = j) > 0$, we have

$$\alpha \equiv P_0\{\beta_1^0 \leqslant s+t\} \geqslant P_0\{\tau < s, x(s+t) = 0\} \geqslant P_0\{\tau < s, x(s) = j\} p_{j0}(t) > 0$$

However, on the set $\{x(0) = i, \beta_1^n \leqslant s+t\}$, for η_i^* defined by (6.7.7), we have $\eta_{i+1}^* < \infty$, and $\beta_1^0 = \eta_{i+1}^* + \theta_{\eta_{i+1}^*} \beta_1^0$. Therefore

$$\{x(0) = i, \beta_1^0 \leqslant s+t\} \subset \{x(0) = i, \eta_{i+1}^* < \infty, \theta_{\eta_{i+1}^*}(\beta_1^0 \leqslant s+t)\}$$

Hence $P_i\{\beta_1^0 \leqslant s+t\}$ increases with i. If we choose $T=s+t$, then $P_i(\beta_1^0 \leqslant T) \geqslant \alpha > 0$ $(i \in E)$. Thus the condition (ii) in Lemma 3.1 is satisfied; obviously, the condition (i) in Lemma 3.1 is satisfied. By Lemma 3.1, this theorem holds for $n=0$. Again noting $\beta_1^n \leqslant \beta_1^0$, we obtain the proof of the theorem. The proof is terminated.

QED

Theorem 2. All honest processes $X \in \mathscr{H}_s(Q)$ are recurrent and ergodic. Furthermore, all order moments $E[\eta_i^*]^l$ of η_i^* defined by (7.7.7) are finite, and the distribution function $P(\eta_i^* \leqslant t)$ is uniquely determined by its moments $E[\eta_i^*]^l$ $(l \geqslant 0)$.

Proof. Since $\eta_i^* \leqslant \beta_1^n$ $(n \geqslant i)$, quoting Theorem 1, we can prove this theorem.

QED

12.5 TWO LEMMAS

Let $X = \{x(t), t < \sigma\} \in \mathscr{H}_s(Q)$. If X is stopping, we may take $\Delta = -1$, and transform X into the honest process $\tilde{X} = \{\tilde{x}(t), t < \infty\} \in \mathscr{H}_s(\tilde{Q})$, in the same way as (7.6.6.), where $\tilde{Q} = (\tilde{q}_{ij})$ $(i, j \in (-1) \cup E)$, and $\tilde{q}_i = q_i$, $\tilde{q}_{i,-1} = \tilde{q}_{-1,-1} = \tilde{q}_{-1,j} = 0$ $(i, j \in E)$. Henceforth we shall adopt the convention above.

Let τ still be the first leaping point, and τ_n be the nth jumping point, and ξ_i be defined by (2.6). For arbitrary $0 \leqslant \varepsilon \leqslant \infty$, put

$$f_\varepsilon(x) = \begin{cases} x & \text{if } 0 \leqslant x \leqslant \varepsilon \\ \varepsilon & \text{if } x > \varepsilon \end{cases} \tag{1}$$

Lemma 1. For $k \geqslant i \geqslant 0$, put

$$H_{ki}^\varepsilon = E_k\left\{ \sum_{0 \leqslant \tau_j < \min(\xi_i, \tau)} f_\varepsilon(\tau_{i+1} - \tau) \right\} \tag{2}$$

In particular,

$$H_{ki}^\infty = E_k\{\min(\xi_i, \tau)\} \tag{3}$$

Then when $R < \infty$, we have

$$H_{ki}^\varepsilon = \frac{z - z_k}{z - z_i} \sum_{j=i+1}^{k-1} (z - z_j)\{1 - \exp[-(a_j + b_j)\varepsilon]\}\mu_j$$

$$+ \frac{z_k - z_i}{z - z_i} \sum_{j=k}^{\infty} (z - z_j)\{1 - \exp[-(a_j + b_j)\varepsilon]\}\mu_j \leqslant N_k \tag{4}$$

and

$$\lim_{\varepsilon \downarrow 0} H_{ki}^\varepsilon = 0 \tag{5}$$

If we also have $s < \infty$, then

$$\lim_{k\to\infty} \frac{H^{\varepsilon}_{ki}}{C_{k0}} = \frac{1}{C_{i0}} \sum_{j=i+1}^{\infty} (z_j - z_i)\{1 - \exp[-(a_j + b_j)\varepsilon]\} \tag{6}$$

$$\lim_{i\to\infty} \lim_{k\to\infty} \frac{H^{\varepsilon}_{ki}}{C_{k0}} = 0 \tag{7}$$

$$\lim_{\varepsilon\downarrow 0} \lim_{k\to\infty} \frac{H^{\varepsilon}_{ki}}{C_{k0}} = 0 \tag{8}$$

where C_{ki} is determined by (2.16).

Proof. Let $i \leqslant k \leqslant n$. Put

$$H^{\varepsilon}_{kin} = E_k\left\{ \sum_{0 \leqslant \tau_j < \min(\xi_i, \xi_n)} f_{\varepsilon}(\tau_{i+1} - \tau) \right\} \tag{9}$$

Obviously, $H^{\varepsilon}_{kin} \uparrow H^{\varepsilon}_{ki}$ $(n \uparrow \infty)$. We see easily that

$$E_k f_{\varepsilon}(\tau_1) = \frac{1}{a_k + b_k}\{1 - \exp[-(a_k + b_k)\varepsilon]\}$$

By an application of Theorem 7.8.2 it follows that $u_k = H^{\varepsilon}_{kin}$ satisfies equation (5.4.11) for $f_i = f_n = 0, f_k = 1 - \exp[-(a_k + b_k)\varepsilon]$ $(i < k < n)$. From Theorem 6.2.3, we obtain

$$\begin{aligned} H^{\varepsilon}_{kin} = {} & \frac{z_n - z_k}{z_n - z_i} \sum_{j=i+1}^{k-1} (z - z_i)\{1 - \exp[-(a_j + b_j)\varepsilon]\}\mu_j \\ & + \frac{z_k - z_i}{z_n - z_i} \sum_{j=k}^{n-1} (z_n - z_j)\{1 - \exp[-(a_j + b_j)\varepsilon]\}\mu_j \end{aligned} \tag{10}$$

Letting $n \to \infty$, we obtain the equality in (4). In comparison with (6.2.6), the inequality in (4) follows.

Since $R < \infty$, by using the dominated convergence theorem, (5) follows from (4).

When $s < \infty$, we have $\sum_{i=0}^{\infty} \mu_i < \infty$, whereas

$$\frac{1}{C_{k0}} \frac{z_k - z_i}{z - z_i} \sum_{j=k}^{\infty} (z - z_j)\{1 - \exp[-(a_j + b_j)\varepsilon\}\mu_j \leqslant z \sum_{j=k}^{\infty} \mu \to 0 \qquad (k \to \infty)$$

$$\frac{1}{C_{i0}} \sum_{j=i+1}^{\infty} (z - z_i)\{1 - \exp[-(a_j + b_j)\varepsilon]\}\mu_j \leqslant z \sum_{j=i+1}^{\infty} \mu_j \to 0 \qquad (i \to \infty)$$

Therefore, (6) follows from (4) and, thus, we obtain (7). Using the dominated convergence theorem, we derive (8) from (6). The proof is completed. QED

Lemma 2. Let $X \in \mathscr{H}_s(Q)$ be a non-minimal Q-process, τ be the first leaping point, and β_1^n be determined by (4.1). Then

$$P\left\{\lim_{n\to\infty} \beta_1^n = \tau\right\} = 1 \tag{11}$$

Proof. Obviously, $\beta_1^n \downarrow (n\uparrow)$, hence $\lim_{n\to\infty} \beta_1^n \geqslant \tau$. On the other hand, for arbitrary positive $\varepsilon_m \downarrow 0$, by (ii) in Theorem 7.6.2, it follows that $P\{x(\tau+\varepsilon_m)=\infty\}=0$. If $\tau=\sigma$, then, certainly, $\lim_{n\to\infty}\beta_1^n=\tau$. If $\tau<\sigma$ then when m is sufficiently large, we have $\tau+\varepsilon_m<\sigma$ and $x(\tau+\varepsilon_m)\in E$. Therefore when $n \geqslant x(\tau+\varepsilon_m)$, we have $\beta_1^n \leqslant \tau+\varepsilon_m$, and $\lim_{n\to\infty}\beta_1^n \leqslant \tau+\varepsilon_m$. Since m is arbitrary, it follows that $\lim_{n\to\infty}\beta_1^n \leqslant \tau$. The proof is concluded. QED

12.6 CHARACTERISTIC SEQUENCE

Let $X=\{x(t), t<\sigma\}\in\mathscr{H}_s(Q)$ be a non-minimal process. We consider the transformation g_n in Definition 10.5.1. Let $\beta_0^n=0$, τ_1^n be the first leaping point for X, and

$$\beta_1^n = \inf\{t \mid \tau_1^n \leqslant t < \sigma, x(t) \leqslant n\} \tag{1}$$

If the above set is empty, we take σ as 'inf'. Suppose that τ_{m-1}^n, β_{m-1}^n are already defined. If $\beta_{m-1}^n=\sigma$, then we define $\tau_m^n=\beta_m^n=\sigma$; otherwise, we define τ_m^n as the first leaping point after β_{m-1}^n, and

$$\beta_m^n = \inf\{t \mid \tau_m^n \leqslant t < \sigma, x(t) \leqslant n\}. \tag{2}$$

Transformation W_{τ^n,β^n} becomes transformation g_n and $X^n = W_{\tau^n,\beta^n}(X)$ becomes

$$X^n = g_n(X) \tag{3}$$

We have

$$g_n(X^{n+1}) = X^n \tag{4}$$

In particular, if X is a (Q,π) Doob process and $\pi_j=0$ $(j>n)$, then $g_n(X)=X$.

Theorem 1. Let $X\in\mathscr{H}_s(Q)$ be a non-minimal process. Then $X^n=\{x^n(t), t<\sigma^n\}\in\mathscr{H}_s(Q)$ given by (3) is a (Q, V^n) Doob process, satisfying (4), where $V^n=(v_j^n, 0\leqslant j\leqslant n)$, with

$$v_j^n = P\{x(\beta_1^n)=j\} \qquad (-1\leqslant j\leqslant n) \tag{5}$$

satisfies

$$v_j^n = v_j^{n+1}\left(\sum_{i=-1}^{n} v_i^{n+1} + v_{n+1}^{n+1}C_{n+1,n}\right)^{-1} \qquad -1\leqslant j<n$$

$$v_n^n = (v_n^{n+1} + v_{n+1}^{n+1} C_{n+1,n}) \left(\sum_{i=-1}^{n} v_1^{n+1} + v_{n+1}^{n+1} C_{n+1,n} \right)^{-1}$$

$$\sum_{j=-1}^{n} v_j^n = 1 \qquad \sum_{i=0}^{n} v_j^n > 0 \tag{6}$$

For X to be honest it is necessary and sufficient that $v_{-1}^n = 0$ $(n \geqslant 0)$.

Proof. Noticing that β_k^n is a Markov time and $P\{x(\beta_k^n) > n\} = 0$, we now proceed to prove that V^n defined by (5) satisfies (6).

Let the quantity $\bar{\beta}_1^n$ be determined by the process $\bar{X} = \{x(\beta_1^{n+1} + t), t < \sigma - \beta_1^{n+1}\}$, in fashion (1). Suppose that Λ_n is the event that the process $\bar{X}$ goes from $n+1$ to n in finite ($\geqslant 0$) steps, and $\bar{\Lambda}_n$ is the complementary event of Λ_n. We can easily identify $\{x(\beta_1^{n+1}) = n+1\}$, Λ_n, $\bar{\Lambda}_n$ all belong to $\mathscr{F}_{\tau_{2-}^{n+1}}$, and $\{x(\bar{\beta}_1^n) = j\} \in \mathscr{F}'_{\tau_2^{n+1}}$. By the corollary to Theorem 7.17.1 (note that the conditional independence in this corollary becomes independence in the case of birth–death processes), for $-1 \leqslant j \leqslant n$,

$$\begin{aligned}
\Delta_j^n &\equiv P\{x(\beta_1^{n+1}) = n+1, \bar{\Lambda}_n, \bar{x}(\bar{\beta}_1^n) = j\} \\
&= P\{x(\beta_1^{n+1}) = n+1, \bar{\Lambda}_n\} P\{\bar{x}(\beta_1^n) = j \mid x(\beta_1^{n+1}) = n+1\} \\
&= P\{x(\beta_1^{n+1}) = n+1\} P\{\bar{\Lambda}_n \mid x(\beta_1^{n+1}) = n+1\} v_j^n \\
&= v_{n+1}^{n+1}(1 - C_{n+1,n}) v_j^n
\end{aligned}$$

Consequently,

$$\begin{aligned}
v_n^n &= P\{x(\beta_1^{n+1}) = n\} + P\{x(\beta_1^{n+1}) = n+1, \Lambda_n\} + \Delta_n^n \\
&= v_n^{n+1} + v_{n+1}^{n+1} C_{n+1,n} + v_{n+1}^{n+1}(1 - C_{n+1,n}) v_n^n \\
v_j^n &= P\{x(\beta_1^{n+1}) = j\} + \Delta_j^n = v_j^{n+1} + v_{n+1}^{n+1}(1 - C_{n+1,n}) v_j^n \qquad (-1 \leqslant j < n)
\end{aligned}$$

Hence

$$\begin{aligned}
v_j^n &= v_j^{n+1} / [1 - v_{n+1}^{n+1}(1 - C_{n+1,n})] \qquad -1 \leqslant j < n \\
v_n^n &= (v_n^{n+1} + v_{n+1}^{n+1} C_{n+1,n}) / [1 - v_{n+1}^{n+1}(1 - C_{n+1,n})]
\end{aligned} \tag{7}$$

From the above expression, we know that either $P\{\beta_1^n < \sigma\} > 0$, for all n, or $P\{\beta_1^n = \sigma\} = 0$ for all n. If the latter holds, then by Lemma 5.2 we have $P\{\tau = \sigma\} = 1$, that is, X is a minimal process, which is contradictory to the hypothesis of this theorem. Hence

$$\sum_{j=0}^{n} v_j^n = P\{\beta_1^n < \sigma\} > 0 \qquad (n \geqslant 0)$$

Next we shall prove $\sum_{j=-1}^{n} v_j^n = 1$. From this and (7) follows (6). When X is honest, by Theorem 4.1, we have $P\{\beta_1^n < \infty\} = 1$. Therefore $v_{-1}^n = 0$, and $\sum_{j=0}^{n} v_j^n = 1$. When X is stopping, we taking $\pi_0 = 1$, $\pi_j = 0$ $(j > 0)$. By Theorem 11.2.2, we may take into account the πD-type extension process

$\bar{X}=\{\bar{X}(t), t<\infty\}$ of X, that is,

$$\begin{aligned} \bar{X}(t) &= x(t) \qquad t<\sigma \\ P\{\bar{x}(\sigma)=0 \mid \sigma<\infty\} &= 1 \end{aligned} \tag{8}$$

Letting $\bar{\beta}_1^n$ be the quantity for $\bar{X}$, obviously, $\beta_1^n=\bar{\beta}_1^n$. Since $\bar{X}$ is honest, and $P(\bar{\beta}_1^n<\infty)=1$, it follows that $\sum_{j=-1}^{n} v_j^n = P\{\beta_1^n<\infty\}=1$.

Suppose $v_{-1}^n=0$, that is, $P\{\beta_1^n<\sigma\}=\sum_{j=0}^{n} v_j^n=1$. By the strong Markov property, we have $P\{\beta_1^n<\beta_2^n<\cdots<\sigma\}=1$. This shows that there exist infinitely many $j=j(\omega)$ $(\leqslant n)$ intervals in $[0,\sigma(\omega))$. By Theorem 7.7.1, we have $P\{\sigma=\infty\}=1$, that is, X is honest.

We now prove that $X^n=g_n(X)$ is a (Q, V^n) Doob process. Set $X_m=\{x(\beta_m^n+t), t<\tau_{m+1}^n-\beta_m^n\}$ $(m\geqslant 0)$. We can easily see that $X_m\in\mathscr{H}_s(Q)$ is a minimal Q process, satisfying (ii*) in Lemma 11.3.1. By $R<\infty$, i.e. $P(\tau<\infty)=1$, we know that $\beta_m^n<\sigma$ if and only if $0<\tau_{m+1}^n-\beta_m^n<\infty$. By the corollary to Theorem 7.17.1, we have

$$P\{x^{m+1}(0)=j \mid 0<\tau_{m+1}^n-\beta_m^n<\infty\}=P\{x(\beta_{m+1}^n)=j \mid \beta_m^n<\sigma\}=v_j^n \qquad 0\leqslant j\leqslant n$$

Therefore (iii*) in Lemma 11.3.1 is satisfied, too. By the strong Markov property of X, we may directly deduce that (iv*) in Lemma 11.3.1 is satisfied for $\Delta=\{x^{m+1}(0)=i\}$. It is not difficult to verify that X_m $(m\leqslant k)$ is $\mathscr{F}_{\tau_{k+1}^n}$-measurable, and X_m $(m>k)$ is $\mathscr{F}'_{\tau_{k+1}^n}$-measurable. We have already pointed out $\{0<\tau_{k+1}^n-\beta_k^n<\infty\}=(\beta_k^n<\sigma)$. By the corollary to Theorem 7.17.1, (iv*) in Lemma 11.3.1 is satisfied. According to Theorem 11.4.2, the process determined by (11.2.5) and (11.2.6) for X_m $(m\geqslant 0)$ is a (Q, V^n) Doob process. But this process is precisely $X^n=g_n(X)$. The proof is completed. QED

Theorem 2. For any $X=\{x(t), t<\sigma\}\in\mathscr{H}_s(Q)$, we have

$$P\{\sigma<\infty\}=0 \text{ or } 1$$

Proof. The conclusion of this theorem is valid for each minimal process or honest process X. Assume x is stopping and non-minimal.

By Theorem 1, we have

$$P\{\beta_1^n=\sigma=\infty\}=1-\sum_{j=-1}^{n} v_j^n=0$$

that is,

$$P\{\beta_1^n<\sigma=\infty\}+P\{\sigma<\infty\}=1 \tag{9}$$

By the strong Markov property, we have

$$P\{\beta_1^n<\sigma=\infty\}=\sum_{j=0}^{n} v_j^n P_j\{\sigma=\infty\}$$

But $u_j \equiv P_j\{\sigma = \infty\}$ satisfies the equation $Qu = 0$, so that u_j is a constant; that is, $P_j(\sigma = \infty) = P(\sigma = \infty)$ is independent of j. Therefore, (9) becomes

$$\left(\sum_{j=0}^{n} v_j^n\right) P\{\sigma = \infty\} + P\{\sigma < \infty\} = 1$$

By Theorem 1, we have $0 < \sum_{j=0}^{n} v_j^n < 1$. Therefore, $P\{\sigma = \infty\} = 0$ or $P\{\sigma < \infty\} = 1$ must follow from the above formula. The proof is concluded.

QED

Theorem 3. Let $X = \{x(t), t < \sigma\} \in \mathscr{H}_s(Q)$ be a non-minimal Q process. Then there exists a non-negative sequence of numbers p, q and r_n ($n \geqslant -1$) satisfying

$$\begin{aligned} &p + q = 1 && \\ &q = 0 && \text{if } S = \infty \\ &r_n = 0\,(n \geqslant 0) && \text{if } p = 0 \\ &0 < \sum_{n=0}^{\infty} r_n N_n < \infty && \text{if } p > 0 \end{aligned} \tag{10}$$

such that

$$v_j^n = P\{x(\beta_1^n) = j\} \qquad -1 \leqslant j \leqslant n \tag{11}$$

which may be expressed as

$$\begin{aligned} &v_i^n = (X_n/A_n) r_j && -1 \leqslant j < n \\ &v_n^n = Y_n + (X_n/A_n) \sum_{l=n}^{\infty} r_l C_{ln} && \end{aligned} \tag{12}$$

where

$$A_n = \sum_{l=0}^{\infty} r_l C_{ln} \qquad d = \begin{cases} (q/p) A_0 & \text{if } p > 0 \\ 1 & \text{if } p = 0 \end{cases}$$

$$\begin{aligned} X_n &= \frac{A_n C_{n0}}{(r_{-1} + A_n) C_{n0} + d} \\ Y_n &= \frac{d}{(r_{-1} + A_n) C_{n0} + d} \\ \frac{X_n}{A_n} &= \frac{C_{n0}}{r_{-1} C_{n0} + 1} \qquad \text{if } p = 0 \end{aligned} \tag{13}$$

If we let η be the last leaping point before β_1^0, then

$$P\{x(\eta) = j\} = \begin{cases} r_{-1}/(r_{-1} + A_0 + d) & \text{if } j = -1 \\ r_j C_{j0}/(r_{-1} + A_0 + d) & \text{if } 0 \leqslant j < \infty \\ d/(r_{-1} + A_0 + d) & \text{if } j = \infty \end{cases} \tag{14}$$

Above, p and q are uniquely determined by X. If $p>0$, then γ_n $(n\geqslant -1)$ is uniquely determined by X except for a constant factor. If $p=0$, then γ_n $(n\geqslant -1)$ is uniquely determined by X

Proof. The proof is to be established in several steps.

(a) Put

$$R_n=\sum_{i=0}^{n-1} v_i^n C_{i0} \qquad S_n=v_n^n C_{n0} \qquad \Delta_n=R_n+S_n \tag{15}$$

Then by (6), we have

$$0<\Delta_n=\frac{\Delta_{n+1}}{\delta_{n+1}} \qquad S_n=\frac{v_n^{n+1}C_{n0}+S_{n+1}}{\delta_{n+1}} \tag{16}$$

$$\delta_{n+1}=\sum_{j=-1}^{n} v_j^{n+1}+v_{n+1}^{n+1}C_{n+1,n} \tag{17}$$

Therefore, v_j^n/Δ_n is independent of $n>j$ $(j\geqslant -1)$, and there exist limits

$$\frac{S_n}{\Delta_n}\downarrow p\geqslant 0 \qquad \frac{R_n}{\Delta_n}\uparrow q\geqslant 0 \tag{18}$$

If $p=0$, we take $r=1$; if $p>0$, we arbitrarily take $\gamma>0$. Put

$$r_j=\frac{v_j^n}{\Delta_n}r \qquad n>j\geqslant -1 \tag{19}$$

Thus we obtain the non-negative sequence of numbers p, q and r_n $(n\geqslant -1)$.

(b) Obviously, $p+q=1$. Since when $p=0$, we have $r_n=0$ $(n\geqslant 0)$. If $p>0$, then there exists a $r_k>0$ $(k\geqslant 0)$ at least. Therefore, we need only prove

$$\sum_{n=0}^{\infty} r_n N_n<\infty \tag{20}$$

Note that $E\beta_1^0<\infty$. When X is honest, (20) follows from Theorem 4.1. When X is stopping, there exists an honest Q process $\bar{X}$ satisfying (8). Hence we also have $E\beta_1^0=E\bar{\beta}_1^0<\infty$. Whence

$$E\tau_2^0=E\beta_1^0+E\{\tau_2^0-\beta_1^0\}=E\beta_1^0+v_0^0E_0\tau=E\beta_1^0+v_0^0R<\infty$$

Put

$$M_i^n=\begin{cases}\tau_{i+1}^n-\beta_i^n & \text{if } \beta_i^n<\beta_1^0\\ 0 & \text{otherwise}\end{cases}$$

Obviously,

$$\tau_2^0\geqslant \sum_{i=1}^{\infty} M_i^n$$

$$\infty>E\tau_2^0\geqslant \sum_{i=1}^{\infty} EM_i^n=\sum_{i=1}^{\infty} E\{\tau_{i+1}^n-\beta_i^n, \beta_i^n<\beta_1^0\}$$

But

$$E\{\tau_2^n - \beta_1^n, \beta_1^n < \beta_1^0\} = \sum_{j=1}^{n} v_j^n E_j \tau_1^n = \sum_{j=1}^{n} v_j^n N_j$$

$$E\{\tau_{i+1}^n - \beta_i^n, \beta_i^n < \beta_1^0\} = \sum_{j=1}^{n} P\{\beta_{i-1}^n < \beta_1^0, x(\beta_i^n) = j\} E_j \tau_1^n$$

$$= \sum_{j=1}^{n} \left\{ \sum_{k=1}^{n} v_k^n (1 - C_{k0}) \right\}^{i-1} v_j^n N_j$$

Therefore,

$$\infty > E\tau_2^0 \geqslant \frac{\sum_{j=1}^{n} v_j^n N_j}{1 - \sum_{k=1}^{n} v_k^n (1 - C_{k0})} \geqslant \frac{\sum_{j=1}^{n-1} v_j^n N_j}{v_{-1}^n + \Delta_n}$$

By (19) it follows that

$$\frac{\sum_{j=1}^{n-1} r_j N_j}{r_{-1} + r} \leqslant E\tau_2^0 < \infty$$

Letting $n \to \infty$, we obtain (20).

(c) We now proceed to prove (12). By (18) and (19) we have

$$p = \lim_{n \to \infty} \frac{R_n}{\Delta_n} = \lim_{n \to \infty} \frac{1}{r} \sum_{j=0}^{n-1} r_i C_{i0} = \frac{A_0}{r}$$

so that

$$r = \begin{cases} A_0/p & \text{if } p > 0 \\ 1 & \text{if } p = 0 \end{cases} \tag{21}$$

Using induction on (6) we easily obtain that: $m > n$,

$$\begin{aligned} v_j^n &= v_j^m / \delta_{m,n} \qquad -1 \leqslant j < n \\ v_n^n &= \left(\sum_{j=n}^{m} v_j^m C_{jn} \right) \Big/ \delta_{mn} \\ 0 &< \Delta_n = \Delta_m / \delta_{mn} \\ \delta_{mn} &= \sum_{j=-1}^{m} v_j^m + \sum_{j=n+1}^{m} v_j^m C_{jn} \end{aligned} \tag{22}$$

From the above, (18) and (19), we have

$$\frac{v_n^n}{\Delta_n} = \frac{\sum_{j=n}^{m} v_j^m C_{jm}}{\Delta_m} = \frac{1}{r} \sum_{j=n}^{m-1} r_j C_{jn} + \frac{S_m}{\Delta_m C_{n0}}$$

so that when $m \to \infty$, we have

$$\frac{v_n^n}{\Delta_n} = \frac{1}{r} \sum_{j=n}^{\infty} r_j C_{jn} + \frac{q}{C_{n0}} \tag{23}$$

By (19) and (23), we obtain

$$1 = \sum_{j=-1}^{n} v_j^n = \Delta_n \left(\frac{\sum_{j=-1}^{n-1} r_j}{r} + \frac{\sum_{j=n}^{\infty} r_j C_{jn}}{r} + \frac{q}{C_{n0}} \right)$$

hence,

$$\Delta_n = \frac{rC_{n0}}{(r_{-1} + A_n)C_{n0} + qr}$$

Substituting the above formula into (19) and (23) and noticing (21), we have (12).

(d) We now prove (14). If X is honest, then $1 = P\{\beta_1^n < \infty\}$. By the strong Markov property, we have $P(\beta_1^n < \beta_2^n < \cdots) = 1$; therefore, for almost all $\omega \in \Omega$, there exist infinitely many $j(\omega)(\leqslant n)$ intervals in $[0, \lim_{l\to\infty} \beta_l^n(\omega))$ for X, whence $P(\lim_{l\to\infty} \beta_l^n = \infty) = 0$. If X is stopping, by Theorem 2, we have $P(\sigma < \infty) = 1$. Since if $\beta_l^n(\omega) < \sigma(\omega)$, then $\beta_l^n(\omega) < \tau_{l+1}^n(\omega) \leqslant \beta_{l+1}^n(\omega)$, by Theorem 7.7.1, we have $P\{$there exists l such that $\beta_l^n = \sigma\} = 1$. Thus we always have

$$P\left\{ \lim_{l\to\infty} \beta_l^n = \sigma \right\} = 1 \tag{24}$$

whether X is stopping or honest. Moreover, for almost all ω, there exists unique l such that $\beta_{l-1}^n < \eta \leqslant \beta_l^n \leqslant \beta_1^0$. Write l as l_n, that is,

$$l_n = \min\{l \mid \beta_l^n \geqslant \eta\} \tag{25}$$

$$\beta_{l_n}^n = \inf\{t \mid \eta \leqslant t < \sigma, x(t) \leqslant n\} \tag{26}$$

If $x(\beta_{l_n}^n) = j$ for some $n > j \geqslant -1$, then it necessarily follows that $\beta_{l_n}^n = \eta$. Since if $\eta < \beta_{l_n}^n$ by definition we have $n < x(t) < \infty$ for $t \in (\eta, \beta_{l_n}^n)$, while $x(\beta_{l_n}^n) = j < n$. As the jump of birth–death processes at the jump point is 1, this is not possible. Hence for $n > j \geqslant -1$,

$$\{x(\beta_{l_n}^n) = j\} \subset \{\beta_{l_n}^n = \eta, x(\eta) = j\} \subset \{x(\eta) = j\} \tag{27}$$

Again it is quite clear that, if $x(\eta) = j$, then for $n > j$ we have $\beta_{l_n}^n = \eta$, so that

$$\{x(\eta) = j\} = \lim_{n\to\infty} \{x(\beta_{l_n}^n) = j\} \qquad (-1 \leqslant j < \infty) \tag{28}$$

We shall now prove

$$\{x(\eta) = \infty\} = \lim_{n\to\infty} \{x(\beta_{l_n}^n) = n\} \tag{29}$$

In fact, assuming that ω belongs to the right-hand side in (29), we surely find

$\beta^n_{l_n}\downarrow\bar{\eta}\geqslant\eta$. Hence by the right-continuity, we have $x(\bar{\eta})=\lim_{n\to\infty}x(\beta^n_{l_n})=\lim_{n\to\infty}n=\infty$. But by the definition of η it follows that for any $\eta<t<\beta^0_1$, $x(t)\in E$. Hence $\bar{\eta}=\eta$, whence $x(\eta)=\infty$, i.e. ω belongs to the left-hand side in (29). Letting ω belong to the left-hand side in (29) for arbitrary n, we certainly have $x(\beta^n_{l_n})=n$. Otherwise, (27) will yield $x(\eta)=j\neq\infty$. Thus (29) holds.

For $0\leqslant j\leqslant n$, we have

$$
\begin{aligned}
P\{x(\beta^n_{l_n})=j\}&=\sum_{l=1}^{\infty}P\{x(\beta^n_l)=j,l_n=l\}\\
&=\sum_{l=1}^{\infty}P\{\beta^n_{l-1}<\beta^0_1,x(\beta^n_l)=j,\beta^n_l\leqslant\beta^0_1<\tau^n_{l+1}\}\\
&=\sum_{l=1}^{\infty}\left\{\sum_{k=1}^{n}v^n_k(1-C_{k0})\right\}^{l-1}v^n_jC_{j0}=\frac{v^n_jC_{j0}}{v^n_{-1}+\Delta_n}
\end{aligned}\tag{30}
$$

and

$$
\begin{aligned}
P\{x(\beta^n_{l_n})=-1\}&=\sum_{l=1}^{\infty}P\{x(\beta^n_l)=-1,l_n=l\}\\
&=\sum_{l=1}^{\infty}P\{\beta^n_{l-1}<\beta^0_1,x(\beta^n_l)=-1\}\\
&=\sum_{l=1}^{\infty}\left\{\sum_{k=1}^{n}v^n_k(1-C_{k0})\right\}^{l-1}v^n_{-1}=\frac{v^n_{-1}}{v^n_{-1}+\Delta_n}
\end{aligned}\tag{31}
$$

Substituting (30) and (31) into (28) and (29) and, moreover, noticing (18) and (19), we obtain

$$
P\{x(\eta)=j\}=\begin{cases}r_{-1}/(r_{-1}+r) & \text{if } j=-1\\ r_jC_{j0}/(r_{-1}+r) & \text{if } 0\leqslant j<\infty\\ rq/(r_{-1}+r) & \text{if } r=\infty\end{cases}
$$

Again noticing (21), we find that the above expression is (14).

(e) Suppose $S=\infty$. By Theorem 2.8, we have $P\{\xi_{\infty 0}<\sigma\}=0$. But $\{x(\eta)=\infty\}\subset\{\xi_{\infty 0}<\sigma\}$. Hence $P\{x(\eta)=\infty\}=0$. Then $q=0$ follows from (13) and (14).

(f) Suppose that there exists a non-negative sequence of numbers $\bar{p},\bar{q}$ and $\bar{r}_n\,(n\geqslant-1)$ such that (10)–(14) hold. By (12) we know the following: Either both p and $\bar{p}$ are zero; hence $\bar{r}_n=r_n=0\,(n\geqslant 0)$ and $q=\bar{q}=1$; and again from

$$
v^n_{-1}=\frac{C_{n0}}{r_{-1}C_{n0}+1}r_{-1}=\frac{C_{n0}}{\bar{r}_{-1}C_{n0}+1}\bar{r}_{-1}
$$

it follows that $r_{-1}=\bar{r}_{-1}$. Or both p and $\bar{p}$ are positive, and (18) follows from (11) and (12), so that $p=\bar{p}$ and $q=\bar{q}$; and again by (19) we have $r_j/r_k=v^n_j/v^n_k=\bar{r}_j/\bar{r}_k$. The proof is completed. QED

Definition 1. The non-negative sequence of numbers p, q and $r_n\,(n \geqslant -1)$ is called the characteristic sequence of the Q process X.

12.7 PROBABILITY STRUCTURE OF PROCESSES

Suppose that a sequence of numbers p, q and $r_n\,(n \geqslant -1)$ satisfying (6.10) is given.

Lemma 1. $v_j^n(-1 \leqslant j \leqslant n)$ defined by (6.12) and (6.13) satisfies (6.6).

Proof. When $p=0$ we have $v_j^n = 0\,(0 \leqslant j < n)$,

$$v_{-1}^n = \frac{r_{-1}C_{n0}}{r_{-1}C_{n0}+1} \qquad v_n^n = \frac{1}{r_{-1}C_{n0}+1}$$

Equation (6.6) follows from direct identification.

Let $p > 0$. By direct identification, we know that the last expression in (6.6) is valid. By (6.12) and (6.13) we have

$$\begin{aligned} 1 - v_{n+1}^{n+1}(1 - C_{n+1,n}) &= \sum_{j=-1}^{n} v_j^{n+1} + v_{n+1}^{n+1}C_{n+1,n} \\ &= \frac{X_{n+1}}{A_{n+1}} \sum_{j=-1}^{n} r_j + Y_{n+1}C_{n+1,n} + \frac{X_n}{A_n}\sum_{j=n+1}^{\infty} r_j C_{jn} \\ &= \frac{X_{n+1}}{A_{n+1}}\left(r_{-1} + A_n + \frac{Y_{n+1}A_{n+1}C_{n+1,n}}{X_n}\right) \\ &= \frac{X_{n+1}}{A_{n+1}}\left(r_{-1} + A_n + \frac{d}{C_{n0}}\right) = \frac{X_{n+1}}{A_{n+1}}\frac{A_n}{X_n} \end{aligned} \tag{1}$$

$$\begin{aligned} v_n^{n+1} + v_{n+1}^{n+1}C_{n+1,n} &= \frac{X_{n+1}}{A_{n+1}} r_n + Y_{n+1}C_{n+1,n} + \frac{X_{n+1}}{A_{n+1}} \sum_{l=n+1}^{\infty} r_l C_{ln} \\ &= \frac{X_{n+1}}{A_{n+1}}\frac{A_n}{X_n}\left(Y_n + \frac{X_n}{A_n}\sum_{l=n}^{\infty} r_l C_{ln}\right) \end{aligned} \tag{2}$$

Therefore, (6) holds. QED

Lemma 2. Suppose that $V^n = \{v_j^n, 0 \leqslant j \leqslant n\}\,(n \geqslant 0)$ satisfies (6.6). Then there exists a probability space $(\Omega, \mathscr{F}, P)$ on which we may define a sequence of (Q, V^n) Doob processes $X^n = \{x^n(t), t < \sigma^n\} \in \mathscr{H}_s(Q)(n \geqslant 0)$ satisfying (6.4).

Proof. Fix a distribution (v_i) as an initial distribution. For each n, there exists a probability space $(\Omega_n, \mathscr{F}_n, P_n)$ on which a (Q, V^n) Doob process $\bar{X}_n = \{\bar{x}_n(t, \omega_n), t < \bar{\sigma}_n(\omega_n)\}\,(\omega_n \in \Omega_n)$ is defined. By Theorem 6.1, it follows that for $m<n, Z_m=g_m(\bar{X}_n)$ is a (Q, V^n) Doob process on $(\Omega_n, \mathscr{F}_n, P_n)$. Let $k\geqslant 1$. For non-negative integers $n_i (1\leqslant i\leqslant k)$ and non-negative real numbers $t_i\geqslant 0$ $(1\leqslant i\leqslant k)$ and $j_i\in E (1\leqslant i\leqslant k)$, we may choose $n>\max(n_1, n_2, \ldots, n_k)$ and define the k-dimensional distribution

$$F_{n_1t_1,\ldots,n_kt_k}(j_1,\ldots,j_k)=P_n\{Z_{n_i}(t_i,\omega_n)=j_i, 1\leqslant i\leqslant k\} \tag{3}$$

It is obvious that this distribution is independent of the choice of n and, furthermore, the family of finite-dimensional distributions $\{F_{n_1t_1,\ldots,n_kt_k}\}$ is consistent. According to the Kolmogorov theorem (see Zi-kun Wang, 1965a, section 1.1, Theorem 1), there exists a probability space $(\Omega, \mathscr{F}, P)$ on which a sequence of processes $X^n=\{x^n(t), t<\sigma^n\}$ is defined and, moreover,

$$P\{x^n{}_i(t_i)=j_i, 1\leqslant i\leqslant k\}=F_{n_1t_1,\ldots,n_kt_k}(j_1,\ldots,j_k) \tag{4}$$

From the above and (3), $X^n\in\mathscr{H}_s(Q)$ is a (Q, V^n) Doob process.

Secondly, according to the theorem quoted above, we may choose $\Omega=(\omega)$, where $\omega=\omega(n,t)$ is a bivariate function taking values in E, $(n=0,1,2,\ldots, t\in [0,\sigma_n), \sigma_n\leqslant\infty)$, and $x^n(t,\omega)=\omega(n,t), \sigma^n(\omega)=\sigma_n$. Therefore, (6.4) holds. The proof is terminated. QED

Let $X^n=\{x^n(t), t<\sigma^n\}$ be the sequence of processes in Lemma 2. From (6.4) it follows that $\sigma^n\leqslant\sigma^{n+1}$. Hence we may set $\sigma^n\uparrow\sigma$.

The quantities defined by the fashion in (6.1) and (6.2) for X^k are written as τ_m^{kn} and β_m^{kn}. From (6.4) it follows that $\beta_m^{n0}\leqslant\beta_m^{n+1,0}\leqslant\sigma^{n+1}$. Therefore, there exists the limit $\beta_m^0=\lim_{n\to\infty}\beta_m^{n0}\leqslant\sigma$.

On account of (7.2.4) $\lim_{m\to\infty}\beta_m^0\geqslant\lim_{m\to\infty}\beta_m^{n0}=\sigma^n$. Therefore

$$P\left\{\lim_{m\to\infty}\beta_m^0=\sigma\right\}=1 \tag{5}$$

For $n>m$, put

$$L_{nm}^i=\begin{cases}\beta_i^{nm}-\tau_i^{nm} & \text{if } \beta_i^{nm}<\beta_1^{n0}\\ 0 & \text{otherwise}\end{cases} \tag{6}$$

and

$$T_\varepsilon^{nm}=\sum_{\tau_1^n\leqslant\tau_{ij}^n<\beta_1^{nm}} f_\varepsilon(\tau_{i,j+1}^n-\tau_{ij}^n) \tag{7}$$

where τ_{ij}^n is the jth jump point after the ith leaping point of X^n. The process $f_\varepsilon(X)$ is defined as (5.1). By (6.4), we easily obtain

$$\begin{gathered} L_{nm}=\sum_{i=1}^{\infty}L_{nm}^i\leqslant L_{n+1,m}\\ T_\varepsilon^{n0}\leqslant T_\varepsilon^{n+1,0} \qquad T_{\varepsilon_1}^{n0}\leqslant T_{\varepsilon_2}^{n0}\,(\varepsilon_1<\varepsilon_2)\end{gathered} \tag{8}$$

Consequently we may set

$$\begin{array}{ll} L_{nm}\uparrow L_m(n\uparrow\infty) & L_m\downarrow L(m\uparrow\infty) \\ T_\varepsilon^{n0}\uparrow T_\varepsilon(n\uparrow\infty) & T_\varepsilon\downarrow T(\varepsilon\downarrow 0) \end{array} \tag{9}$$

Lemma 3. $P\{L=0\}=P\{T=0\}=1$.

Proof. Put

$$\bar{\beta}_i^{nm}=\inf\{t\,|\,t\geqslant\tau_{i0}^n,\ x^n(t)\leqslant m\}$$

We consider

$$\bar{L}_{nm}^i=\begin{cases}\min(\tau_{i+1,0}^n,\bar{\beta}_i^{nm})-\tau_{i0}^n & \text{if } \tau_{i0}^n<\beta_1^{n0} \text{ and } m<x^n(\tau_{i0}^n)\\ 0 & \text{otherwise}\end{cases}$$

and

$$\bar{T}_{\varepsilon_i}^{n0}=\begin{cases}\displaystyle\sum_{\tau_{i0}^n\leqslant\tau_{ij}^n<\min(\tau_{i+1}^n,\bar{\beta}_i^{n0})} f_\varepsilon(\tau_{i,j+1}^n-\tau_{ij}^n) & \text{if } \tau_{i0}^n<\beta_1^{n0}\\ 0 & \text{otherwise}\end{cases}$$

By the definition of L_{nm} and T_ε^{n0} we have

$$L_{nm}=\sum_{i=1}^\infty\bar{L}_{nm}^i\qquad T_\varepsilon^{n0}=\sum_{i=1}^\infty\bar{T}_{\varepsilon_i}^{n0}$$

But by Lemma 5.1, (6.11) and (6.12), we have

$$\begin{aligned}EL_{nm}&=\sum_{i=1}^\infty E\bar{L}_{nm}^i\\ &=\sum_{i=1}^\infty\sum_{j=m+1}^n P\{\tau_{i0}^n<\beta_1^{n0},x^n\{\tau_{i0}^n)=j\}E\{\min(\tau_{i+1,0}^n,\bar{\beta}_i^{nm})-\tau_{i0}^n|x^n(\tau_{i0}^n)=j\}\\ &=\sum_{i=1}^\infty\sum_{j=m+1}^n\left\{\sum_{k=1}^n v_k^n(1-C_{k0})\right\}^{i-1}v_j^nH_{jm}^\infty\\ &=\frac{\sum_{j=m+1}^n v_j^nH_{jm}^\infty}{v_{-1}^n+\sum_{k=0}^n v_k^nC_{k0}}\\ &=\frac{\sum_{j=m+1}^{n-1}r_jH_{jm}^\infty+(d/C_{n0}+\sum_{l=n}^\infty r_lC_{ln})H_{nm}^\infty}{r_{-1}+A_0+d}\end{aligned}$$

and

$$\begin{aligned}ET_\varepsilon^{n0}&=\sum_{i=1}^\infty E\bar{T}_{\varepsilon_i}^{n0}\\ &=\sum_{i=1}^\infty\sum_{k=1}^n P\{\tau_{i0}^n<\beta_1^{n0},x^n(\tau_{i0}^n)=k\}\\ &\quad\times E\left\{\sum_{\tau_{i0}^n\leqslant\tau_{ij}^n<\min(\tau_{i+1,0}^n,\bar{\beta}_i^{n0})} f_\varepsilon(\tau_{i,j+1}^n-\tau_{ij}^n)x_n(\tau_{i0}^n)=k\right\}\end{aligned}$$

$$= \sum_{i=1}^{\infty} \sum_{k=1}^{n} \left\{ \sum_{j=1}^{n} v_j^n (1 - C_{j0}) \right\}^{i-1} v_k^n H_{k0}^{\varepsilon}$$

$$= \frac{\sum_{k=1}^{n} v_k^n H_{k0}^{\varepsilon}}{v_{-1}^n + \sum_{j=0}^{n} v_j^n C_{j0}}$$

$$= \frac{\sum_{k=1}^{n-1} r_k H_{k0}^{\varepsilon} + (d/C_{n0} + \sum_{l=n}^{\infty} r_l C_{ln}) H_{n0}^{\varepsilon}}{r_{-1} + A_0 + d}$$

By (6.2.6) we may verify $C_{ln} N_n \leqslant N_l (l \geqslant n)$. By Lemma 5.1, (6.10) and (6.12), it follows that $H_{jm}^{\varepsilon} \leqslant N_j, \sum_{j=0}^{\infty} r_j N_j < \infty$ and $\lim_{\varepsilon \downarrow 0} H_{k0}^{\varepsilon} = 0$. If $s = \infty$, then $d = 0$; if $s < \infty$, then

$$\lim_{m \to \infty} \lim_{n \to \infty} \frac{H_{nm}^{\infty}}{C_{n0}} = \lim_{\varepsilon \downarrow 0} \lim_{n \to \infty} \frac{H_{n0}^{\varepsilon}}{C_{n0}} = 0$$

Therefore, by the above two expressions, we obtain

$$EL = \lim_{m \to \infty} \lim_{n \to \infty} EL_{nm}$$

$$= \frac{\lim_{m \to \infty} \sum_{j=m+1}^{\infty} r_j H_{jm}^{\infty} + \lim_{m \to \infty} \lim_{n \to \infty} (d/C_{n0} + \sum_{l=n}^{\infty} r_l C_{ln}) H_{nm}^{\infty}}{r_{-1} + A_0 + d} = 0$$

and

$$ET = \lim_{\varepsilon \downarrow 0} \lim_{n \to 0} ET_{\varepsilon}^{n0}$$

$$= \frac{\lim_{\varepsilon \downarrow 0} \sum_{k=1}^{\infty} r_k H_{k0}^{\varepsilon} + \lim_{\varepsilon \downarrow 0} \lim_{n \to \infty} (d/C_{n0} + \sum_{l=n}^{\infty} r_l C_{ln}) H_{n0}^{\varepsilon}}{r_{-1} + A_0 + d} = 0$$

On the basis of this we have proved this lemma, and the proof is completed.

QED

Theorem 4. For the sequence of (Q, V^n) Doob processes $X^n = \{x^n(t), t < \sigma^n\}$ in Lemma 2, its strong limit process $X = \{x(t), t < \sigma\}$ exists. $X \in \mathscr{H}_s(Q)$ is a non-minimal process. X is the unique Q process satisfying (6.11), that is, X is the unique Q process having the characteristic sequence p, q and $\gamma_n (n \geqslant -1)$.

Proof. (a) Obviously the limit $\sigma^n(\omega) \uparrow \sigma(\omega)$ exists. We now proceed to prove that for almost all $\omega \in \Omega$, and almost all $t \in [0, \sigma(\omega))$ in Lebesgue measure L, $x^n(t, \omega)$ is convergent to some state in E; for other $t \in [0, \sigma(\omega)), x^n(t, \omega)$ it is convergent to ∞.

Without loss of generality let us suppose that, for each $\omega \in \Omega, X^n(\omega)$ has only a finite number of i intervals $(i \in E, n \geqslant 0)$ in any finite interval $[0, t)(t \leqslant \sigma(\omega))$. If every constant interval (i.e. general i interval) of each X^n is translated towards the left and, furthermore, the distance covered by translation of every interval is not greater than ε, then the total length of the intervals composed of the points t which are in the interval $[0, \beta_1^{n0}(\omega))$ and make $x_n(t, \omega) \neq x_m(t, \omega)(n > m)$ is not more than $\varepsilon + T_\varepsilon^n(\omega) < \varepsilon + T_\varepsilon(\omega)$. Fixing k, taking $n > m > l(> k)$, on account of $\beta_1^{n0}(\omega) > \beta_1^{k0}(\omega)$ we obtain

$$L\{t \mid t \in [0, \beta_1^{k0}(\omega)), x_n(t, \omega) \neq x_m(t, \omega)\} \leqslant L_l(\omega) + T_{L_l(\omega)}(\omega) \tag{10}$$

Set $\Omega_0 = \{L_l + T_{L_l} \downarrow 0, l \uparrow \infty\}$. From Lemma 3 it follows that $P\{\Omega_0\} = 1$. Given $\omega \in \Omega_0$, from (10) we know that $x_n(t, \omega)$ converges in $[0, \beta_1^{k0}(\omega))$ in accordance with the Lebesgue measure L and hence there exists a subsequence $n_i \to \infty$ such that $x_{n_i}(t, \omega)$ converges for almost all t in $[0, \beta_1^{k0}(\omega))$. Fixing a convergence point t_0, since the Doob processes do not take the value '∞', consequently there exists $M \in E$ such that $x_{ni}(t_0, \omega) \to M(i \to \infty)$. As E is discrete, there is a positive number N such that

$$n_{n_i}(t_0, \omega) = M \qquad (i \geqslant N) \tag{11}$$

Now we start to prove that there exists a positive number N' such that, if $n > N'$, $x_n(t_0, \omega) = M$; hence $x_n(t_0, \omega)$ converges to M. Otherwise, there must exist $m_i \to \infty$ such that

$$x_{m_i}(t_0, \omega) \neq M \tag{12}$$

From this formula and (11) and by $g_m(X^n) = X^m (m < n)$ we know that in $[0, t_0]$, X^m has infinitely many M intervals. And this contradicts the hypothesis at the beginning of the proof.

Consequently, provided that $\omega \in \Omega_0$, then for almost all $t \in [0, \beta_1^{k0}(\omega)), x_n(t, \omega)$ is convergent to the states in E. Setting $k \to \infty$ we find that the same conclusion is valid for $[0, \beta_1^0(\omega))$. We can verify in the same way that the same conclusion is true for $[0, \beta_i^0(\omega))$. From (5) we know that $x^n(t, \omega)$ converges to the states in E for almost all $t \in [0, \sigma(\omega))$. As for the exceptional $t \in [0, \sigma(\omega))$, if $x^n(t, \omega)$ does not converge to ∞, then there surely exist two subsequences n_i and m_i and $M \in E$ so that (11) and (12) hold, which will likewise lead to contradictions.

(b) We are going to prove that $P\{x(t) = \infty\} = 0 (t \geqslant 0)$. On account of (a), $L\{t \mid x(t, \omega) = \infty\} = 0$. By the Fubini theorem there exists a set T, $L(T) = 0$, such that, if $t \notin T$, $P\{x(t) = \infty\} = 0$. Evidently $0 \notin T$.

Assume that $t_0 \in T$. Then $t_0 > 0$. We may take t_1 such that $t_1 \notin T, t_0 - t_1 \notin T$. Hence,

$$\begin{aligned} P\{x(t_0) \geqslant N\} &= \lim_{n \to \infty} P\{x^n(t_0) \geqslant N\} \\ &= \lim_{n \to \infty} E\{P x^n(t_1)[x^n(t_0 - t_1) \geqslant N]\} \end{aligned}$$

$$= E\left\{\lim_{n\to\infty} Px^n(t_1)[x^n(t_0 - t_1) \geqslant N]\right\}$$

$$= E\left\{\lim_{n\to\infty} Px(t_1)[x^n(t_0 - t_1) \geqslant N]\right\}$$

$$= E\{Px(t_1)[x(t_0 - t_1) \geqslant N]\}$$

Letting $N \to \infty$ we have $P\{x(t_0) = \infty\} = E\{Px(t_1)[x(t_0 - t_1) = \infty]\} = 0$.

(c) From (a) and (b) it is easily shown that $X \in \mathscr{H}_s$ and that it is a non-minimal Q process.

(d) We proceed to prove that X satisfies (6.11). In actual fact, let β_1^{kn} be defined for X^k in fashion (6.1) and (6.2). Because of (6.4), $\beta_1^{kn}\uparrow(n \leqslant k\uparrow)$. It is easily seen that there exist limits $\lim_{k\to\infty}\beta_1^{kn} = \beta_1^n, \lim_{k\to\infty} x^n(\beta_1^{kn}) = x(\beta_1^n)$. Furthermore on account of (6.4), $x^k(\beta_1^{kn}) = x^n(\beta_1^{nn})$. Thus, for $0 \leqslant j \leqslant n$,

$$P\{x(\beta_1^n) = j\} = P\{x^n(\beta_1^{nn}) = j\} = v_j^n$$

Hence the above formula holds true for $j = -1$, too.

(e) Suppose that $\bar{X} = \{\bar{x}(t), t < \bar{\sigma}\} \in \mathscr{H}_s(Q)$ also satisfies (6.11). By Corollary 2 to Theorem 10.5.2, $\bar{X}$ is the strong limit of the (Q, V^n) Doob processes $\bar{X}^n = g_n(\bar{X})$.

It follows that both $\bar{X}$ and X take the limit of the transition probability $p_{ij}^n(t)$ of the (Q, V^n) Doob processes as their transition probability, that is $\bar{X}$ and X have the same transition probability, and so they belong to the same process. The proof is terminated. QED

12.8 SUMMARY

Theorem 1. Assume that $X \in \mathscr{H}_s(Q)$ is a non-minimal Q process. Then its characteristic sequence $p, q, \gamma_n (n \geqslant -1)$ satisfies (6.10).

Conversely, given a sequence of non-negative numbers $p, q, \gamma_n (n \geqslant -1)$ satisfying (6.10), then there exists the unique non-minimal process $X \in \mathscr{H}_s(Q)$ whose characteristic sequence is just the given sequence $p, q, \gamma_n (n \geqslant -1)$. Moreover, we may take X as the strong limit of a sequence of (Q, V^n) Doob processes X^n, where $V^n = (v_j^n, 0 \leqslant j \leqslant n)$ is determined by (6.12) and (6.13) under $p, q, \gamma_n (n \geqslant -1)$.

For X to be honest it is necessary and sufficient that $\gamma_{-1} = 0$. X satisfies the system of forward equations if and only if $p = 0$.

It is still necessary to prove the last sentence.

If X satisfies the system of forward equations, by Theorem 10.4.4, the U intervals in $[\tau, \sigma)$ are all ${}_\infty U$ intervals and hence $P\{x(\eta) = j\} = 0$ and from (6.14) it follows that $\gamma_j = 0$ $(j \geqslant 0)$. Consequently $p = 0$. Conservely, we assume $p = 0$. From (6.14) follows $P\{\xi_{k0} < \sigma\} = 0 (k \in E)$, where ξ_{kj} is determined by (10.4.1).

In addition, for $j \in E$,

$$P\{\xi_{kj} < \sigma\} \prod_{i=1}^{j} \frac{a_i}{a_i + b_i} \leqslant P\{\xi_{k0} < \sigma\}$$

Therefore, on account of Lemma 10.4.2,

$$P\{\xi_{EE} < \sigma\} \leqslant \sum_{k,j \in E} P\{\xi_{kj} < \sigma\} = 0$$

By Theorem 10.4.4, X satisfies the system of forward equations.

CHAPTER 13

Relation Between Two Kinds of Construction Theories of Birth–Death Processes

13.1 INTRODUCTION

In Chapter 6 and Chapter 12, we have constructed conservative ($a_0 = 0$) birth–death processes by using analytical methods and probability methods respectively. We naturally ask: What is the relationship between the results for the two methods? We have completely solved the problem (Xiang-qun Yang, 1965b).

In this chapter, we suppose that $a_0 = 0$ and the boundary point z is regular or exit. Moreover, we write $X^2(\lambda)$ in (6.5.1) as $X(\lambda)$. Let $X_i(\lambda) = E_i(e^{-\lambda\tau})$, τ being the first leaping point for the Q process.

13.2 CORRESPONDING THEOREMS

Theorem 1. Let the characteristic sequence of the non-minimal process $X \in \mathscr{H}_s(Q)$ be p, q and r_n ($n \geqslant -1$). Then the resolvent operators of X are

$$\psi_{ij}(\lambda) = \phi_{ij}(\lambda) + X_i(\lambda)\frac{\sum_k r_k\phi_{kj}(\lambda) + dzX_j(\lambda)\mu_j}{r_{-1} + \sum_k r_k[1 - X_k(\lambda)] + dz\lambda\sum_k X_k(\lambda)\mu_k} \tag{1}$$

where d is determined by (12.6.13).

Proof. Let $\psi^n_{ij}(\lambda)$ be the resolvent operators of (Q, v^n) Doob process X^n. Then by (11.4.5), we have

$$\psi^n_{ij}(\lambda) = \phi_{ij}(\lambda) + X_i(\lambda)\frac{\sum_{k=0}^n v^n_k\phi_{kj}(\lambda)}{1 - \sum_{k=0}^n v^n_k + \sum_{k=0}^n v^n_k[1 - X_k(\lambda)]} \tag{2}$$

We denote the fraction in the above expression as $H^n_j(\lambda)$. Substituting (12.6.12) into $H_j^n(\lambda)$ and noticing $\sum_{k=-1}^{n} v_k^n = 1$, we have

$$H_j^n(\lambda) = \frac{(X_n/A_n)\sum_{k=0}^{n-1} r_k\phi_{kj}(\lambda) + [Y_n + (X_n/A_n)\sum_{l=n}^{\infty} r_l C_{ln}]\phi_{nj}(\lambda)}{(X_n/A_n)r_{-1} + (X_n/A_n)\sum_{k=0}^{n-1} r_k[1 - X_k(\lambda)] + [Y_n + (X_n/A_n)\sum_{l=n}^{\infty} r_l C_{ln}][1 - X_n(\lambda)]}$$

$$= \frac{\sum_{k=0}^{n-1} r_k\phi_{kj}(\lambda) + (d/C_{n0} + \sum_{l=n}^{\infty} r_l C_{ln})\phi_{nj}(\lambda)}{r_{-1} + \sum_{k=0}^{n-1} r_k[1 - X_k(\lambda)] + (d/C_{n0} + \sum_{l=n}^{\infty} r_l C_{ln})[1 - X_n(\lambda)]}$$

$$= \frac{\sum_{k=0}^{n-1} r_k\phi_{kj}(\lambda) + (d + \sum_{l=n}^{\infty} r_l C_{l0})\phi_{nj}(\lambda)/C_{n0}}{r_{-1} + \sum_{k=0}^{n-1} r_k[1 - X_k(\lambda)] + (d + \sum_{l=n}^{\infty} r_l C_{l0})[1 - X_n(\lambda)]/C_{n0}}$$

Note that $A_0 = \sum_{l=0}^{\infty} r_l C_{l0} < \infty$, and $d = 0$ if z is exit. By Lemmas 6.9.1 and 6.9.2, we obtain

$$\lim_{n\to\infty} H_j^n(\lambda) = \frac{\sum_k r_k\phi_{kj}(\lambda) + dzX_j(\lambda)\mu_j}{r_{-1} + \sum_k r_k[1 - X_k(\lambda)] + dz\lambda\sum_k X_k(\lambda)\mu_k}.$$

Since $X = \lim_{n\to\infty} X^n$, it follows that $\psi_{ij}(\lambda) = \lim_{n\to\infty} \psi_{ij}^n(\lambda)$, hence the theorem holds. And the proof is terminated. QED

By Theorems 6.6.1, 6.9.3 and 6.9.4, it follows that each non-minimal Q process $\psi(\lambda)$ has the following representation:

$$\psi_{ij}(\lambda) = \phi_{ij}(\lambda) + X_i(\lambda)\frac{\sum_k \alpha_k\phi_{kj}(\lambda) + DX_j(\lambda)\mu_j}{c + \sum_i \alpha_k[1 - X_k(\lambda)] + D\lambda\sum_k X_k(\lambda)\mu_k} \tag{3}$$

where the row vector $\alpha \geqslant 0$ satisfies (6.9.5), constant $D \geqslant 0$, and furthermore $D = 0$ if z is exist. Moreover, $\sum_k \alpha_k\phi_{kj}(\lambda) + DX_j(\lambda)\mu_j \not\equiv 0$, and the constant $c \geqslant 0$.

We point out that, except for a constant factor, the vector α, the constants c and D are uniquely determined by the process. In fact, suppose that α, c, D and $\bar{\alpha}, \bar{c}, \bar{D}$ correspond to the same process, then

$$\frac{\alpha\phi(\lambda) + DX(\lambda)\mu}{A_\lambda} = \frac{\bar{\alpha}\phi(\lambda) + \bar{D}X(\lambda)\mu}{\bar{A}_\lambda} \tag{4}$$

where $A_\lambda = c + [\alpha, 1 - X(\lambda)] + D\lambda[X(\lambda)\mu, 1]$, $\bar{A}_\lambda$ being the quantity corresponding to $\bar{\alpha}$, $\bar{c}$ and $\bar{D}$. Multiplying both sides of (4) by $\lambda I - Q$, we obtain

$$\alpha/A_\lambda = \bar{\alpha}/\bar{A}_\lambda$$

hence

$$D/A_\lambda = \bar{D}/\bar{A}_\lambda$$

Therefore $K = A_\lambda/\bar{A}_\lambda > 0$ and is independent of λ. Hence $\alpha = K\bar{\alpha}$ and $D = K\bar{D}$. Again substituting into (4) we get $c = K\bar{c}$.

Definition 1. The process in (3) is called a (Q, α, c, D) process.

Theorem 2. The characteristic sequence p, q and r_n $(n \geqslant -1)$ of a (Q, α, c, D) process is

$$r_{-1} = c \qquad r_n = \alpha_n (n \geqslant 0)$$
$$p = \begin{cases} 0 & \text{if } \alpha = 0 \\ A_0 z/(A_0 z + D), & \text{if } \alpha \neq 0 \end{cases}$$
$$q = \begin{cases} 1 & \text{if } \alpha = 0 \\ D/(A_0 z + D) & \text{if } \alpha \neq 0 \end{cases} \tag{5}$$

where $A_0 = \sum_{l=0}^{\infty} r_l C_{l0}$.

Proof. Comparing (1) and (3), we know that r_n and α_n differ by a constant factor. We may as well consider $r_n = \alpha_n$ $(n \geqslant 0)$, so that $r_{-1} = c$ and $dz = D$. From this (5) follows. The proof is over. QED

13.3 PROPERTIES OF THE PROCESS AT THE FIRST LEAPING POINT

On account of the corresponding theorem, it becomes clear how the (Q, α, c, D) process constructed by means of the analytical method moves and, moreover, some probability quantities of the process can be computed.

Theorem 1. Let $X \in \mathscr{H}_s(Q)$ be a (Q, α, c, D) process, and β_1^n be determined by (12.6.1). Then the probability $v_j^n = P\{x(\beta_1^n) = j\}$ $(-1 \leqslant j \leqslant n)$ is calculated as follows:

$$v_{-1}^n = (X_n/A_n)c \qquad v_j^n = (X_n/A_n)\alpha_j \, (0 \leqslant j < n)$$
$$v_n^n = Y_n + (X_n/A_n) \sum_{l=n}^{\infty} \alpha_l C_{ln} \tag{1}$$

where

$$A_n = \sum_{l=0}^{\infty} \alpha_l C_{ln} \qquad d = \begin{cases} D/z & \text{if } \alpha \neq 0 \\ 1 & \text{if } \alpha = 0 \end{cases}$$
$$X_n = \frac{A_n C_{n0}}{(c + A_n) C_{n0} + d} \qquad Y_n = \frac{d}{(c + A_n) C_{n0} + d} \tag{2}$$
$$\frac{X_n}{A_n} = \frac{C_{n0}}{c C_{n0} + d} \qquad \text{if } \alpha = 0$$

Proof. The conclusion of this theorem follows from Theorem 2.2 and (12.6.3). QED

Theorem 2. Let $X \in \mathscr{H}_\delta(Q)$ be a (Q, α, c, D) process, τ be the first leaping point. That $D > 0$ or $[\alpha, 1] = \infty$ is called case A; that $D = 0$ and $[\alpha, 1] < \infty$ is called case B. Then for $0 \leqslant i < \infty$,

$$P\{x(\tau) = i\} = \begin{cases} 0 & \text{case A} \\ \alpha_i/(c + [\alpha, 1]) & \text{case B} \end{cases}$$

$$P\{\tau = \sigma < \infty\} = P\{x(\tau) = -1\} = \begin{cases} 0 & \text{case A} \\ c/(c + [\alpha, 1]) & \text{case B} \end{cases}$$

$$P\{x(\tau) = \infty\} = \begin{cases} 1 & \text{case A} \\ 0 & \text{case B} \end{cases}$$

Proof. By (12.5.11) and the right-continuity of X, we obtain

$$\{x(\tau) = i\} = \lim_{n \to \infty} \{x(\beta_1^n) = i\} \qquad -1 \leqslant i < \infty$$

By Theorem 1, we have

$$P\{x(\tau) = i\} = \lim_{n \to \infty} \frac{X_n}{A_n} a_i \qquad i \geqslant 0$$

$$P\{x(\tau) = -1\} = P\{\tau = \sigma < \infty\} = \lim_{n \to \infty} \frac{X_n}{A_n} c$$

But

$$\frac{X_n}{A_n} = \frac{zC_{n0}}{z(c + A_n)C_{n0} + D}$$

If $D > 0$, obviously, $X_n/A_n \to 0$. If $D = 0$, then

$$\frac{X_n}{A_n} = \frac{1}{c + A_n} \to \frac{1}{c + [\alpha, 1]} \tag{3}$$

In fact, since $A_n = \sum_{l=0}^{n} \alpha_l + \sum_{l=n}^{\infty} \alpha_l C_{ln}$, it follows that, if $[\alpha, 1] = \infty$, obviously (3) holds. If $[\alpha, 1] < \infty$, then (3) follows from $\sum_{l=n}^{\infty} \alpha_l C_{ln} \leqslant \sum_{l=n}^{\infty} \alpha_l \to 0$. Again noticing $P\{x(\tau) = \infty\} = 1 - \sum_{i=-1}^{\infty} P\{x(\tau) = i\}$ we obtain the conclusion of this theorem. The proof is concluded. QED

PART VI PROPERTIES OF MARKOV PROCESSES RELATED TO CONSTRUCTION THEORY

CHAPTER 14

Properties of Birth–Death Processes

14.1 INTRODUCTION

Birth–death processes constructed with the probability method have clear path structures. Every birth–death process is the strong limit of a sequence of Doob processes. Therefore, in order to study the properties of a Q process we only need to study the limits. In this chapter we shall carry out this procedure by studying distributions of first return times. We assume $a_0 = 0$ and adopt the notations given in Chapter 6, such as the increasing solutions $u(\lambda)$ and the decreasing solutions $v(\lambda)$; we also write

$$X(\lambda) = X^2(\lambda) = \frac{u(\lambda)}{u(z, \lambda)}$$

14.2 SOME FINE RESULTS OF THE MINIMAL PROCESS

Assume that $X = \{x(t), t < \sigma\} \in \mathscr{H}_s(Q)$, and that τ is the first leaping point and τ_1 is the first jumping point. Let

$$\eta_i = \begin{cases} \inf\{t \mid \tau_1 \leqslant t \leqslant \sigma, x(t) = i\} \\ \sigma \qquad \text{if the above set is empty} \end{cases} \tag{1}$$

be the time of the process returning to i for the first time. Zi-kun Wang and Xiang-qun Yang (1988, section 5.2) have pointed out that there exists $h = h(j) > 0$ such that for $\lambda > -h$, $E_k\{e^{-\lambda\eta_j}\} (k < j)$ are finite and are the unique solutions of the equations

$$\begin{aligned} D_\mu u_k^+ &= \lambda \mu_k \qquad k < j \\ u_j &= 1 \end{aligned} \tag{2}$$

For $k < j$ the moments ${}_1N_{kj}^l = E_k\{\eta_j^l\} = E_k\{\eta_j^l, \eta_j < \tau\}$ $(k < j)$ are finite and satisfy the relation

$$\begin{aligned} {}_1N_{kj}^l &= l \sum_{i=k}^{j-1} (z_{j+1} - z_j) \sum_{s=0}^{i} {}_1N_{sj}^{l-1} \mu_s \qquad (k < j) \\ {}_1N_{kj}^0 &= 1 \end{aligned} \tag{3}$$

Moreover $N_k^l = E_k\{\tau^l\}$ satisfy

$$N_k^l = l\sum_{i=k}^{\infty}(z_{i+1}-z_i)\sum_{s=0}^{i}N_s^{l-1}\mu_s \leqslant l!R^l \qquad k\geqslant 0$$
$$N_k^0 = 1 \tag{4}$$

Hence, one has

$$E_k\{e^{-\lambda\eta_j}\} = \frac{u_k(\lambda)}{u_j(\lambda)} \qquad (k<j)$$
$$E_k\{e^{-\lambda\tau}\} = X_k(\lambda) \qquad \lambda>0 \tag{5}$$

and

$$E_k\{e^{-\lambda\eta_j}\} = \sum_{i=0}^{\infty}(-\lambda)^l\frac{{}_1N_{kj}^l}{l!} \qquad (|\lambda|<h, k<j) \tag{6}$$

When $R<\infty$,

$$E_k\{e^{-1\tau}\} = \sum_{l=0}^{\infty}(-\lambda)^l\frac{N_k^l}{l!} \qquad |\lambda|<\frac{1}{R} \tag{7}$$

Now, for $i\leqslant k<j$ and $i+1<j$, let

$${}_1\varphi_{kij}(\lambda) = E_k\{e^{-\lambda\eta_i}, \eta_i<\eta_j\}$$
$${}_2\varphi_{kij}(\lambda) = E_k\{e^{-\lambda\eta_j}, \eta_j<\eta_i\} \tag{8}$$

$${}_1N_{kij}^l = E_k\{\eta_i^l, \eta_i<\eta_j\} \qquad {}_2N_{kij}^l = E_i\{\eta_j^l, \eta_j<\eta_i\} \tag{9}$$

$${}_1\varphi_{kj}(\lambda) = E_k\{e^{-\lambda\eta_i}, \eta_i<\tau\} \qquad {}_2\varphi_{ki}(\lambda) = E_k\{e^{-\lambda\tau}, \tau\leqslant\eta_i\} \tag{10}$$

$${}_1N_{ki}^l = E_k\{\eta_i^l, \eta_i<\tau\} \qquad {}_2N_{ki}^l = E_k\{\tau^l, \tau\leqslant\eta_i\} \tag{11}$$

Clearly, $P_k\{\eta_j\uparrow\tau\}=1$ as $j\uparrow\infty$ and, for $i\leqslant k$ as $j\uparrow\infty$

$${}_1\varphi_{kij}(\lambda)\uparrow{}_1\varphi_{ki}(\lambda) \qquad {}_1\varphi_{kij}(\lambda)+{}_2\varphi_{kij}(\lambda)\uparrow{}_1\varphi_{ki}(\lambda)+{}_2\varphi_{ki}(\lambda) \tag{12}$$

$${}_1N_{kij}^l\uparrow{}_1N_{ki}^l \qquad {}_1N_{kij}^l+{}_2N_{kij}^l\uparrow{}_1N_{ki}^l+{}_2N_{ki}^l \tag{13}$$

Theorem 1. (i) There exists $h=h(j)>0$ such that for $\lambda>-h$ ${}_a\varphi_{kij}(\lambda)$ $(a=1,2, i\leqslant k<j)$ are finite, all moments ${}_aN_{kij}^l$ $(a=1, 2, l\geqslant 0, i\leqslant k<j)$ are finite and they satisfy the relations:

$${}_aN_{kij}^l = l\left(\frac{z_j-z_k}{z_j-z_i}\sum_{s=i+1}^{k}(z_s-z_i){}_aN_{sij}^{l-1}\mu_s + \frac{z_k-z_i}{z_j-z_i}\sum_{s=k+1}^{j-1}(z_j-z_s){}_aN_{sij}^{l-1}\mu_s\right)$$
$$i<k<j$$

$${}_1N_{kij}^0 = \frac{z_j-z_k}{z_j-z_i} \qquad {}_2N_{kij}^0 = \frac{z_k-z_i}{z_j-z_i} \qquad i<k<j \tag{14}$$

$$
\begin{aligned}
{}_1N^l_{iij} &= \frac{1}{q_i}(l_1 N^{l-1}_{iij} + a_{i1} N^l_{i-1,i} + b_{i1} N^l_{i+1,ij}) \\
{}_1N^0_{iij} &= \frac{a_i}{q_i} + \frac{b_i}{q_i}\frac{z_j - z_{i+1}}{z_j - z_i}
\end{aligned}
\tag{15}
$$

$$
\begin{aligned}
{}_2N^l_{iij} &= \frac{1}{q_i}(l_2 N^{l-1}_{iij} + b_{i2} N^l_{i+1,ij}) \\
{}_2N^0_{iij} &= \frac{b_i}{q_i}\frac{z_{i+1} - z_i}{z_j - z_i}
\end{aligned}
\tag{16}
$$

$$
{}_a\varphi_{kij}(\lambda) = \sum_{l=0}^{\infty} (-\lambda)^l \frac{{}_aN^l_{kij}}{l!} \qquad (i \leqslant k < j, |\lambda| < h) \tag{17}
$$

$$
\begin{aligned}
{}_1\varphi_{kij}(\lambda) &= \frac{u_j(\lambda)v_k(\lambda) - u_k(\lambda)v_j(\lambda)}{u_j(\lambda)v_k(\lambda) - u_i(\lambda)v_j(\lambda)} \\
& \qquad\qquad (\lambda > 0, i < k < j) \\
{}_2\varphi_{kij}(\lambda) &= \frac{u_k(\lambda)v_i(\lambda) - u_i(\lambda)v_k(\lambda)}{u_j(\lambda)v_i(\lambda) - u_i(\lambda)v_j(\lambda)}
\end{aligned}
\tag{18}
$$

$$
\begin{aligned}
{}_1\varphi_{iij}(\lambda) &= \frac{1}{\lambda + q_i}\left(a_i \frac{u_{i-1}(\lambda)}{u_i(\lambda)} + b_{i1}\varphi_{i+1,ij}(\lambda) \right) \\
& \qquad\qquad (\lambda > 0) \\
{}_2\varphi_{iij}(\lambda) &= \frac{1}{\lambda + q_i} b_{i2}\varphi_{i+1,ij}(\lambda)
\end{aligned}
\tag{19}
$$

(ii) The following also hold:

$$
{}_1\varphi_{ki}(\lambda) = \frac{v_k(\lambda)}{v_i(\lambda)} \qquad {}_2\varphi_{ki}(\lambda) = X_k(\lambda) - X_i(\lambda)\frac{v_k(\lambda)}{v_i(\lambda)} \qquad (i < k, \lambda > 0) \tag{20}
$$

$$
\begin{aligned}
{}_1\varphi_{ii}(\lambda) &= \frac{1}{\lambda + q_i}\left(a_i \frac{u_{i-1}(\lambda)}{u_i(\lambda)} + b_i \frac{v_{i+1}(\lambda)}{v_i(\lambda)} \right) \qquad \lambda > 0 \\
{}_2\varphi_{ii}(\lambda) &= \frac{b_i}{\lambda + q_i}\left(X_{i+1}(\lambda) - X_i(\lambda)\frac{v_{i+1}(\lambda)}{v_i(\lambda)} \right) \qquad \lambda > 0
\end{aligned}
\tag{21}
$$

$$
\begin{aligned}
{}_aN^l_{ki} &= l\left(\frac{z - z_k}{z - z_i} \sum_{s=i+1}^{k} (z_s - z_i)_a N^{l-1}_{si}\mu_s + \frac{z_k - z_i}{z - z_i} \sum_{s=k+1}^{\infty} (z - z_s)_a N^{l-1}_{si}\mu_s \right) \\
&\leqslant l! R^{l-1} N_k \leqslant l R^l \qquad (i < k,\ a = 1 \text{ or } R < \infty \text{ as } a = 2)
\end{aligned}
\tag{22}
$$

$$
{}_1N^0_{ki} = \frac{z - z_k}{z - z_i} \qquad {}_2N^0_{ki} = \frac{z_k - z_i}{z - z_i} \qquad (i < k)
$$

$$
\begin{aligned}
{}_1N^l_{ii} &= \frac{1}{q_i}(l_1 N^{l-1}_{ii} + a_{i1} N^l_{i-1,i} + b_{i1} N^l_{i+1,i}) \leqslant l! R^l \\
{}_1N^0_{ii} &= \frac{a_i}{q_i} + \frac{b_i}{q_i}\frac{z - z_{i+1}}{z - z_i}
\end{aligned}
\tag{23}
$$

$$
\begin{aligned}
{}_2N^l_{ii} &= \frac{1}{a_i}(l_2 N^{l-1}_{ii} + b_{i2} N^l_{i+1,i}) \leqslant l! R^l \\
{}_2N^0_{ii} &= \frac{b_i}{q_i}\frac{z_{i+1} - z_i}{z - z_i}
\end{aligned}
\tag{24}
$$

When $R < \infty$,

$$
{}_a\varphi_{kl}(\lambda) = \sum_{l=0}^{\infty} (-\lambda)^l \frac{{}_aN^l_{ki}}{l!} \qquad (i \geqslant k, |\lambda| < 1/R) \tag{25}
$$

Proof. (i) Since for $\lambda > -h$, $E_k\{e^{-\lambda \eta_j}\}$ $(k > j)$ and $E_k\{\eta^l_j\}$ $(k < j)$ are finite, it follows that ${}_a\varphi_{kij}(\lambda)$ and ${}_aN^l_{kij}$ are finite, and by Wilks (1962, p. 14),

$$
{}_aN^l_{kij} = (-1)^l \frac{d^l}{d\lambda^l} {}_a\varphi_{kij}(\lambda)\bigg|_{\lambda=0} \tag{26}
$$

and (17) holds.

The formula (19) follows from the strong Markov property. Multiplying both sides of (19) by $(\lambda + q_i)$, differentiating them l times and noting (26) and (5)–(6), we obtain the first expressions of (15) and (16). Obviously,

$$
{}_1N^0_{iij} = P_i\{\eta_i < \eta_j\} = \frac{a_i}{q_i} P_{i-1}\{\eta_i < \eta_j\} + \frac{b_i}{q_i} P_{i+1}\{\eta_i < \eta_j\}
$$

$$
{}_2N^0_{iij} = \frac{b_i}{q_i} P_{i+1}\{\eta_j < \eta_i\}
$$

Therefore the second expressions of (15)–(16) follow from (12.2.8).

Making use of the strong Markov property we find that ${}_a\varphi_{kij}(\lambda)$ $(i < k < j,$ $\lambda > 0)$ satisfy the following equations:

$$
\begin{aligned}
u_i &= [1 - (-1)^a]/2 \\
(\lambda + a_k + b_k)u_k - a_k u_{k-1} - b_k u_{k+1} &= 0 \qquad (i < k < j) \\
u_j &= [1 + (-1)^a]/2
\end{aligned}
$$

Solve the above equations to obtain (18). Differentiating the above equations and noting (26), we see that $u_k = {}_aN^l_{kij}$ $(i < k < j, l \geqslant 1)$ satisfy equation (5.4.11) with $f_i = 0$, $f_k = l_a N^{l-1}_{kij}$ $(i < k < j)$ and $f_j = 0$. According to Theorem 6.2.3 and (5.4.12) the first expression of (14) is justified. The second one follows from (12.2.8).

(ii) On account of (12), let $j\uparrow\infty$ in (18)–(19) to obtain (20)–(21).

When $R<\infty$, by (7) for $k\geqslant i$ we have

$$ {}_1N^l_{ki}+{}_2N^l_{ki}=E_k\{\min(\eta_i,\tau)\}^l\leqslant E_k\tau^l\leqslant l!R^l<\infty $$

Hence (25) holds.

The second expression in (22) follows from (12.2.16). From (13) we can obtain (23) and (24) merely by taking limits in the corresponding expressions in (i). Let $j\uparrow\infty$ in the first expression of (14); then according to the monotone convergence theorem we find that (22) holds for $a=1$ and that if we substitute ${}_1N^l_{ki}+{}_2N^l_{ki}$, ${}_1N^{l-1}_{si}+{}_2N^{l-1}_{si}$ for ${}_1N^l_{ki}$, ${}_1N^{l-1}_{si}$ respectively in (22), the corresponding formulae still hold and ${}_1N^l_{ki}+{}_2N^l_{ki}\leqslant l!R^l<\infty$. Therefore for $a=2$, (22) also holds. The proof is completed. QED

Theorem 2. The minimal solution is recurrent if and only if $z=\infty$; more precisely,

$$ \int_0^\infty f_{ij}(t)\,\mathrm{d}t=\lim_{\lambda\downarrow 0}\phi_{ij}(\lambda)=\Gamma_{ij}=\begin{cases}(z-z_j)\mu_j & \text{if } j\geqslant i\\ (z-z_i)\mu_j & \text{if } j<i\end{cases} \tag{27} $$

Proof. It follows from (6.3.2) that

$$ u_i(\lambda)\downarrow 1 \qquad (\lambda\downarrow 0) \tag{28} $$

By (6.3.4) $v_i(\lambda)\to\sum_{j=i}^{\infty}(z_{j+1}-z_j)=z-z_i$ as $\lambda\downarrow 0$. So (27) follows and the proof is completed. QED

Theorem 3. Suppose $z=\infty$. Then the minimal solution is ergodic if and only if $\sum_{i=0}^{\infty}\mu_i<\infty$.

Proof. According to (12.2.17) we have ${}_1N^1_{i-1,i}=m_{i-1}$. When $z=\infty$, by (22)–(23) it follows that ${}_1N^0_{ii}=1$ and

$$ E_i\eta_i={}_1N^i_{1i}=\frac{1}{q_i}(1+a_im_{i-1}+b_{i1}N^1_{i+1,i}) $$

$$ {}_1N^1_{i+1,i}=(z_{i+1}-z_i)\sum_{s=i+1}^{\infty}u_s $$

Hence $E_i\{\eta_i\}<\infty$ is equivalent to $\sum_{s=0}^{\infty}u_s<\infty$. QED

Theorem 4. Suppose $R<\infty$ and $S<\infty$. Then

$$ \lim_{n\to\infty}\frac{{}_1\varphi_{nj}(\lambda)}{z-z_n}=\frac{X_0(\lambda)}{v_j(\lambda)} \qquad \lambda>0 \tag{29} $$

$$ \lim_{n\to\infty}\frac{1-{}_2\varphi_{nj}(\lambda)}{z-z_n}=\lambda\sum_k X_k(\lambda)\mu_k+\frac{X_0(\lambda)X_j(\lambda)}{v_j(\lambda)} \qquad \lambda>0 \tag{30} $$

Proof. According to Lemmas 6.9.1 and 6.9.2, (29)–(30) are deduced from (20) immediately. QED

14.3 INVARIANT MEASURES OF THE MINIMAL PROCESSES

Suppose $R < \infty$. If a process is recurrent, it must be honest. Conversely, according to Theorem 12.4.2 an honest process is recurrent and ergodic. We will find the invariant measure.

Theorem 1. Suppose X is a $(Q, a, 0, D)$ process. Then its invariant measure is

$$\pi_j = \lim_{t\to\infty} p_{ij}(t) = \frac{\sum_k a_k \Gamma_{kj} + D\mu_j}{\sum_k a_k N_k + D\sum_k \mu_k} \tag{1}$$

and

$$m_{ii} = E_i \eta_i = \frac{\sum_k a_k \Gamma_{ki} + D\mu_i}{q_i\left(\sum_k a_k N_k + D\sum_k \mu_k\right)} \tag{2}$$

Here Γ_{kj} are determined by (2.27) and N_k by (6.2.6).

Proof. By (7.7.10) we know that the limits in (1) exist. By Tauber's theorem (Hardy, 1949, Theorem 98), $\lim_{\lambda\downarrow 0} \lambda\psi_{ij}(\lambda) = \lim_{t\to\infty} p_{ij}(t)$. Because $P_i\{\tau < \infty\} = 1$, $X_i(\lambda) = E_i\{e^{-\lambda\tau}\} \uparrow 1$ as $\lambda \downarrow 0$ and

$$\frac{1 - X_i(\lambda)}{\lambda} \sum_j \phi_{ij}(\lambda) \uparrow \sum_j \Gamma_{ij} = N_i \qquad \text{as } \lambda\downarrow 0$$

Note that $D = 0$ if z is exist, and $\sum_k u_k < \infty$ if z is regular. Therefore (1) follows from (13.2.3), and (2) follows from (1) and (7.7.10). We conclude the proof. QED

Theorem 2. Suppose that $X \in \mathscr{X}_s(Q)$ is a $(Q, a, 0, D)$ process and π is its invariant measure. Suppose that functions f and g satisfy $\sum_i |f(i)|\pi_i < \infty$, $\sum_i |g(i)|\pi_i < \infty$ and $\sum_i g(i)\pi_i \neq 0$. Then

$$P\left\{\lim_{t\to\infty} \frac{\int_0^t f[x(u)]\,du}{\int_0^t g[x(u)]\,du} = \frac{\sum_i f(i)\pi_i}{\sum_i g(i)\pi_i}\right\} = 1 \tag{3}$$

Proof. See Zhang-nan Li nd Rong Wu (1964), Theorem 3.1). QED

14.4 DISTRIBUTION OF THE FIRST RETURNING TIME

Theorem 1. Suppose that $X \in \mathscr{H}_s(Q)$ is a $(Q, \alpha, 0, D)$ process and η_j is defined by (2.1). Then for $\lambda > 0$,

$$E_i\{e^{-\lambda\eta_j}\} = \frac{u_i(\lambda)}{u_j(\lambda)} \qquad (i < j) \tag{1}$$

$$E_i\{e^{-\lambda\eta_i}\} = 1 - m(\lambda)(\lambda + q_i)^{-1}\left[m(\lambda)\phi_{ii}(\lambda) + X_i(\lambda)\left(\sum_k a_k\phi_{ki}(\lambda) + DX_i(\lambda)u_i\right)\right]^{-1} \tag{2}$$

$$E_i\{e^{-\lambda\eta_j}\} = \left[m(\lambda)\phi_{ij}(\lambda) + X_i(\lambda)\left(\sum_k a_k\phi_{kj}(\lambda) + DX_j(\lambda)\mu_j\right)\right]$$
$$\cdot\left[m(\lambda)\phi_{jj}(\lambda) + X_j(\lambda)\left(\sum_k a_k\phi_{kj}(\lambda) + DX_j(\lambda)u_j\right)\right]^{-1} \qquad (i > j) \tag{3}$$

where

$$m(\lambda) = \sum_k a_k[1 - X_k(\lambda)] + D\lambda\sum_k X_k(\lambda)\mu_k \tag{4}$$

or equivalently,

$$E_i\{e^{-\lambda\eta_j}\} = {}_1\varphi_{ij}(\lambda) + {}_2\varphi_{ij}(\lambda)\left(\sum_{k=0}^{j} a_k\frac{u_k(\lambda)}{u_j(\lambda)} + \sum_{k=j+1}^{\infty} a_k\frac{v_k(\lambda)}{v_i(\lambda)} + D\frac{X_0(\lambda)}{v_j(\lambda)}\right)$$
$$\cdot\left[\sum_{k=0}^{j} a_k + \sum_{k=j+1}^{\infty} a_k\left(1 - X_k(\lambda) + X_j(\lambda)\frac{X_0(\lambda)}{v_j(\lambda)}\right)\right]^{-1} \qquad (i \geqslant j) \tag{5}$$

where ${}_a\varphi_{ij}(\lambda)$ are determined by (2.20) and (2.21).

Proof. The formula (1) is just (2.5). By the well known formulae in Chung (1969a, pp. 192–3),

$$p_{ii}(t) = e^{-q_i t} + \int_0^t p_{ii}(t-u)\,dP_i \qquad (\eta_i \leqslant u)$$

$$p_{ij}(t) = \int_0^t p_{jj}(t-u)\,dP_i \qquad (\eta_i \leqslant u, i \neq j)$$

Taking Laplace transformations, noting (13.2.3) and making some simple rearrangements, we can obtain (2)–(3).

Now consider (Q, V^n) Doob processes. Let σ_h be the hth leaping point of X. By recurrence $P_i\{\eta_j < \infty\} = 1$ for $i \geqslant j$. It follows from the structure of Doob processes that

$$E_i\{e^{-\lambda\eta_j}\} = E_i\{e^{-\lambda\eta_j}, \eta_j < \sigma_1\} + \sum_{h=1}^{\infty} E_i\{e^{-\lambda\eta_j}, \sigma_h \leqslant \eta_j < \sigma_{h+1}\}$$

$$= {}_1\varphi_{ij}(\lambda) + \sum_{h=1}^{\infty} E_i\{e^{-\lambda\sigma_1}, \sigma_1 \leqslant \eta_j\}\left(\sum_{k=i+1}^{\infty} v_k E_k(e^{-\lambda\sigma_1}, \sigma_1 \leqslant \eta_j)\right)^{h-1}$$

$$\cdot\left(\sum_{h=0}^{j-1} v_k E_k(e^{-\lambda\eta_j}) + v_i + \sum_{k=j+1}^{\infty} v_k E_k(e^{-\lambda\eta_j}, \eta_j \leqslant \sigma_1)\right)$$

$$= {}_1\varphi_{ij}(\lambda) + {}_2\varphi_{ij}(\lambda)\left(\sum_{k=0}^{j} v_k \frac{u_k(\lambda)}{u_j(\lambda)} + \sum_{k=j+1}^{\infty} v_k \frac{v_k(\lambda)}{v_j(\lambda)}\right)\left(1 - \sum_{k=j+1}^{\infty} v_{k\,2}\varphi_{kj}(\lambda)\right)^{-1}$$

$$= {}_1\varphi_{ij}(\lambda) + {}_2\varphi_{ij}(\lambda)\left(\sum_{k=0}^{j} v_k \frac{u_k(\lambda)}{u_j(\lambda)} + \sum_{k=j+1}^{\infty} v_k \frac{v_k(\lambda)}{v_j(\lambda)}\right)$$

$$\cdot\left(\sum_{k=0}^{j} v_k + \sum_{k=j+1}^{\infty} v_k(1 - {}_2\varphi_{kj}(\lambda))\right)^{-1} \tag{6}$$

For a general $(Q, a, 0, D)$ process X, it is the strong limit of a sequence of honest (Q, V^n) Doob processes X^n. Define η_i and η_i^n for X and X^n, respectively. Then $\eta_i^n \uparrow \eta_i$ as $n \uparrow \infty$.

Substituting V^n in (13.3.1) into (6), letting $n \to \infty$ and noting (6.9.2)–(6.9.3), (2.30) and (2.20) we can obtain (5) by a simple calculation. The proof is terminated.

QED

CHAPTER 15

Recurrence and Ergodic Properties

15.1 INTRODUCTION

The recurrence and ergodic properties of processes are extremely useful for the study of approximation properties of their transition functions at infinity. They are also important for the study of excessive functions, zero–one law (see Zi-kun Wang, 1964, 1965c, 1966, 1980) and reversibility (see Min Qian and Zhen-ting Hou, 1979). General studies for the classification of Markov processes have been summed up in Chung (1967, II, section 10). Li-de Wu (1965) and Miller (1963) studied the classification of states of minimal Q processes and found the necessary and sufficient conditions related to the Q matrix, under which Q processes are recurrent or ergodic.

This chapter studies the classification of states of Q processes, which is closely related to construction theory. This kind of classification depends not only on Q but also on the construction of processes. Hence, different constructions need different treatments. The content of this chapter is derived from Xiang-qun Yang (1980d).

15.2 TWO LEMMAS

We call i, $\psi(\lambda)$-recurrent or ergodic if the state i is recurrent or ergodic with respect to $\psi(\lambda)$. We say a process is recurrent or ergodic if all its states are recurrent or ergodic. Obviously, i is $\psi(\lambda)$-recurrent if and only if $\lim_{\lambda \downarrow 0} \psi_{ii}(\lambda) = \infty$. By (7.7.9) and Tauber's theorem, i is $\psi(\lambda)$-ergodic if and only if $\lim_{\lambda \downarrow 0} \lambda \psi_{ii}(\lambda) = \pi_i > 0$. And in terms of m_{ii} defined in (7.7.8), we have

$$\pi_i = 1/q_i m_{ii} \tag{1}$$

In the following, the notation $i \overset{\psi(\lambda)}{\Rightarrow} j$ means that i may reach j with respect to $\psi(\lambda)$ (i.e. there exists $\lambda > 0$ such that $\psi_{ij}(\lambda) > 0$). We always have $i \overset{\psi(\lambda)}{\Rightarrow} i$.

Suppose $X \in \mathscr{H}_s(Q)$, τ_1 and τ are the first point and leaping point, respectively.

Lemma 1. As $\lambda \downarrow 0$,

$$\phi_{ij}(\lambda) \uparrow \Gamma_{ij} = \sum_{n=0}^{\infty} \pi_{ij}^{n} \frac{1}{q_j} \leqslant \Gamma_{jj} \tag{2}$$

$$\xi_i(\lambda) \equiv E_i\{e^{-\lambda\tau}\} \uparrow \xi_i = P_i\{\tau < \infty\} \tag{3}$$

$$\frac{1 - \xi_i(\lambda)}{\lambda} \uparrow N_i = \sum_j \Gamma_{ij} = E_i \tau \tag{4}$$

where (π_{ij}) is defined as in (2.9.7) and

$$\xi_i(\lambda) = 1 - \lambda \sum_j \phi_{ij}(\lambda) \tag{5}$$

Proof. The formula (2) is just (2.10.28). The inequality $\Gamma_{ij} \leqslant \Gamma_{jj}$ follows from Theorem 7.2.1. Imitate the proof of Theorem 7.12.3, and we obtain

$$\lambda \sum_j \phi_{ij}(\lambda) = 1 - E_i\{e^{-\lambda\tau}\} \tag{6}$$

Comparing this with (5), we obtain $\xi_i(\lambda) = E_i\{e^{-\lambda\tau}\}$. Moreover (3) is proved. Finally,

$$\frac{1 - \xi_j(\lambda)}{\lambda} = \sum_i \phi_{ij}(\lambda) = E_i\left\{\int_0^{\tau} e^{-\lambda t}\, dt\right\} \tag{7}$$

From this (4) follows. QED

Lemma 2. Suppose X is a $\psi(\lambda)$-process and i is $\psi(\lambda)$-recurrent. Then $P_i\{\sigma = \infty\} = 1$. If i is $\psi(\lambda)$-recurrent and $\phi(\lambda)$-non-recurrent, then $P_i\{\tau < \infty\} = 1$.

Proof. The first claim is clear because if i is $\psi(\lambda)$-recurrent, then on $\{x(0) = i\}$ x returns to i for infinitely many times. But according to Theorem 7.7.3, the nth sojourn times at i, ρ_i^n $(n \geqslant 0)$, are independent of each other. So $\sigma \geqslant \sum_{n=0}^{\infty} \rho_i^n = \infty$.

By Theorem 7.7.4, if i is $\psi(\lambda)$-recurrent then

$$P_i\{X(\omega) \text{ has infinitely many } i\text{-intervals in } [0, \infty)\} = 1$$

If i is $\phi(\lambda)$-non-recurrent, then

$$P_i\{X(\omega) \text{ has only finitely many } i\text{-intervals in } [0, \tau(\omega))\} = 1$$

Therefore $P_i\{\tau < \infty\} = 1$. We conclude the proof. QED

15.3 DOOB PROCESSES

Theorem 1. Suppose $\psi(\lambda)$ is a (Q, π) Doob process and i is $\phi(\lambda)$- non-recurrent. Then i is $\psi(\lambda)$-recurrent if and only if $\xi_i = 1$, $\sum_k \pi_k \xi_k = 1$ and there exists k such that $\pi_k > 0$ and

$$k \overset{\phi(\lambda)}{\Rightarrow} i \tag{1}$$

If i is $\psi(\lambda)$-recurrent, then i is ergodic if and only if

$$\sum_k \pi_k N_k < \infty \tag{2}$$

Remark

Because $\sum_k \pi_k \leqslant 1$ and $\xi_k \leqslant 1$, if (1) holds, necessarily $\sum_k \pi_k = 1$, i.e. $\psi(\lambda)$ is honest.

Proof. From the hypothesis, $\Gamma_{ii} < \infty$. By Lemma 2.1 let $\lambda \downarrow 0$ in (11.4.5) so that we find that

$$\lim_{\lambda \downarrow 0} \psi_{ij}(\lambda) = \Gamma_{ii} + \xi_i \frac{\sum_k \pi_k \Gamma_{kl}}{1 - \sum_k \pi_k \xi_k} \tag{3}$$

Suppose that i is $\psi(\lambda)$-recurrent. Then the left-hand side of (3) is infinite. According to Lemma 2.2, $\xi_i = 1$. Moreover, surely

$$\sum_k \pi_k \Gamma_{k_i} \leqslant (\textstyle\sum_k \pi_k)\Gamma_{ii} < \infty$$

and is positive, for otherwise (3) becomes $\lim_{\lambda \downarrow 0} \psi_{ii}(\lambda) = \Gamma_{ii} < \infty$, a contradiction. Since $\sum_k \pi_k \Gamma_{ki} > 0$, it follows that there exist $\pi_k > 0$ and $\Gamma_{ki} > 0$. But the latter is equivalent to $k \overset{\phi(\lambda)}{\Rightarrow} i$. Hence (1) holds. Conversely, by (1) it follows that the right-hand sde of (3) is infinite.

Suppose (1) hold. It is derived from Lemma 2.1, the remark of this theorem and (11.4.5) that

$$\lim_{\lambda \downarrow 0} \lambda \psi_{ii}(\lambda) = \lim_{\lambda \downarrow 0} \lambda \phi_{ii}(\lambda) + \lim_{\lambda \downarrow 0} \xi_i(\lambda) \frac{\sum_k \pi_k \phi_{ki}(\lambda)}{\sum_k \pi_k [1 - \xi_k(\lambda)]/\lambda}$$

It has been pointed out that the denominator of the right-hand side of the above formula is finite and positive. Hence the $\psi(\lambda)$-ergodic property of i is equivalent to (2). The proof is finished. QED

Corollary 1

Suppose for any $i, j, i \overset{\phi(\lambda)}{\Rightarrow} j$ and $\phi(\lambda)$ is non-recurrent. Then a (Q, π)Doob process is recurrent if and only if $\sum_k a_k \xi_k = 1$. If the process is recurrent,then it is ergodic if and only if (2) holds.

Proof. The necessity for recurrence is clear. We now prove the sufficiency. As $\sum_k \pi_k \xi_k = 1$, there surely exists some i such that $\pi_i > 0$ and $\xi_i = 1$. Hence for this i, (1) is satisfied, and so i is $\psi(\lambda)$-recurrent. On the other hand, obviously $i \overset{\phi(\lambda)}{\Leftrightarrow} j$ for any i, j for any i, j; therefore $\psi(\lambda)$ is recurrent.

Corollary 2

Suppose for any i, j, $i \overset{\phi(\lambda)}{\Leftrightarrow} j$ and Q is single exit. Then a (Q, π) Doobprocess is recurrent if and only if it is honest, i.e. Q is conservative and $\sum_k \pi_k = 1$.

Proof. For single exit Q, $P_i\{\tau < \infty \mid x(\tau - 0) \in B_e\} = 1$. Therefore by Theorem 7.12.8, $\xi_i = P_i\{\tau < \infty\} = 1$ and the condition $\sum_k \pi_k \xi_k = 1$ becomes $\sum_k \pi_k = 1$.

QED

15.4 SINGLE EXIT PROCESSES

Lemma 1. For $\eta(\lambda)$ and ξ given in Lemma 2.11.4

$$\lambda[\eta(\lambda), \xi] \downarrow 0 \qquad \text{as } \lambda \downarrow 0 \tag{1}$$

Proof. It follows from (2.11.40) and the property $\xi(\lambda) \uparrow \xi$ as $\lambda \downarrow 0$. QED

Theorem 2. Suppose $\psi(\lambda)$ is a process in Theorem 3.2.1 and i is $\phi(\lambda)$-non-recurrent. Then i is $\psi(\lambda)$-recurrent if and only if

$$\bar{X}_i > 0 \qquad \sum_k \alpha_k \Gamma_{kj} + \bar{\eta}_j > 0 \qquad c = [a, X^0] + \bar{\sigma}^0 = 0 \tag{2}$$

If i is $\psi(\lambda)$-recurrent, then i is $\psi(\lambda)$-ergodic if and only if

$$\sum_k \left(\alpha_k \sum_j \Gamma_{kj} \bar{X}_j + \bar{\eta}_k \bar{X}_k \right) < \infty \tag{3}$$

where $\bar{\eta}_j(\lambda) \uparrow \bar{\eta}_j$, as $\lambda \downarrow 0$.

Proof. Imitate the proof of Theorem 3.1. It suffices to note that from (2.11.7)

$$[\bar{X}_i - \bar{X}_i(\lambda)]/\lambda = \sum_j \phi_{ij}(\lambda) \bar{X}_j \uparrow \sum_j \Gamma_{ij} \bar{X}_j \qquad (\lambda \downarrow 0)$$

and to apply Lemma 1. QED

Corollary

Assume that Q is conservative and single exit. If for any $i, j \in E$, $i \overset{\phi(\lambda)}{\Leftrightarrow} j$and $\phi(\lambda)$ is non-recurrent, then $\psi(\lambda)$ is recurrent if and only if$c = [a, X^0] + \bar{\sigma}^0 = 0$.

15.5 FIRST-ORDER PROCESSES

Suppose $X = \{x(t), t < \sigma\} \in \mathscr{H}_1(Q)$, i.e. X is a first-order process. It is a $\{\pi(a,\cdot), a \in B_e\}$-generalized D-type extension process of the minimal Q process $X^0 = \{x(t), t < \tau\}$, where $\pi(a,\cdot)$ satisfy (11.6.2). For this reason we call X a $\{Q, \pi(a,\cdot), a \in B_e\}$ first-order process.

For simplicity, in this section we assume that the non-atomic exit boundary B_{e_2} induced by Q is empty. That is, the exit boundary B_e entirely consists of atomic boundary points; simply put $\mathscr{A} = B_e$.

By (11.5.26) the resolvent operator $\psi(\lambda)$ of a $\{Q, \pi(a,\cdot), a \in \mathscr{A}\}$ first-order process has the following representation:

$$\psi_{ij}(\lambda) = \phi_{ij}(\lambda) + \sum_{a \in \mathscr{A}} \sum_{b \in \mathscr{A}} X_i^a(\lambda) G_{ab}(\lambda) A_j^b(\lambda) \tag{1}$$

where

$$X_i^a(\lambda) = E_i\{e^{-\lambda\tau}, x(\tau - 0) = a\} \qquad A_j^b(\lambda) = \sum_k \pi(b,k)\phi_{kj}(\lambda).$$

$$\left.\begin{aligned} &V_{ab}(\lambda) = \sum_k \pi(a,k) X_k^b(\lambda) \\ &\mathscr{V}(\lambda) = \{V_{ab}(\lambda)\} \qquad \{\mathscr{V}(\lambda)\}^l = \{V_{ab}^l(\lambda)\} \qquad \text{are } \mathscr{A} \times \mathscr{A} \text{ matrices} \\ &G_{ab}(\lambda) = \sum_{l=0}^{\infty} V_{ab}^l(\lambda) \end{aligned}\right\} \tag{2}$$

If we set

$$A = \{i \mid i \in E, \text{ there exists } a \in \mathscr{A} \text{ such that } \pi(a,i) > 0\} \tag{3}$$

then obviously

$$\pi(a, E - A) = 0 \qquad a \in \mathscr{A} \tag{4}$$

and the formula (1) can be rewritten as

$$\psi_{ij}(\lambda) = \phi_{ij}(\lambda) + \sum_{r \in A} \sum_{q \in A} Z_{ir}(\lambda) D_{rq}(\lambda) \phi_{j}(\lambda) \tag{5}$$

where

$$\left.\begin{aligned} &Z_{iq}(\lambda) = \sum_{a \in \mathscr{A}} X_i^a(\lambda)\pi(a,q) \qquad i \in E, q \in A \\ &\mathscr{Z}(\lambda) = \{Z_{rq}(\lambda)\} \qquad \{\mathscr{Z}(\lambda)\}^l = \{Z_{rq}^l(\lambda)\} \qquad \text{are } A \times A \text{ matrices} \\ &D_{rq}(\lambda) = \sum_{l=0}^{\infty} Z_{rq}^l(\lambda) \end{aligned}\right\} \tag{6}$$

Clearly, as $\lambda\downarrow 0$

$$
\begin{aligned}
&X_i^a(\lambda)\uparrow X_i^a = P_i\{x(\tau-0)=a\} \qquad && A_j^a(\lambda)\uparrow A_j^a = \sum_k \pi(a,k)\Gamma_{kj} \\
&\mathscr{V}(\lambda)\uparrow\mathscr{V} = \left\{\sum_k \pi(a,k)X_k^b\right\} && \{\mathscr{V}(\lambda)\}^l\uparrow\mathscr{V}^l \\
&Z_{iq}(\lambda)\uparrow Z_{iq} = \sum_{a\in A} X_i^a\pi(a,q) && \{\mathscr{Z}(\lambda)\}^l\uparrow\mathscr{Z}^l \\
&\mathscr{G}(\lambda)\uparrow\mathscr{G} = \sum_{l=0}^{\infty}\mathscr{V}^l && \mathscr{D}(\lambda)\uparrow\mathscr{D} = \sum_{l=0}^{\infty}\mathscr{Z}^l
\end{aligned}
\tag{7}
$$

Again obviously $\mathscr{Z}(\lambda)$, $\mathscr{Z}$ and $\mathscr{V}(\lambda)$, $\mathscr{V}$ may be considered as one-step transition matrices of Markov chains on the state spaces $\mathscr{A}$ and A, respectively. Hence we can say a is $\mathscr{V}(\lambda)$-recurrent, and so on. Thus it follows from (2.2.10) that

$$
\begin{aligned}
G_{ab}(\lambda) &\leqslant G_{bb}(\lambda) \qquad & G_{ab} &\leqslant G_{bb} \\
D_{rq}(\lambda) &\leqslant D_{qq}(\lambda) & D_{rq} &\leqslant D_{qq}
\end{aligned}
\tag{8}
$$

Lemma 1. (i) Suppose for some $a, b\in\mathscr{A}$, $\lim_{\lambda\downarrow 0}\lambda G_{bb}(\lambda)=0$ and $a\overset{\mathscr{V}}{\Rightarrow}b$. Then

$$\lim_{\lambda\downarrow 0}\lambda G_{ab}(\lambda) = \lim_{\lambda\downarrow 0}\lambda G_{ba}(\lambda) = \lim_{\lambda\downarrow 0}\lambda G_{aa}(\lambda) = 0$$

(ii) Suppose for some $r, q\in A$, $\lim_{\lambda\downarrow 0}\lambda D_{qq}(\lambda)=0$ and $r\overset{\mathscr{Z}}{\Rightarrow}q$. Then

$$\lim_{\lambda\downarrow 0}\lambda D_{rq}(\lambda) = \lim_{\lambda\downarrow 0}\lambda D_{qr}(\lambda) = \lim_{\lambda\downarrow 0}\lambda D_{rr}(\lambda) = 0$$

Proof. It suffices to prove (i). By the hypothesis, there exist α, β such that $V_{ab}^{\alpha}V_{ba}^{\beta}>0$. However, by Chung (1966a, remark on p. 22)

$$V_{bb}^{\alpha+\beta+l}(\lambda)\geqslant V_{ba}^{\beta}(\lambda)V_{aa}^{l}(\lambda)V_{ab}^{\alpha}(\lambda)$$

It follows that

$$\lambda\left(G_{bb}(\lambda)-\sum_{l=0}^{\alpha+\beta}V_{bb}^{l}(\lambda)\right)\geqslant V_{ba}^{\beta}(\lambda)\lambda G_{aa}(\lambda)V_{ab}^{\alpha}(\lambda)$$

Noting (7) we have $V_{ba}^{\beta}(\lambda)V_{ab}^{\alpha}(\lambda)\uparrow V_{ba}^{\beta}V_{ab}^{\alpha}$ $(\lambda\downarrow 0)$. Therefore

$$\lim_{\lambda\downarrow 0}\lambda G_{bb}(\lambda)\geqslant V_{ba}^{\beta}\lim_{\lambda\downarrow 0}\lambda G_{aa}(\lambda)V_{ab}^{\alpha}$$

And so $\lim_{\lambda\downarrow 0}\lambda G_{aa}(\lambda)=0$. Then by (8) we have completed the proof. QED

Lemma 2. Suppose $\psi(\lambda)$ is a $\{Q, \pi(a,\cdot), a\in\mathscr{A}\}$ first-order process and i is

$\phi(\lambda)$-non-recurrent and $\psi(\lambda)$-recurrent. Then:

(i) if $X_i^a > 0$, a is $\mathscr{V}$-recurrent;

(ii) if $Z_{ir} > 0$, r is $\mathscr{Z}$-recurrent.

Proof. (i) Suppose X is a $\psi(\lambda)$ process and τ^n is its nth leaping point for each n. It follows from the structure of the first-order processes that $x(\tau^n - 0)$ $(n \geqslant 1)$ and $x(\tau^n)$ are $\mathscr{V}$-chains and $\mathscr{Z}$-chains, respectively. By the hypothesis and Lemma 2.2

$$P_i\{\tau < \infty\} = P_i\{a = \infty\} = 1 \tag{9}$$

Let $\delta_0 = 0, \beta_0 = \tau$, δ_1 be the time that X first returns to i after β_0, and β_1 be the first leaping point after δ_1. Then by (9) and the $\psi(\lambda)$-recurrence of $i, P_i\{\delta_1 < \infty\} = P_i\{\beta_1 < \infty\} = 1$. Let δ_n be the time after β_{n-1} that X returns to i for the first time. Let β_n be the first leaping point after δ_n. Then it can be proved that $P_i\{\delta_n < \infty$ for all $n\} = 1$.

It is easy to see that $\{x(\beta_n - 0) = a\} \in \mathscr{F}'_{\delta_n} \cap \mathscr{F}_{\delta_{n+1}}$. By Theorem 7.7.3 they are independent. By the strong Markov property

$$P_i\{x(\beta_l - 0) = a\} = P_i\{\delta_l < \infty, \theta_{\delta_l}[x(\tau - 0) = a]\} = X_i^a > 0 \tag{10}$$

Hence it follows from the Borel–Cantelli lemma that

$$P_i\{x(\beta_l - 0) = a \text{ for infinitely many } l\} = 1$$

and of course

$$\begin{aligned} &P_i\{x(\tau^l - 0) = a \text{ for infinitely many } l\} \\ &\quad = \sum_{b \in \mathscr{A}} X_i^b P\{x(\tau^l - 0) = a \text{ for infinitely many } l / x(\tau^1) = b\} = 1 \end{aligned}$$

Since $X_i^a > 0$, it follows that $P\{x(\tau^l - 0) = a$ for infinitely many $l / x(\tau^l - 0) = a\} = 1$, i.e. a is $\mathscr{V}$-recurrent.

(ii) The proof is similar to (i). The left-hand side of (10) is replaced by $P_i\{x(\beta_l) = r\} = Z_{ir} > 0$. Hence $P_i\{x(\beta_l) = r$ for infinitely many $l\} = 1$ and it goes without saying that

$$\begin{aligned} &P_i\{x(\tau^l) = r \text{ for infinitely many } l\} \\ &\quad = \sum_{q \in A} Z_{iq} P_i\{x(\tau^l) = r \text{ for infinitely many } l | x(\tau^1) = q\} = 1 \end{aligned}$$

Since $Z_{ir} > 0$, it follows that $P_i\{x(\tau^l) = r$ for infinitely many $l | x(\tau^1) = r\} = 1$, i.e. r is $\mathscr{Z}$-recurrent. And the proof is completed. QED

Theorem 3. Suppose $\psi(\lambda)$ is a $\{Q, \pi(a, \cdot), a \in \mathscr{A}\}$ first-order process and i is

$\phi(\lambda)$-non-recurrent. Then:

(i) i is $\psi(\lambda)$-recurrent if and only if there exist $a, b \in \mathscr{A}$ such that $X_i^a > 0$, $A_i^b > 0$, a is $\mathscr{V}$-recurrent and $a \overset{\mathscr{V}}{\Rightarrow} b$[1].

(ii) i is $\psi(\lambda)$-recurrent if and only if there exist $r, q \in A$ such that $Z_{ir} > 0$, $r \overset{\mathscr{Z}}{\Rightarrow} q \overset{\phi(\lambda)}{\Rightarrow} i$ and r is $\mathscr{Z}$-recurrent.

Proof. We need only prove (i). Let $\lambda \downarrow 0$ in (1) so that

$$\begin{aligned}\lim_{\lambda \downarrow 0} \psi_{ii}(\lambda) &= \Gamma_{ii} + \sum_{a \in \mathscr{A}} \sum_{b \in \mathscr{A}} X_i^a G_{ab} A_i^b \\ &= \Gamma_{ii} + \sum_{a, b \in \mathscr{A}}{}^{+} X_i^a g_{ab} G_{bb} A_i^b \qquad (11)\end{aligned}$$

where $\sum^+$ means that the summation is taken over positive summands and g_{ab} is the probability that a $\mathscr{V}$-chain starting from a reaches b after finite steps (≥ 0).

If i is $\psi(\lambda)$-recurrent, then the left-hand side of (11) is infinite, and so there must exist $a, b \in \mathscr{A}$ such that $X_i^a > 0, g_{ab} > 0$ and $A_i^b > 0$. By Lemma 2, a is $\mathscr{V}$-recurrent. But $g_{ab} > 0$ precisely demonstrates $a \overset{\mathscr{V}}{\Rightarrow} b$.

Suppose the sufficient conditions are satisfied. If $a = b$, then $g_{ab} = 1$, $G_{aa} = \infty$ and the right-hand side of (11) is infinite. If $a \neq b$, then since a is $\mathscr{V}$-recurrent and $a \overset{\mathscr{V}}{\Rightarrow} b$, i.e. $g_{ab} > 0$, it follows that b is also recurrent, i.e. $G_{bb} = \infty$. Therefore the right-hand side of (11) is also finite. We conclude the proof. QED

Theorem 4. Suppose for any i, j, $i \overset{\phi(\lambda)}{\Leftrightarrow} j$ and $\phi(\lambda)$ is non-recurrent. And suppose $\psi(\lambda)$ is a $\{Q, \pi(a, \cdot), a \in \mathscr{A}\}$ first-order process.

(i) If $\psi(\lambda)$ is recurrent, then naturally

$$Q \text{ is conservative} \qquad \xi_i = 1 \ (i \in E) \qquad \pi(a, E) = 1 \ (a \in \mathscr{A}) \qquad (12)$$

where $\xi_i = P_i\{\tau < \infty\}$ is the same as (2.3).

(ii) If (12) holds and so does one of the following: (1°) $\mathscr{A}$ is finite; (2°) A is finite; (3°) $\mathscr{A}$ has at least one $\mathscr{V}$-recurrent state; (4°) A has at least one $\mathscr{Z}$-recurrent state; then $\psi(\lambda)$ is recurrent.

Proof. (i) By Lemma 2.2 $\xi_i = P_i\{\tau < \infty\} = 1$ $(i \in E)$ and $P_i\{\sigma = \infty\} = 1$. Hence Q must be conservative. Therefore, for $a \in \mathscr{A}$, by (11.6.2) we have

$$\sum_k \pi(a, k) = \sum_k P\{x(\tau) = k \mid x(\tau - 0) = a\}$$

[1] a and b may be the same, $a \overset{\mathscr{V}}{\Rightarrow} a$.

(ii) Suppose (12) holds. Under the hypotheses of the theorem, surely $X_i^a > 0$ and $A_i^a > 0$ for all $a \in \mathscr{A}, i \in E$, and so $Z_{ir} > 0$ for all $i \in E, r \in A$. Because

$$\sum_{b \in \mathscr{A}} V_{ab} = \sum_k \pi(a,k) \sum_{b \in \mathscr{A}} X_k^b = \sum_k \pi(a,k)\xi_k = 1$$

$$\sum_{q \in A} Z_{rq} = \sum_{a \in \mathscr{A}} X_r^a \sum_{q \in A} \pi(a,q) = \sum_{a \in \mathscr{A}} X_r^a = \xi_r = 1$$

if one of the conditions (1°)–(4°) holds, we can deduce that $a\ (\in A)$ is $\mathscr{V}$-recurrent and $\gamma\ (\in A)$ is $\mathscr{Z}$-recurrent. Thus the sufficient conditions in Theorem 3 hold. Hence $\psi(\lambda)$ is recurrent. The proof is completed. QED

Theorem 5. Suppose i is $\phi(\lambda)$-non-recurrent and $\psi(\lambda)$ is a $\{Q, \pi(a,\cdot), a \in \mathscr{A}\}$ first-order process. Suppose $\delta \equiv \sup_{a \in \mathscr{A}} \pi(a,E) < 1$, then i is $\psi(\lambda)$-non-recurrent.

Proof. By the hypothesis we know that $\Gamma_{ki} \leqslant \Gamma_{ii} < \infty$. Since

$$\sum_{a \in \mathscr{A}} X_i^a(\lambda) \leqslant \sum_{a \in \mathscr{A}} X_i^a \leqslant 1$$

it easily follows by (7) that

$$A_i^b(\lambda) = \sum_k \pi(b,k)\Gamma_{ki} \leqslant \delta \Gamma_{ii}$$

$$\sum_{b \in \mathscr{A}} T_{ab}^l(\lambda) \leqslant \delta^l \qquad \sum_{b \in \mathscr{A}} G_{ab}(\lambda) \leqslant \sum_{l=0}^{\infty} \delta^l = \frac{1}{1-\delta} < \infty$$

It follows from (1) that

$$\psi_{ii}(\lambda) \leqslant \Gamma_{ii} + \sum_{a \in \mathscr{A}} X_i^a(\lambda) \frac{1}{1-\delta} \delta \Gamma \leqslant \frac{\Gamma_{ii}}{1-\delta} < \infty$$

From this we know that i is $\psi(\lambda)$-non-recurrent. The proof is finished. QED

Lemma 6. Suppose $\psi(\lambda)$ is a $\{Q, \pi(a,\cdot), a \in \mathscr{A}\}$ first-order process and for any i, j, $i \overset{\psi(\lambda)}{\Leftrightarrow} j$. Then

(i) if $\pi(a,E) > 0$, then $a \overset{\mathscr{V}}{\Rightarrow} b\ (b \in \mathscr{A})$;

(ii) for any $r, q \in A$, $r \overset{\mathscr{Z}}{\Rightarrow} q$.

Proof. Suppose X is a $\psi(\lambda)$ process and η_i^* is defined as in (7.6.7). Because i and j communicate with each other, with respect to $\psi(\lambda)$, we have

$$u_i = P_i\{\eta_j^* < \sigma\} > 0 \tag{13}$$

(i) By the hypothesis there exists i such that $\pi(a,i) > 0$ and for $b \in \mathscr{A}$ there exists j such that $X_j^b > 0$. Thus

$$P\{\text{there exists } l \geqslant 1 \text{ such that } x(\tau^l - 0) = b \mid x(\tau^1 - 0) = a\}$$
$$\geqslant \pi(a, i) u_{ij} X_i^b > 0$$

i.e. $a \overset{\mathscr{Y}}{\Rightarrow} b$.

Claim (ii) can be proved similarly and the proof is completed. QED

Theorem 7. Suppose that $\psi(\lambda)$ is a $\{Q, \pi(a, \cdot), a \in \mathscr{A}\}$ first-order process, for any $i, j, i \overset{\phi(\lambda)}{\Leftrightarrow} j$, $\phi(\lambda)$ is non-recurrent and $\psi(\lambda)$ is recurrent. Then

(i) if A is finite, $\psi(\lambda)$ is ergodic if and only if

$$N_k = E_k \tau < \infty \qquad k \in A \tag{14}$$

(ii) if $\mathscr{A}$ is finite, $\psi(\lambda)$ is ergodic if and only if

$$\sum_k \pi(a, k) N_k < \infty \qquad a \in \mathscr{A} \tag{15}$$

Proof. We only prove (i). Claim (ii) can be proved similarly. Since $\psi(\lambda)$ is recurrent, by Lemma 2.2 $\psi(\lambda)$ is honest, i.e.

$$\lambda \sum_j \psi_i(\lambda) = 1 \qquad i \in E \tag{16}$$

Q is conservative since $\psi(\lambda)$ satisfies the system of backward equations. According to Theorem 3 there exists $r \in A$ such that r is $\mathscr{L}$-recurrent. By Lemma 6 A is a $\mathscr{L}$-recurrent class. Substituting (5) into (16), considering (1.5) and noting that $\xi_i(\lambda) = \sum_{a \in \mathscr{A}} X_i^a(\lambda)$ if Q is conservative, we arrive at

$$\sum_{a \in \mathscr{A}} X_i^a(\lambda) = \sum_{a \in \mathscr{A}} X_i^a(\lambda) \sum_{r, q \in \mathscr{A}} \pi(a, r) \lambda D_{rq}(\lambda) \sum_j \phi_q(\lambda)$$

It follows from the linear independence of $X^a(\lambda)$ that

$$1 = \sum_{r, q \in \mathscr{A}} \pi(a, r) \lambda D_{rq}(\lambda) \sum_j \phi_g(\lambda) \qquad a \in \mathscr{A} \tag{17}$$

(a) If there exists some $q \in A$ such that $N_q = \infty$, then for this q there exists $a \in \mathscr{A}$ such that $\pi(a, q) > 0$. By (17)

$$\pi(a, q) \lambda D_{qq}(\lambda) \sum_j \phi_{qj}(\lambda) \leqslant 1$$

$$\lambda D_{qq}(\lambda) \leqslant \frac{1}{\pi(a, q) \sum_j \phi_{qj}(\lambda)}$$

It follows that $\lim_{\lambda \downarrow 0} \lambda D_{qq}(\lambda) = 0$. By Lemma 1 for any $r, q \in A$, $\lim_{\lambda \downarrow 0} \lambda D_{rq}(\lambda) = 0$. Hence by (5) and the finiteness of A we have $\lim_{\lambda \downarrow 0} \lambda \psi_{ii}(\lambda) = 0$, i.e. $\psi(\lambda)$ is non-ergodic.

(b) Suppose $N_k < \infty$ for all k. We might as well assume that $\lim_{\lambda\downarrow 0}\lambda D_{rq}(\lambda)$ exist for all $r, q \in A$. Otherwise we can always choose a suitable subsequence of λ. Then by (17) and the finiteness of A,

$$1 = \sum_{r,q\in A} \pi(a,r) \lim_{\lambda\downarrow 0} \lambda D_{rq}(\lambda) \sum_j \Gamma_{qj}$$

$$= \sum_j \sum_{r,q\in A} \pi(a,r) \lim_{\lambda\downarrow 0} \lambda D_{rq}(\lambda)\Gamma_{qj} \qquad a\in\mathscr{A}$$

Consequently for fixed $a\in\mathscr{A}$ there exist $j\in E$ and $r, q\in A$ such that

$$\pi(a,r) \lim_{\lambda\downarrow 0} \lambda D_{rq}(\lambda)\Gamma_{qj} > 0$$

But for $a\in\mathscr{A}$ there must exist i such that $X_i^a > 0$. Therefore by (5)

$$\pi_i = \lim_{\lambda\downarrow 0} \lambda\psi_{ij}(\lambda) \geqslant \lim_{\lambda\downarrow 0} z_{ir}(\lambda)\lambda D_{iq}(\lambda)\phi_{qj}(\lambda)$$

$$\geqslant X_i^a \pi(a,r)\left(\lim_{\lambda\downarrow 0} \lambda D_{rq}(\lambda)\right)\Gamma_{qj} > 0$$

i.e. $\psi(\lambda)$ is ergodic. The proof is concluded. QED

Theorem 8. Suppose that the non-atomic exit boundary B_{e2} induced by Q is an empty set and $\mathscr{A} = B_e$ is a finite set. Then any D-type Q process $X\in\mathscr{H}_D(Q)$ is a $\{Q, \pi(a,\cdot), a\in\mathscr{A}\}$ first-order process, where

$$P\{x(\tau) = j \mid x(\tau - 0) = a\} = \pi(a,j) \tag{18}$$

Here τ is the first leaping point.

Proof. Since $X\in\mathscr{H}_D$, it follows that

$$P_{ij}(t) = P_{ij}\{x(t) = j, t < \tau\} + P_i\{\tau \leqslant t, x(t) = j\}$$

$$= f_{ij}(t) + \sum_{a\in\mathscr{A}} \sum_k P_i\{x(\tau - 0) = a, \tau \leqslant t, x(\tau) = k, x(t) = j\}$$

Making use of the strong Markov property and applying Theorem 7.17.1, it is easy to prove that the summand of the above is equal to

$$\int_{\substack{x(\tau-0)=a\\ x(\tau)=k,\tau\leqslant t}} p_{kj}(t-\tau)\mathrm{d}P_i = \int_0^t p_{kj}(t-s)\mathrm{d}P_i\{x(\tau - 0) = a, x(\tau) = k, \tau \leqslant s\}$$

$$\times P_i\{x(\tau - 0) = a, x(\tau) = k, \tau \leqslant s\}$$

$$= P_i\{x(\tau - 0) = a, \tau \leqslant s\} P\{x(\tau) = k \mid x(\tau - 0) = a\}$$

$$= L_i^a(s)\pi(a,k) \tag{19}$$

Thus (19) becomes

$$p_{ij}(t) = f_{ij}(t) + \sum_{a\in\mathscr{A}} \sum_{k} \pi(a,k) \int_0^t p_{kj}(t-s)\,\mathrm{d}L_i^a(s)$$

Taking Laplace transforms we have

$$\psi_{ij}(\lambda) = \phi_{ij}(\lambda) + \sum_{a\in\mathscr{A}} X_i^a(\lambda) \sum_k \pi(a,k)\psi_{kj}(\lambda) \tag{20}$$

or

$$\psi(\lambda) = \phi(\lambda) + \{X^a(\lambda)\}'\{B^a(\lambda)\} \tag{21}$$

where $B_j^a(\lambda) = \sum_k \pi(a,k)\psi_{kj}(\lambda)$. Upon multiplying the above on the lefts by $\pi(a,\cdot)$ we obtain

$$\begin{aligned} \{B^a(\lambda)\} &= \{A^a(\lambda)\} + \mathscr{V}(\lambda)\{B^a(\lambda)\} \\ \{I - \mathscr{V}(\lambda)\}\{B^a(\lambda)\} &= \{A^a(\lambda)\} \end{aligned} \tag{22}$$

where $A^a(\lambda)$ is defined as (7) and $\mathscr{A}\times\mathscr{A}$ matrix $\mathscr{V}(\lambda)$ as (4.2). Because the matrices $\mathscr{V}(\lambda)$ have row sums $\sum_{b\in\mathscr{A}}\sum_k \pi(a,k)X_k^b(\lambda) < 1$ and $\mathscr{A}$ is finite, for each $\lambda > 0$, $\{I - \mathscr{V}(\lambda)\}^{-1}$ exists and is equal to $\mathscr{G}(\lambda) = \sum_{l=0}^{\infty}(\mathscr{V}(\lambda))^l$. Therefore by (22)

$$\{B^a(\lambda)\} = \{I - \mathscr{V}(\lambda)\}^{-1}\{A^a(\lambda)\} = \mathscr{G}(\lambda)\{A^a(\lambda)\}$$

Substituting this into (21) we find that $\psi(\lambda)$ have the form (1), i.e. $\psi(\lambda)$ is a $\{Q, \pi(a,\cdot), a\in\mathscr{A}\}$ first-order process. The proof is complete. QED

Remark

Many results in this section can also be extended to kth-order processes.

15.6 BILATERAL BIRTH–DEATH PROCESSES

In this section we shall utilize the notations given in Chapter 5.

Suppose that X is a bilateral birth–death process and τ is its first leaping point. Define the first hitting times ξ_i according to (12.2.6).

Lemma 1. Suppose $i \leqslant k \leqslant n$. Then

$$P_k\{\xi_i < \xi_n\} = \frac{z_n - z_k}{z_n - z_i} \qquad P_k\{\xi_n < \xi_i\} = \frac{z_k - z_i}{z_n - z_i} \tag{1}$$

Proof. Imitate the proof of Theorem 12.2.2. QED

Theorem 2. Let C_{kn} denote the probability that the process X starting from k

reaches n after finite ($\geqslant 0$) jumps, i.e. $C_{kn} = P_k\{\xi_n < \tau\}$. Then

$$C_{kn} = \begin{cases} (r_2 - z_k)/(r_2 - z_n) & \text{if } n \leqslant k \\ (z_k - r_1)/(z_n - r_1) & \text{if } n > k \end{cases} \tag{2}$$

Proof. It follows by letting $n \to +\infty$ or letting $i \to -\infty$ respectively in (1).
QED

Theorem 3. The minimal Q process is recurrent if and only if r_1 and r_2 are both infinite.

Proof. The probability that the process X starting from 0 returns to 0 is

$$f_0^* = \frac{a_0}{a_0 + b_0} C_{-10} + \frac{b_0}{a_0 + b_0} C_{10}$$

Hence by (2) we can see that $f_0^* = 1$ if and only if r_1 and r_2 are both infinite. The proof is completed.
QED

Theorem 4. Suppose that the minimal Q process is recurrent. Then it is ergodic if and only if

$$\sum_k \mu_k < \infty \tag{3}$$

and in this case

$$m_{ii} = \left(\sum_k \mu_k\right) \Big/ q_i \mu_i \tag{4}$$

$$m_{ij} = (z_j - z_i) \sum_{s \leqslant t} \mu_s + \sum_{i < s < j} (z_j - z_s)\mu_s \qquad \text{if } j > i \tag{5}$$

$$m_{ij} = \sum_{j < s < i} (z_s - z_j)\mu_s + (z_i - z_j) \sum_{s \geqslant i} \mu_s \qquad \text{if } j < i \tag{6}$$

where $m_{ij} = E_i \eta_j^*$, where η_i^* is determined by (7.7.7).

Proof. Consider a minimal Q process $X = \{x(t), t < \infty\}$. Note that by recurrence we have $P_i\{\eta_i^* < \infty\} = P_i\{\xi_n < \infty\} = 1$. It is easy to see that for $i < k < n, u_k = \min(\xi_i, \xi_n)$ satisfy the equations (5.4.11) with $f_i = f_n = 0$ and $f_k = 1\,(i < k < n)$. Hence it follows from (5.4.12) that

$$E_k \min(\xi_i, \xi_n) = \sum_{i < s \leqslant k} \frac{(z_s - z_i)(z_n - z_k)}{z_n - z_i} \mu_s + \sum_{k < i < n} \frac{(z_k - z_i)(z_n - z_s)}{z_n - z_i} \mu_s \tag{7}$$

Letting $i \to -\infty, n \to +\infty$ in the above expression and noting that r_1 and r_2

are infinite, we obtain (5)–(6). By the strong Markov property

$$m_{ii} = \frac{1}{q_i} + \frac{a_i}{q_i} m_{i-1,i} + \frac{b_i}{q_i} m_{i+1,i}$$

Noting that $a_i(z_i - z_{i-1}) = b_i(z_{i+1} - z_i) = \mu_i^{-1}$ and substituting (5) and (6) into the above, we obtain (4). The proof is completed. QED

Lemma 5. If the minimal Q process is non-recurrent, then

$$N_k = E_k\tau = \sum_{s \leqslant k} \frac{(z_s - r_1)(r_2 - z_k)}{r_2 - r_1} \mu_s + \sum_{k<s} \frac{(z_k - r_1)(r_2 - z_s)}{r_2 - r_1} \mu_s$$

Proof. It follows by letting $i \to -\infty$ and $n \to +\infty$ in (7). QED

Theorem 6. Suppose that r_1 is entrance or natural and r_2 is exit or regular. Then a Q process $\psi(\lambda)$ is recurrent if and only if r_1 is inifinite and $\psi(\lambda)$ is honest.

Proof. This theorem is a special case of Theorem 4.2. Under the hypothesis, $\bar{X}_i = \bar{X}_i^2$ and $X_i^0 = X_i^1$. Thus the first two inequalities in (4.2) hold. Therefore (4.2) is equivalent to $c = 0$ and $X^0 = X^1 = 0$. By noting (5.7.2) our proof is finished. QED

Theorem 7. Suppose that r_1 is entrance or natural, r_2 is exit or regular, and $\psi(\lambda)$ is recurrent. Then

(i) if r_1 is exit, then $\psi(\lambda)$ is ergodic;

(ii) if r_1 is natural, (1*) when $\sum_{s \leqslant 0} \mu_s < \infty$, then $\psi(s)$ is non-ergodic; (2*) when $\sum_{s \leqslant 0} \mu_s < \infty$, and $\psi(\lambda)$ has the representation (5.8.3), then $\psi(\lambda)$ is ergodic if and only if

$$\sum_{s \leqslant 0} a_s N_s < \infty \tag{9}$$

In particular, if $\sum_{s \leqslant 0} \mu_s < \infty$ and $\sum_{s \leqslant 0} a_s(r_2 - z_s) < \infty$, then $\psi(\lambda)$ is ergodic.

Proof. Letting $\lambda \downarrow 0$ in (5.8.12) we have

$$\bar{\eta}_i = P_1(r_2 - z_j)\mu_j + P_2\mu_j \qquad \text{in which } P_1 = 0 \text{ if } r_1 \text{ is natural and } P_2 = 0 \text{ if } r_2 \text{ is exit} \tag{10}$$

But r_2 is regular or exit and, moreover,

$$\sum_{j \leqslant 0} (r_2 - z_j)\mu_j < \infty \qquad \text{if } r_1 \text{ is entrance}$$
$$\sum_{j \geqslant 0} \mu_j < \infty \qquad \text{if } r_2 \text{ is regular} \tag{11}$$

Hence $\sum_i \bar{\eta}_i < \infty$ is equivalent to $P_2 \sum_{s<0} \mu_s < \infty$. Note $\bar{X} = X^2 = 1$, so the ergodicity condition (4.3) becomes

$$\sum_s a_s N_s + P_2 \sum_{s<0} \mu_s < \infty \tag{12}$$

Because, by Theorem 6, r_1 is infinite, by (8)

$$N_k = (r_2 - z_k) \sum_{k \leqslant s} \mu_s + \sum_{s<k} (r_2 - z_s)\mu_s \tag{13}$$

We now proceed to prove that if $\sum_{s \leqslant 0} \mu_s < \infty$, then

$$\sum_{s \geqslant 0} a_s N_s < \infty \tag{14}$$

In fact, when $k \geqslant 0$

$$N_k = (r_2 - z_k) \sum_{s<0} \mu_s + M_k \tag{15}$$

where M_k is just N_k in (5.11.11). It follows from Theorems 5.11.3 and 5.11.4 that

$$\sum_{k \geqslant 0} a_k N_k < \infty$$

(i) If r_1 is entrance, then surely $\sum_{s \leqslant 0} \mu_s < \infty$. By (14) the ergodic property is equivalent to (9). But when $i \leqslant 0$, by (11) we have

$$N_i \leqslant \sum_{s \leqslant k} (r_2 - z_s)\mu_s + \sum_{k<s} (r_2 - z_s)\mu_s = \sum_s (r_2 - z_s)\mu_s < \infty$$

Hence by Theorem 5.11.5

$$\sum_{i \leqslant 0} a_i N_i \leqslant \left(\sum_{i \leqslant 0} a_i \right) \sum_s (r_2 - z_s)\mu_s < \infty$$

Therefore $\psi(\lambda)$ is ergodic.

(ii) Suppose r_1 is natural. (1*) Suppose $\sum_{s \leqslant 0} \mu_s < \infty$. By (13) $N_i = \infty$. If $a \neq 0$, then (12) is not satisfied. If $a = 0$, then necessarily $P_2 > 0$. Hence (12) is not satisfied, and so $\psi(\lambda)$ is not ergodic.

(2*) Suppose $\sum_{s \leqslant 0} \mu_s < \infty$. By (14), the ergodicity condition (12) becomes (9). Furthermore, suppose $\sum_{s \leqslant 0} a_s (r_2 - z_s) < \infty$. Then by (13) for $i \leqslant s$,

$$N_i \leqslant (r_2 - z_i) \sum_{s<i} \mu_s + \sum_{i<s<0} (r_2 - z_s)\mu_s + \sum_{s \geqslant 0} (r_2 - z_s)\mu_s$$

$$\leqslant (r_2 - z_i) \sum_{s<0} \mu_s + \sum_{s \geqslant 0} (r_2 - z_s)\mu_s$$

By Theorem 5.11.5 we have $\sum_{k\leqslant 0} a_k < \infty$. Therefore $\sum_{s\leqslant 0} a_s N_s < \infty$, and so $\psi(\lambda)$ is ergodic. The proof is terminated. QED

Theorem 8. Suppose that r_1 and r_2 are regular or exit. Then:

(i) the Q process $\psi(\lambda)$ is recurrent if and only if it is honest;

(ii) any honest Q processes are ergodic.

Proof. By Lemma 2.2 $\psi(\lambda)$ is recurrent, so it must be honest.

First, we prove that if the vector $\alpha \geqslant 0$ satisfies $\alpha\phi(\lambda)\in l$, then

$$\sum_k a_k N_k < \infty \tag{16}$$

where N_k is determined acording to (8). When $k \geqslant 0$ from (8) it follows that

$$N_k \leqslant (r_2 - z_k) \sum_{s<0} \frac{z_s - r_1}{r_2 - r_1} \mu_s + M_k \tag{17}$$

where M_k is precisely N_k in (5.11.11). By Theorems 5.11.3 and 5.11.4 we have $\sum_{k\geqslant 0} a_k N_k < \infty$. Similarly $\sum_{k\leqslant 0} a_k N_k < \infty$ can be proved.

Next, for $\bar{\eta}_j(\lambda)$ defined in (5.7.12), let $\lambda \downarrow 0$. We obtain

$$\bar{\eta} = P_1 X^1 \mu + P_2 X^2 \mu \qquad \text{where } P_a = 0 \text{ if } r_a \text{ is exit} \tag{18}$$

However, when r_2 is regular

$$\sum_j X_j^2 \mu_j \leqslant \frac{1}{r_2 - r_1} \sum_{j<0} (z_j - r_1)\mu_j + \sum_{j\geqslant 0} \mu_j < \infty$$

Similarly when r_1 is regular

$$\sum_j X_j^1 \mu_j < \infty$$

Therefore we always have

$$\sum_j \bar{\eta}_j < \infty \tag{19}$$

Furthermore, it follows from Lemma 4.1 that

$$[\alpha^a, X^a - X^a(\lambda)] \downarrow 0 \qquad \lambda[\bar{\eta}(\lambda), X^a] \downarrow 0 \qquad (\lambda \downarrow 0) \tag{20}$$

If $\psi(\lambda)$ has the representation (5.10.16) ($c = 0$ and $\alpha_1 = \alpha_2$), then by (16) and (19), imitating Theorem 4.2, we can find that $\psi(\lambda)$ is recurrent and ergodic and, furthermore,

$$\pi_j = \lim_{\lambda\downarrow 0} \lambda\psi_{ij}(\lambda) = \frac{\sum_k a_k \Gamma_{kj} + P_1 X_j^1 \mu_j + P_2 X_j^2 \mu_j}{\sum_k a_k N_k + \sum_k (P_1 X_k^1 + P_2 X_k^2)\mu_k} \tag{21}$$

Suppose that $\psi(\lambda)$ has the representation (5.10.1), (5.10.4) and (5.10.33) and that $\bar{S}^{12} = \bar{S}^{21} = 1$. Write

$$\mathscr{G}(\lambda) = \mathscr{L}_\lambda^{-1} = (I - \bar{\mathscr{S}} + \bar{\mathscr{H}}_\lambda + \bar{\mathscr{M}}\,\mathscr{U}_\lambda)^{-1}$$

$$= \sum_{l=0}^{\infty} \{\mathscr{F}(\lambda)\}^l \operatorname{diag}\left(\frac{1}{1 - e_\lambda^{aa}}\right) \tag{22}$$

where e_λ^{ab} are elements of the matrix $\bar{\mathscr{S}} - \bar{\mathscr{H}}_\lambda - \bar{\mathscr{M}}\,\mathscr{U}_\lambda$, and $\mathscr{F}(\lambda) = \{f^{ab}(\lambda)\}$ with

$$f^{aa}(\lambda) = 0 \qquad f^{ab}(\lambda) = \frac{e_\lambda^{ab}}{1 - e_\lambda^{aa}} \qquad (b \neq a) \tag{23}$$

It follows from (20) and $\bar{S}^{aa} = 0$ that

$$1 - e_\lambda^{aa} \uparrow 1 \qquad f^{ab}(\lambda) \uparrow \bar{S}^{ab} \qquad (\lambda \downarrow 0) \tag{24}$$

Thus

$$\mathscr{F}(\lambda) \uparrow \bar{\mathscr{S}} \qquad (\lambda \downarrow 0) \qquad \mathscr{G}(\lambda) \to \mathscr{G} = \sum_{l=0}^{\infty} \bar{\mathscr{S}}^l$$

and $\psi(\lambda)$ can be expressed as

$$\psi_{ij}(\lambda) = \phi_{ij}(\lambda) + \sum_{a=1}^{2} X_i^a(\lambda) G_{ab}(\lambda) \eta_i^b(\lambda) \tag{25}$$

where $\eta^b(\lambda) = \bar{\alpha}^b \phi(\lambda) + \bar{M}^{bb} X^b(\lambda)\mu$, in which $\bar{M}^{bb} = 0$ when r_b is exit.

Because

$$\bar{\mathscr{S}} = \begin{pmatrix} 0 & 1 \\ 1 & 0 \end{pmatrix} \tag{26}$$

let $\lambda \downarrow 0$ in (25) so that $\psi(\lambda)$ is recurrent. As $\psi(\lambda)$ is honest, it follows that

$$\lambda G_{ab}(\lambda) \sum_j \eta_j^b(\lambda) = 1 \qquad a = 1, 2 \tag{27}$$

Paying attention to (16) and (18)–(19), and letting $\lambda \downarrow 0$ we obtain

$$\sum_j \eta_j^b(\lambda) \uparrow \sum_k (\bar{\alpha}_k^b N_k + \bar{M}^{bb} X_k^b \mu_k) < \infty$$

From (27) $\lim_{\lambda \downarrow 0} \lambda G_{ab}(\lambda) \sum_j \eta_j^b = 1$. Hence there exists j such that $\lim_{\lambda \downarrow 0} \lambda G_{ab}(\lambda) \eta_j^b > 0$.

From (25) we have

$$\lim_{\lambda \downarrow 0} \lambda \psi_{ij}(\lambda) \geqslant X_i^a \left(\lim_{\lambda \downarrow 0} \lambda G_{ab}(\lambda) \right) \eta_j^b > 0$$

Therefore $\psi(\lambda)$ is ergodic. We conclude the proof. QED

References

Chung, K. L. (1963) On the boundary theory for Markov chains, *Acta. Math.*, **110**, 19–77
——(1966) On the boundary theory for Markov chains II *Acta Math.*, **115**, 111–63
——(1967) *Markov Chains with Stationary Transition Probabilities*, 2nd edn, Springer-Verlag
Cramer, H. (1946) *Mathematical Methods of Statistics*, Princeton University Press
Doob, J. L. (1945) Markov chains—denumerable case, *Trans. Am. Math. Soc.*, **58**, 455–73
——(1953) *Stochastic Processes*, John Wiley & Sons
——(1959) Discrete potential theory and boundaries, *J. Math. Mech.*, **8**, 433–58
Dynkin, E. B. (1959) *Foundations of the Theory of Markov Processes*, Moscow (in Russian), English translation, Prentice-Hall, 1961
——(1963) *Markov Processes*, Moscow (in Russian) (English translation in two volumes, Springer-Verlag, 1965)
——(1969) Boundary theory of Markov processes, *Uspehi Mat, Nauk SSSR*, **24**, 2(146), 3–42 (in Russian)
Feller, W. (1940) On the integro-differential equations of purely discontinuous Markov process, *Trans. Am. Math. Soc.*, **48**, 488–515
——(1945) *Trans. Am. Math. Soc.*, **58**, 474
——(1956) Boundaries induced by non-negative matrices, *Trans. Am. Math. Soc.*, **83**, 19–54
——(1957a) On boundaries and lateral conditions for the Kolmogorov differential equations, *Ann. Math.*, **65**, 527–70
——(1957b) *An Introduction to Probability Theory and Its Applications*, vol. 1 John Wiley & Sons
——(1958) Notes on my paper 'On boundaries and lateral conditions for Kolmogorov differential equations', *Ann. Math.*, **68**, 735–6
——(1971) *An Introduction to Probability Theory and Its Applications*, vol. 2 John Wiley & Sons
Guan, Zhao-zhi (1958) *Lecture on Functional Analysis*, Higher Educational Publishing House, Beijing (in Chinese)
Hardy, G. H. (1949) *Divergent Series*, Oxford University Press
Hou, Zhen-ting (1974) Criteria for the uniqueness of Q-processes, *Scientia Sinica*, **2**, 115–30
——(1975) The construction of sample functions for homogeneous and denumerable Markov process, *Scientia Sinica*, **3**, 259–66
——(1982) *Criteria for the Uniqueness of Q-Processes*, Hunan Science and Technology Publishing House (in Chinese)
Hou, Zhen-ting and Guo, Qin-fen (1976) The quantitative theory of construction theory for homogeneous and denumerable Markov processes, *Acta Math. Sinica*, **19**, 239–62 (in Chinese)

——(1978) *Homogeneous and Denumerable Markov Processes*, Science Publising House (in Chinese)
Hu, Di-he (1965) The construction theory of denumerable Markov Processes, *J. Beijing Uiniv.* (*Natural Sci.*), **2**, 111–43 (in Chinese)
——(1966) The construction theory of Q-processes in abstract space, *Acta Math. Sinica*, **16**, 150–65 (in Chinese)
——(1983) *Theory of Markov Processes with Denumerable States*, Wuhan University Press (in Chinese)
Hunt, G. A. (1960) Markoff chains and Martin boundaries, *Illnois J. Math.* **4**, 66–100
Ito, K. (1961) *Stochastic Processes*, Shanghai Science and Technology Publishing House (in Chinese)
Karlin, S. and McGregor, J. L. (1957) The classification of birth and death processes, *Trans. Am. Math. Soc.*, **86**, 366–400
Kemeny, J. G., Snell, J. L. and Knapp, A. W. (1966) *Denumerable Markov Chains*, D.V. *Nostrand, Princeton*
Kendall, D. G. (1960) Hyperstonian spaces associated with Markov chains, *Proc. London Math. Soc.*, **10**, 67–87
Kolmogorov, A. (1931) Uber die analytischen Methoden in der Wahrscheinlichkeitsrechnung, *Math. Ann.*, **104**, 415–58
Kunita, H. (1962) Applications of Martin boundaries to instantaneous return Markov processes over a denumerable space, *J. Math. Soc. Japan*, **14**, 66–100
Lamb, C. W. (1971) On the construction of certain transition functions, *Ann. Math. Stat.*, **42**, 439–50
Li, Zhang-nan and Wu, Rong (1964) Some limit theorems of additive functionals for Markov chains with denumerable states, *J. Nankai Univ.* (*Natural Sci.*) **5**(5), 121–40 (in Chinese)
Miller, R. G. (1963) Stationarity equations in continuous time Markov chains, *Trans. Am. Math. Soc.*, **109**, 35–44
Qian, Min and Hou, Zhen-ting (1979) *Reversible Markov Processes*, Hunan Science and Technology Publishing House (in Chinese)
Reuter, G. E. H. (1957) Denumerable Markov processes and the associated semi-group on *l*, *Acta Math.*, **97**, 1–46
——(1959) Denumerable Markov processes, II, *J. London Math. Soc.*, **34**, 81–91
——(1962) Denumerable Markov processes, III, *J. London Math. Soc.*, **37**, 64–73
——(1976) Denumerable Markov processes (IV): On C. T. Hou's uniqueness theorem for Q semigroups, *Z. Wahrsch.*, **33**, 309–15
Saks, S. (1937) *Theory of the Integral*, Warzawa Lwow
Shi, Ren-jie (1964) The time substitution of denumerable Markov processes, *J. Nankai Univ.* (*Natural Sci.*), **5**(5), 51–88 (in Chinese)
Sun, Zhen-Zu (1962) General expressions for a class of Markov processes, *J. Zhengzhou Univ.*, **2**, 17–23 (in Chinese)
Titchmarsh, E. C. (1939) *The Theory of Functions*, 2nd edn, Oxford University Press
Wang, Zi-kun (1958) Classification of all birth–death processes, *Hayku*, **4**, 19–25 (in Russian)
————(1961) On distributions of functionals of birth and death processes and their applications in theory of queues, *Scientia sinica*, *X* (2), 160–70
——(1962) The construction theory of birth–death processes, *Chinese Adv. Math.*, **5**, 137–87 (in Chinese)
——(1964) The ergodic property and zero-one laws of birth–death processes, *J. Nankai Univ.* (*Natural Sci.*), **5**(5), 89–94

——(1965a) *Theory of Stochastic Processes*, Science Publishing House, Beijing (in Chinese)

——(1965b) The Martin boundary and limit theorems for excessive functions, *Scientia sinica*, **XIV**(8), 1118–29

——(1965c) The zero-one laws of Markov processes, *Acta Math. Sinica*, **15**, 342–53 (in Chinese)

——(1966) Some properties of recurrent Markov processes, *Acta Math. Sinica*, **16**, 166–78 (in Chinese)

——(1980) *Birth–Death Processes and Markov Chains*, Science Publishing House, Beijing (in Chinese)

Wang, Zi-kun and Yang, Xiang-qun (1978) The construction of stopping birth–death processes, *Acta Math. Sinica*, **21**, 66–71 (in Chinese)

——(1979) The probability-analysis method in construction theory of stopping birth–death processes, *J. Nankai Univ.*(*Natural Sci.*), **3**, 1–32 (in Chinese)

——(1988) *The Birth–Death Processes and Markov Chains*, Springer-Verlag

Watanabe, T. (1960a) Some topics related to Martin boundaries induced by countable Markov processes, *32nd Session of ISI*

——(1960b) On the theory of Martin boundaries induced by countable Markov processes, *Mem. Coll. Sci. Univ. Kyoto, Ser. A, Math.*, **XXXIII**, 39–108

Wilks, S. S. (1962) *Mathematical statistics*, John Wiley & Sons

Williams, D. (1964) On the construction problem for Markov chains, *Z. Wahrsch.*, **3**, 227–46

——(1966) *Z. Wahrsch.*, **5**, 296–9

——(1976) The *Q* matrix problem, *Séminaire de Probabilitiés X* (Lecture Notes in Mathematics, 511), Springer, pp. 216–34

Wu, Li-de (1963) The distributions of functionals of integral type for homogeneous and denumerable Markov processes, *Acta Math. Sinica*, **13**, 86–93 (in Chinese)

——(1965) The classification of states for denumerable Markov processes, *Acta Math. Sinica*, **15**, 32–41 (in Chinese)

Xiong, Da-guo (1980) Markov processes with density matrix *Q. J. Beijing Industry Univ.*, **1**, 1–10 (in Chinese)

——(1981) (*Q*,*b*) process, *Kexue Tongbao*, **26**, 446–7 (in Chinese)

Yang, Xiang-qun (1964a) The functionals of integral type for denumerable Markov processes and lateral properties of bilateral birth and death processes, *Chinese Adv. Math.*, **7**, 397–424 (in Chinese)

——(1964b) The bilateral birth and death processes, *J. Nankai Univ.* (*Natural Sci*), **5**(5), 9–40 (in Chinese)

——(1965a) A class of birth and death processes, *Acta Math. Sinica*, **15**, 9–31 (in Chinese); or see *Chinese Math.*, **6**, 305–29

——(1965b) Notes on the construction theory of birth and death processes, *Acta Math. Sinica*, **15**, 173–87 (in Chinese); or see *Chinese Math.*, **6**, 479–94

——(1966a) The lateral conditions of Kolmogorov's backward equations, *Acta Math. Sinica*, **16**, 429–52 (in Chinese); or see *Chinese Math.* **6**, 449–74

——(1966b) The properties of birth and death processes, *Chinese Adv. Math.*, **9**, 365–80 (in Chinese)

——(1973) Selected translation of *Math. Stat. Prob.*, **12**, 209–48

——(1978) W transformation of denumerable Markov processes, *J. Xiangtan Univ.* (*Natural Sci.*), **1**, 29–43 (in Chinese)

——(1979) *U* intervals of denumerable Markov processes, *J. Xiangtan Univ.* (*Natural Sci.*), **1**, 1–8 (in Chinese)

——(1980a) W transformation and strong limit for denumerable Markov processes, *Scientia Sinica*, **XXIII** (9), 1092–109

——(1980b) *Chinese Ann. Math.*, **1**, 131–8 (in Chinese)

——(1980c) The strong limits and the entrance decomposition of denumerable Markov processes, *Chinese Ann. Math.*, **1**, 255–60 (in Chinese)

——(1980d) The classification of states for nonregular denumerable Markov processes, *Acta Math. Sinica*, **23**, 583–608 (in Chinese)

——(1981a) Construction of Q processes satisfying Kolmogorov's backward equations in single exit case or forward equations in single entrance case, *Kexue Tongbao*. **26**, 390–4 (in Chinese)

——(1981b) Martin boundaries and Q processes, *J. Xiangtan Univ.* **1**, 10–35 (in Chinese)

——(1982) The construction theory of Q processes for the Q matrix with a finite set of non-conservative states and finite exit boundary, *Scientia Sinica, Ser. A*, **XXV**, 476–91

——(1983a) Construction of Q processes when non-conservative state set and exit boundary set being finite, *Kexue Tongbao*, Special Issue, 58–61

——(1983b) The approximating Markov chains and the structure of sample paths for non-sticky return processes, (I), (II), *J. Xiangtan Univ.* (*Natural Sci.*), **1**, 43–55; **2**, 43–55 (in Chinese)

——(1984a) Notes on the construction theory of bifinite Q processes, *J. Xiangtan Univ.* (*Natural Sci.*), **1**, 9–23 (in Chinese)

——(1984b) Construction of paths for a class of non-sticky Q processes, *Kexue Tongbao*, **29**, 1036–8

Index

The section numbers given in the following mean that the entry appears for the first time or its definition is given in that section